Complex Brain Functions

Conceptual Advances in Brain Research

A series of books focusing on brain dynamics and information processing systems of the brain.

Edited by *Robert Miller, University of Otago, New Zealand (Editor-in-chief), Günter Palm, Universität Ulm, Germany* and *Gordon Shaw, University of California at Irvine, USA.*

Volume 1

Brain Dynamics and the Striatal Complex
edited by *R. Miller* and *J.R. Wickens*

Volume 2

Complex Brain Functions: Conceptual Advances in Russian Neuroscience
edited by *R. Miller, A.M. Ivanitsky* and *P.M. Balaban*

Forthcoming Volumes

Time and the Brain
edited by *R. Miller*

Sex Differences in Animal Brain Lateralization
V.L. Bianki and *E.B. Filipova*

Cortical Areas: Unity and Diversity
edited by *A. Schüz* and *R. Miller*

Volumes in Preparation

The Female Brain
Neural Determinism
Functional Memory and Brain Oscillations

This book is part of a series. The publisher will accept continuation orders which may be cancelled at any time and which provide for automatic billing and shipping of each title in the series upon publication. Please write for details.

Complex Brain Functions
Conceptual Advances in Russian Neuroscience

Edited by

R. Miller

School of Medical Sciences
University of Otago
New Zealand

A.M. Ivanitsky

and

P.M. Balaban

Institute of Higher Nervous Activity and Neurophysiology
Russian Academy of Sciences
Moscow

harwood academic publishers
Australia • Canada • France • Germany • India • Japan
Luxembourg • Malaysia • The Netherlands • Russia • Singapore
Switzerland

Amsteldijk 166
1st Floor
1079 LH Amsterdam
The Netherlands

British Library Cataloguing in Publication Data
Complex brain functions : conceptual advances in Russian
neuroscience. – (Conceptual advances in brain research ; v. 2)
1. Neurosciences – Soviet Union 2. Neurosciences – Soviet
Union – History
I. Miller, R. II. Ivanitsky, A. III. Balaban, P.
612.8′0947′0904

ISBN 90-5823-021-X
ISSN: 1029-2136

CONTENTS

SERIES PREFACE

The workings of the brain, including the human brain are a source of endless fascination. In the last generation, experimental approaches to brain research have expanded massively, partly as a result of the development of powerful new techniques. However, the development of concepts which integrate and make sense of the wealth of available empirical data has lagged far behind the experimental investigation of the brain. This series of books entitled Conceptual Advances in Brain Research (CABR) is intended to provide a forum in which new and interesting conceptual advances can be presented to a wide readership in a coherent and lucid way.

The series will encompass all aspects of the sciences of brain and behaviour, including anatomy, physiology, biochemistry and pharmacology, together with psychological approaches to defining the function of the intact brain. In particular, the series will emphasise modern attempts to forge links between the biological and the psychological levels of describing brain function. It will explore new cybernetic interpretations of the structure of nervous tissue; and it will consider the dynamics of brain activity, integrated across wide areas of the brain and involving vast numbers of nerve cells. These are all subjects which are expanding rapidly at present. Subjects relating to the human nervous system as well as clinical topics related to neurological or psychiatric illnesses will also make important contributions to the series.

These volumes will be aimed at a wide readership within the neurosciences. However, brain research impinges on many other areas of knowledge. Therefore, some volumes may appeal to a readership, extending beyond the neurosciences. Books suitable for the series are monographs, edited multiauthor collections or books deriving from conferences, provided they have a clear underlying conceptual theme. In order to make these books widely accessible within the neurosciences and beyond, the style will emphasise broad scholarship comprehensible by readers in many fields, rather than descriptions in which technical detail of a particular speciality is dominant.

The next decades promise to provide major new revelations about brain function, with far-reaching impact on the way we view ourselves. These great breakthroughs will require a broad interchange of ideas across many fields. We hope that the CABR series plays a significant part in the exploration of this important frontier of knowledge.

LIST OF CONTRIBUTORS

Aleksandrov, I.O.
Institute of Psychology
Russian Academy of Sciences
Yaroslavskaya 13
Moscow 129366
Russia

Alexandrov, L.I.
Institute of Higher Nervous Activity and Neurophysiology
Russian Academy of Sciences
Butlerova St. 5A
Moscow 117865
Russia

Alexandrov, Yu.I.
Institute of Psychology
Russian Academy of Sciences
Yaroslavskaya 13
Moscow 129366
Russia

Anokhin, K.V.
PK Anokhin Institute of Normal Physiology
Russian Academy of Medical Sciences
Leninskii Pr. 14
Moscow 117901
Russia

Balaban, P.M.
Institute of Higher Nervous Activity and Neurophysiology
Russian Academy of Sciences
Butlerova St. 5A
Moscow 117865
Russia

Bezdenezhnych, B.N.
Institute of Psychology
Russian Academy of Sciences
Yaroslavskaya 13
Moscow 129366
Russia

Bodunov, M.V.
Institute of Psychology
Russian Academy of Sciences
Yaroslavskaya 13
Moscow 129366
Russia

Dubrovinskaya, N.V.
Institute of Developmental Physiology
Russian Academy of Education
Pogodinskaya ul. 8
Moscow 119905
Russia

Dumenko, V.N.
Institute of Higher Nervous Activity and Neurophysiology
Russian Academy of Sciences
Butlerova St. 5A
Moscow 117865
Russia

Evtikhin, D.V.
Biology Faculty
M V Lomonosov State University
Vorobjevy Gory
Moscow 117234
Russia

Gavrilov, V.V.
Institute of Psychology
Russian Academy of Sciences
Yaroslavskaya 13
Moscow 129366
Russia

Golubeva, T.B.
Institute of Higher Nervous Activity and Neurophysiology
Russian Academy of Sciences
Butlerova St. 5A
Moscow 117865
Russia

Gorkin, A.G.
Institute of Psychology
Russian Academy of Sciences
Yaroslavskaya 13
Moscow 129366
Russia

Grechenko, T.N.
Institute of Psychology
Russian Academy of Sciences
Yaroslavskaya 13
Moscow 129366
Russia

Grinchenko, Yu.V.
Institute of Psychology
Russian Academy of Sciences
Yaroslavskaya 13
Moscow 129366
Russia

Gulyaeva, N.V.
Institute of Higher Nervous Activity and Neurophysiology
Russian Academy of Sciences
Butlerova St. 5A
Moscow 117865
Russia

Ioffe, M.E.
Institute of Higher Nervous Activity and Neurophysiology
Russian Academy of Sciences
Butlerova St. 5A
Moscow 117865
Russia

Ivanitsky, A.M.
Institute of Higher Nervous Activity and Neurophysiology
Russian Academy of Sciences
Butlerova St. 5A
Moscow 117865
Russia

Korneeva, E.V.
Institute of Higher Nervous Activity and Neurophysiology
Russian Academy of Sciences
Butlerova St. 5A
Moscow 117865
Russia

Kudriashov, I.E.
Institute of Higher Nervous Activity and Neurophysiology
Russian Academy of Sciences
Butlerov St. 5A
Moscow 117865
Russia

Kudriashova, I.V.
Institute of Higher Nervous Activity and Neurophysiology
Russian Academy of Sciences
Butlerova St. 5A
Moscow 117865
Russia

Kulakovsky, Yu.V.
Institute of Developmental Physiology
Russian Academy of Education
Pogodinskaya ul. 8
Moscow 119905
Russia

Latanov, A.V.
Biology Faculty
M V Lomonosov State University
Vorobjevy Gory
Moscow 117234
Russia

Leonova, A.Yu.
Biology Faculty
M V Lomonosov State University
Vorobjevy Gory
Moscow 117234
Russia

Machinskaya, R.I.
Institute of Developmental Physiology
Russian Academy of Education
Pogodinskaya ul. 8
Moscow 119905
Russia

Maksimova, N.E.
Institute of Psychology
Russian Academy of Sciences
Yaroslavskaya 13
Moscow 129366
Russia

Miller, R.
Department of Anatomy and Structural Biology
University of Otago
PO Box 913
Dunedin
New Zealand

Popova, N.K.
Institute of Cytology and Genetics
Siberian Branch of Russian Academy of Sciences
Pr. akademika Lavrenteva 10
Novosibirsk 630090
Russia

Raevsky, V.V.
Institute of Higher Nervous Activity and Neurophysiology
Russian Academy of Sciences
Butlerova St. 5A
Moscow 117865
Russia

Saulskaya, N.B.
I.P. Pavlov Institute of Physiology
Russian Academy of Sciences
6 Makarova Quay
V-34, St Petersburg 199034
Russia

Sergin, V.Ya.
Neuroinformatics Laboratory
Russian Academy of Sciences Far East Division
9 Piyp Avenue
Petropavlovsk-Kamchatsky 683006
Russia

Shapovalova, K.B.
I.P. Pavlov Institute of Physiology
Russian Academy of Sciences
6 Makarova Quay
V-34, St Petersburg 199034
Russia

Shevchenko, D.G.
Institute of Psychology
Russian Academy of Sciences
Yaroslavskaya 13
Moscow 129366
Russia

Silkis, I.G.
Institute of Higher Nervous Activity and Neurophysiology
Russian Academy of Sciences
Butlerova St. 5A
Moscow 117865
Russia

Simonov, P.V.
Institute of Higher Nervous Activity and Neurophysiology
Russian Academy of Sciences
Butlerova St. 5A
Moscow 117865
Russia

Sokolov, E.N.
Department of Psychophysiology
Lomonosov Moscow State University
8/5 Mokhovaya
Moscow 103009
Russia

Stepanichev, M.Yu.
Institute of Higher Nervous Activity and Neurophysiology
Russian Academy of Sciences
Butlerova St. 5A
Moscow 117865
Russia

Strelets, V.B.
Institute of Higher Nervous Activity and Neurophysiology
Russian Academy of Sciences
Butlerova St. 5A
Moscow 117865
Russia

INTRODUCTION

There are many difficulties hindering Western scientists from getting to know Russian scientific literature: these have included political barriers, as well as the barrier of language. Translation of Russian papers often leaves the reader with much hard work to do, because the translations seldom give much attention to capturing the nuances of the original. Apart from these difficulties, the conceptual language of Russian papers is often unfamiliar, which creates an additional and substantial barrier. However, Russian science has a long and proud tradition going back to Peter the Great and the founding of the Russian Academy of Science in 1725. Since then there have been many famous names, including figures such as Lomonosov, Lobachevsky and Mendeleef. In neuroscience, the name of Pavlov (the 150th anniversary of whose birth coincides with the publication of the present book) is familar to Western scientists. However, there were major figures in Russian neuroscience in generations before this, who are little known in the West, notably Sechenov, working in the middle years of last century, a pioneer in the study of reflexes. In the twentieth century, further major influences on Russian neuroscience came from the work of Vvedensky and Ukhtomsky. The writings of these physiologists are hardly known at all in the West.

This book, initiated by Professor Ivanitsky, contains a collection of chapters representing many avenues of contemporary research in Russian neuroscience. The chapters range from basic research at the cellular level, to studies of higher nervous function in animals and humans, including innovative analyses of the EEG, comparative studies, psychopharmacology and neurochemistry, as well as papers with a more philosophical content. Some of the chapters describe particular research projects, others are wider-ranging reviews of work that has been in progress for many years. The chapters vary in their conceptual difficulty. However, it is hoped that the editing of the original versions has made the meaning as clear as possible, so that the difficulties for the reader are only those inevitably associated with novel concepts, rather than the subsidiary and unnecessary ones of language.

Having carefully read all eighteen chapters in this book I have gained several strong impressions about the characteristics of Russian neuroscience which I will attempt to summarise.

Firstly, in many of the chapters in this book there is a rich sense of history. Authors are cited from the early years of this century, from the nineteenth century, and in one case from the eighteenth century, as though they are part of a living tradition. Some of these older authors are Russian, but it is also clear that many of the chapter authors are familiar with early works in English, German and French.

The second point is that all the chapters really are concerned with conceptual advance. One sees many issues being addressed in these chapters which are so fundamental that they might slip by without one realizing what the main question is. Examples are the discussion of fundamental concepts of neurotransmitter function by Saulskaya, or of the mechanisms of synaptic plasticity by Silkis, the explicit discussion of different explanations of Gestalt representation by Sokolov, comparisons of radically different models of memory by Anokhin, as well as issues of psychobiological development by Raevsky, and by Dubrovinskaya and her colleagues. At the level of the higher human faculties, the reader will be challenged by the concise account of emotions by Simonov, and by reviews of the physiological correlates of conscious awareness in chapters by Ivantisky and Sergin. Always there is the attempt to go beyond immediate data to discover general principles.

The strong sense of history and the concern with basic conceptual questions are closely related. It has been said (in another context), that those who are ignorant of history are condemned to repeat it. In the context of science, researchers who are ignorant of the history of their subject are condemned to repeat old experiments, using ever more elaborate and explicit techniques perhaps, but without major conceptual advance. Only when one knows the history of one's subject well does one have a clear idea of what the real conceptual issues are.

Several of the chapters in this book focus on rhythmic aspects of brain activity. Study of oscillatory processes has been a prominent theme in Russian science as a whole, not only in brain research. In Russian neuroscience this theme was prominent in the work of Vvedensky. In more recent times the idea that coincidence of rhythms is of major significance in information processing was been developed some time before it became a focus for Western neuroscience.

Although I seldom read papers in ethology, another feature which strikes me is the concern for studying behaviour in settings as close as possible to the natural environment. I am fascinated by (but can only imagine) what it must be like to study the genetics of behaviour in silver foxes, in Novosibirsk (Popova's paper). In the same connection, I also sense a keen awareness of the continuity between animal species and humans.

Underlying many of these chapters is a sense that the functions of the brain are an integrity, which can actually be understood. In so much neuroscience literature one senses that the researchers regard the highest nervous functions (such as those which define a 'person') as something which is metaphysically separate from what can actually be studied. This leads to an emphasis on the sensory and motor pathways, the 'way in' and the 'way out', without ever really confronting as a major scientific issue the neurobiology of that central entity lying between the way in and the way out. This way of thinking seems to be preserved in old (but still often unchallenged) views of a homunculus, and in more recent times, by ideas that there is a 'central executive' controlling attentional processes. From reading works of Alexander Luria, it has become clear to me that he at least could surmount this difficulty, and think of the integration of the whole person as a fascinating scientific question. From reading the chapters in the present book, it is evident that this is a characteristic not only of Luria, but of a whole tradition of neuroscience and psychology in Russia. Therefore, this book is recommended to anyone who wants a view of the brain which goes beyond the more simple-minded reductionism of human psychological faculties.

Several of the chapters raise issues of the relation between mind and brain, either explicitly, or as underlying assumptions. These chapters obviously give importance to the process of the investigator 'looking inwards' to discover his or her own mental processes. This has also become a respectable approach in Western neuroscience in recent decades. It is an interesting question to ask whether this aspect of Russian neuroscience is as recent as that in the West, or has been prominent in earlier times, when behaviorism was dominant in North America. Luria, at least was interested in a cognitive approach when, in the West, behaviorism held sway. Beyond this issue, it is clear that the scientists whose work appears in this book discussing the status of mind are fundamentally materialist in orientation, as are most of their Western colleagues. There is however a difference in tone to these discussions from those by Western neuroscientists with a philosophical orientation. For Russian neuroscientists, there is a step-by-step progression from the physiological phenomena to mental phenomena and then to social phenomena (and back again, if one has in mind a reflex arc). A distinctive part of this sequence, also a part of Luria's thinking, is that 'mind' is mainly a social creation. These ways of thinking about mind are sometimes found amongst Western writers, but are a much more consistent feature in the chapters of this book.

For myself, a materialist view is inadequate.The physiological events associated with conscious awareness, to my way of thinking, are only correlates of consciousness, not its explanation. If terms are defined carefully, there is no common language crossing between physiological events and inner subjective experience, and therefore there is no way of constructing a true explanatory argument which crosses between these two modes of description. Therefore, I would prefer some sort of dualist or parallelist approach (though not one which allows causal interaction *sensu stricto* between the two levels). Nevertheless, both Ivanitsky and Sergin use the word 'mystery' in connection with mind-brain relationships. Maybe these authors have some sympathy for such an approach.

There is a final question I raise: to what extent are the distinctive features I have noticed in these chapters a product of Soviet science, or are they part of a longer tradition which predates the Soviet era? I do not know the answer to this question, but my suspicion is that much of this tradition is pre-Soviet, and also that there have been strong efforts to keep alive the older tradition of neuroscience. Thus, L.R. Graham (*Science in Russia and the Soviet Union*, Cambridge University Press, 1993) comments that as far back as the eighteenth century, Russian's were attracted to the idea, stemming from Locke, that environmental influences form the mind. Luria (*The Making of Mind,* Harvard University Press, 1979) mentions that Sechenov's work—'Reflexes of the Brain'—written in the middle of last century included an explicit program for explaining mental phenomena as the central link in a reflex arc. Similar ways of thinking are to be found in several of the chapters in this book. Apart from these specific examples, my overall impression of this book is that the sense of history it conveys is so strong that many of the underlying habits of thought by the scientists who write here derive from a tradition which predates the Soviet era.

Although I have travelled twice in Russia, the barriers (mentioned above) in the way of fully understanding Russian neuroscience mean that my own knowledge is still rather superficial. There is a real danger that major programs of scientific work, and important conceptual developments in Russia in the last generation will be known to

Western scientists only if they are Russian speakers, and have access to Russian publications. At the same time, Russian scientists are very much aware that in the past they have been separated from the scientific developments in the West, and hope for much closer links with Western scientists. The present work is just a small contribution to making the large body of Russian neuroscience research better known in the West. It is hoped that it may lead to improved dialogue between Russian brain researchers and their colleagues in the West. There is undoubtedly scope for many projects similar to the present book. Such work is demanding in time, but is nevertheless rewarding. I thank the contributors to this book for their papers, and hope that our editing of them is sufficiently lucid that their work will be read widely in the West. I also thank the following scientists at Otago University for their assistance in editing these chapters: Professor W.C. Abraham (Department of Psychology), Professor K.G. White (Department of Psychology) and Dr. J.R. Wickens (Department of Anatomy and Structural Biology).

R. Miller

1 Volume Transmission in the Striatum as Constituting Information Processing

N.B. Saulskaya

I.P. Pavlov Institute of Physiology, Russian Academy of Science, St. Petersburg, Russia
nbs@infran.ru

In this paper we review experimental evidence from the literature and our own studies in favour of a special mode of intercellular communication in the striatum—volume transmission. It is characterised by diffusion of neurotransmitters from release points through the extracellular fluid of the brain to distant non-synaptic receptor sites. We discuss the evidence for the existence of volume transmission of three important classical neurotransmitters of the striatum (dopamine, glutamate and GABA), the tentative mechanisms underlying this means of neuronal communication, receptors involved, and the role of volume transmission in the striatum in controlling behavioural functions.

KEYWORDS: synaptic transmission, interneuronal communication, diffusion

1. INTRODUCTION

Synaptic transmission, based on precise neurone-to-neurone signalling, is proposed to be the basic tenet of the neurone doctrine. Over the last decade, however, another mode for interneuronal communication in the central nervous system (CNS) has been advanced and has gained experimental support (Otellin and Arushanian, 1989; Sakharov, 1990; Agnati *et al.*, 1995; Bach-y-Rita, 1993; Grace, 1991; Zigmond *et al.*, 1990). This new concept is based on diffusion of neurotransmitters and other biologically active compounds through the brain extracellular fluid to distant receptors. Agnati *et al.* suggested the term "volume transmission" to define this complementary means of intercellular communication (Agnati *et al.*, 1995). By volume transmission, neurotransmitters may spread for distances beyond the point of release through the extracellular space and exert their activity at multiple receptor sites within a brain area; this may permit the area to operate as a unified whole.

Several investigators have undertaken an historical analysis of this idea (Agnati *et al.*, 1995; Bach-y-Rita, 1993). The concept can be traced to Golgi's reticular tenet, postulating that the CNS operates as a global neuronal continuity in which all elements are connected to others (see Agnati *et al.*, 1995). In the Russian physiological school, D.A. Sakharov has proposed a critical revision of synaptic theory in his concept of "heteron" (Sakharov, 1990). Using morphological studies, V.A. Otellin has provided evidence in favour of a non-synaptic nature of interactions between different neurotransmitter systems in the CNS (Otellin and Arushanian, 1989).

The first indications that neurotransmission by diffusion may exist in the CNS were provided by morphological studies showing receptors for glutamate, GABA, monoamines and neuropeptides outside classical synapses in many brain areas (Otellin and Arushanian, 1989; Agnati *et al.*, 1995; Bjorklund and Lindvall, 1986; Groves *et al.*, 1994; La Gamma *et al.*, 1994; Levey *et al.*, 1993; Martin *et al.*, 1993; Petralia *et al.*, 1996; Yung *et al.*, 1995). This hypothesis gained further support from studies using *in vivo* microdialysis, which provided an opportunity to describe directly the processes occurring in extracellular space of the brain. These studies have revealed that, among the classical neurotransmitters found in the extracellular fluid at concentrations within the range required for activation of at least metabotropic non-synaptic receptors, are monoamines, glutamate and GABA (Abercrombie *et al.*, 1990; Conor *et al.*, 1991; Gonon *et al.*, 1994; Imperato *et al.*, 1992; Parsons *et al.*, 1991; Westerink *et al.*, 1987). The efflux of the neurotransmitter dopamine from the synaptic cleft (Garris *et al.*, 1994) and its diffusion through the extracellular fluid over a long distance (Parsons *et al.*, 1991; Stamford *et al.*, 1988) has been demonstrated in direct experiments. Taken together, these results strongly imply that release and spread of classical neurotransmitters within the extracellular fluid is a mode of information-handling in the CNS, rather than a non-functional component of the synaptic release of neurotransmitters that have already exerted their physiological action. This hypothesis has raised a number of important questions which currently remain unanswered. For example, what information is conveyed by volume transmission? How important is this information for the expression of adaptive behaviour? What mechanisms underlie this mode of interneuronal communication? Most of these questions represent a major challenge to future research in the field.

Here we discuss current ideas concerning some of the problems mentioned above. This paper focuses on three important classical neurotransmitters of the striatum (dopamine, glutamate and GABA) which may act via the volume transmission mode in this brain area. We review the evidence in favour of the physiological importance of volume transmission of these neurotransmitters in controlling striatal functions and in expression of behaviour regulated by this brain area.

2. THE EXISTENCE OF VOLUME TRANSMISSION IN THE STRIATUM

The striatum is a large forebrain structure, which is implicated in the normal control of motor functions as well as emotional and motivational processes (Otellin and Arushanian, 1989; Shapovalova *et al.*, 1992). The main function of the striatum is proposed to be the gathering of information from different cortical areas, and then conveying the integrated signals to the brainstem pre-motor area (i.e. the pedunculo-pontine nucleus), on the one hand, and back to the cortex, on the other hand (Carlsson and Carlsson, 1990; Goldman-Rakic and Selemon, 1990; Shapovalova *et al.*, 1992).

The structural units of the striatum are considered to be the principal GABA-ergic medium-sized spiny neurones; these are (simultaneously) the target cells for the major afferent systems to the striatum and the projection neurones of this brain area (Smith and Bolam, 1990). In addition to a cortical glutamatergic input, all regions of the striatum receive topographical dopaminergic input from the ventral tegmental area and the substantia nigra (Shapovalova *et al.*, 1992; Bjorklund and Lindvall, 1986). Morphological studies have revealed a convergence of cortical (glutamatergic) and dopaminergic inputs on the same dendritic spines of principal GABA-ergic striatal neurones (Smith and Bolam, 1990) that underlies interactions between dopaminergic, glutamatergic and GABA-ergic systems in this brain region.

During the past two decades, several models of striatal function have been advanced (Shapovalova *et al.*, 1992; Carlsson and Carlsson, 1990; Goldman Rakic and Selemon, 1990). All of them are based solely on synaptic transmission as the means of inter-neuronal communication in this brain area. Nonetheless, convincing arguments have been made that volume transmission may also underlie cross-talk between axon terminals and neurones in the striatum.

In this respect, most investigators have concentrated their studies on the diffuse action of dopamine in the striatum. Since the discovery of dopaminergic systems in the CNS, the idea that dopamine might be considered as a classical neurotransmitter has been called into question several times. Some studies have revealed non-classical morphological features of dopamine synapses, such as the formation of multiple presynaptic varicosities *en passage*, and a lack of postsynaptic densities in some of these synaptic contacts (Bjorklund and Lindvall, 1986; Smith and Bolam, 1990). Nevertheless, subsequent analysis using electron microscopic preparations, has revealed evidence of classic pre- and postsynaptic densities in at least some dopamine synapses (Smith and Bolam, 1990).

A very interesting analysis of processes underlying dopamine efflux from the synaptic cleft in the ventral striatum has been undertaken by Garris *et al.* (Garris and Wightman, 1994; Garris *et al.*, 1994). Dopamine synapses in the striatum have been described as two parallel, thickened membranes, 300 nm in length, with a synaptic cleft of 15 nm (Garris *et al.*, 1994; Groves *et al.*, 1994). Synaptic vesicles are densely packed not only in the presynaptic region but also in the adjacent axon segments. As originally stressed by Garris *et al.* (Garris and Wightman, 1994; Garris *et al.*, 1994) high-affinity dopamine uptake (Near *et al.*, 1985) was suggested to be the means of terminating the action of dopamine in the synaptic cleft. Moreover, the extracellular space was considered to be a separating zone between synapses (Gonon *et al.*, 1987). Nevertheless Garris *et al.* (Garris and Wightman, 1994; Garris *et al.*, 1994) have shown, that striatal membranes express dopamine uptake sites underlying dopamine reuptake at a concentration of 5.9 pmol/mg protein. Taking into account the density of dopaminergic synapses (one synapse per 4 μm), the number of uptake sites per dopamine synapse is calculated to be 1750 uptake sites/synapse (Garris *et al.*, 1994). However, in spite of such a high density of dopamine uptake sites, and approximately the same calculated density of dopamine receptors (1655 D1 receptor sites and 433 D2 receptor sites per synapse), recent investigations using fast-scan cyclic voltammetry have revealed that dopamine released in response to a single stimulus pulse (approximately 1000 molecules) escapes from a synaptic cleft and penetrates into the extracellular space (Garris *et al.*, 1994). These calculations imply that the majority of dopamine reuptake sites and dopamine receptors are located outside dopamine synapses (Garris *et al.*, 1994). Garris *et al.*, postulated that the dopamine synapse is designed for the effective efflux of dopamine from the synaptic cleft to the extracellular space. The reuptake system is proposed to regulate extrasynaptic dopamine levels and the distance that dopamine can diffuse from the synapse (Garris and Wightman, 1994; Garris *et al.*, 1994).

The hypothesis of non-synaptic action of dopamine in the striatum has been further substantiated by studies showing that, in the rat striatum, there is no spatial correspondence between sites of dopamine release and sites of dopamine receptor concentration. Indeed, morphological evidence has been obtained that dopamine terminals in the striatum form synaptic contact on the neck of dendritic spines of principal striatal neurones (Groves *et al.*, 1994; Smith and Bolam., 1990) whereas the majority of striatal D1 and D2 receptors are located on spine heads, i.e. far from sites of dopamine release (Levey *et al.*, 1993). Interestingly, glutamate terminals that originate from cortical areas also form synaptic contact on spine heads of principal striatal neurones (Smith and Bolam, 1990) (Figure 1.1).

The presynaptic interactions in the striatum might be another important piece of evidence for the existence of volume transmission that involves not only dopamine but also glutamate, GABA and acetylcholine (Abercrombie and Keefe, 1991; Connor *et al.*, 1991; Zigmond *et al.*, 1990) in this brain region. As mentioned, in these and other studies, these neurotransmitters exert a direct tetrodotoxin-insensitive presynaptic action on each other's release, despite an apparent absence of axo-axonic synaptic contacts (only 7 axo-axonic synapses per 100,000 synaptic junctions)(Otellin and Arushanian, 1989; Kornhuber and Kornhuber, 1986; Zigmond *et al.*, 1990). These data have led some investigators to conclude that presynaptic interactions in the striatum are based primarily on diffuse action of neurotransmitters (Otellin and Arushanian, 1989; Grace, 1991; Zigmond *et al.*, 1990). Diffusion of dopamine throughout the extracellular space in dorsal and ventral striatum has been demonstrated in direct experiments (Parsons *et al.*, 1991; Stamford *et al.*, 1988). The length of the diffusion path for a dopamine molecule in the striatum is about 100 μm (Parsons *et al.*, 1991). However, in the dopamine-denervated striatum, this length normally increases up to 1 mm (Stamford *et al.*, 1988), a distance that corresponds to the size of striatal cell clusters—a small group of medium-sized spiny principal neurones lying around a large aspiny cholinergic interneurone (Goldman-Rakic and Selemon, 1990). Recent studies using electron microscopy combined with immunostaining with a monoclonal antibody against choline acetyltransferase have revealed that only 8% of cholinergic axon terminals in the rat striatum form classical synaptic junctions (Contant *et al.*, 1996), whereas striatal neurones express a high amount of acetylcholine receptors (Shapovalova *et al.*, 1992; Smith and Bolam, 1990). This finding makes it possible that a cholinergic interneurone located at the centre of a cell cluster of the striatum interacts with other neurones in the cluster, presumably via volume transmission. Irrespective of the role that cell clusters play in the striatum, it is likely that the diffusion of extracellular dopamine and acetylcholine within the cluster serves as an important mechanism of integration between cells in the cluster, which influences the functional state of the cluster as a whole.

Molecular diffusion of glutamate and GABA in the striatal extracellular space has not been investigated. However, studies using microdialysis have shown that glutamate and GABA are permanently present in the striatal extracellular fluid at concentrations of 10^{-6} M and 10^{-7} M respectively (Saulskaya and Marsden, 1995a, 1996; Connor *et al.*, 1991; Saulskaya and Marsden, 1995c).

A current belief is that in the striatum, $GABA_B$ receptors serve as presynaptic ones. Indeed, as demonstrated in morphological and electrophysiological studies, striatal neurones express both $GABA_A$ and $GABA_B$ receptors that appear to be non-uniformly dis-

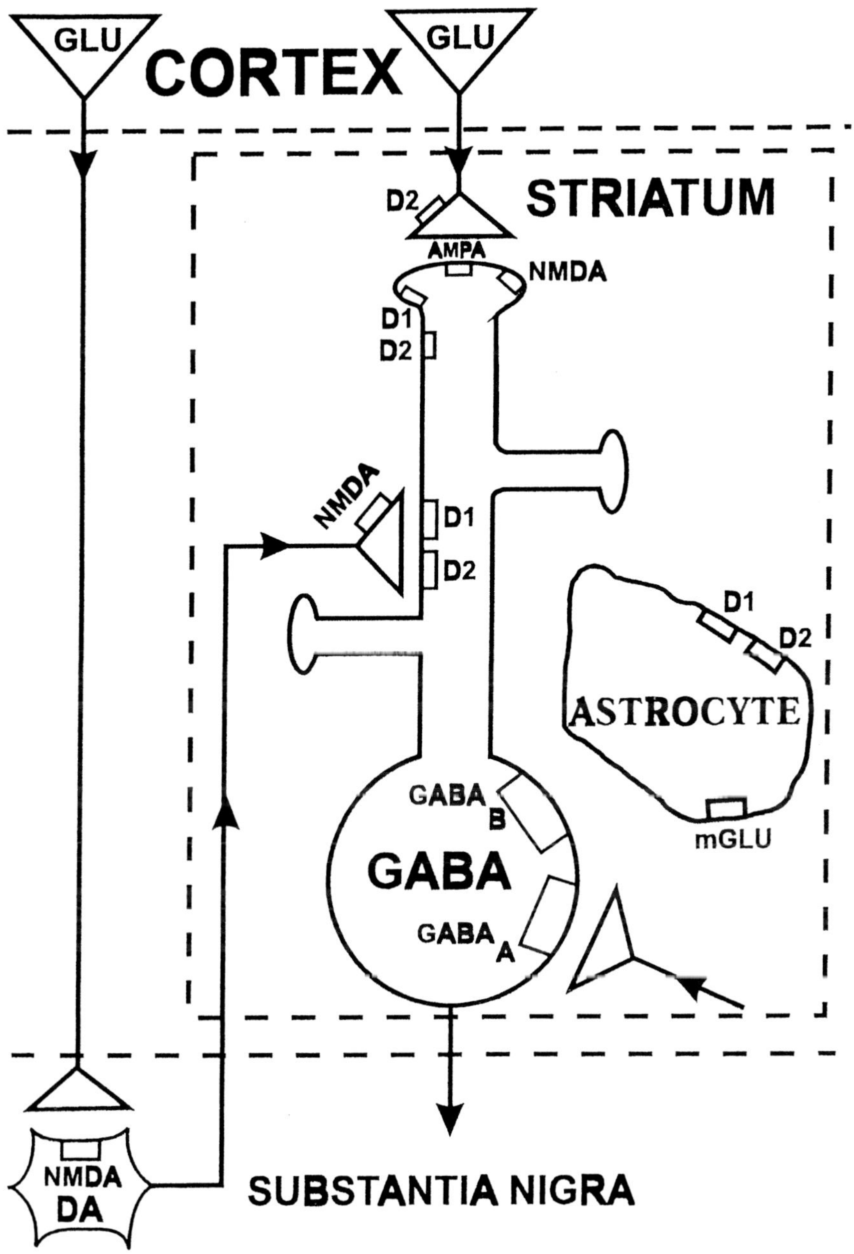

Figure 1.1. Tentative receptor mechanisms involved in volume transmission of dopamine, glutamate and GABA in the striatum (Otellin and Arushanian, 1989; Agnati *et al.,* 1995; Calabresi *et al.*, 1990; Grace, 1991; Groves *et al.*, 1994; Levey *et al.*, 1993; Martin *et al.*,1993; Petralia *et al.*, 1996; Shi and Rayport, 1994, Smith and Bolam., 1990; Yung *et al.*, 1995). GABA: Principal GABA-ergic medium-sized spiny neurones of the striatum that provide integration of the input information to form output signals; D1, D2: dopamine receptors, DA: dopamine; NMDA: NMDA receptor; AMPA: AMPA/kainate receptors. mGLU: metabotropic glutamate receptor. GLU: glutamate. $GABA_A$, $GABA_B$: GABA receptors.

tributed (Calabresi *et al.,* 1990; Shi and Rayport, 1994). In particular, $GABA_A$ receptors appear to show preferential localisation to postsynaptic sites, and they are responsible exclusively for synaptic action of GABA, whereas $GABA_B$ receptors show non-synaptic and presynaptic localisation (Shi and Rayport, 1994). Taken together these data have led to the suggestion that $GABA_B$ receptors in the striatum account for most of the GABA receptors for volume transmission in this brain area.

Non-synaptic glutamate receptor have been revealed in many brain regions (Agnati *et al.,* 1995). Striatal neurones express a high density of NMDA, AMPA/kainate and metabotropic glutamate receptor subtypes. However, the postsynaptic actions of glutamate in the striatum appear to be mediated solely via AMPA/kainate receptors (Herrling, 1985) that are normally located at postsynaptic sites (Martin *et al.*, 1993). This receptor subtype has not been observed at presynaptic sites in the striatum (Martin *et al.*, 1993). These morphological findings suggest, but of course do not prove, that in the striatum, extracellular glutamate exerts its action presumably via presynaptic and extrasynaptic NMDA and metabotropic glutamate receptors. Future research needs to address more specifically the question of the glutamate receptor subtypes involved in volume transmission in this brain area.

Membrane receptors of astroglia have been proposed to be another important target for non-synaptic action of extracellular neurotransmitters (Agnati *et al.,* 1995; Bach-y-Rita, 1993; La Gamma *et al.*, 1994; Martin *et al.,* 1993). At the present time, studies using electron microscopy combined with immunostaining have revealed that striatal astrocytes express dopamine, metabotropic glutamate, but not AMPA/kainate receptors (La Gamma *et al.*, 1994; Petralia *et al.*, 1996; Martin *et al.*, 1993).

3. MECHANISMS UNDERLYING VOLUME TRANSMISSION IN THE STRIATUM

The level of dopamine in the striatal extracellular fluid is proposed to be regulated by several independent mechanisms. The first one is a synaptic vesicular release, due to exocytosis. This process is impulse- and Ca^{++}-dependent (Abercrombie and Keefe, 1991). The second process is high affinity uptake of released dopamine, which is the primary mechanism by which dopamine is inactivated (Abercrombie and Keefe, 1991). The third mechanism is the non-vesicular carrier-mediated release of dopamine, due to reversal of the dopamine re-uptake mechanism. This process is independent of impulses and Ca^{++} (Levi and Raiteri, 1993). The balance between these processes is suggested to be the major determinant of the extracellular dopamine level in the striatum (Abercrombie and Keefe, 1991). As has been shown recently, enzymatic degradation of released dopamine does not play a role in determining the basal dopamine level in this brain area (Justice *et al.*, 1994). The origin of extracellular glutamate and GABA in the striatum appears to be vesicular and nonvesicular release, limited by re-uptake (Smolders *et al.*, 1994). A significant proportion of glutamate and GABA (60–80%) in the striatal extracellular space arises via vesicular and non-vesicular release from neurones (Smolders *et al.*, 1994). In addition, extracellular glutamate and GABA may arise from glial cells.

Electrophysiological recording from identified dopaminergic neurones has shown that under physiological conditions, these cells typically exhibit two patterns of discharge activity (Figure 1.2): either single spikes at frequencies averaging 3–4 Hz (pacemaker-like firing) or bursts of action potentials (2 to 6 action potentials at a frequency of 15 Hz) (Grace, 1991). Electrical stimulation of the nigro-striatal dopaminergic pathway, mimicking the spontaneous bursting pattern, is several times more potent in effecting dopamine release than regularly spaced ones having the same average frequency (Gonon, 1994). These data have led to the conclusion that the physiological significance of burst-like firing of dopaminergic neurones in the striatum is to initiate volume transmission. In contrast, pacemaker-like firing of dopaminergic neurones results in synaptic transmission, and under these conditions, dopamine does not escape from the synaptic cleft, and can exert its action on synaptic receptors in close proximity to sites of release. Since burst firing is never observed in experiments *in vitro* (Sanghera *et al.*, 1984), this pattern appears to depend on the activity of afferent fibres (Kalivas, 1993). A detailed analysis of this phenomenon undertaken by Kalivas, reveals that glutamatergic cortical inputs of dopaminergic cells are responsible, at least partly, for converting pacemaker-like firing in dopaminergic cells of the ventral tegmental area into burst-firing patterns (Kalivas, 1993). Electrical stimulation of the prefrontal cortex converts dopamine neuronal activity into bursting patterns (Gariano and Groves, 1988; Kalivas, 1993) and cooling the prefrontal cortex converts spontaneous burst firing of dopamine cells back to pacemaker-like firing (Svensson and Tung, 1989; Kalivas, 1993). Moreover, the local administration of NMDA to the ventral tegmental area induces burst firing accompanied by dopamine release in the ventral striatum (Saud-Chagny *et al.*, 1991).

These data allow a conclusion to be drawn, that in the striatum, dopaminergic volume transmission is typically initiated by burst firing of dopaminergic neurones induced by glutamatergic cortical signals through NMDA receptor activation.

In addition, the prefrontal cortex (as well as other glutamatergic afferent sites that project to the striatum) may influence dopaminergic volume transmission in this brain area via presynaptic mechanisms. Administration of excitatory amino acids or their analogues into the striatum elicits an increase in extracellular striatal dopamine levels in a tetrodotoxin-insensitive manner (Shapovalova *et al.*, 1992; Abercrombie and Keefe, 1991). Interestingly, experiments have shown that neither local application of glutamate antagonists into the striatum via the microdialysis probe (Shapovalova *et al.*, 1992; Abercrombie and Keefe, 1991) nor damage to the cortical area projecting to the striatum (our own studies: Saulskaya and Gorbachevskaya, 1997; Saulskaya *et al.*, 1996), influences basal dopamine release into the striatal extracellular space. Therefore, this data suggests that glutamatergic inputs to the striatum only exert a transient influence on dopamine extracellular outflow in this brain area, which does not occur under resting conditions. In contrast, impulse activity in nigrostriatal dopaminergic neurones appears to be the principal determinant of extracellular dopamine concentration under basal conditions, as evidenced by the dramatic decreases in basal extracellular dopamine levels in the striatum after injections of tetrodotoxin into the medial forebrain bundle (Abercrombie and Keefe, 1991). However, in our recent study, we have demonstrated that learning causes long-lasting changes in the mechanisms involved in the presynaptic glutamatergic control of basal dopamine release into the extracellular space in the striatum (Saulskaya and Marsden, 1995b). Using microdialysis, we revealed an NMDA-dependent component of basal dopamine release in the ventral striatum, that appeared two hours after the acquisition of a conditioned emotional response in rats after learning

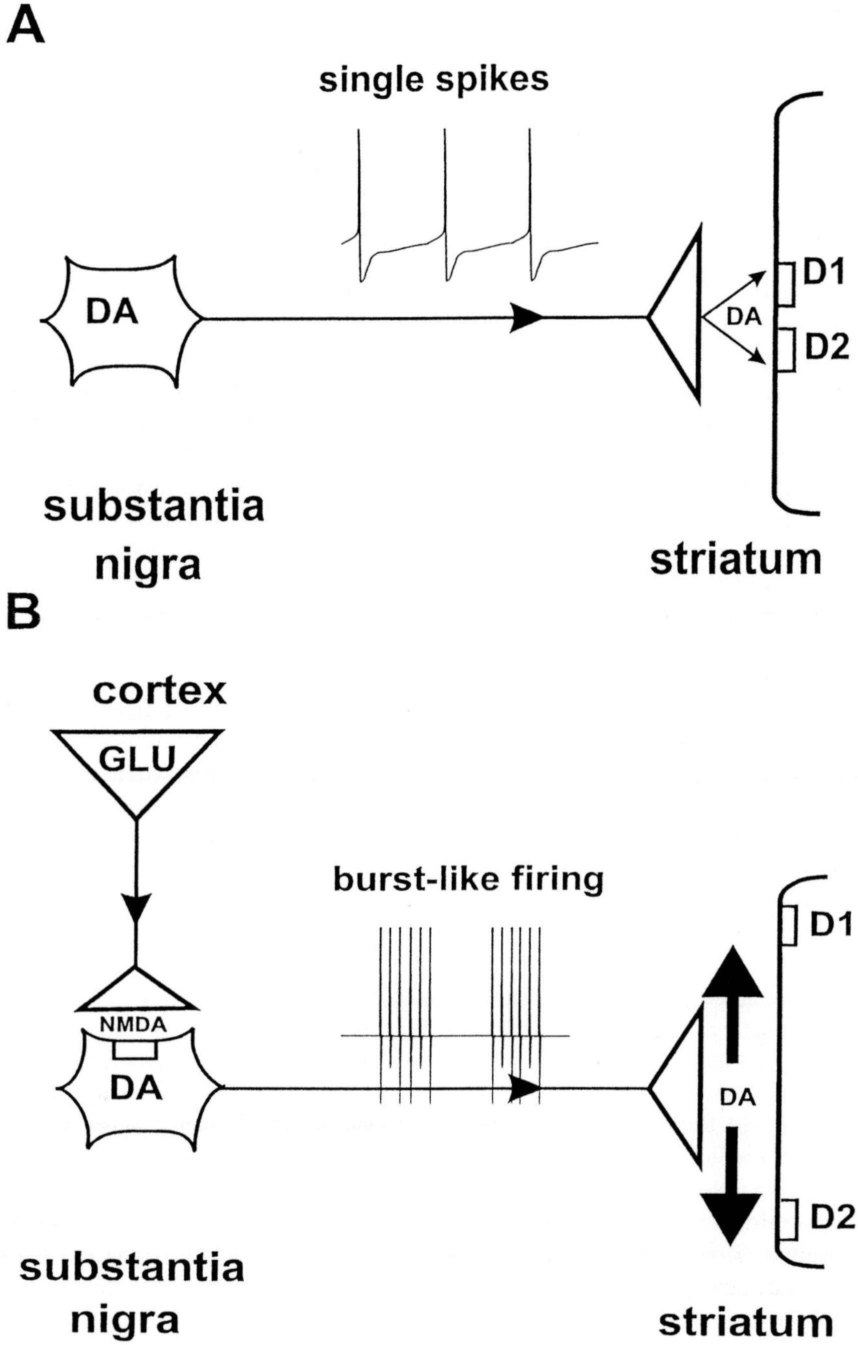

Figure 1.2. Tentative neuronal mechanisms underlying synaptic (A) and volume (B) transmission of dopamine in the striatum. Two patterns of activity of dopaminergic neurones, single spikes and burst firing, result in synaptic and volume transmission of dopamine in the striatum respectively (Garris *et al.*, 1994; Gonon, 1994; Gonon *et al.*, 1987; Grace, 1991) Burst firing of dopaminergic neurones is induced by glutamatergic cortical signals through NMDA receptor activation (Gariano and Groves, 1988; Sanghera *et al.*, 1984; Saud-Chagny *et al.*, 1991; Svensson and Tung, 1989). D1, D2: dopamine receptors; DA: dopamine; NMDA: NMDA receptor; GLU: glutamate.

(but not in untrained animals), although the apparent "basal" dopamine release had returned to normal.

Therefore, dopaminergic volume transmission in the striatum appears to be under the double control of cortical glutamatergic areas. Dopamine release into the striatal extracellular space may be induced either by burst firing of dopaminergic neurones initiated by cortico-nigral glutamatergic signals, or by a presynaptic cortico-striatal influence.

4. THE FUNCTIONAL ROLE OF VOLUME TRANSMISSION IN THE STRIATUM

Although a number of investigations using *in vivo* microdialysis, have provided evidence that extracellular levels of striatal dopamine, glutamate and GABA change in response to behavioural challenge (Saulskaya, 1993; Saulskaya and Marsden, 1994, 1995a,b, 1996; Imperato *et al.*, 1992; McCullough *et al.*, 1993; Phillips *et al.*, 1991; Shi and Rayport, 1994), until recently it has not been established whether these changes are essential for the expression of behavioural activity.

The first direct evidence for the functional importance of dopamine diffusion within the striatal extracellular fluid has been obtained in studies that revealed compensations after lesions of central dopaminergic neurones. These studies, using the neurotoxin, 6-hydroxydopamine, suggest that the subtotal loss of nigral dopaminergic neurones is compensated, to some extent, by increased release of dopamine from residual dopaminergic neurones (Zigmond *et al.*, 1990). This maintains the constancy of the striatal extracellular dopamine level, correlated with the absence of functional impairments (Abercrombie *et al.*, 1990; Parsons *et al.*, 1991; Zigmond *et al.*, 1990). Significant decrease in the striatal extracellular dopamine level accompanied by akinesia, develops only if degeneration of the dopaminergic neurones is almost complete (not less then 90%) (Abercrombie *et al.*, 1990; Zigmond *et al.*, 1990).

Moreover, we have observed a comparable phenomenon in studies of glutamatergic systems of the striatum. In particular, using microdialysis and measurements of glutamate release, we have revealed that partial excitotoxic lesions of the hippocampal glutamatergic inputs to the ventral striatum, made by infusion of ibotenic acid, influence neither basal glutamate extracellular level in this area nor open-field activity. Only a complete loss of neurones in the ventral subiculum and CA1 projecting to the ventral striatum is able to cause both a decrease in striatal extracellular glutamate levels and hyperlocomotion.

Therefore, extensive impairment of synaptic transmission dependent on impulse-flow in the striatum, due to damage to dopaminergic and glutamatergic inputs does not influence locomotion, whereas impairment of volume transmission influences this type of behavioural activity, which is known to be under the control of the striatum.

Taken together, these data suggest firstly that compensatory events occur to maintain the extracellular level of dopamine and glutamate after subtotal degeneration of dopaminergic and glutamatergic inputs to the striatum. Secondly, they indicate the functional significance of volume transmission of dopamine and glutamate for the expression of behavioural activity regulated by the striatum.

Regulation of locomotion by dopaminergic and glutamatergic systems in the striatum involving volume transmission, is thought to be an integral part of more global functions operating via this brain area. As postulated by Stricker and Zigmond, "...motivated behaviours can be described in terms of two components: one that directs activities toward a distinctive goal and one that energises activities regardless of their goal" (Stricker and Zigmond, 1989). Important functions of meso-striatal dopaminergic systems are considered to provide non-specific activational component of motivated behaviours (Stricker and Zigmond, 1989). The way they operate is by modulation of the striatal responsiveness to cortical glutamatergic inputs and, via stria-pallido-thalamo-cortical loop, the responsiveness of cortical areas to sensory inputs that influence the degree of arousal (Carlsson and Carlsson, 1990; Stricker and Zigmond, 1986). Various lines of evidence have indicated that it is volume transmission that underlies these functions of the dopaminergic systems (Phillips *et al.*, 1991; Joseph *et al.*, 1991). Data from the literature and our own studies have shown that polymodal environmental stimuli (e.g. mild stress, food and water intake, social and sexual contacts, novelty) are associated with prolonged dopamine release into the extracellular space of the striatum (Saulskaya, 1993; Saulskaya and Marsden, 1994, 1995b; Imperato *et al.*, 1992; Joseph *et al.*, 1991; McCullough *et al.*, 1993; Phillips *et al.*, 1991). Studies utilising brain microdialyses have observed increased extracellular levels of dopamine in the striatum during all behavioural situations studied. These include those involving movements (McCullough *et al.*, 1993) as well as inhibition of movements (Saulskaya, 1993; Saulskaya and Marsden, 1994, 1995b) as well as behaviours accompanied by either reward (Phillips *et al.*, 1991), or punishment (Saulskaya, 1993; Saulskaya and Marsden, 1994, 1995b). Both stress and relief from stress causes dopamine release into the ventral striatum (Imperato *et al.*, 1992). Therefore, initiation of volume transmission, reflected in dopamine release into the extracellular space of the striatum, is hypothesised to be a non-specific response occurring during situations that require enhanced behavioural responsiveness (Joseph *et al.*, 1991). This response depends neither on motivation underlying ongoing behaviour, nor on a pattern of movement activity.

The standard view holds that the action of dopamine in the striatum is mediated by two families of G-protein-coupled receptors, whose stimulation exerts both immediate and delayed action on striatal neurones. One of the most important consequences of dopamine's action in the region of the striatum is the control of signal-to-noise ratio. Enhanced dopamine levels appear to be associated with inhibition of spontaneous activity of principal striatal neurones, and with an increase in their response elicited by glutamatergic inputs (Rolls *et al.*, 1984; Stamford *et al.*, 1988). Furthermore, extracellular dopamine decreases the degree of electrotonic coupling between striatal neurones via gap junctions (Donnel and Grace, 1995). These temporary anatomical changes make the activity of individual striatal neurones more segregated and, therefore, more selective for input signals. These findings suggest an important contribution of extracellular actions of dopamine to behaviour.

In addition, dopamine release into the striatal extracellular space occurs slowly with a delayed peak and extended duration. The entire process occurs over periods of tens of minutes and reaches a maximum after the behavioural session (Imperato *et al.*, 1992; Joseph *et al.*, 1991; Phillips *et al.*, 1991; Saulskaya, 1993; Saulskaya and Marsden, 1994, 1995b). Two questions arise: Through what mechanisms does such prolonged activation of volume transmission occur in response to behaviour? What could be the function of such prolonged activation?

We believe that the necessity of prolonged dopamine release is accounted for by the importance of extended extracellular dopaminergic stimulation, in particular, for the expression of immediate-early genes, encoding transcriptional factors that regulate the expression of late specific target genes providing coupling of short-term stimulus-response cascades to long-term changes in neurones (Berretta *et al.*, 1992).

In addition to an influence of volume transmission on the processes underlying plasticity of striatal neurones, regulatory mechanisms of volume transmission as such also involve memory processes. In particular, findings from our studies have shown that dopamine and glutamate release into the striatal extracellular space can be caused by conditioned stimuli previously paired with reinforcement (Saulskaya and Marsden, 1994; 1995a,b,c).

In studies of the mechanisms underlying prolonged dopamine release into the striatal extracellular space in response to behaviour, we have revealed that, at least under certain conditions, the presynaptic glutamatergic influence (through activation of NMDA presynaptic receptors) may increase the duration of dopamine release after the behavioural session (Saulskaya and Marsden, 1994, 1995b). In particular, the NMDA antagonist MK-801 (but not the AMPA/kainate antagonist CNQX) administered into the extracellular space of the ventral striatum, completely prevented the later phase of dopamine release, which lasted for an hour after the behavioural session (Saulskaya and Marsden, 1994, 1995b) while having no significant effect on the immediate increase in dopamine release induced by a conditioned emotional response. This supposition was later substantiated by our studies of glutamate release in the striatal extracellular space, which have demonstrated further the existence of the delayed increase in extracellular glutamate that might be responsible for the maintained delayed phase of the increase in dopamine release following a conditioned emotional response (Saulskaya and Marsden, 1995a,c). We suggest that during a conditioned emotional response, glutamate, released by terminals located on the head of a dendritic spine of principal striatal neurones, escapes from the synaptic cleft and diffuses through the extracellular space to the dopaminergic terminals located on a neck of the dendritic spine. Thus, cortical glutamatergic inputs may be capable of modulating the extracellular concentration of glutamate in the vicinity of the dopaminergic terminals so that glutamate may exert an additional action on NMDA presynaptic receptors located there, and therefore causes an additional dopamine release registered as the NMDA-dependent delayed component.

Therefore, glutamatergic volume transmission may also be capable of setting the level of responsiveness of principal striatal neurones by increasing the duration of extracellular dopamine release in response to behaviour.

There is also evidence that extracellular GABA would be expected to regulate the responsiveness of striatal neurones. Our data have indicated that all behavioural situations studied (exploratory activity, mild stress, response to, as well as relief from danger) are associated with increased extracellular GABA release in the ventral striatum (Saulskaya and Marsden, 1995c; 1996). Consistent with these observations, recent investigations of cultured striatal neurones have provided evidence for the involvement of the intrinsic striatal GABA-ergic system (which is known to set the level of extracellular GABA in

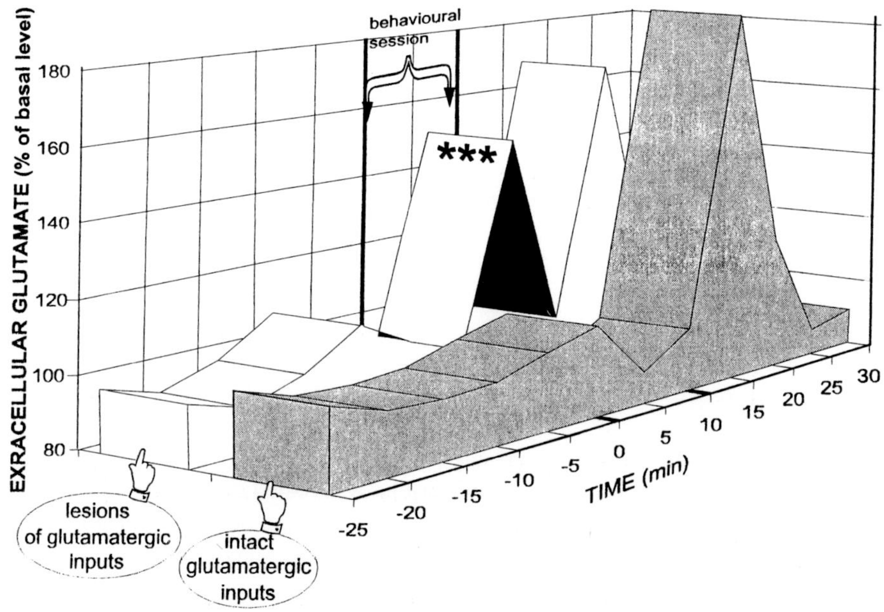

Figure 1.3. Extracellular glutamate levels in the ventral striatum, following expression of conditioned emotional responses (exposure of rats to the box where the footshock was given previously) in hippocampal-lesioned, and sham-operated rats. Results are expressed as percentage of basal (pre-testing) mean. Note that the increase in glutamate release during the behavioural session only occurs after lesions of glutamatergic hippocampal input to the ventral striatum. *** $p < 0.001$, compared with sham operated rats.

the striatum [Smolders *et al.*, 1994]), into the regulation of signal-to-noise ratio in striatal information processing (Shi and Rayport, 1994).

The contribution of volume transmission to pathological states of the brain is even more striking than the role it plays in physiological conditions. As mentioned above, in animal models of Parkinson's disease, lesions of the nigral dopaminergic neurones cause increased dopamine release from residual dopaminergic neurones, resulting in maintenance of extracellular dopamine levels in the striatum, and compensation of behavioural impairments (Abercrombie *et al.*, 1990; Parsons *et al.*, 1991; Zigmond *et al.*, 1990). Based on this observation, it was postulated that the preclinical phase of Parkinson's disease is prolonged, due to this compensatory mechanism involving enhanced volume transmission of dopamine (Zigmond *et al.*, 1990). However, the involvement of volume transmission in brain pathology is not restricted to positive behavioural effects. In some cases, an injury-induced homeostatic increase in neurotransmitter release into the extracellular space, rather than injury *per se*, might be the cause of behavioural impairments. Using *in vivo* microdialysis, combined with measurement of glutamate and dopamine release, we have shown that an excitotoxic lesion of the hippocampal glutamatergic input to the ventral striatum, itself having no effect on basal glutamate and dopamine release within this brain area under resting conditions, causes a paradoxical increase in dopamine (Saulskaya *et al.*, 1996; Saulskaya and Gorbachevskaya, 1997) and glutamate release (Figure 1.3) in response to behavioural challenge (conditioned emotional response). Behavioural observations showed that lesioning the hippocampal formation impaired the expression of the conditioned emotional response, as shown by increased ambulation and rearing—symptoms known to be the behavioural consequences of local injections of dopaminergic and glutamatergic agonists into the ventral striatum (Arnt, 1981; Boldry *et al.,* 1991; Shapovalova *et al.*, 1992). We believe that, similar to lesions of dopaminergic inputs, hippocampal lesions may result in compensatory mechanisms intended to maintain the basal level of glutamate-related activity within the ventral striatum, even in the absence of a large portion of the glutamatergic input (Figure 4). These may include up-regulation of glutamate receptor sensitivity in the microenvironment of the destroyed hippocampal glutamatergic terminals, including presynaptic receptors located on dopaminergic terminals, and increased glutamate release from residual glutamatergic terminals derived from non-hippocampal sources, with increased diffusion of glutamate through the extracellular space, due to the loss of reuptake sites on hippocampal afferents. These processes restore, to some extent, the basal glutamate-related activity under resting conditions (including presynaptic regulation of dopamine release), but could cause hyper-responsiveness during behavioural activity, accompanied by functional activation of other glutamatergic inputs to the ventral striatum. With the observed increase of release and diffusion of glutamate, this neurotransmitter may diffuse within a region of supersensitivity causing additional dopamine release and behavioural impairment. This idea is supported by the similarity of the effects of hippocampal lesions (Hannigan *et al.,* 1984) and acute stimulation (Wu and Brudzynski, 1995) of the hippocampal glutamatergic input to the ventral striatum on dopamine-related hyperlocomotion. Furthermore, the observed enhanced volume transmission of glutamate and dopamine after a lesion of the hippocampal glutamatergic input may explain hyperdopaminergic states in schizophrenia, a disease characterised by lowered activity within some cortical structures including the hippocampus (Tamminga *et al.*, 1992).

5. CONCLUSIONS

In this paper, we review evidence that, in addition to synaptic transmission, another neurotransmitter mechanism, named volume transmission, exists in the striatum. This complementary neurotransmitter mechanism, operating independently and differently from synaptic transmission, is based on diffusion of dopamine, glutamate, GABA and other compounds through the striatal extracellular fluid. Neurones, axon terminals and astroglia might be candidate targets for volume transmission signals. D1, D2 dopamine receptors, NMDA and, probably, metabotropic glutamate receptors as well as GABA-B receptors are expected to be involved as receptors for volume transmission in the striatum (Figure 1).

One of the primary functions of volume transmission in the striatum is to maintain the levels of striatal and cortical responsiveness, that influence the degree of arousal, depending on the on-going behavioural situation. At present, it can safely be said that under resting conditions, extracellular dopamine sets the background level of neuronal excitability in the striatum, via D2 nonsynaptic, tonically-active receptors (Connor *et al.*, 1991). D1 receptors, whose activation requires higher concentrations of dopamine than

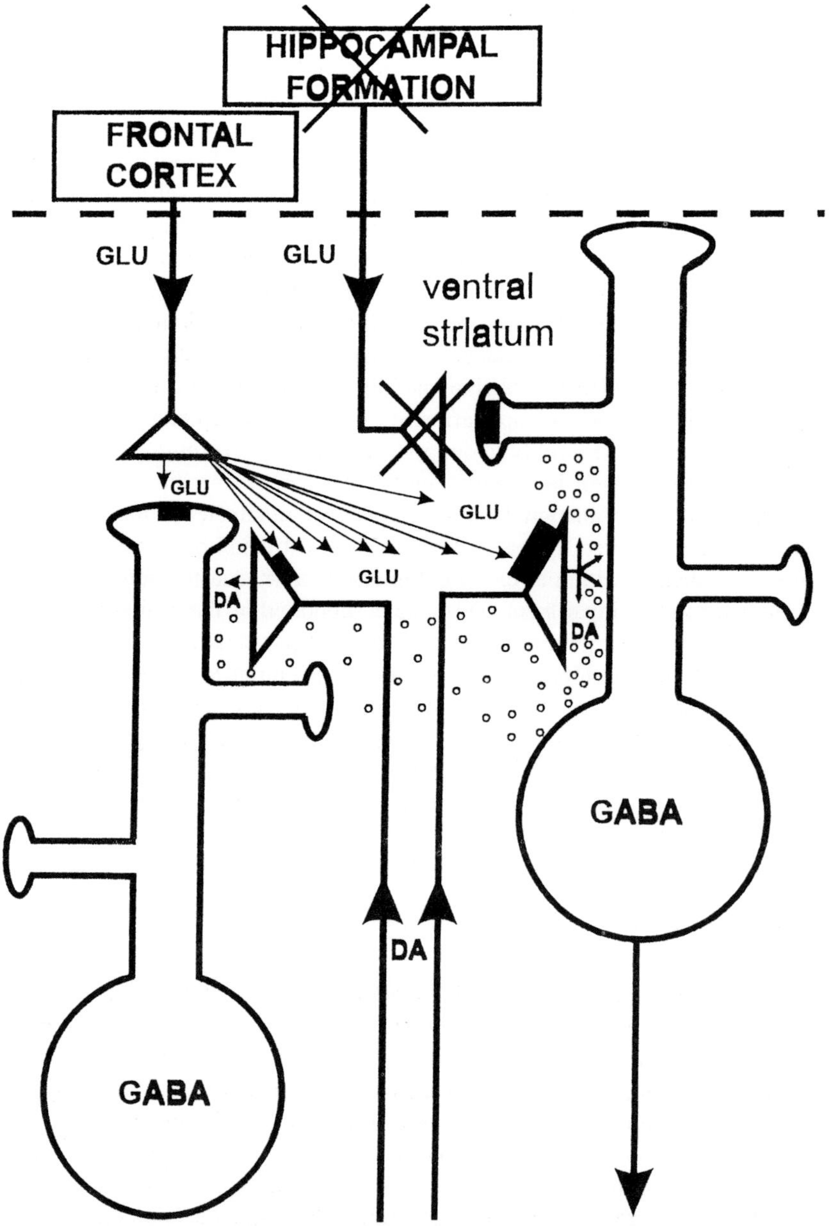

Figure 1.4. A tentative scheme explaining volume transmission hyper-response in the ventral striatum following behavioural challenge (conditioned emotional response) in hippocampal lesioned rats. The subtotal loss of glutamatergic input to the ventral striatum results in compensatory mechanisms, intended to maintain the basal level of glutamate-related activity. These include up-regulation of glutamate receptors, and increased glutamate release from those glutamate terminals that remain, together with increased diffusion of glutamate within the extracellular space, due to the loss of reuptake sites on hippocampal afferent (see text). GLU: glutamate; DA: dopamine. Small black blocks: normal glutamate receptors; large black blocks: supersensitive glutamate receptors.

those that exist in the striatal extracellular space under basal conditions, as well as NMDA receptors (because of Mg^{++} block), are not tonically active (Abercrombie and Keefe, 1991; Connor *et al.*, 1991; Saulskaya and Marsden, 1994; 1995b). There is no evidence available in favour of tonic activation of GABA nonsynaptic receptors and metabotropic glutamate receptors in the striatum.

One of the most important properties of volume transmission in the striatum appears to be to produce a switch of receptor subtypes involved in the transition from resting situations to situations when the animal starts behavioural activity. Indeed, relevant environmental stimuli have been shown to cause additional dopamine release in the striatal extracellular space (Imperato *et al.*, 1992; McCullough *et al.*, 1993; Phillips *et al.*, 1991; Saulskaya and Marsden, 1994, 1995b). Studies using voltammetry have shown that under these conditions, of additional dopamine release, local increases in extracellular dopamine level in the striatum can be higher than those global levels obtained by microdialysis, and high enough to activate D1 dopamine receptors. It should be mentioned that this increase in extracellular dopamine in the striatum causes qualitative rather than quantitative changes. Thus, rather than the D2 receptor, another receptor subtype (D1) exhibiting a different electrophysiological and pharmacological profile and located at output striatal neurones of another type (Gerfen *et al.*, 1990), becomes involved in the response. This view has gained further support from our studies showing that behavioural activity causes tonic activation of NMDA presynaptic receptors, which results in the appearance of an NMDA-dependent component of basal dopamine release in trained animals (Saulskaya and Marsden, 1994, 1995b). While the precise mechanisms underlying the activation of NMDA receptors without AMPA/kainate receptor activation still remains to be investigated, it is clear that NMDA receptors become available for glutamate released in response to behaviour. Therefore, one could suggest the existence of two receptor systems involved in volume transmission: the system for rest and the system for functional activity. We believe that a switch from one receptor system to another is an important component in the organisation of behaviour.

It should be stressed that volume transmission is not a unique mechanism of mammalian CNS (Agnati *et al.*, 1995; Arshavsky *et al.*, 1988; Bach-y-Rita, 1993). Arshavsky *et al.* (1988) provided direct evidence for the functional significance of volume transmission in ganglia of invertebrates. After complete disruption of synaptic contacts, a single neurone of isolated pedal ganglia was still able to express its functional activity (Arshavsky *et al.*, 1988). V.A. Otellin (Otellin and Arushanian, 1989) and Agnati *et al.* (1995) have suggested that, in phylogeny, volume transmission is a more ancient mechanism of neuronal communication in the CNS than synaptic transmission. Nevertheless, volume transmission still plays a very important role throughout phylogeny. In fact, neurones in the youngest brain areas, such as the cortex, express a maximum of nonsynaptic receptors for acetylcholine, excitatory amino acids, monoamines (see: Agnati *et al.*, 1995; Contant *et al.*, 1996; Martin *et al.*, 1993). It would be wrong to put volume and synaptic transmission in opposition, it being better to consider them as alternative mechanisms of neuronal communication. As suggested by D.A. Sakharov (Sakharov, 1990), synaptic transmission may be defined as a particular case of volume transmission, occurring if a diffusion area is restricted within a synaptic cleft. We believe that in the CNS, both neurotransmitter mechanisms operate as complementary processes. Volume transmission plays a role in "tuning", whereas synaptic transmission is responsible for quick "executive" processes.

In general, it must be emphasised that in spite of the accumulating data in this field, our ideas concerning the functional role of volume transmission in the CNS are mostly speculative. Future research needs to address more specifically some questions mentioned here. In particular, our knowledge about the physiological relevance of extracellular GABA in the striatum is far from complete. We currently know very little of glutamate and GABA receptor mechanisms for volume transmission in the striatum. It still remains to be seen how glial cells are involved in volume transmission in the striatum. These and perhaps other questions are going to be high on the agenda in the near future.

ACKNOWLEDGEMENT

The research is supported by Russian Fund for Fundamental Research (grant N 95-04-11524a)

REFERENCES

Abercrombie, E.D., Bonatz, A.E. and Zigmond M.J. (1990) Effects of L-DOPA on extracellular dopamine in striatum of normal and 6-hydroxydopamine-treated rats. *Brain Research* **525**, 36–44.

Abercrombie, E.D. and Keefe, K.A. (1991) Determinants of extracellular dopamine concentration as measured by microdialysis. In: H. Rollema, B. Westerink and J.D. Drijfhout (eds). *Monitoring Molecules in Neuroscience*, University Centre for Pharmacy, Groningen. pp. 308–311.

Agnati, L.F., Zoli, M., Stromberg, I. and Fuxe K. (1995) Intercellular communication in the brain: Wiring versus volume transmission. *Neuroscience*, **69**, 711–726.

Arnt, J. (1981) Hyperactivity following injection of a glutamate agonist and 6.7.ADIN into the rat nucleus accumbens and its inhibition by THIP. *Life Science*, **28**, 1597–1603.

Arshavsky, Yu.I., Deliagina, T.G., Gelfand, I.M., Orlovsky, G.N., Panchin, Yu.V., Pavlova G.A. and Popova G.A. (1988) Non-synaptic interaction between neurons in molluscs. *Comparative Biochemistry and Physiology*, **91C**, 199–203.

Bach-y-Rita, P. (1993) Neurotransmission in the brain by diffusion through the extracellular fluid: A review. *NeuroReport*,. **4**, 343–350.

Berretta, S., Robertson, H.A. and Graybiel, A.M. (1992) Dopamine and glutamate agonists stimulate neurone-specific expression of fos-like protein in the striatum. *Journal of Neurophysiology*, **68**, 767–777.

Bjorklund, A. and Lindvall, O. (1986) Dopamine-containing systems in the CNS. In: A. Bjorklund, T. Hokfelt, and M.J. Kuhar, *Handbook of Chemical Neuroanatomy. Classical Transmitters and Transmitter Receptors in the CNS., Part II, 3.* Elsevier, Amsterdam, New York, Oxford, pp. 55–111.

Boldry, R.C., Willins, D.L., Wallace, L.J. and Uretsky, N.J. (1991) The role of endogenous dopamine in the hypermotility response to intra-accumbens AMPA. *Brain Research*, **559**, 100–108.

Calabresi, P., Mercuri, N.B., DeMurtas, M. and Bernardi, G. (1990) Endogenous GABA mediates presynaptic inhibition of spontaneous and evoked excitatory synaptic potentials in the rat neostriatum. *Neuroscience Letters*, **118**, 99–102.

Carlsson, M. and Carlsson, A. (1990) Interactions between glutamatergic and monoaminergic systems within the basal ganglia—implications for schizophrenia and Parkinson's disease. *Trends in Neuroscience*, **13**, 272–276.

Connor, W.T.O., Herrera-Marschitz, M., Lindefors, N., Osborne, P.G., Drew, K.L., Reid, M. and Ungerstedt, U. (1991) Dopamine-GABA interactions in the striatum. In: H. Rollema, B. Westerink and J.D. Drijfhout (eds) *Monitoring Molecules in Neuroscience.* University Centre for Pharmacy, Groningen. pp. 93–95.

Contant, C., Umbriaco, D., Garcia, S., Watkins, K.S. and Descarries L. (1996) Ultrastructural characterization of the acetylcholine innervation in adult rat neostriatum. *Neurosci*ence, **7**, 937–947.

Donnel, P.O and Grace, A. (1995) Different effects of subchronic clozapine and haloperidol on dye-coupling between neurons in the rat striatal complex. *Neuroscience*, **66**, 763–767.

Gariano, R.F. and Groves, P.M. (1988) Burst firing induced in midbrain dopamine neurons by stimulation of the medial prefrontal and anterior cingulate cortex. *Brain Research*, **462**, 194–198.

Garris, P.A., Ciolkowski, E.L., Pastore, P. and Wightman R.M. (1994) Efflux of dopamine from the synaptic cleft in the nucleus accumbens of the rat brain. *Journal of Neuroscience*, **14**, 6084–6093.

Garris, P.A. and Wightman, R.M. (1994) Efflux of dopamine from the synaptic cleft in the nucleus accumbens of the rat brain: evidence for an extrasynaptic mode of dopamine transmission. In: A. Louilot, T. Durkin, U. Spampinato, M. Cador (eds) *Monitoring Molecules in Neuroscience*, Bordeaux University, Bordeaux. pp. 62–63.

Gerfen, C.R., Engber, T.M., Mahan, L.C., Susel, Z. and Chase, T.N. (1990) D_1 and D_2 dopamine receptor-regulated gene expression of striatonigral and srtiatopallidal neurons. *Science, New York*, **250**, 1429–1432.

Goldman-Rakic, P.S. and Selemon, L. (1990) New frontiers in basal ganglia research. *Trends in Neuroscience*, **13**, 241–244.

Gonon, F. (1994) Kinetics of catecholamine release and elimination in the peripheral and central nervous system. In: A. Louilot, T. Durkin, U. Spampinato, M. Cador (eds) *Monitoring Molecules in Neuroscience*. Bordeaux University. Bordeaux, pp. 60–61.

Gonon, F.G., Mermet, C.C. and Marcenac, F.H. (1987) *In vivo* voltammetry with carbon fiber electrodes: a tool for studying dopaminergic neurons of the basal ganglia. In: M. Sandler, C. Feuerstein and B. Scatton (eds), *Neurotransmitter Interactions in the Basal Ganglia*, Raven Press, New York, pp. 101–110.

Grace, A.A. (1991) Phasic versus tonic dopamine release and the modulation of dopamine system responsivity: A hypothesis for the aetiology of schizophrenia. *Neuroscience*, **41**, 1–24.

Groves, P.M., Linder, J.C. and Young, S.J. (1994) 5-hydroxydopamine-labelled dopaminergic axons: Three dimensional reconstructions of axons, synapses and postsynaptic targets in the neostriatum. *Neuroscience*, **58**, 593–604.

Hannigan, J.H., Springer, J.E. and Isaacson, R.L. (1984) Differentiation of basal ganglia dopaminergic involvement in the behaviour after hippocampectomy. *Brain Research*, **291**, 83–91.

Herrling, P.L. (1985) Pharmacology of corticocaudate excitatory postsynaptic potential in the cat: Evidence for its mediation by quisqualate or kainate-receptors. *Neuroscience*, **14**, 417–426.

Imperato, A., Angelucci, L., Casolini, P., Zocchi, A. and Puglisi-Allegra, S. (1992) Repeated stressful experience differentially affect limbic dopamine release during and following stress. *Brain Research*, **577**, 194–199.

Joseph, M.H., Young, A.M.J. and Gray, J.A. (1991) Reinforcement, conditioning and dopamine function in the nucleus accumbens. In: H. Rollema, B. Westerink, J.D. Drijfhout (eds) *Monitoring Molecules in Neuroscience*. University Centre for Pharmacy. Groningen, pp. 200–203.

Justice, J.B., Smith, A.D. and Sam, P.M. (1994) Characterization of dopamine microdialysis and application to regulation of the dopamine transport. In: A. Louilot, T. Durkin, U. Spampinato and M. Cador, *Monitoring Molecules in Neuroscience*, Bordeaux University. Bordeaux. pp. 73–74.

Kalivas, P.T. (1993) Neurotransmitter regulation of dopamine neurons in the ventral tegmental area. *Brain Research Reviews*, **18**, 75–113.

Kornhuber, J. and Kornhuber, M.E. (1986) Presynaptic dopaminergic modulation of cortical input to the striatum. *Life Science*, **39**, 669–674.

La Gamma, E.F., Strecker, E., Lenn, N.J., DeCristofaro, J.D. and Weisinger, G. (1994) Dopamine regulation of transfected preproenkephaline promoter in primary rat astrocytes *in vitro* and *in vivo*. *Experimental Neurology*, **130**, 304–310.

Levey, A.I., Hersch, H., Rue, D.B., Sunahara, R.K, Niznik, H.B., Kitt, C.A., Price, D.L., Maggio, R., Brann, M.R. and Ciliax, B.J. (1993) Localization of D1 and D2 dopamine receptors with subtype-specific antibodies. *Proceedings of the.National Academy of Sciences, U.S.A.*, **90**, 8861–8865.

Levi, G. and Raiteri, M. (1993) Carrier-mediated release of neurotransmitters. *Trends in Neuroscience*, **16**, 415–419.

Martin, L.J., Blackstone, C.D., Levey, A.I., Huganir, R.L. and Price, D.L. (1993) AMPA glutamate receptor subunits are differentially distributed in rat brain. *Neuroscience,* **53**. 327–358.

McCullough, L.D., Sokolowski, J.D. and Salamone, J.D. (1993) A neurochemical and behavioral investigation of the involvement of nucleus accumbens dopamine in instrumental avoidance. *Neuroscience*, **52**, 919–925.

Near, J.A., Bigelow, J.B. and Wightman, R.M. (1985) Comparison of uptake of dopamine in rat striatal chopped tissue and synaptosomes. *Journal of Pharmacology and Experimental Therapeutics*, **245**, 921–927.

Otellin, V.A. and Arushanian, E.B. (1989) *The Nigro-Strio-Nigral System* (In Russian). Medicina, Moskow.

Parsons, L.H., Smith, A.D. and Justice, J.B. (1991) The *in vivo* microdialysis recovery of dopamine is altered independently of basal level by 6-hydroxydopamine lesions to the n. accumbens. *Journal of Neuroscience Methods*, **40**, 139–147.

Phillips, A.G., Atkinson, L.J., Blackburn, J.R. and Blaha, C.D. (1991) *In vivo* analyses of dopamine neurotransmission during anticipatory and consummatory phases of feeding behaviour. In: H. Rollema, B. Westerink, J.D. Drijfhout (eds) *Monitoring Molecules in Neuroscience*, University Centre for Pharmacy, Groningen. pp. 204–297.

Petralia, R.S., Wang, Y.X., Niedzielski, A.S. and Wenthold, R.J. (1996) The metabotropic glutamate receptors, MGLUR2 and MGLUR3, show unique postsynaptic, presynaptic and glial localizations. *Neuroscience*, **71**, 949–956.
Rolls, E.T., Thorpe, S.J., Boytim, M., Szabo, I. and Perrett, D.I. (1984) Responses of striatal neurons in the behaving monkey. 3. Effects of iontophoretically applied dopamine on normal responsiveness. *Neuroscience*, **12**, 1201–1212.
Sakharov, D.A. (1990) A diversity of neurotransmitters: functional meaning. (In Russian) *Zhurnal Evoliutsionni Biokhimii i Fisiologii*, **26**, 733–741.
Sanghera, M.K., Trulson, M.E. and German, D.C. (1984) Electrophysiological properties of mouse dopamine neurons: *In vivo* and *in vitro* studies. *Neuroscience*, **12**, 793–801.
Saud-Chagny, M.F., Ponec, J. and Gonon, F. (1991) Presynaptic autoinhibition of the electrically evoked dopamine release studied in the rat olfactory tubercule by *in vivo* electrochemistry. *Neuroscience*, **45**, 641–652.
Saulskaya, N.B. Synaptic release of dopamine during an acquisition of passive avoidance task. (1993) (In Russian) *Zhurnal Vysshey Nervnoy Dejatelnosty*, **43**, 1025–1029.
Saulskaya, N.B. and Marsden, C.A. (1994) The role of glutamatergic inputs to the n. accumbens in control of synaptic dopamine release during association learning. (In Russian) *Fiziolicheskii Zhurnal i I.M.Sechenova* **80**, 45–54.
Saulskaya, N.B. and Marsden, C.A. (1995a) Extracellular glutamate levels in the n. accumbens during conditioned emotional response. *Fiziolicheskii Zhurnal i I.M.Sechenova* (In Russian) **81**, 161–165.
Saulskaya, N. and Marsden, C.A. (1995b) Conditioned dopamine release: Dependence upon N-methyl-D-aspartate receptors. *Neuroscience*, **67**, 57–63.
Saulskaya, N. and Marsden, C.A. (1995c) Extracellular glutamate in the nucleus accumbens during a conditioned emotional response in the rat. *Brain Research*, **698**, 114–120.
Saulskaya, N.B. and Marsden, C.A. (1996) The GABA extracellular level in the n. accumbens in the course of a conditioned nociceptive response in rats. (In Russian) *Fiziolicheskii Zhurnal i I.M. Sechenova,* **82**, 37–43.
Saulskaya, N., Marsden, C.A. and Gorbachevskaya, A.I. (1996) Conditioned emotional response and glutamate–dopamine interaction. In: J.L. Gonzales-Mora, R. Borges and M. Mas. *Monitoring Molecules in Neuroscience*. University of La Laguna, Santa Cruz de Tenerife, pp. 329–330.
Saulskaya, N.B. and Gorbachevskaya, A.I. (1997) Conditioned dopamine release in the nucleus accumbens in rats with lesions of the hippocampal formation. (In Russian) *Fiziolicheskii Zhurnal i I.M. Sechenova*, **83**, 76–82.
Shapovalova, K.B., Gorbachevskaya, A.I. and Saulskaya, N.B. (1992) Structural organisation and neurochemical mechanisms of participation of the nucleus accumbens in the interaction of limbic and motor systems and in the regulation of motor behaviour. (In Russian) *Zhurnal Vysshey Nervnoy Dejatelnosty*, **42**, 226–276
Shi, W.-X. and Rayport, S. (1994) GABA synapses formed *in vitro* by local axon collaterals of nucleus accumbens neurons. *Neuroscience* **14**, 4548–4560.
Smith, A.D. and Bolam, J.P. (1990) The neural network of the basal ganglia as revealed by the study of synaptic connections of identified neurones. *Trends in Neuroscience*, **13**, 259–265.
Smolders, I., Herregodts, P., Michotte, Y. and Ebinger, G. (1994) Modulation of the extracellular levels of GABA and glutamate in rat striatum by means of local administration of uptake inhibitors and after decortication. In: A. Louilot, T. Durkin, U. Spampinato, M. Cador, (eds) *Monitoring Molecules in Neuroscience.* Bordeaux University, Bordeaux, p. 291.
Stamford, J.A., Zurmunt, L. and Kruh, P. (1988) Diffusion and uptake of dopamine in rat caudate and nucleus accumbens compared using fast cyclic voltammetry. *Brain. Res*earch, **448**, 381–386.
Stricker, E.M. and Zigmond, M.J. (1986) Brain monoamines, homeostasis and adaptive behaviour. In: V.B. Mountcastle., F.E. Bloom and S.R. Geiger (eds) *Handbook of Physiology. The Nervous System, Vol IV.* American Physiological Society, Bethesda, Maryland, pp. 677–700.
Svensson, T.H. and Tung, C.-S. (1989) Local cooling of prefrontal cortex induced pacemaker-like firing of dopaminergic neurons in rat ventral tegmental area *in vivo. Acta Physiologica Scandinavica*, **136**, 135–136.
Tamminga, C.A., Thaker, G.K., Buchanan, R., Kirkpatrick, B., Alphs, L.D., Chase, T.N. and Carpenter, W.T. (1992) Limbic systems abnormalities identified in schizophrenia using positron emission tomography with fluorodeoxyglucose and neocortical alteration with deficit syndrome. *Archives of General Psychiatry*, **49**, 522–530.
Westerink, B.H.C., Tuntler, J., Damsma, G., Rollema, H. and De Vries, J.B. (1987) The use of tetrodotoxin for the characterization of drug-enhanced dopamine release in conscious rats studied by brain dialysis. *Naunyn-Schmiedeberg's Archives of Pharmacology*, **336**, 502–507.
Wu, M. and Brudzynski, M. (1995) Mesolimbic dopamine terminals and locomotor activity induced from the subiculum. *Neuroreport*, **6**, 1601–1604.

Yung, K.K.L., Bolam, J.P., Smith, A.D., Hersch, S.M., Ciliax, J. and Levey, A. (1995) Immunocytochemical localization of D1 and D2 dopamine receptors in the basal ganglia of the rat: light and electron microscopy. *Neuroscience*, **65**, 709–730.

Zigmond, M., Abercrombie, E.D., Berger, T.W., Grace, A.A. and Stricker, E.M. (1990) Compensations after lesions of central dopaminergic neurons: some clinical and basic implications. *Trends in Neuroscience*, **13**, 290–295.

2 Unitary Postsynaptic Mechanisms of LTP and LTD in the Neocortex, Hippocampus and Cerebellum

I.G. Silkis

Institute of Higher Nervous Activity and Neurophysiology, Russian Academy of Sciences, Moscow, Russia
mlno@mlno.msk.ru

A unitary postsynaptic model of excitatory and inhibitory synaptic plasticity for neocortical, hippocampal and cerebellar Purkinje cells is proposed. To provide for the fulfilment of the Hebbian rule, it is postulated that only synapses activated by neurotransmitter are modifiable. It is proposed that heterosynaptic effects occur if homosynaptic and heterosynaptic afferents activate both a target cell and a "common" inhibitory neurone, which forms modifiable synapses on the target cell. According to the proposed model, unitary mechanisms underlie homosynaptic, associative and heterosynaptic plasticity. It was revealed by computational modeling of post-tetanic processes in a neocortical/hippocampal cell under stationary conditions that: (i) excitatory (inhibitory) synaptic efficacy depends on the ratio between active protein kinases and protein phosphatases; (ii) it monotonically increases (decreases) with the intracellular Ca^{2+} rise; and (iii) it does not depend on the initial synaptic efficacy. The necessary and sufficient conditions for synaptic modifications are: the coincidence of pre- and postsynaptic cell activity; and the change of pre- and/or postsynaptic cell activity that causes a shift in the ratio between active protein kinases and protein phosphatases. LTP (LTD) of excitation and LTD (LTP) of inhibition occur due to the positive (negative) post-tetanic Ca^{2+} shift relative to the rise in Ca^{2+} produced by prior stimulation. A dependence on the previous history of activity is an intrinsic property of LTP and LTD. It is also postulated that the properties of receptors on neocortical/hippocampal and Purkinje cells are similar. The opposite Ca^{2+}-dependent modification rules in these structures are conceivably the result of an up-regulation of cAMP levels by Ca^{2+} ions in neocortical/hippocampal cells and a down-regulation of cGMP levels by Ca^{2+} ions in Purkinje cells.

KEYWORDS: LTP, LTD, excitation, inhibition, pyramidal cell, Purkinje cell

1. COMMONLY ACCEPTED MECHANISMS OF EXCITATORY SYNAPTIC PLASTICITY IN THE NEOCORTEX, HIPPOCAMPUS AND CEREBELLUM

1.1. Mechanisms of Homosynaptic LTP and LTD of Excitatory Transmission in the Neocortex and Hippocampus

The modification of the efficacy of synaptic transmission is considered to be one of the mechanisms underlying the storage of information. Long-term potentiation (LTP) and long-term depression (LTD) in the efficacy of excitatory synaptic transmission have been observed in the neocortex, hippocampus, cerebellum, thalamus and others structures of the central nervous system (CNS) (Bliss and Collingridge, 1993; Bear and Malenka, 1994; Weber *et al.,*1984; Silkis, 1994b, 1996b, 1997a; Ito and Karachot, 1992). Properties of LTP and LTD such as their duration, input specificity, associativity and dependence on previous synaptic efficacy (Abraham and Bear, 1996; Linden, 1994; Tsumoto, 1992) are most important with respect to their participation in the learning. The necessity for pre- and postsynaptic cells to be coactive for an increase in synaptic efficacy to occur was postulated by Hebb (1949), and is known as the Hebbian rule.

During recent years, the properties and mechanisms of LTP and LTD of excitatory transmission have been extensively studied in the hippocampus. It has been found that presynaptic cell activity causes a rise in the postsynaptic Ca^{2+} ion concentration and changes in the activity of intracellular molecules. Detailed analysis of the participation of Ca^{2+}-dependent intracellular substances in the induction of LTP/LTD has shown that a key role is played by protein kinases (PKs) and protein phosphatases (PPs), which determine the phosphorylation state of ionotropic AMPA (alpha-amino-3-hydroxy-5-methyl-4-isoxazole propionic acid) and NMDA (N-methyl-D-aspartate) glutamate receptors. The greater (less) the proportion of phosphorylated receptors (Rph), the more (less) the synaptic sensitivity to glutamate. It was found that LTP and LTD can be induced in the same synapse (Abraham and Bear, 1996; Weber *et al.,* 1984; Silkis, 1996b) and the same intracellular substances, but in different relations, can result in LTP, as well as in LTD (Bear and Malenka, 1994; Bliss and Collingridge, 1993; Lisman, 1994). These data were of great importance to studies of plasticity mechanisms. The sequence of postsynaptic Ca^{2+}-dependent processes causing LTP or LTD of excitatory synaptic transmission is schematically shown in Figure 2.1. Rhythmic stimulation activates AMPA, NMDA and metabotropic glutamate (mGlu) receptors, and causes an increase in postsynaptic Ca^{2+} levels due to Ca^{2+} influx through NMDA and voltage-dependent Ca^{2+}-channels (VDCCs) and due to Ca^{2+} release from intracellular stores. The tetanization also causes an increase in cAMP concentration as well as an enhancement of cAMP-dependent protein kinase A (PKA) and protein kinase C (PKC) activity. Ca^{2+} activates calmodulin (CaM), which in turn activates Ca^{2+}-calmodulin-dependent protein kinase II (CaMKII) and protein phosphatase 2B (PP2B). The phosphorylation state of AMPA and NMDA receptors is determined, on the one hand, by the phosphorylating action of PKA, PKC and CaMKII, and on the other hand, by the dephosphorylating action of protein phosphatase 1 (PP1). PP1 activity is controlled by its inhibitor (I1), which is activated by PKA and inhibited by PP2B. At high Ca^{2+} concentrations, phosphorylation usually prevails and LTP is produced, while at low Ca^{2+} concentrations the predominance of dephosphorylation results in LTD (Bear and Malenka, 1994; Lisman, 1994).

1.2. Heterosynaptic and Associative LTP and LTD of Excitatory Transmission in the Neocortex and Hippocampus

The study of simultaneous changes in the efficacy of two excitatory inputs to the same postsynaptic cell has demonstrated the existence of associative and heterosynaptic plasticity. It is possible to induce associative LTP (LTPa) and LTD (LTDa) by combining low-frequency stimulation (LFS) with membrane depolarization and hyperpolarization, respectively. Heterosynaptic LTD (LTDh) is observed in a non-activated synaptic input simultaneously with homosynaptic LTP in the tetanized input to a single target cell

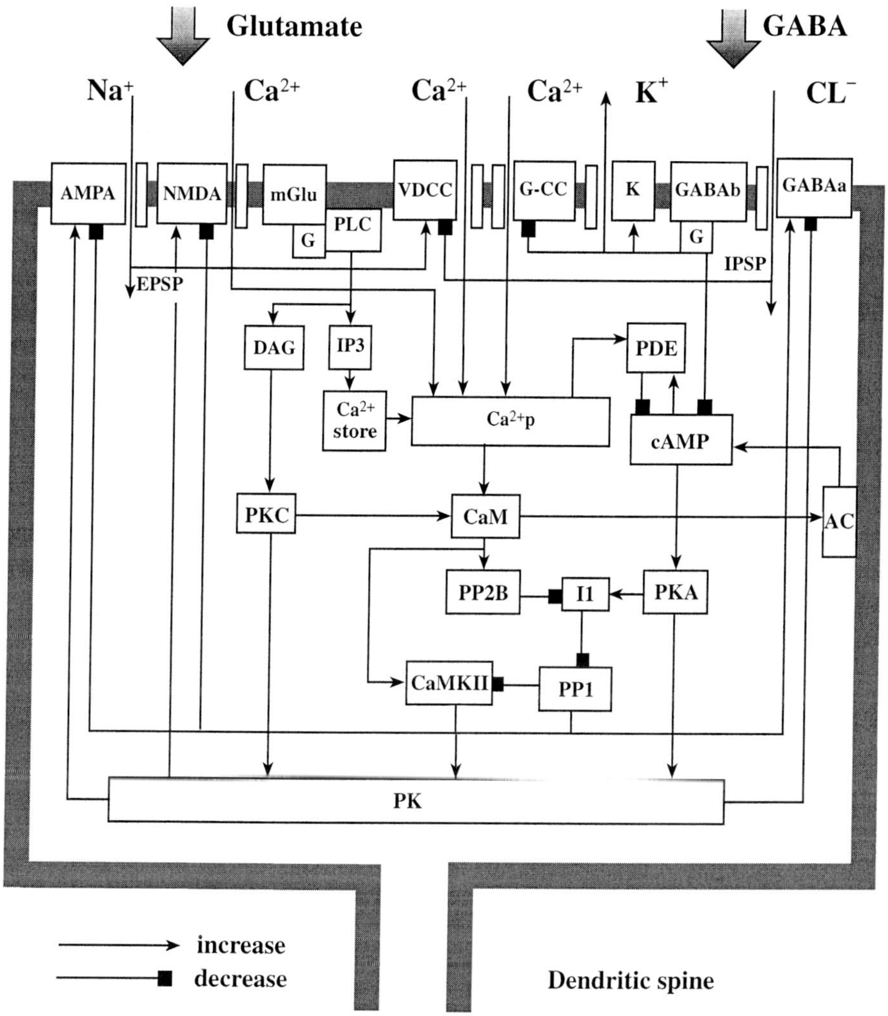

Figure 2.1. Posttetanic processes underlying the modification of synaptic transmission in a neocortical/hippocampal pyramidal neurone. G—G protein; PLC—phospholypase C; IP3—inositol-1,4,5-triphosphate; DAG—diacylglycerol; PDE—phosphodiesterase ; PP2B—calcineurin; AC—adenylate cyclase; I1—inhibitor of protein phosphatases 1; Ca^{2+}p—postsynaptic calcium level. The other designations are given in the text. (See also complete list of abbreviations at the end of this chapter.)

(Linden, 1994). Further investigations have demonstrated that LTDh can occur without homosynaptic LTP or with homosynaptic LTD (Linden, 1994; Otani and Connor, 1996; Silkis *et al.,* 1994b). According to the present view, fulfilment of the Hebbian rule is not required for LTPh induction, since a postsynaptic cell is active in the absence of presynaptic activity (Linden, 1994). Although it has been believed that it is impossible to induce heterosynaptic long-term potentiation (LTPh), since LTP is an input specific effect, nevertheless LTPh has been reported (Misgeld *et al.,* 1979; Otani *et al.*, 1995; Silkis *et al.*, 1994b).

It has been found that a rise of Ca^{2+} is also required for the induction of associative and heterosynaptic effects. Thus, LTDa was not obtained in the presence of NMDA receptor antagonists, and LTDh was not induced in the absence of extracellular Ca^{2+} (Linden, 1994). It was proposed, therefore, that a depolarizing potential propagates from the activated synapse to the inactive ones, where it promotes the opening of VDCCs (Jaffe *et al.*, 1994), and the ensuing Ca^{2+} influx postsynaptically triggers heterosynaptic plasticity.

1.3. Mechanisms of LTD of Excitatory Inputs to the Cerebellar Purkinje Cell

LTD of the efficacy of excitatory synaptic connections between parallel fibers (PFs) and Purkinje cells (PC), termed cerebellar LTD (LTDc), is obtained when PFs and a climbing fiber (CF) are activated in phase. LTDc is input specific (Ekerot and Kano, 1985). We assume therefore that LTDc can be considered as an associative effect and is mediated postsynaptically by changes in AMPA receptor sensitivity. Thus, pulses of glutamate that were insufficient to substitute for PFs activation during LTDc induction, were sufficient as test stimuli to detect LTDc once it has been induced (Crepel and Krupa, 1988). A Ca^{2+} rise is necessary for LTDc induction, since LTDc is blocked by postsynaptic application of Ca^{2+} chelators or removal of external Ca^{2+} (Linden and Connor, 1991). LTDc is usually observed when conjunctive stimulation of PFs and CF results in a high Ca^{2+} level, while a moderate Ca^{2+} rise produced by PFs simulation alone or caused by Ca^{2+} chelators leads to cerebellar LTP (LTPc) (Hartell, 1994a; Hirano, 1990; Kasono and Hirano, 1994; Shibuki and Okada, 1992). This result is in contrast to the neocortex/hippocampus, where a large rise in Ca^{2+} causes LTP, while a lower one causes LTD (Bear and Malenka, 1994; Lisman, 1994). Activation of mGlu receptors, which promote Ca^{2+} rise and PKC activation, is required for LTDc occurrence (Crepel and Krupa, 1988; Linden and Connor, 1991). Cerebellar PCs contain mGlu receptors in unusually high quantities, particularly in the distal dendritic spines where the PFs terminate (Martin *et al.*, 1992). cGMP-dependent protein kinase G (PKG) is also involved in cerebellar synaptic plasticity (Hartell, 1994b; Linden, 1994). It was proposed that a rise in cGMP concentrations is provided by the action of a gaseous messenger molecule, nitric oxide (NO), on soluble guanylyl cyclase (GCs) after PF or CF stimulation (Linden, 1994). According to commonly accepted opinion, the mechanisms of LTDc are distinct from the mechanisms of neocortical/hippocampal plasticity (Linden, 1994). It is assumed that phosphorylation of AMPA receptors on the PC produced by a Ca^{2+} rise is essential for LTDc induction (Ito and Karachot, 1992).

1.4. Homosynaptic LTP and LTD of Inhibitory Transmission in the Neocortex, Hippocampus and Cerebellum

LTP and LTD of the efficacy of inhibitory transmission (LTPi, LTDi) was recently demonstrated in the neocortex (Komatsu, 1994; Silkis, 1994a; 1996b; 1997a), hippocampus (McLean *et al.*, 1996), and cerebellum (LTPic, LTDic) (Kano *et al.*, 1992; Llano *et al.*, 1991). Neocortical LTPi is an input-specific effect since it was induced only in the tetanized pathway (Komatsu, 1994). The mechanisms of LTPi and LTDi were not specifically studied but it was demonstrated that an intracellular Ca^{2+} rise is also essential for modifications of inhibition. In the neocortex and hippocampus, LTDi induction required an additional Ca^{2+} increase compared with that which causes LTPi (Komatsu, 1994; Mclean *et al.*, 1996). *Vice versa*, the large Ca^{2+} rise in PCs, produced by PF tetanization paired with CFs stimulation or PC depolarization, resulted in LTPic (Llano *et al.*, 1991), while the use of a Ca^{2+} chelator led to LTDic in this pathway (Kano *et al.*, 1992). It was demonstrated that LTPi and LTDi could be sequentially expressed at the same synapses (Mclean *et al.*, 1996).

2. CONTRADICTIONS IN THE COMMONLY ACCEPTED MODELS OF EXCITATORY SYNAPTIC PLASTICITY

2.1. Is It Possible to Induce Non-Hebbian Types of LTD and LTP?

There are some weak points in the commonly accepted models of excitatory synaptic plasticity. Thus, according to the present view, the Hebbian rule is not universal (Linden, 1994). However, the fulfilment of this rule is necessary for the increase in information storage during learning (Frolov and Muraviev, 1987).

It was proposed that the fulfilment of the Hebbian rule is not required for LTDh induction, because a postsynaptic cell is active in the absence of presynaptic cell activity (Linden, 1994). However, it has recently been demonstrated that postsynaptic cell depolarization does not cause a modification of an untetanized input without its synaptic activation (Otani *et al.*, 1995; Otani and Connor, 1996; Otsu *et al.*, 1995). In addition, recent experimental data provide evidence that activation of one input to a postsynaptic cell does not affect identically all unstimulated inputs to this cell (Muller *et al.*, 1995; Silkis *et al.*, 1994). Thus, heterosynaptic modifications do seem to be related to the tested input.

LTDa is also considered to be non-Hebbian (Linden, 1994), since it is develops when discharges of the presynaptic cell are not accompanied by spikes of a hyperpolarized postsynaptic cell. However, it must be noted that, while such hyperpolarization excludes Ca^{2+} entry through VDCCs, it does not prevent the rise of intracellular Ca^{2+} level due to activation of postsynaptic mGlu receptors and the release of Ca^{2+} from intracellular stores. The necessity of this last condition for LTDa induction was recently demonstrated (Wang *et al.*, 1997). Thus, LTDa is also the result of a conjunction of presynaptic cell discharge and a postsynaptic cell response. For these reasons, we assume that LTDh and LTDa cannot be assigned unconditionally as non-Hebbian types of synaptic plasticity.

It is interesting that LTP in synapses formed by mossy fibres on hippocampal CA3 pyramidal cells (which synapses do not contain NMDA receptors) was earlier explained only by presynaptic mechanisms. However, it has now been demonstrated that this type of LTP also requires postsynaptic responses, and thus conforms to the Hebbian principle (Derric and Martinez, 1996; Urban and Barrionuevo, 1996).

2.2. Is the Magnitude of the Ca^{2+} Level which Leads to LTP or LTD Absolute or Relative?

It follows from the analysis of existing experimental data that the magnitude of the Ca^{2+} concentration which leads to LTP or LTD is not absolute (not necessarily high and low), as assumed earlier (Artola and Singer, 1993; Linden, 1994), but relative. Usually, high-frequency stimulation (HFS) and NMDA receptor activation are required for LTP occurrence, but LTP can also be obtained without NMDA receptor activation (Linden, 1994; Tsumoto, 1992) or after LFS (Grassi *et al.*, 1996; O'Dell and Kandel, 1994). LFS or NMDA receptor blockade during HFS is often used for LTD induction. However, in some cases the opening of NMDA channels is required for LTD occurrence (Dudek and Bear, 1993; Linden, 1994). In addition, HFS can simultaneously cause LTP and LTD in different cells of the same group. Moreover, the recycling of HFS episodes permitted the sequential induction of LTP and LTD in the same synapse (Weber *et al.*, 1984; Linden, 1994). Finally, the magnitude of the Ca^{2+} level at which LTP or LTD occurs cannot be absolute, since the sign of synaptic modification depends not only on the post-tetanic Ca^{2+} rise, but also on the initial synaptic efficacy (Abraham and Bear, 1996; O'Dell and Kandel, 1994).

2.3. Does a Sliding Modification Threshold and Metaplasticity Actually Exist?

The existence of a variable (sliding) threshold for conversion from LTD to LTP has been proposed to explain the experimentally observed dependence of synaptic plasticity on initial synaptic efficacy. It was speculated that this threshold is determined by the level of membrane depolarization (Bear, 1995). However, at least in the neocortex, cell depolarization alone is not sufficient for synaptic modification to occur (Otsu *et al.*, 1995). In addition, it has been assumed that the stimulation frequency at which the conversion from LTD to LTP occurs, also slides. This frequency is bidirectionally shifted in relation to the previously used frequency, in the range between HFS and LFS (Abraham and Bear, 1996; Bear, 1995). The advancement of this theory has led to the assumption that a new effect, "plasticity of synaptic plasticity" (termed "metaplasticity"), determines the direction of modification depending on the history of synaptic activity (Abraham and Bear, 1996). Ca^{2+}-dependent variations of intracellular substances, as well as the changes in gene expression, have been suggested as possible metaplasticity mechanisms. However, the last assumption seems improbable, because the necessary time for gene expression far exceeds those tens of minutes that are sufficient for LTP or LTD production. Besides, the effects explained by "metaplasticity" are input-specific (Abraham and Bear, 1996; Bear, 1995; Huang *et al.*, 1992), while the specificity of gene expression for a stimulated input has not been established. It was proposed that "metaplasticity" might not be obtained independently by special experiments, because the Ca^{2+}-dependent processes underlying this phenomena do not differ from those underlying LTP and LTD. However, it was not excluded that metaplasticity could occur without concurrent changes in synaptic efficacy (Abraham and Bear, 1996). Nonetheless the existence of such a new effect as "metaplasticity" cannot yet be considered as proven.

It is pertinent to note that synaptic modification (M) is, by definition, the difference between post-tetanic (E_p) and initial (E_o) synaptic efficacy: $M = E_p - E_o$. Thus, the sign and magnitude of synaptic modification (LTP or LTD) cannot be determined unless the initial synaptic efficacy is known, regardless of whether metaplasticity exists or not. As mentioned above, the efficacy of the synaptic input (E) is determined by the number of highly sensitive phosphorylated receptors (R_{ph}). So, (E) is a function of R_{ph} [$E = F(R_{ph})$], where R_{ph} is determined by the Ca^{2+}-dependent balance between the postsynaptic concentration of active protein kinases (PK_a) and protein phosphatases (PP_a) (Lisman, 1994). Since the post-tetanic Ca^{2+} rise depends on stimulation frequency (f), R_{ph} must also depend on this frequency. Therefore, E_p and E_o can be represented respectively as functions of f [$E_p = F(f)$], and of prior stimulation frequency (f_o) [$E_o = F(f_o)$], while M is a function of f and f_o [$M = E_p - E_o = F(f) - F(f_o)$].

The question could be raised whether the post-tetanic activity of PKs and PP might be regulated not only by the postsynaptic Ca^{2+} rise produced by a stimulation with instantaneous frequency f, but also by the prior synaptic activation with a frequency f_o. In the last case E_p will be a function of f and f_o [$E_p = F'(f, f_o)$]. In experimental conditions, post-tetanic and initial EPSPs ($EPSP_p$ and $EPSP_o$) that are proportional to E_p and E_o are usually estimated, and synaptic modification is determined by the difference between $EPSP_p$ and $EPSP_o$ ($EPSP_p$–$EPSP_o$). If E_p actually depends on E_o, the experimentally evoked $EPSP_p$ will be a function of f and $EPSP_o$ [$EPSP_p = F''(f, EPSP_o)$].

2.4. Does Receptor Phosphorylation Underlie LTD in Cerebellar Purkinje Cells?

It has been proposed that phosphorylation of AMPA receptors underlies LTDc (Ito and Karachot, 1992; Nakazawa *et al.,* 1995). If this proposed mechanism of cerebellar LTD is correct, then the properties of AMPA receptors on PCs are distinctive from those on hippocampal/neocortical cells, where AMPA receptor phosphorylation underlies LTP (Bear and Malenka, 1994; Lisman, 1994). Phosphorylation of AMPA receptors on the PC was obtained after the application of AMPA together with 8-Br-cGMP (which causes activation of PKG in PCs) (Nakazawa *et al.*, 1995). However, this experimental protocol does not necessarily cause LTDc. In fact, we assume that it is more likely to result in LTPc. Actually, the Ca^{2+} level due to AMPA application must be low, since in this case mGlu receptors, that promote a Ca^{2+} rise in PCs up to 50% (Ross *et al.*, 1990), are inactive. According to known experimental data, the low Ca^{2+} level must lead to LTPc, but not to LTDc, because a high Ca^{2+} level is necessary for LTDc induction (Hartell, 1994a,b; Kasono and Hirano, 1994). Moreover, AMPA application may additionally promote LTPc occurrence due to the parallel activation of PCs and inhibitory interneurones and a consequent reduction of Ca^{2+} levels in the PCs. Thus it was found that, owing to GABA action and a reduced level of Ca^{2+}, LTPc could be induced by a strong stimulation protocol that usually results in LTDc (Shibuki and Okada, 1992). It follows from the above-mentioned data that experimentally obtained phosphorylation of AMPA receptors on PCs is more likely to be correlated with induction of LTPc than LTDc. Thus, the properties of these receptors on PCs appear similar to those on neocortical/hippocampal neurones. We have not found in the literature any distinction between AMPA receptor properties in different CNS structures. If the existing model of cerebellar plasticity is correct, and AMPA receptor phosphorylation causes LTDc at high Ca^{2+} levels, then at this level, PK activity must dominate. However, at large Ca^{2+} concentrations PKG activity is low (Olson *et al.*, 1976) and an inhibitor of PP1 (G-substrate) becomes inactive (Kennedy, 1992). We assume that both these effects—PK inactivation and PP1 activation must promote the dephosphorylation of AMPA receptors on PCs, not their phosphorylation as is commonly proposed.

Clearly, existing models of synaptic plasticity do not explain the differential dependence of synaptic modifications on Ca^{2+} levels between neocortical/hippocampal cells and

PCs. In neocortical/hippocampal cells, a large post-tetanic Ca^{2+} elevation triggers LTP and LTDi, while a moderate increase in Ca^{2+} concentration leads to LTD and LTPi (Komatsu, 1994; Linden, 1994). In contrast, LTDc and LTPic are observed in PCs when the Ca^{2+} level was high, while a moderate Ca^{2+} rise results in LTPc and LTDic (Hirano, 1990; Kasono and Hirano, 1994).

2.5. Is NO Action on Soluble Guanylyl Cyclase the Only Mechanism of cGMP Production in Cerebellar Purkinje Cells?

The accepted mechanism for cGMP production in PCs during LTD contains an inherent contradiction. On the one hand, it has been suggested that NO action on GCs causes a cGMP rise and thus participates in LTDc induction. On the other hand, it has been shown that NO or NO-synthase (NOS) influence neither the rise of cGMP levels in PCs, nor LTDc induction (Linden, 1994). The cGMP level in a PC is high, but the amount of GCs is low (Luo *et al.*, 1994). Moreover, GCs are inhibited at high Ca^{2+} concentrations (Luo *et al.*, 1994), at which LTDc is usually developed. With regard to NO, the source of its formation is also problematic. One suggestion has been that NO is formed outside of PCs, while the "target" of NO is inside PCs (Tsien, 1996). In one study, NO was suggested to be formed as a result of CF stimulation (Linden, 1994); in another study, NO was considered to be the result of PF activation (Tsien, 1996). PF and CF stimulation both result in cGMP formation in the PC, but NOS is not contained in the terminals of the CFs, although it is possibly contained in PF terminals (Linden, 1994; Ross *et al.*, 1990).

2.6. The Basis of the Suggested Unitary Model of Synaptic Plasticity

We have developed a unitary model of synaptic plasticity in neocortical/hippocampal and PCs, that removes the above-mentioned contradictions. The model is based on the postulate that unified Hebbian modification rules should be fulfilled for all types of synaptic plasticity if excitatory and inhibitory inputs to a neurone are stimulated simultaneously (Silkis, 1995a,b). The suggested mechanism for modifying inhibitory transmission (Silkis, 1996a) is based on changes in the activity of the same intracellular PKs and PP1 that influence the modification of excitatory transmission (Sigel, 1995).

We have investigated some features of excitatory and inhibitory transmission modifications using a mathematical model of postsynaptic biochemical processes in CA3 hippocampal pyramidal neurones, triggered by rhythmic stimulation (Murzina and Silkis, 1997b; 1997c). It follows from the results of this modeling that a Ca^{2+} rise, which leads to synaptic modification, is relative, and that the sign of the synaptic modification depends on the initial synaptic efficacy. The initial synaptic efficacy can be considered as a modification threshold, making metaplasticity a redundant mechanism.

The suggested model for synaptic plasticity in PCs is based on the postulate that the properties of receptors on neocortical/hippocampal and Purkinje cells are similar and therefore that dephosphorylation underlies LTDc. We have proposed that there is another mechanism for cGMP production, which is triggered by activation of inhibitory inputs to PCs (Silkis, 1996d,f).

3. THE PROPOSED POSTSYNAPTIC MODEL OF SIMULTANEOUS EXCITATORY AND INHIBITORY PLASTICITY IN THE NEOCORTEX AND HIPPOCAMPUS

3.1. The Participation of Ca^{2+}-dependent Processes in the Modification of Inhibitory Synaptic Transmission

Although the mechanisms underlying the modification of inhibitory transmission have not been studied in detail, there is enough experimental data to elaborate a model of inhibitory synaptic plasticity. On the one hand, rhythmic discharges of inhibitory cells and consequent $GABA_a$ receptor activation cause a decrease of Ca^{2+} influx through VDCCs, owing to cell hyperpolarization. In addition, the activation of G protein-coupled $GABA_b$ receptors results in the inhibition of G protein-coupled Ca^{2+}-channels, a decrease in cAMP levels and a lowering of Ca^{2+} efflux from intracellular stores (Kuriyama *et al.*, 1993). On the other hand, it has been found that Ca^{2+} and PKs influence the modification of IPSP amplitude. It is particularly remarkable that PKA, PKC and CaMKII phosphorylate not only glutamate-, but also GABA-sensitive receptors. As a result of such phosphorylation, the sensitivity of AMPA and $GABA_a$ receptors varies in opposite directions (McDonald and Moss, 1994; Sigel and Baur, 1988). Thus, an increase in Ca^{2+} levels and PK activity results in the reduction of $GABA_a$ receptor sensitivity and IPSP amplitude, while the inactivation of PKs causes a long-term increase of IPSP amplitude (Moss *et al.*, 1992; Szente *et al.*, 1990).

When examining the possible mechanisms of mechanisms of modification of inhibitory transmission, we focused our attention on several features. First, this modification is input-specific, since LTPi developed as a result of the stimulation of inhibitory inputs, but was not observed on unstimulated inhibitory inputs (Komatsu, 1994). Second, we have shown that LTP of inhibitory inputs to a target cell could be induced simultaneously with LTP of one or two excitatory inputs to the same target cell. Therefore, the firing rate of some target cells does not necessarily decrease (increase) during LTPi (LTDi). Moreover, LTPi was obtained simultaneously with an increase in postsynaptic cell spontaneous frequency. Therefore, the modification of inhibitory inputs does not lead to changes in postsynaptic cell excitability. These results suggest that modifiable inhibitory synapses are probably located on the dendritic spines of a target cell, but not on its soma (Silkis, 1994; 1996a, c). Indeed, metabotropic $GABA_b$ receptors, together with ionotropic $GABA_a$ receptors, are found on dendritic spines (Kanter and Haberly, 1993).

We proposed that LTPi occurs as a result of $GABA_b$ receptor activation, a decrease of Ca^{2+} and cAMP concentrations in the dendritic spine, an inactivation of PKA, PKC, CaMKII, and a consequent dephosphorylation of GABA receptors (Silkis, 1996a). In fact, the necessity of $GABA_b$ receptor activation for LTPi induction was recently demonstrated in neocortical slices (Komatsu, 1996). The role of $GABA_b$ receptors in the modification of inhibition is supported also by the data that a depression of the IPSP occurs as a result of G protein inactivation (Lambert and Wilson, 1993). The proposed sequence of neocortical/hippocampal processes, underlying inhibitory transmission modification, is schematically shown in Figure 2.1.

3.2. Mechanisms of Simultaneous Modifications of Excitatory and Inhibitory Synaptic Efficacy

Inhibitory synapses located on dendritic spines only occur together with excitatory ones, and the fraction of such spines is about 5–20% of the total (Dehay *et al.*,1991). It is important to note that glutamate- and GABA-sensitive receptors can be activated almost simultaneously, since the latency of monosynaptic EPSPs is not far removed from the latency of disynaptic IPSPs (Miles, 1990). This fact is possibly a result of the lower threshold for inhibitory cell activation and a greater rising slope of inhibitory cell EPSPs in comparison with pyramidal neurones. During tetanization, an enhanced release of glutamate causes mGlu- and NMDA receptor activation, a rise in Ca^{2+} levels and PK activation. In contrast, an increase in GABA release and its action on $GABA_a$ and $GABA_b$ receptors causes a reduction of Ca^{2+} levels and PK inactivation (Figure 2.1). If, in the end, PK activity prevails, then ionotropic receptors should be phosphorylated and LTP and LTDi should develop simultaneously. If the activity of PP1 prevails, then receptor dephosphorylation should result in the simultaneous induction of LTD and LTPi (Silkis, 1996a). These predicted results are in the accord with our experimental data (Silkis, 1997a).

It follows from the suggested model, that HFS of excitatory inputs alone can result only in a significant Ca^{2+} rise. However, if one wishes to convert LTP into LTD while using HFS, it is necessary to decrease Ca^{2+} levels and reduce PK activity. Therefore, we assume that the induction of LTD by HFS will be hindered in those pathways in which inhibitory inputs are not activated as well, and in those pathways where only $GABA_a$ receptors are activated. These assumptions are supported by the recent data that $GABA_b$ receptor antagonists prevented LTD induction, while LTD occurred in the presence of $GABA_a$ receptor antagonists (Wagner and Alger, 1995).

3.3. Proposed Participation of Inhibition in Heterosynaptic LTD Induction

Distinct from the existing opinion that LTDh is developed in the absence of presynaptic activity (Linden, 1994), our model holds that all types of synaptic plasticity are Hebbian (Silkis, 1995a,b). The evidence for this assumption comes from the recent finding that postsynaptic cell depolarization does not cause a modification of an untetanized input without its synaptic activation by LFS (Otani *et al.*, 1995). An intracellular Ca^{2+} rise also does not lead to synaptic modification if the synapse is not activated (Christie *et al.*, 1996). To provide the fulfilment of the Hebbian rule, we postulated that only receptors activated by a transmitter are modifiable. This feature will provide LTP, LTD, LTPi and LTDi with properties such as use-dependence and input specificity. In accordance with this suggested postulate, untetanized excitatory inputs are not modifiable (Figure 2.2). We have proposed that the observed changes in heterosynaptic PSPs reflect IPSP variations, while the EPSP amplitude is invariant. In accordance with this assumption, the following conditions have been suggested as necessary and sufficient for LTDh induction: 1) the convergence of homo- and heterosynaptic afferents not only on the target cell, but also on a "common" inhibitory interneurone, which inhibits this target cell (Figure 2.2); and 2) the modification of synaptic transmission in "common inhibitory pathways" (Silkis, 1995a). The availability of "common" inhibitory neurones is confirmed by known morphological and electrophysiological data (Miles, 1990; Misgeld *et al.*, 1979).

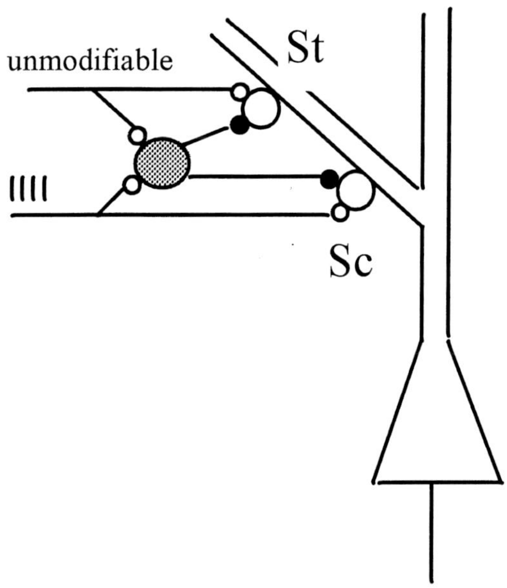

Figure 2.2. The proposed scheme for Hebbian modification of heterosynaptic input efficacy. St, test spine; Sc, conditioned spine.

In order to modify an inhibitory input to the test spine (St), where heterosynaptic afferents terminate (Figure 2.2), the Ca^{2+} concentration in this spine must be changed. Since Ca^{2+} molecules cannot pass through the spine neck (Koch and Zador, 1993), a Ca^{2+} rise in the St could be provided by extending the depolarizing potential from the conditioned spine (Sc) to the St, and subsequent opening of VDCCs (Figure 2.2). In this condition, a heterosynaptic test signal will evoke a changed IPSP, and a modified PSP will be observed despite an unchanged EPSP. LTPi of inhibitory synapses in a heterosynaptic pathway requires $GABA_a$ receptor dephosphorylation, which must be provided by PP1 activation. If the suggested mechanism of LTDh is correct, the blockade of VDCCs should prevent the development of LTDh, but should not influence the induction of homosynaptic LTP, since the high Ca^{2+} level in Sc can be provided by the opening of NMDA channels due to the large depolarization of this spine. This prediction of the model has experimental support (Wickens and Abraham, 1991). Moreover, it was demonstrated recently that heterosynaptic input stimulation influences LTDh induction, as well as homosynaptic LTD occurrence (Otani and Connor, 1996). We assume that this effect is possibly a consequence of the presence of a "common" inhibitory neurone, activated by the two groups of afferents, whereas excitatory homo- and heterosynaptic afferents activate the target cell independently. The intimate interaction between homo- and heterosynaptic modification was also demonstrated in an earlier experiment (Muller *et al.*, 1995).

In some cases, the Ca^{2+} enlargement in St can be so high that LTDi will be induced together with LTPh, owing to the prevalence of PK activity. However, the Ca^{2+} level must be low and the activity of PP1 must dominate for LTPi and LTDh production. The participation of PP1 in LTDh induction is supported by experimental data (Scanziani *et al.*, 1996). The presence of $GABA_b$ receptors on the heterosynaptic input could prevent excessive increase of Ca^{2+} concentration in St during tetanization. The involvement of $GABA_b$ receptors in LTDh induction was confirmed by the data that $GABA_b$ receptor antagonists hindered LTDh occurrence (Davies and Collingridge, 1989).

It has been shown that different synaptic inputs must be located closely enough on a dendritic tree for the heterosynaptic effects to occur (White *et al.*, 1990). Otherwise, an EPSP extending from a point of excitation would decay and the depolarization levels of other dendritic sites would be insufficient for the opening of VDCCs. If simultaneous activation of heterosynaptic inhibitory inputs to St, and homosynaptic excitatory inputs to Sc is required for heterosynaptic plasticity, LTDh is an associative effect. Similar processes possibly occur during LTDa induction. The complete occlusion of LTD and LTDa, found recently (Christie *et al.*, 1995) can be considered as evidence for unitary mechanisms underlying LTD and LTDa.

4. THE STUDY OF LTP AND LTD PROPERTIES BY COMPUTATIONAL MODELING OF POSTSYNAPTIC PROCESSES IN THE HIPPOCAMPAL PYRAMIDAL CELL

4.1. The Dependence of the Number of Phosphorylated Receptors on the Parameters of Rhythmic Stimulation and the Post-tetanic Ca^{2+} Rise

We developed a computational model of synaptic plasticity on the basis of the suggested mechanism of synaptic plasticity in neocortical/hippocampal cells (Figure 2.1). This model enabled us to investigate simultaneously various interconnected post-tetanic metabolic processes which occur in a dendritic spine of the pyramidal neurone of hippocampal area CA3 (Murzina and Silkis, 1996a; 1997b,c). In accordance with the suggested postulate, which fulfils the Hebbian rule, it is accepted in the computational model that the connection or disconnection of a phosphate group with a receptor becomes possible only if the receptor is activated by transmitter. This assumption is now directly supported by the data that PKG activation causes AMPA receptor phosphorylation only in the presence of a receptor agonist (Nakazawa *et al.*, 1995).

The computational model represents a system of equations of chemical kinetics (Murzina and Silkis, 1996a, 1997b,c). It follows from the result of these equations that the number of phosphorylated ionotropic AMPA or NMDA receptors (R_{ph}) depends on the ratio between active protein kinases and protein phosphatase 1 (PKs/PP1), and is completely determined by the amount of transmitter. The time-averaged concentration of the excitatory transmitter released during rhythmic stimulation (M) is proportional to the stimulation frequency f and the amount of transmitter (M_o) released per presynaptic spike. Varying the magnitude of M_o in our mathematical treatments, we simulated the "force" of a synaptic input (Murzina and Silkis, 1996b). Using the outcome of these equations, we studied the dependence of the number of R_{ph} on such parameters as the stimulation frequency f, the input "force", and the types of activated postsynaptic receptors. The dependence of R_{ph} on frequency f for a "middle force" input, on the condition that all types of glutamate- and GABA-sensitive receptors are activated, is shown in Figure 2.3a.

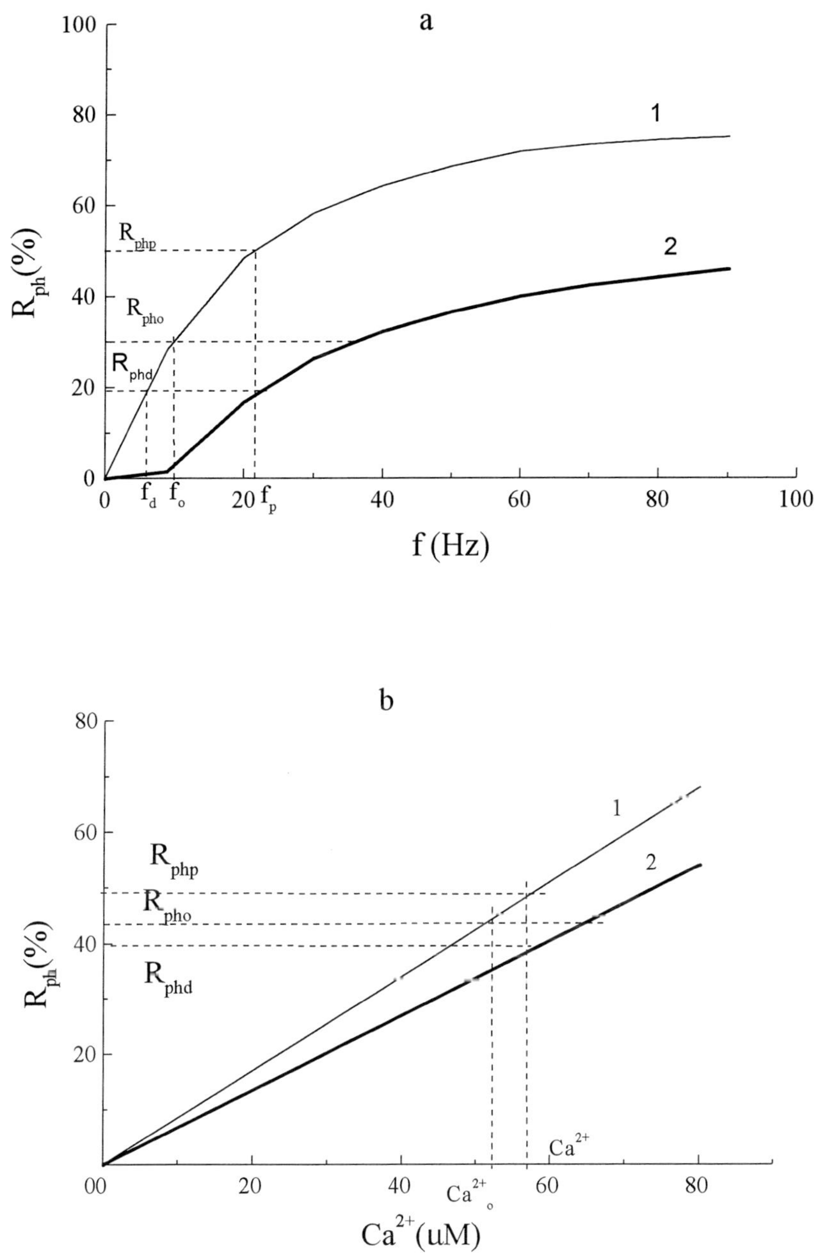

Figure 2.3. The dependence of the number of phosphorylated receptors, that is proportional to excitatory synaptic transmission efficacy, on stimulation frequency (a) and intracellular Ca^{2+} concentration (b).
a) Ordinate—the number of phosphorylated AMPA and NMDA receptors Rph (%), abscissa—the frequency of rhythmic stimulation f (Hz); Rpho, Rphp, Rphd—initial, potentiated, depressed numbers of phosphorylated receptors.
b) Ordinate—the number of phosphorylated AMPA and NMDA receptors Rph (%), abscissa—postsynaptic Ca^{2+} concentration (μM); 1—activation of excitatory input alone; 2—simultaneous activation of excitatory and inhibitory inputs.

The number of R_{ph} is maximal if HFS is applied to an isolated excitatory input (curve 1, Figure 2.3a). Stimulation of additional inhibitory inputs and $GABA_b$ receptor activation essentially reduces *Rph* (curve 2, Figure 2.3a). The Ca^{2+} concentration also depends on the stimulation frequency and availability of inhibition (Murzina and Silkis, 1996c,d). The expression of the modification of excitatory transmission following a postsynaptic Ca^{2+} rise is shown in Figure 2.3b. It follows from the calculations that if an excitatory "middle force" input is stimulated alone, and the stimulation frequency varies in the usually-used range of 5–100 Hz, the Ca^{2+} concentration increases from 15 μM up to 100 μM. In this range, the Ca^{2+}-dependence of the number of phosphorylated receptors *Rph* (i.e. efficacy of synapse) is close to linear.

As mentioned above, the efficacy of inhibitory synaptic inputs is proportional to the number of dephosphorylated $GABA_a$ receptors. We have found that the number of dephosphorylated $GABA_a$ receptors decreases monotonically with the Ca^{2+} rise (Figure 2.4).

4.2. Comparative Analysis of Different Conditions of LTP Induction

It should be mentioned that computational modeling permits one to exclude some indeterminate events, that occur for reasons beyond the control of the experimenter. These reasons are: a) the non-specific action of a number of applied substances; b) the influence of applied substances on both pre- and postsynaptic cells, thus not allowing one to differentiate between pre- and postsynaptic effects; and c) the action of applied substances

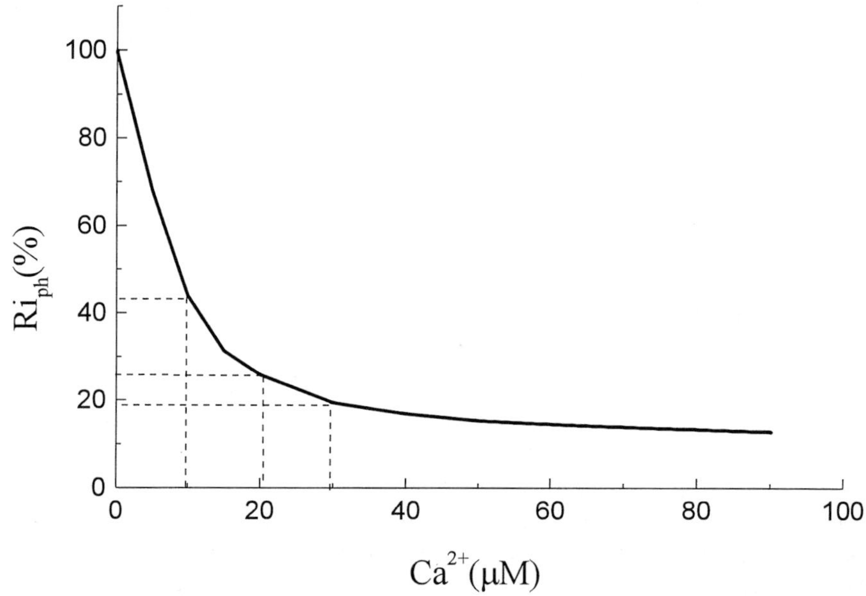

Figure 2.4. The dependence of inhibitory synaptic transmission efficacy on intracellular Ca^{2+} concentration. Ordinate—the number of dephosphorylated GABAa receptors Ri_{ph} (%), abscissa—postsynaptic Ca^{2+} concentration (mM);

on a set of neighbouring spines, that adds the associative and heterosynaptic effects to the homosynaptic one. In experimental conditions, the influence of each intracellular substance on synaptic plasticity is usually investigated separately, whereas the mathematical model allows one to study the participation of individual substances, taking into account their interactions (Murzina and Silkis, 1997a).

For the investigation of some features of excitatory and inhibitory plasticity in various cells, allowance must be made in the model parameters for the functional properties of these cells. For example, AMPA receptors located on PCs and neocortical inhibitory interneurones have higher Ca^{2+} permeability than AMPA receptors on neocortical/hippocampal pyramidal neurones (Brorson *et al.*, 1995). In addition, there are no NMDA receptors on the PCs, and so the essential Ca^{2+} rise is provided by Ca^{2+} efflux from intracellular stores (Ross *et al.*, 1990). In our mathematical treatments, we did not take into account that Ca^{2+} entry through VDCCs can be strengthened by generation of an action potential and subsequent active propagation of a spike into the dendritic tree. In such cases, the Ca^{2+} current into distal dendrites is maximal (Jaffe *et al.*, 1994). This mechanism possibly provides for a more effective modification of synaptic efficacy in those experiments where spikes of postsynaptic cells follow the discharge of presynaptic neurones.

In fact, our experiments *in vivo* used a strong stimulation current, which resulted in the discharge of postsynaptic cells, and obtained homosynaptic LTP in "weak" thalamo-cortical inputs (Silkis, 1996b,1997a). In contrast, in other experiments that evoked EPSPs in postsynaptic cells, only LTPa was observed as a result of conjoint stimulation of thalamo-cortical and cortico-cortical inputs (Ivakiri *et al.*, 1991). It must be noted that the majority of synaptic plasticity studies have been conducted using cortical slices, where synaptic inhibition is considerably stronger and NMDA receptor-mediated activity is lower than in similar neurones of intact cortex (Armstrong-James *et al.*, 1993). Furthermore, a higher level of incoming signal is required for NMDA receptor activation in slices (Fox and Daw, 1993). The level of synaptic excitation in the slices is possibly low, due to a disruption of excitatory afferents, whereas the level of inhibition does not vary strongly, because inhibitory interneurones and recorded cells are located in close proximity. Therefore, in comparison with slices, the intact neocortex seems to be more favorable for LTP induction.

4.3. The Participation of Modifiable Inhibition in Long-Term Strengthening of an Excitatory Signal

Using the computational model we described in our previous articles (Murzina and Silkis 1996a,b,c,d), the Ca^{2+} concentration, cAMP level, PP1 and PKs activity, and R_{ph} proportion as well as changes in EPSP/IPSP amplitude and latency depend on various conditions of pre- and postsynaptic cell activation. Using this model, we have demonstrated that modifiable inhibition leads to long-term regulation of the strength of excitatory signals (Murzina and Silkis, 1996d). For example, if a presynaptic neurone, that monosynaptically excites and disynaptically inhibits a certain postsynaptic cell, fires rhythmically for a long time, and if the rise in Ca^{2+} in the postsynaptic cell is sufficiently large, the LTD of the inhibitory input which occurs simultaneously with LTP of the excitatory input, selectively promotes expression of this LTP. If other excitatory inputs to the same postsynaptic cell simultaneously activate "common" inhibitory interneurones, LTPh of inhibitory transmission may be induced. This effect, together with unmodifiable excitatory heterosynaptic transmission, will lead to LTDh of excitatory connections between various presynaptic cells and a given postsynaptic cell. Thus, a significant selective increase of the efficacy of one input will be supplemented by a decrease in the efficacy of other inputs. We have shown that LTD in the efficacy of disynaptic inhibitory input underlies this contrasting effect, even in the case when the efficiency of the excitatory input was previously weak and is only weakly potentiated due to isolated activation.

4.4. The Main Features of Synaptic Plasticity Arising from the Model

Two main features of synaptic plasticity are derived from the mathematical simulation. First, in a stationary state the number of R_{ph} does not depend on the initial number of phosphorylated receptors (R_{pho}), but is completely defined by the amount of transmitter released during instantaneous tetanization. R_{pho} affects only the time (T) of achievement of the stationary state; T is not less than several tens of minutes. According to this result, R_{ph} must be the same for a given stimulation frequency f, regardless of whether the previous stimulation frequency f_0 was higher or lower than f. If so, then E_p is clearly a function of f [$E_p = F'(f)$] and does not depend on E_o. Second, as a consequence of the monotonic rise of the function $R_{ph} = F(f)$, the sign of excitatory transmission modification (LTP or LTD) can be obtained at once by the difference (positive or negative) between R_{ph} and R_{pho}. Each successive increase or decrease in stimulation frequency must lead to LTP or LTD induction. Thus, the stimulation frequency at which LTP or LTD occurs is not absolute but relative, and depends on the value of f_o. The initial synaptic efficacy E_o can be considered as a point at which LTD reverses into LTP (LTP reverses into LTD). This point is commonly accepted as the LTD/LTP crossover point (Kirkwood *et al.*, 1996).

4.5. Dependence of the Character of Homosynaptic and Heterosynaptic LTP and LTD on Previous Stimulation

It follows from the results of our calculations that previous stimulation has differential influences on the possibility of LTP or LTD induction. For example, it is difficult to produce LTP if R_{pho} is large due to prior HFS. However, previous LFS and/or additional activation of inhibitory inputs, that cause a decrease of R_{pho}, must promote LTP expression after subsequent HFS (Figure 2.3a). These computational results are confirmed by experimental data that LTP is obtained mostly if a synapse was depressed previously (Abraham and Bear, 1996). In contrast, it has been impossible in some experiments to produce LTD without prior LTP (Dudek and Bear, 1993). This effect has been named depotentiation (O'Dell and Kandel, 1994; Wagner and Alger, 1995; Yang and Faber, 1991).

It is perhaps of interest to note here that an analogous effect of prior stimulation was also observed for heterosynaptic plasticity. Thus, it was found that LTP induced in the medial perforant path can be depotentiated heterosynaptically (Doyere *et al.*, 1997). Moreover, depotentiation of the medial perforant path synapses was found to be linearly correlated with the magnitude of LTP induced in the lateral perforant path synapses.

Moreover, it has been demonstrated that induction of LTD (LTP) on one input to CA1 hippocampal cells leads to heterosynaptic reversal of LTPh (LTDh) previously induced on a separate pathway by HFS (LFS) without modification of naive heterosynaptic responses (Muller *et al.,* 1995). Similarly, it has been shown that tetanization, which failed to induce LTDh, did cause a heterosynaptic depotentiation if the pathway has been tetanized earlier (Scanziani *et al.*, 1996). With regard to the suggested mechanism of heterosynaptic plasticity, the obtained LTDh (LTPh) could be a consequence of LTPih (LTDih). It follows from these experiments that the modification of inhibitory transmission in a heterosynaptic pathway, such as LTPih (LTDih), could be easily induced if LTDi together with LTP (LTPi together with LTD) were previously induced in this pathway.

It has also been found that previous activation of a heterosynaptic input influenced the sign of modification of both homo- and heterosynaptic inputs (Otani and Connor, 1996). This result can be related to the availability of a "common" inhibitory neurone and its simultaneous influence on the initial efficiency of inhibitory inputs to different spines of the target cell.

It follows from our modeling results (Figure 2.4) that the character of inhibitory input modifications also depends on previous activation, or initial efficacy, of inhibition. This result is supported now by the finding that LTPi in hippocampal slices is induced more easily following previous induction of LTDi than in naive slices (McLean *et al.*, 1996).

4.6. A Similarity between Theoretical and Experimental Features of Synaptic Efficacy

The properties of synaptic plasticity that are evident from the monotonic rise of the $R_{ph}(f)$-curve can be related to those experimentally obtained. We have shown that a successive increase (decrease) in stimulation frequency must lead to a consequent rise (decline) in the number of R_{ph} and therefore in EPSP amplitude. LTP is observed if $f > f_o$, while LTD is induced if $f < f_o$. The relative nature of the frequency-dependence of LTP and LTD is supported by the finding that successive stimulation with frequencies 1 Hz, 2 Hz, 5 Hz, 100 Hz leads to sequential enlargement of EPSP amplitude (Grassi *et al.*, 1996). It was concluded in this article that the effective frequency for LTP induction is not necessarily a high frequency, but probably any frequency higher than the test frequency, whereas the most effective frequency to counteract the induced LTP should be the lowest one. It can be seen from curve 1 (Figure 2.3a) that if f_o is equal to 10 Hz then successive 3 Hz stimulations must cause a decrease of R_{ph} (LTD induction), while the succeeding stimulations with increasing frequencies such as 20 Hz, 50 Hz, 100 Hz must lead to the enlargement of R_{ph}, and thus cause LTP of increasing amplitude for each subsequent tetanus. Such a feature of synaptic modifications was recently described (Malen and Chapman, 1997). Since the spontaneous rate of most neocortical cells ranges from 5 to 10 Hz, HFS (> 25 Hz) must usually lead to LTP, while a LFS (1–5 Hz) must result in LTD. Just such effects of HFS and LFS are often described, while a tetanus at an intermediate frequency (10 Hz) does not result in a change in synaptic efficacy (Dudek and Bear, 1992). However, inhibition of protein kinase or protein phosphatase activity during a 10-Hz tetanus results in LTD or LTP, respectively (Coussens and Teyler, 1996).

It is clear from our results that any one of many stimulation frequencies may cause LTP if it exceeds the prior stimulation frequency f_o (Figure 2.3a). In the stationary state, synaptic efficacy is proportional to the maximal number of R_{ph} (for the stimulation frequency used). Therefore, after achievement of the stationary state any repetitive stimulation with the same frequency cannot cause a change in the existing synaptic efficacy. This conclusion is confirmed by the saturation property of LTP.

It can be seen from the comparison of curves 1 and 2 in Figure 2.3a that for a given E_o the same stimulation frequency f can lead to LTP, if an excitatory input is activated alone (curve 1), or can promote LTD if an inhibitory input is activated simultaneously (curve 2). Electrophysiological data also have shown that HFS together with GABA application may lead to LTD (Linden, 1994).

From the $R_{ph}(f)$-curve it follows that the greater the difference between the prior and successive stimulation frequency, the greater the modification effect. Thus, prior HFS (large f_o), on the one hand, must reduce subsequent LTP or even prevent its induction, but on the other hand, must enhance subsequent LTD. Indeed, it has been demonstrated in different experiments that prior HFS leads to less LTP and more LTD (for review see Abraham and Bear, 1996). It is obvious also that the more (less) the value of f_o, the more (less) the value of subsequent stimulation frequency f might be used for LTP induction. In other words, prior HFS (LFS) shifts to the right (left) the previously existing LTD/LTP crossover point (modification threshold for LTP). If the initial spontaneous firing rate of the presynaptic cell was low (for example, due to the absence of afferent signals) then the LTD/LTP crossover point should be shifted to the left as compared with the crossover point associated with the normal level of spontaneous activity. This prediction of our model is confirmed by the experimental data that the LTD/LTP crossover point is shifted to the right (left) by HFS (LFS) (Bear, 1995), and by the finding that the LTD/LTP crossover point in the light-deprived visual cortex is shifted to the left as compared with normal cortex (Kirkwood *et al.*, 1996).

The expression of LTP or LTD is determined by the difference between post-tetanic and initial synaptic efficacy ($E_p - E_o$), i.e., the difference between f and f_o. Thus, the more (less) the value of E_o, the less (more) the effect of LTP produced by a tetanization with the given frequency f. Likewise LTD, which is determined by the difference ($E_o - E_p$), is more (less) expressed if E_o is high (low). There is the analogous assumption that a low initial level of synaptic efficacy would shift the threshold in favour of greater LTP and less LTD (Stanton, 1996). Such an effect is convenient to study in light-deprived animals, since deprivation leads to an activity-dependent decrease in initial synaptic efficacy (Bear, 1995). Due to the low initial synaptic efficacy in these animals, LFS must be less effective in producing LTD, while HFS must be more advantageous in LTP induction. This prediction of our model was also confirmed by the experimental data (Kirkwood *et al.*, 1996).

4.7. A Comparison with a Model Based on Metaplasticity

The other feature of synaptic plasticity that is the consequence of our model is that the post-tetanic number of R_{ph}, and therefore E_p, does not depend on E_o. It follows from this result that, in a stationary state, post-tetanic EPSP amplitude depends only on the value of f, and must be the same for naive, previously potentiated or previously depressed synapses. Several experimental data support this conclusion. Thus, it was demonstrated (Heynen *et al.,* 1996) that HFS causes the same rise of EPSP amplitude in CA1 pyramidal cells, regardless of whether the synapse was naive or previously depressed by LFS. In other experiments, the same EPSP amplitude (the same LTP effect) was observed after HFS of naive inputs and inputs that has been previously activated by LFS (Staubli and Chun, 1996). It was found also that even though prior LFS caused LTD of the EPSP slope in CA1, subsequent HFS is able to achieve virtually the same absolute amount of LTP as in control (naive) slices (Dudek and Bear, 1993). The analogous result was obtained when studying associative plasticity. It was demonstrated for unitary CA3-CA1 EPSPs that synchronous pairing of synaptic activation and postsynaptic depolarization resulted in an increase in the amplitude of EPSPs to the same absolute level, regardless of whether the input was naive or had been previously depressed (Debanne *et al.*, 1997). In four of six cells recorded in the rat medial vestibular nucleus, HFS delivered after reducing LTP by LFS enhanced the response again to the same level as it was established by the first HFS (Grassi *et al.*, 1996).

The results of our model are similar to the results of the model based on a variable modification threshold. Thus, a positive (negative) post-tetanic shift in the level of membrane depolarization relative to the modification results in LTP (LTD); the greater the initial synaptic efficacy, the higher the modification threshold (Abraham and Bear, 1996; Kirkwood *et al.*, 1996). Therefore, the same level of membrane depolarization may result in LTP (LTD), if the initial synaptic efficacy is low (high). It must be noted that, despite the similarity of their final effects, our model does not require an additional parameter such as a variable modification threshold. The role of a modification threshold can be played by the initial synaptic efficacy that varies with prior activation.

Ca^{2+}-dependent variations of intracellular substances were suggested as possible mechanisms of metaplasticity (Abraham and Bear, 1996; Kirkwood *et al.*, 1996). We assume that effects of previous activation, such as a change in NMDA receptor sensitivity, or the modification of the threshold of calcium/calmodulin kinase II autophosphorylation, that have been proposed as possible metaplasticity mechanisms (Abraham and Bear, 1996) could only change the time (T) to achievement of the stationary state. We assume also that there is no need to propose that inhibition participates directly in the mechanisms of metaplasticity (Abraham and Bear, 1996). However, inhibition is related to synaptic plasticity. Thus, we have shown that an additional disynaptic inhibitory pathway, when involved during prior activity, decreases initial synaptic efficacy and therefore can promote LTP induction by successive tetanizations, particularly if the last does not activate the inhibitory interneurones.

It follows from our results that the dependence of synaptic plasticity on initial efficacy is not an independent phenomenon (not the result of metaplasticity), but one of the intrinsic properties of the known types of synaptic plasticity. Metaplasticity seems an excessive mechanism. Besides, we have not found any experimental evidence for the existence of a sliding modification threshold. We conclude that a new effect such as metaplasticity cannot be considered as proven. Moreover, metaplasticity cannot occur without changes in synaptic efficacy and the mechanisms of metaplasticity are in much less competition with those of synaptic modification than has been proposed (Abraham and Bear, 1996).

The preceding discussion does not exclude the existence of forms of metaplasticity based on gene expression, or changes in dendritic spine configuration, or any other mechanisms. Such forms of metaplasticity are more prolonged than LTP and LTD, could pertain to other forms of plasticity, and could be fundamental for the long-term storage of information.

4.8. Unified Modification Rules for Different Types of LTP and LTD

A significant result of our model is that synaptic modification (LTP or LTD) can be obtained only if the frequency f of each successive stimulation is not equal to the prior stimulation frequency f_o. Indeed, it is impossible to modify synaptic efficacy using the same stimulation frequency f as used previously. The same is true for the post-tetanic rise in Ca^{2+} (Figure 2.3b).

It is clear from our results that fulfilment of the Hebbian rule (coincidence of pre- and postsynaptic cell activity), which is a necessary condition for synaptic modification, could be insufficient if there is no change in pre- or postsynaptic cell activity. We formulated modification rules in the following way: Both activation of a synapse by the transmitter, and changes in pre- and/or postsynaptic cell activity during a time that is large enough for the shift in the ratio of PKs/PP1 activity in the postsynaptic neurone, are necessary and sufficient for the modification of synaptic transmission.

The unified modification rules for homosynaptic, heterosynaptic, associative and cerebellar synaptic plasticity are presented in Table 2.1. These rules are formulated with regard to the simultaneous activation of excitatory and inhibitory inputs. One of the most important conclusions to emerge from our model is that the same postsynaptic mechanisms underlie LTP, LTD and depotentiation. The question of whether the same mechanisms underlie LTD and depotentiation was checked by special experiments, and the similarity of these processes was indicated (Wagner and Alger, 1995), and a hypothetical model that can reconcile the apparent disparities between LTD and depotentiation was suggested (Wagner and Alger, 1996).

The induction of one or another type of modification depends on the following parameters: the amount of transmitter released per presynaptic impulse; the presence of different types of activated postsynaptic receptors such as NMDA, mGlu and $GABA_b$, and the initial phosphorylation state of ionotropic receptors. The last condition should be taken into account for both homosynaptic and heterosynaptic pathways.

5. THE PROPOSED POSTSYNAPTIC MODEL OF EXCITATORY AND INHIBITORY PLASTICITY IN THE CEREBELLAR PURKINJE CELL

5.1. Proposed Mechanisms for the Modification of Excitatory Inputs to a Cerebellar Purkinje Cell

When elaborating the hypothetical postsynaptic model for PC synaptic plasticity, we took into account existing evidence for postsynaptic mechanisms of LTD induction (Linden, 1994). Our postulate that only receptors activated by transmitter are modifiable is experimentally supported. Thus, it was demonstrated that without synaptic activation of the PC the application of the membrane-permeable analogue of cGMP causes neither LTDc (Hartell, 1994a), nor AMPA receptor phosphorylation (Nakazawa *et al.*, 1995). LTDc is an associative effect, in the sense that it is developed only due to conjoint PF and CF

Table 2.1. The unitary modification rules for known types of LTP and LTD

The type of modification	*Activation of modifiable spine*		*The depolarization level of a spine*		*Shift in the Ca^{2+} level*	*Shift in the cAMP and cGMP level*	*Receptors phosphorylation*	*Changes in EPSP and IPSP amplitude*		*Modification of resulting PSP*
	excitatory	*inhibitory*	*tested*	*conditioned*				*EPSP*	*IPSP*	
Homosynaptic	xxx	x	**		+	(+)	Phosphor.	LTP	LTDi	LTP
	xxx	xx	*		–	(–)	Dephosphor.	LTD	LTPi	LTD
Associative	x	x	**	***	+	(+)	Phosphor.	LTPa	LTDia	LTPa
	x	xx	*	***	–	(–)	Dephosphor.	LTDa	LTPia	LTDa
Heterosynaptic	0	x	**	***	–	(+)	Phosphor.	no	LTDih	LTPh
	0	xx	*	***	–	(–)	Dephosphor.	no	LTPih	LTDh
Cereb.(a)	xxx	xx	**		+	[–]	Dephosphor	LTDc	LTPic	LTDc
Cerebel.	x	xx	*		–	[+]	Phosphor.	LTPc	LTDic	LTPc

. Cereb.(a)—PFs and CF are stimulated, Cerebel.—PFs are stimulated; in both cases—PFs-PC input is tested; 0—no activation; xxx, xx, x—large, middle, small amount of transmitter; ***, **, *—high, middle, low level of spine depolarization; +, –; (+), (–); [+], [–]—positive, negative shift of calcium level: cAMP concentration: cGMP concentration in relation to previous rise; phosphor., dephosphor.—phosphorylation, dephosphorylation of receptors; no—no modification stimulation causing the essential rise in Ca^{2+} in a PC. Homosynaptic LTPc was observed after isolated PF stimulation, and heterosynaptic LTPc was induced after CF tetanization (Hirano, 1990).

We have postulated (Silkis, 1996b) that the properties of AMPA receptors are similar in different CNS structures. Therefore, the phosphorylation of AMPA receptors on PCs must lead to LTPc, as was shown for hippocampal/neocortical cells, but cannot cause LTDc assumed earlier (Ito and Karachot, 1992). According to our postulate, LTDc in PCs must be the consequence of AMPA receptor dephosphorylation, that causes a decrease of their sensitivity to glutamate. In our model, we used the data that cGMP prevails in PCs (Kennedy, 1992), as distinct from hippocampal/neocortical cells, where cAMP is dominant. Accordingly, cGMP-dependent PKG but not cAMP-dependent PKA is involved in LTDc induction (Hartell, 1994b; Ito and Karachot, 1992; Linden, 1994). We assumed that PKG, together with PP2B, controls PP1 activity in PCs. Protein phosphatase 1, which dephosphorylates receptors on PCs, is inhibited by the G substrate that has been proposed to be the target for PKG (Kennedy, 1992). The amino acid sequence of the G substrate is similar to the sequence of PP1 inhibitor I1 in neocortical/hippocampal cells. The hypothesized sequence of postsynaptic processes that is triggered by stimulation of excitatory inputs, and leads to LTDc is schematically represented in Figure 2.5. It is agreed that not only PKG, but also PKC influences the sensitivity of glutamate receptors on a PC (Hartell, 1994b; Ito and Karachot, 1992). PKC in the PCs is activated due to the action of glutamate binding to mGlu receptors that are similar in function to neocortical/hippocampal processes. We do not exclude a participation of CaMKII in cerebellar synaptic plasticity, because an increase in Ca^{2+} levels may cause activation of CaM and CaMKII. Active CaMKII could be inhibited by PP1, as is believed to occur in neocortical/hippocampal cells (Lisman, 1994). It is probable that this PK phosphorylates AMPA receptors on the PC, since certain sites of AMPA receptors could not be phosphorylated by PKC and PKG (Nakazawa *et al.*, 1995).

It has been observed that LTDc is developed at high Ca^{2+} concentrations (Linden, 1994). According to our postulate, at high Ca^{2+} concentrations, receptors must be dephosphorylated, i.e. the activity of PP1 must dominate the activity of PKC and PKG. The fulfilment of this condition is possibly provided by the fact that cGMP concentration and PKG activity decreases with an increase in Ca^{2+} levels (Olson *et al.*, 1976). This decrease occurs through the action of phosphodiesterases (PDEs), which are expressed in the PCs (Hartell, 1994a,b; Luo *et al.*, 1994). One of these PDEs is activated by CaM. The efficacy of another is Ca^{2+}-dependent, and strongly increases in the presence of a negligible amount of cGMP. Therefore, the activity of PKG at high Ca^{2+} levels decreases, while the activity of PP1 simultaneously increases, because PKG does not phosphorylate the PP1 inhibitor (G substrate). The high PP1 activity may also cause CaMKII inactivation (Figure 2.5). All these processes at a high Ca^{2+} level must lead to AMPA receptor dephosphorylation and a decrease in their sensitivity to glutamate, i.e. they must induce LTDc (see Table 2.1). In contrast, in neocortical/hippocampal cells, Ca^{2+} and cAMP increase simultaneously and cause LTP at a high Ca^{2+} level (Figure 2.1). Thus, the dependence of the sign of synaptic modification on Ca^{2+} levels differs in PCs as compared with neocortical/hippocampal cells, possibly as a result of the differential involvement of cGMP and cAMP. When the Ca^{2+} level in the PCs is low, for example due to the PF activation alone, PDE activity must also be low, while cGMP concentration and PKG

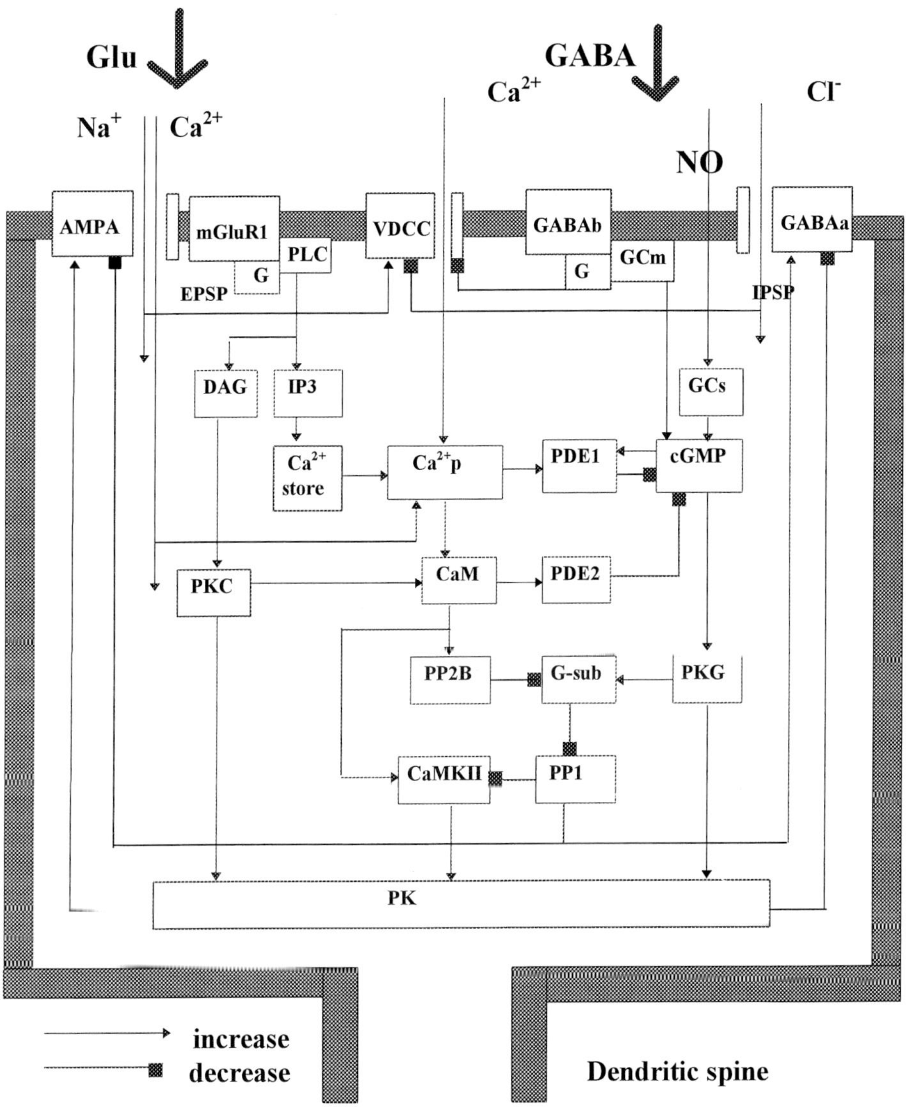

Figure 2.5. The proposed post-tetanic processes that cause the modification of synaptic transmission in a cerebellar Purkinje cell. The designations are given in the text and in Figure 1. (See also complete list of abbreviations at end of this chapter.)

activity must be high. In this case PP1 activity should be low and cannot inactivate CaMKII. Therefore, PKG and CaMKII together with PKC should phosphorylate AMPA receptors and cause LTPc. Indeed, LTPc was observed after isolated PF stimulation (Hartell, 1994a; Hirano, 1990), and under other experimental conditions that down-regulate the Ca^{2+} level (Hartell, 1994a,b; Kasono and Hirano, 1994; Shibuki and Okada, 1992).

5.2. Hypothetical Mechanism of cGMP Production in the Cerebellar Purkinje Cell

As noted above, the existing view that NO participates in cGMP production and subsequent induction of LTDc is inconsistent with some of the experimental data. For example, NOS is not contained in the CF and possibly not in PF terminals (Linden, 1994; Ross *et al.,* 1990). However, the stimulation of these fibres results in LTDc induction (Linden, 1994). Since NOS was found in axon terminals of inhibitory interneurones, we proposed that inhibitory cell discharges can lead to NO production and a cGMP rise in PCs (Silkis, 1996e). The increase in cGMP level after PF and CF stimulation may occur not only because these fibres monosynaptically excite, but also disynaptically inhibit, the PCs (Ross *et al.*, 1990; Vigot *et al.,* 1993). The necessity of inhibitory cells for the cGMP rise is confirmed by experimental data (Wood *et al.*, 1994). Since the level of GCs, the target of NO, is low in the PCs (Luo *et al.*, 1994), we proposed that an important role in cGMP production is probably played by membrane-bound GC (GCm), whose properties are distinctive from those of soluble GCs (Kennedy, 1992). The GCm could be activated through G proteins due to metabotropic receptor activation. There are two types of such receptors on PCs: mGlu1 and $GABA_b$ receptors. Because mGlu1 receptors activate only phospholipase C, we proposed that $GABA_b$ receptors participate in cGMP production (Silkis, 1996e). The amount of GABA required for activation of these receptors can be provided by rhythmic stimulation of inhibitory cells during tetanization of PFs and CFs.

5.3. Proposed Mechanisms of Modification of Inhibitory Inputs to a Cerebellar Purkinje Cell

According to recent data, the efficacy of inhibitory inputs to the PCs is also modifiable (Kano *et al.*, 1992; Llano *et al.*, 1991). This modification is an associative effect, since the inhibitory current in the PCs increases only if GABA application or inhibitory cell stimulation is conjoint with PC depolarization (Llano *et al.*, 1991). It is important that the same PKs, PKC, PKG and CaMKII, which participate in the modification of excitatory inputs to the PC, can also influence long-term changes in the efficiency of inhibition (Krishek *et al.,* 1994; McDonald and Moss, 1994; Sigel, 1995). It was demonstrated that LTPic and LTDic are input-specific and depend on the activity of pre- and postsynaptic cells (Kano *et al.*, 1992; Llano *et al.*, 1991). If so, then the change in balance between PK and PP1 in a PC, which determines the phosphorylation state of GABA receptors, must depend on the discharge of presynaptic inhibitory neurones. As a result of GABA release, only PKG can be activated, while PKC as well as CaMKII are known to become active due to excitatory input stimulation.

By analogy with our assumption regarding the properties of AMPA receptors in different structures, we proposed that the properties of $GABA_a$ receptors on cerebellar PCs and neocortical/hippocampal cells are identical (Silkis, 1996e). If so, then the sensitivity of GABA receptors on the PCs should rise as a result of their dephosphorylation and should decrease owing to their phosphorylation. This proposed feature of receptor modification corresponds to those experimental results in which the increase in PK activity in PCs resulted in a decrease of the Cl^- current through $GABA_a$ receptors (Pasqualotto *et al.*, 1993). It is pertinent to note here that it is unlikely that AMPA and $GABA_a$ receptor sensitivity decrease simultaneously due to a rise in PK activity in the PCs. It can be predicted using our model that inhibitory interneurone tetanization without excitatory input stimulation will cause a rise in the cGMP level, increase PKG activity (Figure 2.5), and lead to a phosphorylation of $GABA_a$ receptors and LTDic occurrence, because in such an experimental conditions the Ca^{2+} level and PDEs activity could not rise. Evidence supporting these two predictions comes from the experiments where rhythmic stimulation of a PC resulted in LTD of the IPSP in neurones of the deep cerebellar nuclei (Morishita and Sastry, 1993). In these experiments the participation of NO in cGMP production can be excluded, since NOS is absent in PC axon terminals (Ross *et al.*, 1990). Owing to the absence of glutamate, mGlu receptors were not activated and PKC was inactive. Neither depolarization of deep cerebellar nuclei neurones nor a rise in intracellular Ca^{2+} were observed in this study (Morishita and Sastry, 1993). In the absence of Ca^{2+} ions, the activation of PDEs, PP1 and CaMKII can be excluded. However, PKG could have been activated due to the action of GABA on $GABA_b$ receptors. Thus, the LTD of the IPSP which was obtained in deep cerebellar nuclei neurones could have been a consequence of cGMP production and the phosphorylation of $GABA_a$ receptors by PKG. It is interesting to note that, according to the suggested model, a Ca^{2+} rise is not required for LTDic induction since cGMP production and PKG activation could be achieved without Ca^{2+} ions. Indeed, it was found in the cerebellum that an increase in cGMP levels can occur in the absence of Ca^{2+} (Luo *et al.*, 1994).

5.4. Simultaneous Modification of Excitatory and Inhibitory Inputs to Cerebellar Purkinje Cells

It follows from the suggested mechanism of cerebellar synaptic plasticity that inhibitory and excitatory input modifications are interrelated (Figure 2.5). Disynaptic inhibition plays an essential role in long-term changes of the efficacy of monosynaptic excitatory transmission from PFs to a PC. Thus, strengthening of the activity of inhibitory cells may cause a decrease in Ca^{2+} levels, an increase in cGMP concentration, and the induction of LTPc and LTDic instead of LTDc and LTPic, which can occur if inhibition is weak. This prediction of the model is confirmed by experiments showing that PF stimulation, together with the application of Bi-cGMP and GABA, resulted in strongly expressed LTPc (Shibuki and Okada, 1992). Decreasing Ca^{2+} levels by the use of a Ca^{2+} chelator or hyperpolarizing current also resulted in LTPc, that was explained by simultaneous LTDic (Hartell, 1994a). On the contrary, the blockade of inhibition may promote LTDc induction, because such an experimental protocol results in a Ca^{2+} rise, reductions of cGMP levels and PKG activity, and a consequent decrease in the phosphorylation state and sensitivity of AMPA receptors on the PCs. Support for this prediction is provided by experiments wherein LTDc was induced only in the presence of a GABA antagonist (Shibuki and Okada, 1992). In turn, synaptic activation or depolarization of the PC that caused enhanced rises of Ca^{2+}, promoted LTPic induction (Llano *et al.*, 1991).

6. CONCLUSION

According to commonly accepted models of synaptic plasticity for the neocortex and hippocampus, the modification rules for homo-, hetero- and associative LTD are distinctive, and the mechanism of synaptic plasticity for cerebellar Purkinje cells is usually considered unique. In contrast, the present model suggests that all known types of excitatory and inhibitory synaptic plasticity in the neocortex, hippocampus and cerebellum conform to common modification rules. We have proposed that the following conditions are necessary for the modification of homo-, hetero- and associative synaptic inputs: the coincidence of pre- and postsynaptic cell activity, as well as changes in pre- and/or postsynaptic cell activity during a time sufficient for a change in the ratio between protein kinases and protein phosphatase 1 in a postsynaptic neurone. Presynaptic activation must include monosynaptic excitation and disynaptic inhibition. Heterosynaptic effects occur if homo- and heterosynaptic afferents form synapses not only on a target cell, but also on a "common" interneurone, which is presynaptic to the same target cell. The induction of all types of LTD is facilitated if monosynaptic excitation is followed by disynaptic inhibition.

Computational modeling of post-tetanic processes in a hippocampal pyramidal neurone has shown that the efficacy of synaptic transmission in stationary conditions is determined by the amount of transmitter released during tetanization, and does not depend on the initial synaptic efficacy. The sign of modification (LTP or LTD) depends on the previous synaptic efficacy, and on the post-tetanic shift in Ca^{2+} levels and the concentration of cyclic nucleotides (cAMP or cGMP). The changes in protein kinase and protein phosphatase activity relative to their previous state must cause simultaneous and opposite modifications of excitatory and inhibitory inputs. We propose that the Ca^{2+}-dependent increase in cAMP levels in neocortical/hippocampal neurones and Ca^{2+}-dependent decrease in the cGMP level in cerebellar Purkinje cells can underlie the different character of Ca^{2+}-dependence of the sign of synaptic modification in these structures.

The suggested unitary model of synaptic plasticity explains and integrates various existing experimental data, and moreover predicts results that can be experimentally tested. The proposed unitary modification rules can be used in models of memory and learning based on artificial neural networks with synaptic plasticity, and containing excitatory and inhibitory elements. We assume that such networks will be marked by a large information capacity.

ACKNOWLEDGEMENTS

I sincerely thank G. Murzina for the mathematical modeling, and A. Frolov for the discussion and critical remarks.

REFERENCES

Abraham, W.C. and Bear, M.F. (1996) Metaplasticity: the plasticity of synaptic plasticity, *Trends in Neurosci.*, **19**, 126–130.

Armstrong-James, M., Welker, E. and Callahan, C.A. (1993). The contribution of NMDA and non-NMDA receptors to fast and slow transmission of sensory information in the rat SI barrel cortex. *Journal of Neuroscience*, **13**, 2149–2160.

Artola, A. and Singer, W. (1993) Long-term depression of excitatory synaptic transmission and its relationship to long-term potentiation. *Trends in Neuroscience*, **16**. 480–487.

Bear, M.F. (1995) Mechanism for a sliding synaptic modification threshold. *Neuron*, **15**, 1–4.

Bear, M.F. and Malenka, R.C. (1994) Synaptic plasticity: LTP and LTD. *Current Opinion in Neurobiology*, **4**, 389–399.

Bliss, T.V.P. and Collingridge, G.L. (1993) A synaptic model of memory: long-term potentiation in the hippocampus. *Nature, London*, **361**, 31–39.

Brorson, J.R., Manzolollo, P.A., Gibbons, S.J. and Miller, R.J. (1995) AMPA receptor desensitization predicts the selective vulnerability of cerebellar Purkinje cells to excitotoxicity. *Journal of Neuroscience*, **15**, 4515–4524.

Christie, B.R., Magee, J.C. and Johnston, D. (1996) The role of dendritic action potentials and Ca^{2+} influx in the induction of homosynaptic long-term depresssion in hippocampal CA1 pyramidal neurons. *Learning and Memory* **3**, 160–169.

Christie, B.R., Stellwagen, D. and Abraham, W.C. (1995) Evidence for common expression mechanisms underlying heterosynaptic and associative long-term depression in the dentate gyrus. *Journal of Neurophysiology*, **74**, 1244–1247.

Coussens, C.M. and Teyler, T.J. (1996) Protein kinase and phosphatase activity regulate the form of synaptic plasticity expressed. *Synapse*, **24**, 97–103.

Crepel, F. and Krupa, M. (1988) Activation of protein kinase C induces a long term depression of glutamate sensitivity of cerebellar Purkinje cells. An *in vitro* study. *Brain Research*, **458**, 397–401.

Davies, S.N. and Collingridge, G.L. (1989) Role of excitatory amino acid receptors in synaptic transmission in area CA3 of rat hippocampus. *Proceedings of the Royal Society of London (B)*, **236**, 373–384.

Debanne, D., Gahwiler, B.H. and Thomson, S.M. (1997) Bidirectional associative plasticity of unitary CA3-CA1 EPSPs in the rat hippocampus *in vitro*. *Journal of Neurophysiology*, **77**, 2851–2855.

Dehay, C., Douglas, R.J., Martin, K.A.C. and Nelson, C. (1991) Excitation by geniculo-cortical synapses is not vetoed at the level of dendritic spines in cat visual cortex, *Journal of Physiology, (London)*, **440**, 723–734.

Derric, B.E. and Martinez, J.J.L. (1996) Associative bidirectional modifications at the hippocampal mossy fibr-CA3 synapse. *Nature, London*, **381**, 429–434.

Doyere, V., Srebro, B. and Laroche, S. (1997) Heterosynaptic LTD and depotentiation in the medial perforant path of the dentate gyrus in the freely moving rat. *Journal of Neurophysiology*, **77**, 571–578.

Dudek, S.M. and Bear, M.F. (1992) Long-term depression in area CA1 of hippocampus and the effects of NMDA receptor blockade *Proceedings of the National Academy of Sciences, U.S.A.*, **89**, 4363–67.

Dudek, S.M. and Bear, M.F. (1993) Bidirectional long-term modification of synaptic effectiveness in the adult and immature hippocampus. *Journal of Neuroscience*, **13**, 2910–2918.

Ekerot, C.F. and Kano, M. (1985) Long-term depression of parallel fibre synapses following stimulation of climbing fibres. *Brain Research*, **342**, 357–360.

Fox, K. and Daw, N.W. (1993) Do NMDA receptors have a critical function in visual cortical plasticity. *Trends in Neuroscience*, **16**, 116–122.

Frolov, A.A. and Muraviev, I.P. (1987) *Neuronal models of associative memory*, Moscow: Nauka, pp. 158 (Russian)

Grassi, S., Pettorossi, V.E. and Zampolini, M. (1996) Low-frequency stimulation cancels the high-frequency induced long lasting effects in the rat medial vestibular nuclei. *Journal of Neuroscience*, **16**, 3373–3380.

Hartell, N. A. (1994a) Induction of cerebellar long-term depression requires activation of glutamate metabotropic receptors. *Neuroreport*, **5**, 913–916.

Hartell, N.A. (1994b) cGMP acts within cerebellar Purkinje cells to produce long term depression via mechanisms involving PKC and PKG. *Neuroreport*, **5**, 833–836.

Hebb, D.O. (1949) *The organization of behavior.* New-York: Wiley.

Heynen, A.J., Abraham, W.C. and Bear, M.F. (1996) Bidirectional modification of CA1 synapses in the adult hippocampus *in vivo*. *Nature, London*, **381**, 163–166.

Huang, Y.Y., Colino, A., Selig, D.K. and Malenka, R.C. (1992) The influence of prior synaptic activity on the induction of long-term potentiation. *Science, New York* **255**, 730–733.

Hirano, T. (1990) Depression and potentiation of the synaptic transmission between a granule cell and a Purkinje cell in rat cerebellar culture. *Neuroscience Letters*, **119**, 141–144.

Ito, M. and Karachot, L. (1992) Protein kinases and phosphatase inhibitors mediating long-term desensitization of glutamate receptors in cerebellar Purkinje cells. *Neuroscience Research*, **14**, 27–38.

Ivakiri, A., Pavlides, C., Keller, A. and Asanuma, H. (1991) Long-term potentiation of thalamic input to the motor cortex induced by coactivation of thalamo-cortical and cortico-cortical afferents. *Journal of Neurophysiology*, **65**, 1435–1441.

Jaffe, D.B., Ross, W.N., Lisman, J.E., Lasser-Ross, N., Miyakawa, H., and Johnston, D. (1994) A model for dendritic Ca^{2+} accumulation in hippocampal pyramidal neurons based on fluorescence imaging measurements. *Journal of Neurophysiology*, **71**, 1065–1077.

Kano, M., Rexhausen, U., Dreessen, J. and Konnerth, A. (1992) Synaptic excitation produces a long-lasting rebound potentiation of inhibitory synaptic signals in cerebellar Purkinje cells. *Nature, London*, **356**, 601–604.

Kanter, E.D. and Haberly, L.B. (1993) Associative long-term potentiation of piriform cortex slices requires GABAa blockade. *Journal of Neuroscience*, **13**, 2477–2482.

Kasono, K. and Hirano, T. (1994) Critical role of postsynaptic calcium in cerebellar long-term depression. *Neuroreport*, **6**, 17–20.

Kennedy, M.B. (1992) Second messengers and neural function, In: Z.W. Hall (ed.) *An introduction to molecular neurobiology*, Sunderland, Massachusetts: Sinauer Associates INC publishers, pp.207–246.

Kirkwood, A., Rioult, M.G. and Bear, M.F. (1996) Experience-dependent modification of synaptic plasticity in visual cortex. *Nature, London*, **381**, 526–528.

Koch, C. and Zador, A. (1993) The function of dendritic spines; devices subserving biochemical rather than electrical compartmentalization. *Journal of Neuroscience*, **13**, 413–422.

Komatsu, Y. (1994) Use-dependent long-term potentiation of inhibitory synaptic transmission in rat visual cortex. *Journal of Neuroscience*, **14**, 6488–6499.

Komatsu, Y. (1996) GABAb receptors, monoamine receptors, and postsynaptic inositol triphosphate induced Ca^{2+} release are involved in the induction of long-term potentiation at visual cortical inhibitory synapses. *Journal of Neuroscience*, **16**, 6342–6352.

Krishek, B.J., Xie, X., Blackstone, C., Huganir, R, L, Moss, S.J., and Smart, T.G. (1994) Regulation of GABAa receptor function by protein kinase C phosphorylation. *Neuron*, **12**, 1081–1092.

Kuriyama, K., Hirouchi, M. and Nakayasu, H. (1993) Structure and function of cerebral GABAa and GABAb receptors. *Neuroscience Research*, **17**. 91–99.

Lambert, N.A. and Wilson, W.A. (1993) Discrimination of post- and presynaptic GABAb receptor-mediated responses by tetrahydroaminoacridine in area CA3 of the rat hippocampus. *Journal of Neurophysiology*, **69**, 630–641.

Linden, D.J. (1994) Long-term synaptic depression of the mammalian brain. *Neuron*, **12**, 457–472.

Linden, D. J. and Connor, J.A. (1991). Participation of postsynaptic PKC in cerebellar long-term depression in culture. *Science, New York*, **254**, 1656–1659.

Lisman, J. (1994) The CaMKII hypothesis for the storage of synaptic memory. *Trends in Neuroscience*, **17**, 406–412.

Llano, I., Leresche, N. and Marty, A. (1991) Calcium entry increases the sensitivity of cerebellar Purkinje cells to applied GABA and decreases inhibitory synaptic current. *Neuron*, **6**, 565–574.

Luo, D., Leung, E. and Vincent, S.R. (1994) Nitric-oxide dependent efflux of cGMP in rat cerebellar cortex: an *in vivo* microdialysis study. *Journal of Neuroscience*, **14**, 263–271.

Malen, P.L. and Chapman, P.F. (1997) Nitric oxide facilitates long-term potentiation, but not long-term depression. *Journal of Neuroscience*, **17**, 2645–2651.

Martin, L.J., Blackstone, C.D., Huganir, R.L. and Price, D.L. (1992) Cellular localization of a metabotropic glutamate receptor in rat brain. *Neuron*, **9**, 259–270.

McDonald, B.J. and Moss, S.J. (1994) Differential phosphorylation of intracellular domains of gamma-aminobutyric acid type A receptor subunits by calcium/calmodulin type 2-dependent protein kinase and cGMP-dependent protein kinase. *Journal of Biological Chemistry*, **269**, 18111–18117.

McLean, H.A., Caillard, O., Ben-Ari, Y. and Gaiarsa, J.L. (1996) Bidirectional plasticity expressed by GABAergic synapses in the neonatal rat hippocampus. *Journal of Physiology, London*, **496**, 471–477.

Miles, R. (1990) Synaptic excitation of inhibitory cells by single CA3 hippocampal pyramidal cells of the guinea-pig *in vitro*. *Journal of Physiology, London*, **428**, 61–77.

Misgeld, U., Sarvey, J.M. and Klee, M.R. (1979) Heterosynaptic postactivation potentiation in hippocampal CA3 neurons. Long-term changes of the postsynaptic potentials. *Experimental Brain Research*, **37**, 217–229.

Morishita, W. and Sastry, B.R. (1993) Long-term depression of IPSPs in rat deep cerebellar nuclei. *Neuroreport*, **4**, 719–722.

Moss, S.J., Smart, T.G., Blackstone, C.D. and Huganir, R.L. (1992) Functional modulation of GABAa receptors by cAMP dependent protein phosphorylation. *Science, New York*, **257**, 661–665.

Muller, D., Heft, S. and Figurov, A. (1995) Heterosynaptic interactions between LTP and LTD in Ca1 hippocampal slices. *Neuron*, **14**, 599–605.

Murzina, G.B. and Silkis, I.G. (1996a) Computational model of simultaneous long-term modifications in the efficacy of excitatory and inhibitory inputs to the hippocampal pyramidal neuron. *Neural Network World*, **6**, 331–338.

Murzina, G.B. and Silkis, I.G. (1996b) The mathematical modelling of Ca^{2+}-dependent postsynaptic processes in the hippocampus (the induction of long-term potentiation and long-term depression). *Zhurnal Vysshey Nervnoy Dejatelnosty*, **46**, 674–688. (Russian)

Murzina, G.B. and Silkis, I.G. (1996c) Long-term potentiation and depression of inhibitory transmission studied by using mathematical modelling of the postsynaptic processes. *Zhurnal Vysshey Nervnoy Dejatelnosty*, **46**, 917–928. (Russian)

Murzina, G.B. and Silkis, I.G. (1996d) The contrasting of synaptic signals resulting from simultaneous excitatory and inhibitory input modifications. *Zhurnal Vysshey Nervnoy Dejatelnosty*, **46,** 1076–1087. (Russian)

Murzina, G.B. and Silkis, I.G. (1997a) The changes of protein kinases and protein phosphatases activity in dendritic spine: a mathematical model. *Neurohimia*, **14**, 48–62. (Russian)

Murzina, G.B. and Silkis, I.G. (1997b) A model of posttetanic efficiency of a neuron dendritic spine *Biophysica*, **42**, 702–710.

Murzina, G.B. and Silkis, I.G. (1997c) The study of LTP and LTD of excitatory transmission by mathematical model of postsynaptic processes. *Doklady Academii Nauk SSSR*, (*Proc. USSR Acad. Sci.)* **352**, 240–243 .

Nakazawa, K., Mikawa, S., Hashikawa, T. and Ito, M. (1995) Transient and persistent phosphorylation of AMPA-type glutamate receptor subunits in cerebellar Purkinje cells. *Neuron*, **15**, 697–709.

O'Dell, T.J. and Kandel, E.R. (1994) Low-frequency stimulation erases LTP through an NMDA receptor mediated activation of protein phosphatases. *Learning and Memory*, **1**, 129–139.

Olson, D.R., Kon, C. and Breckenridge, B. (1976) Calcium ion effects on guanylate cyclase of brain. *Life Science*, **18**, 935–940.

Otani, S., Connor, J.A. and Levy, W.B. (1995) Long-term potentiation and evidence for novel synaptic association in CA1 stratum oriens of rat hippocampus. *Learning and Memory*, **2**, 101–106

Otani, S. and Connor, J.A. (1996) A novel synaptic interaction underlying induction of long-term depression in the area CA1 of adult rat hippocampus. *Journal of Physiology, London*, **492**, 225–230.

Otsu, Y., Kimura, F. and Tsumoto, T. (1995) Hebbian induction of LTP in visual cortex: perforated patch-clamp study in cultured neurons. *Journal of Neurophysiology*, **74**, 2439–2444.

Pasqualotto, B.A., Lanius, R.A. and Shaw, C.A. (1993) Regulation of GABAa and AMPA receptors by agonist and depolarizing stimulation requires phosphatase or kinase activity. *Neuroreport*, **4**, 447–450.

Ross, C.A., Bredt, D. and Snyder, S.H. (1990) Messenger molecules in the cerebellum. *Trends in Neuroscience*, **13**, 216–222.

Scanziani, M., Malenka, R.C. and Nicoll, R.A. (1996) Role of intercellular interactions in heterosynaptic long-term depression. *Nature, London*, **380**, 446–450.

Shibuki, K. and Okada, D. (1992) Cerebellar long-term potentiation under suppressed postsynaptic Ca2+ activity. *Neuroreport*, **3**, 231–234.

Sigel, E. and Baur, R. (1988) Activation of protein kinase C differentially modulates neuronal Na+, Ca2+, and γ aminobutyric type A channels. *Proceedings of the National Academy of Sciences, U.S.A.*, **85**, 6192–6196.

Sigel, E. (1995) Functional modulation of ligand-gated GABAA and NMDA receptor channels by phosphorylation. *Journal of Receptor and Signal Transduction Research*. **15**, 325–332.

Silkis, I.G. (1994) Long-term posttetanic modification of the efficiency of inhibitory connections in the thalamo-cortical circuitry. *Doklady Academii Nauk SSSR*, (*Proc. USSR Acad. Sci.*), **337**, 413–419.

Silkis, I.G. Rapoport, S. Sh., Veber, N.B. and Guschin, A.M. (1994b) Neurobiology of the integrative activity of the brain: some properties of long-term posttetanic heterosynaptic depression in the motor cortex of the cat. *Neuroscience and Behavioral Physiology*, **24**, 500–506.

Silkis, I.G. (1995a) Simultaneous activation of excitatory and inhibitory inputs as a necessary condition for production of homo-, hetero-, and associative LTD of excitatory transmission. *Zhurnal Vysshey Nervnoy Dejatelnosty*, **45**, 1151–1163. (Russian)

Silkis, I.G. (1995b) New modification rules for neural networks with excitatory and inhibitory synaptic connections. *The Second International Symposium on Neuroinformatics and Neurocomputers.* Rostov-on-Don. Russia. pp.129–135.

Silkis, I.G. (1996a) Activation of GABAb receptors, reduction of intracellular concentration of Ca^{++}, and inhibition of protein kinases are possible mechanisms of long-term posttetanic modification of the efficacy of inhibitory transmission in the new cortex. *Neuroscience and Behavioral Physiology*, **26**, 80–87.

Silkis, I.G. (1996b) Long-term changes, induced by microstimulation of the neocortex, in the efficiency of excitatory postsynaptic transmission in the thalamo-cortical networks. *Neuroscience and Behavioral Physiology*, **26**, 301–312.

Silkis, I.G. (1996c) Long-term changes in the efficiency of inhibitory transmission in the thalamo-cortical neuronal networks induced by microstimulation of the cortex. *Neuroscience and Behavioral Physiology*, **26**, 416–427.

Silkis, I.G. (1996d) A role of cyclic nucleotides in neuronal synaptic plasticity. *Neurokhimiia*, **13**, 3–6.

Silkis, I.G. (1996e) A possible role of GABAb receptors activation in cGMP production and in long-term depression of inhibitory synaptic transmission efficacy in cerebellar Purkinje cells *Neurokhimiia*. **13**, 7–11.

Silkis, I.G. (1996f) The model of long-term modifications (LTD, LTP) in the efficacy of excitatory and inhibitory transmission to cerebellar Purkinje neurons. *Neural Network World*, **6**, 371–374.

Silkis, I.G. (1997a) Long-term changes in the efficiency of excitatory and inhibitory connections in neural micronetworks of the motor cortex induced by tetanization of the thalamic nuclei and the sensory cortex. *Neuroscience and Behavioral Physiology*, **27**, 6–16.

Staubli, U. and Chun, D. (1996) Proactive and retrograde effects on LTP produced by theta pulse stimulation: mechanisms and characteristics of LTP reversal *in vitro*. *Learning and memory*, **3**, 96–105.

Stanton, P.K. (1996) LTD, LTP, and the sliding threshold for long-term synaptic plasticity. *Hippocampus*, **6**, 35–42

Szente, M.V., Baranyi, A. and Woody, C.D. (1990) Effects of protein kinase C inhibitor H-7 on membrane properties and synaptic responses of neocortical neurons of awake cats. *Brain Research*, **506**, 281–286.

Tsien, R.Y. (1996) LTD in cerebellar Purkinje neurons results from coincidence of NO and depolarization induced Ca^{2+} transients. In: A. Konnerts *et al. Coincidence Detection in the Nervous System*. Strasbourg: HFSP, pp. 95.

Tsumoto, T. (1992) Long-term potentiation and long-term depression in the neocortex. *Progress in Neurobiology*, **39**, 209–228.

Urban, N.N. and Barrionuevo, G. (1996) Induction of Hebbian and non-Hebbian mossy fiber long-term potentiation by distinct patterns of high-frequency stimulation. *Journal of Neuroscience*, **16**, 4283–4299.

Vigot, R., Billard, J.M. and Batini, C. (1993) Reduction of GABA inhibition in Purkinje and cerebellar nuclei neurons in climbing fibre deafferented cerebella of rat. *Neuroscience Research*, **17**, 249–255.

Wagner, J.J. and Alger, B.E. (1995) GABAergic and developmental influences of homosynaptic LTDe and depotentiation in rat hippocampus. *Journal of Neuroscience*, **15**, 1577–1586.

Wagner, J.J. and Alger, B.E. (1996) Homosynaptic LTD and depotentiation: do they differ in name only? *Hippocampus*, **6**, 24–29.

Wang, Y., Rowan, M.J. and Anwyl, R. (1997) Induction of LTD in the dentate gyrus *in vitro* is NMDA receptor independent, but dependent on Ca^{2+} influx via low-voltage-activated Ca^{2+} channels and release of Ca^{2+} from intracellular stores. *Journal of Neurophysiology*, **77**, 812–825.

Weber, N.V., Rapoport, S. Sh. and Silkis, I.G. (1984) Long-lasting excitability changes in pyramidal tract neurons in cats. *Zhurnal Vysshey Nervnoy Dejatelnosty*, **34**, 572–574. (Russian)

White, G., Levy, W.B. and Steward, O. (1990) Spatial overlap between populations of synapses determines the extent of their associative interaction during the induction of long-term potentiation and depression. *Journal of Neurophysiology*, **64**, 1186–1198.

Wickens, J.R. and Abraham, W.C. (1991) The involvement of L-type calcium channels in heterosynaptic long-term depression in the hippocampus. *Neuroscience Letters*, **130**, 128–132.

Wood, P.L., Emmett, M.R. and Wood, J.A. (1994) Involvement of granule, basket and stellate neurons but not Purkinje or Golgi cells in cerebellar cGMP increases *in vivo*. *Life Science*, **54**, 615–620.

Yang, X.D. and Faber, D.S. (1991) Initial synaptic efficacy influences induction and expression of long-term changes in transmission. *Proceedings of the National Academy of Sciences, U.S.A.*, **88**, 4299–4304.

Abbreviations used:

AMPA: alpha-amino-3-hydroxy-5-methyl-4-isoxazole propionic acid
CaM: calmodulin (CaM)
CaMKII: Ca^{2+}-calmodulin-dependent protein kinase II
cAMP: cyclic adenosine monophosphate
CF: climbing fiber
cGMP: cyclic guanosine monophosphate
GC: guanylyl cyclase
GCm: membrane-bound GC
GCs: soluble GC
E: synaptic efficacy
E_o: initial synaptic efficacy
E_p: post-tetanic synaptic efficacy
f: stimulation frequency
f_0: previous stimulation frequency
$GABA_a$, $GABA_b$: receptors for gamma-amino butyric acid
HFS: high frequency stimulation

I1: inhibitor of PP1
LFS: low frequency stimulation
LTP: long term potentiation of excitatory transmission
LTPa: associative LTP
LTPc: cerebellar LTP
LTPh: heterosynaptic LTP
LTPi: LTP of inhibitory transmission (LTPic: ditto in cerebellum; LTPih: inhibitory heterosynaptic LTP)
LTD: long term depression of excitatory transmission
LTDa: associative LTD
LTDc: cerebellar LTD
LTDh: heterosynaptic LTD
LTDi: LTD of inhibitory transmission (LTPic: ditto in cerebellum; LTDih inhibitory heterosynaptic LTD)
M: synaptic modification
M_o: amount of transmitter released per presynaptic spike
mGlu: metabotropic glutamate (receptors)
NMDA: N-methyl-D-aspartate
NO: nitric oxide
NOS: NO-synthase
PC: Purkinje cell
PDEs: phosphodiesterases
PF: parallel fiber
PK: protein kinases (also PKA, PKC, PKG)
PK_a: concentration of active protein kinases
PP: protein phosphatases (also PP1, PP2B)
PP_a: concentration of active protein phosphatases
R_{ph}: number of highly sensitive phosphorylated receptors
R_{pho}: initial number of phosphorylated receptors
Sc: conditioned spine
St: test spine
T: time of achievement of the stationary state
VDCCs: voltage-dependent Ca^{2+}-channels

3 Memory Consolidation: Narrowing the Gap between Systems and Molecular Genetics Neurosciences

K.V. Anokhin

Institute of Normal Physiology, Russian Academy of Medical Sciences, Moscow, Russia
anokhin@neuro.redline.ru

Memory consolidation is a process of information transfer from a short-term to a long-term store, which results in the establishment of a permanent memory trace. Two alternative approaches have been taken to explain the neural basis of this fundamental phenomenon. A network model suggests that memory consolidation is a function of a particular brain system that supports declarative or explicit memory. According to this model, consolidation involves a transfer of memory from the medial temporal lobe to the neocortex storage sites and takes weeks or years to be completed. The molecular model views consolidation as a switch between short-term and long-term mechanisms of memory storage in the same cell. This process requires new gene expression, is universal for various forms of memory and can be completed within minutes or hours after learning. The present review surveys the main features and limitations of both models and suggests the necessity of their integration into a unified model. Such a model should view consolidation as a multi-level set of parallel processes in multiple memory systems, all activated by the same learning event. Memory consolidation in each system according to such a "parallel draft" model is relatively independent and involves multiple phases of gene expression and reorganization of storage sites.

KEYWORDS: learning, memory, consolidation, gene expression, hippocampus

1. INTRODUCTION

Memory consolidation is a neural process of information transfer from a short-term to a long-term store which results in the establishment of a permanent memory resistant to disruptive treatments (McGaugh and Herz, 1972; Weingartner and Parker, 1984; Alvarez and Squire, 1994). Though the concept of memory consolidation forms a core of current research on information storage in the nervous system, there is still no apparent consensus about the neural mechanisms of this event. This is not due to a vague definition of the process itself. On the contrary, recent advances in systems and molecular neuroscience have produced two clear models of memory consolidation (Alvarez and Squire, 1994; DeZazzo and Tully, 1995; Abel *et al.*, 1995; Bailey *et al.*, 1996). However these two models operate on very different scales of time and space. A network model is based mainly on the studies of declarative memory distortions in humans and effects of brain lesions in mammals (Squire, 1992). It views consolidation as a structural reorganization of memory repository between the hippocampal system and the neocortex. Such a process requires lengthy periods of weeks and even years to be accomplished (Squire *et al.,* 1993; Squire and Zola, 1996). The molecular genetics approach adopts a broader view of long-term memory consolidation as a universal biological phenomenon conserved through the animal kingdom and shared by different forms of nondeclarative and declarative memories (Bailey *et al.*, 1996; Tully *et al.*, 1994). A shift from short-term to long-term memory is understood here as a critical "switch" between mechanisms that support synaptic modifications within the same cell. This transfer of information from the short-term to long-term storage is believed to require activation of gene expression through universal transcriptional mechanisms which are conserved from invertebrates to mammals and operate within minutes to hours after learning (Abel *et al.,* 1995; DeZazzo and Tully, 1995; Mayford *et al.*, 1995; Tully, 1997).

Even this brief exposition makes it clear that further progress in the biological understanding of learning and cognition might be substantially hindered by the profound differences in molecular and systems neuroscience approaches to the issue of memory consolidation. The present article is directed toward narrowing the gap between the two models. It reviews main features and assumptions of each model, exposes the main differences between them and suggests conditions for their integration into a more universal "parallel draft" model of memory consolidation. Such a model should cover a wide spectrum of memory forms that exist in mammalian and non-mammalian species and must be able to explain the molecular bases of the reorganization of memory sites during the course of its long-term storage.

2. A HISTORICAL PERSPECTIVE ON MEMORY CONSOLIDATION

The fact that human memory consists of distinct processes became evident in the very first experimental study of remembering, performed by Hermann Ebbinghaus (1885). Ebbinghaus discovered two important divisions in memory acquisition and recall. First, by learning different lists of nonsense syllables, he found that while remembering six or seven items required only one repetition, a list of twelve items needed up to fifteen learning sessions. This led him to postulate the existence of two different memories (Ebbinghaus, 1885), a suggestion which is sometimes interpreted as an anticipation of the modern distinction between short-term memory (STM) and long-term memory (LTM) (Kandel *et al.*, 1987). However, the Ebbinghaus hypothesis about two types of memories had a more subtle projection on the issue of STM and LTM, since both processes in his classification committed learned·items to the LTM.

The second essential discovery of Ebbinghaus concerned the kinetics of memory formation, and was based on a method that he called "savings". Ebbinghaus found that relearning the list of nonsense syllables took him less time and required fewer repetitions than the original learning. Most importantly, testing of such memory savings at different times after learning revealed two periods during the forgetting of new material (Ebbinghaus, 1885). The retention of learned items rapidly decayed in the first minutes to hours after learning and then remained at approximately the same level for many days.

Based on Ebbinghaus' findings and his own self-observations, William James proposed the now classical distinction between a short-term "primary memory", that constitutes a part of the psychological present, and a longer-lasting "secondary memory" into which items could be stored and consciously retrieved at later times (James, 1890).

James's hypothesis soon received support from the experiments of Müller and Pilzecker on verbal learning (1900). They found that learning a second list of verbal material immediately after the first list potentiated forgetting of the first list, a phenomenon they called retroactive interference (Müller and Pilzecker, 1900). Müller and Pilzecker argued that retroactive interference can be explained by postulating two memory processes: first, a perseverating phase during which memory is open for disruption, and then a more stable phase when memory becomes a permanent physical structure. The transition from an initial disruptible state into a later permanent state constitutes the process of memory "consolidation" (Müller and Pilzecker, 1900).

Though this concept was initially introduced to explain retroactive interference in normal learning, it was soon applied to retrograde or pre-morbid amnesia (Burnham, 1903), a phenomenon of memory loss for past events. Systematic studies of retrograde amnesia were made by Ribot (1882; see also Hacking, 1995) and by Korsakoff who described a syndrome of memory impairment in chronic alcoholic patients (Korsakoff, 1889). However, these ideas about the mechanisms and dynamics of memory consolidation did not receive much attention in neuroscience until Donald Hebb (1949) renewed interest in the physiological bases of memory formation by his hypothesis of a dual trace memory mechanism. Hebb's suggestion that "a reverberatory trace might cooperate with structural change and carry the memory until the growth change is made" (Hebb, 1949, p. 62) offered a physiological explanation for the distinction between sequential mechanisms of activity-dependent short-term and growth-related long-term memory. At the same time, Duncan (1949) found that retrograde amnesia could be reproduced in animals by administration of an electroconvulsive shock (ECS) shortly after learning. He also demonstrated the existence of "the gradient of retrograde amnesia"—the closer in time the ECS was to the learning event, the worse was subsequent memory retention. This phenomenon resembled the famous Ribot's Law of Regression, which stated that in human memory "the new perishes before the old" (Ribot, 1882). ECS thus came to be an experimental tool for the study of memory consolidation in animals. Starting from the late 1940s the "consolidation theory" became immensely influential in memory research in psychology (Atkinson and Shiffrin, 1968) and neurobiology (McGaugh and Herz, 1972; Weingartner and Parker, 1984). It has even been suggested that deciphering the critical events behind memory consolidation may give neuroscience the "Rosetta stone" for understanding biological principles of knowledge acquisition both at systems and molecular levels (Rose, 1991a). However, as it will be seen below, this is not what is happening. The remarkable advances made recently by systems and molecular genetic neuroscience have taken these disciplines even further apart in their account of memory consolidation.

3. MEMORY CONSOLIDATION: A NEURAL SYSTEMS SCENARIO

The term "neural systems approach" will be used here to describe a line of research which distinguishes different "memory systems" in the brain (Tulving, 1985; Squire, 1986; Squire and Zola, 1996; Thompson and Kim, 1996). Each of the multiple memory systems is suggested to serve a different biological function, to depend on a different set of neural structures, to learn something different about the situation, to have its own operational rules and to function relatively independently of other memory systems. Together all these memory systems are thought to cooperate towards an output of what was previously believed to be a single memory entity (Schacter and Tulving, 1994; Willingham, 1997).

3.1. Multiple Memory Systems in the Brain

Though philosophical speculations on the existence of different forms of memory (Bergson, 1911) and anecdotal observations of unusual memory dissociation in human amnesia (Claparède, 1911) had existed for a long time, they received little theoretical attention until approximately 20 years ago. The main impetus for a shift from a single to a multiple memory systems view came from neuropsychological studies of amnestic patients with focal brain lesions and Korsakoff syndrome (Squire *et al.*, 1993; Butters and Delis, 1995). It was noted that even in the most severe cases of anterograde amnesia, the loss of memory and learning abilities in such patients was not complete. Perhaps the best studied case of such dissociation is the patient H.M., who underwent at the age of 27 a bilateral medial temporal lobe removal for the treatment of intractable epilepsy (Scoville and Milner, 1957; Milner *et al.*, 1968; Hilts, 1995). Despite the fact that H.M. had an almost entire loss of capacity to remember new facts and events, rendering him unable to learn the environment and personnel of the nursing house in which he lived for years, he nevertheless had intact learning of new motor skills (Corkin, 1984). H.M. also had a preserved performance in priming tasks, like the capacity to complete an unfinished word or to recognize an ambiguous picture more rapidly if it was viewed some time ago (Corkin, 1984).

Studies of patients like H.M. suggested that lesions of certain brain structures can disrupt memory that supports conscious recollection of facts and events, while leaving intact other learning abilities including skills, habits, categorization and simple conditioning. The brain areas that have proved to be particularly involved in such amnesias are the structures of the midline diencephalon and medial temporal lobe (MTL) which includes the hippocampal formation and adjacent perirhinal, entorhinal and parahippocampal cortices (Squire, 1992). Experimental lesions of MTL in monkeys and rodents appeared to mimic the memory dissociation found in humans (Mishkin, 1978; Zola-Morgan and Squire, 1990; Kim and Fanselow, 1992), indicating that different memory systems and their dissociation exist at least in mammals (Squire, 1992).

A direct functional interpretation of brain lesions has always been a difficult task, particularly because of limitations imposed by the processes of neural and behavioural compensation. However, recent studies with positron emission tomography (PET) and functional magnetic resonance imaging (fMRI) in healthy volunteers have corroborated the main conclusions from the lesion approach. For example, Squire *et al.* (1992) found in a PET study that the hippocampus was activated when people remembered recently presented words after receiving their three-letter beginnings. Schacter *et al.* (1996) have additionally reported that the hippocampus was most active when people were recalling words which were subjected to elaborate processing during encoding task. Similar results were obtained by Nyberg *et al.* (1996) who demonstrated that there was a strong correlation between hippocampal activity and success of memory retrieval within individual subjects.

3.2. Memory Consolidation as a Property of the Declarative Memory System

Though the number of memory systems and demarcations between them is a matter of ongoing discussion (Schacter and Tulving, 1994; Willingham, 1997), there is a relative consensus on the existence of two main memory forms. One is declarative or explicit memory, that requires awareness and conscious intention for recall. The second is non-declarative or implicit memory that includes such different forms of experience as sensory and motor skills, habits, priming, category formation, basic associative and non-associative learning. What is common for all forms of non-declarative memory is that their behavioural expression does not require the participation of consciousness and is not dependent on the integrity of the MTL. The neural substrate of these memory forms is believed to involve the same sensory, motor and associative pathways that were used in the expression of the learning process (Bailey *et al.*, 1996).

One of the major characteristics of the declarative memory system is that storage and retrieval of explicit information depends only temporarily on the MTL structures. With the passage of time, lesions of the MTL can no longer disturb the recall of the initial learning episode, a phenomenon called temporally-graded retrograde amnesia (Squire, 1992). This property of the declarative memory system explains, for example, why damage of the MTL usually results in only time-limited retrograde amnesia as described by Ribot's Law of regression. Importantly, temporally-graded retrograde amnesia has also been reproduced with hippocampal lesions in monkeys and rodents, suggesting that this feature of declarative memory system has a phylogenetic history in the mammalian brain.

The proposed explanation for the development of temporally-graded retrograde amnesia after MTL lesions is that storage of declarative memory is gradually reorganized over time, so that it eventually becomes independent of MTL and is stored in other distributed locations. Clinical and experimental studies show that this process can require a long time to be completed. For example, in the experiments of Squire and Spanis (1984), mice were given a series of ECS treatments at different times after one-trial passive avoidance learning. ECS produced a graded impairment of task retention that covered a period from 1 to 3 weeks after training. Squire *et al.* (1975) have also designed a human memory task which was based on questions about television programs transmitted for a single season at different times before ECS treatment in psychiatric patients with severe depression. Amnesia produced by ECS covered a period of 1–2 years. Lesions of MTL structures produce similar effects. For example, memory in a contextual freezing paradigm in rats is impaired only if lesions of the hippocampus are made within the first week after training (Kim and Fanselow, 1992). In monkeys, bilateral lesions of the hippocampal formation impaired retention scores for objects that were used in an object-discrimination task 2–4 weeks before surgery (Zola-Morgan and Squire, 1990). The degree of retrograde amnesia decreased monotonically from 2 to 12 weeks following learning to surgery (Zola-Morgan and Squire, 1990). Amnestic patients with confirmed MTL damage exhibit temporally graded retrograde amnesia that extends into the distant past and may cover many years, sometimes up to 25 years (Squire and Zola, 1996). It is this lengthy process of rearrangement in memory storage sites which is assumed under the term of memory consolidation in the neural systems approach (Alvarez and Squire, 1994).

3.3. Network Models of Memory Consolidation

A number of models were proposed to explain consolidation as a process of gradual reorganization of memory storage at the systems level. According to one class of models, the hippocampus acts as a temporary memory store, from which information is later transferred to permanent locations in the cerebral cortex (Marr, 1971; Thompson and Kim, 1996). Other models suggest that the hippocampus does not store memory itself, but rather binds disparate neocortical areas which were involved in the learning episode (Mishkin, 1982; Teyler and DiScenna, 1986). In these models, the hippocampus acts as a memory "index", communicating through a broad net of its connections with most areas of the cerebral cortex (Teyler and DiScenna, 1986). Still other models point out that the distinction between indexing and storage is unnecessary (Eichenbaum *et al.*, 1992; Alvarez and Squire, 1994). It is suggested that MTL serves as a temporary store of a simple "indexing" memory, but also activates and gradually binds ensembles of neocortical cells that form representations of the original event (Alvarez and Squire, 1994). The latter view is an illustrative example of the network approach to memory consolidation. According to Alvarez and Squire, MTL acts as a temporary memory store, while the neocortex is a permanent repository of long-term memory. After new information is learned, MTL directs its initial recall by binding together the neocortical cells which participated in the original experience. At each round of such reactivation, the direct connections between geographically separated parts of memory representation are gradually stabilized through simultaneous activity according to a Hebbian synaptic rule. MTL also displays random endogenous activity. It likewise excites and links ensembles of neocortical neurones underlying memory representations. This activity-dependent process constitutes the biological substrate of memory consolidation. As a result, long-term memory is gradually established in the neocortex, where it is stored in distributed networks of neocortical neurones specialized for processing and analysis of area-specific kinds of information (Alvarez and Squire, 1994).

3.4. Main Features of Network Models

Despite certain differences between various network models, they all share the same set of fundamental assumptions about memory consolidation. These assumptions can be summarized in three main points (Figure 3.1):

1. Memory consolidation is a systems level phenomenon. It is based on transfer of representation functions from one set of structures to another.

"The most common current view of the memorial functions of the hippocampal-medial temporal lobe system is that declarative memories are stored there for some period of time and then eventually transferred or consolidated to other brain regions for permanent storage" (Thompson and Kim, 1996, p. 13440).

2. Memory consolidation requires extended periods of time ranging from days in rodents to years in humans.

"Observations of temporally-graded retrograde amnesia led to the idea of memory consolidation: as time passes the neural substrate of memory is gradually changed or reorganized in a way that makes memory resistant to disruption" (Alvarez and Squire, 1994, p. 7041).

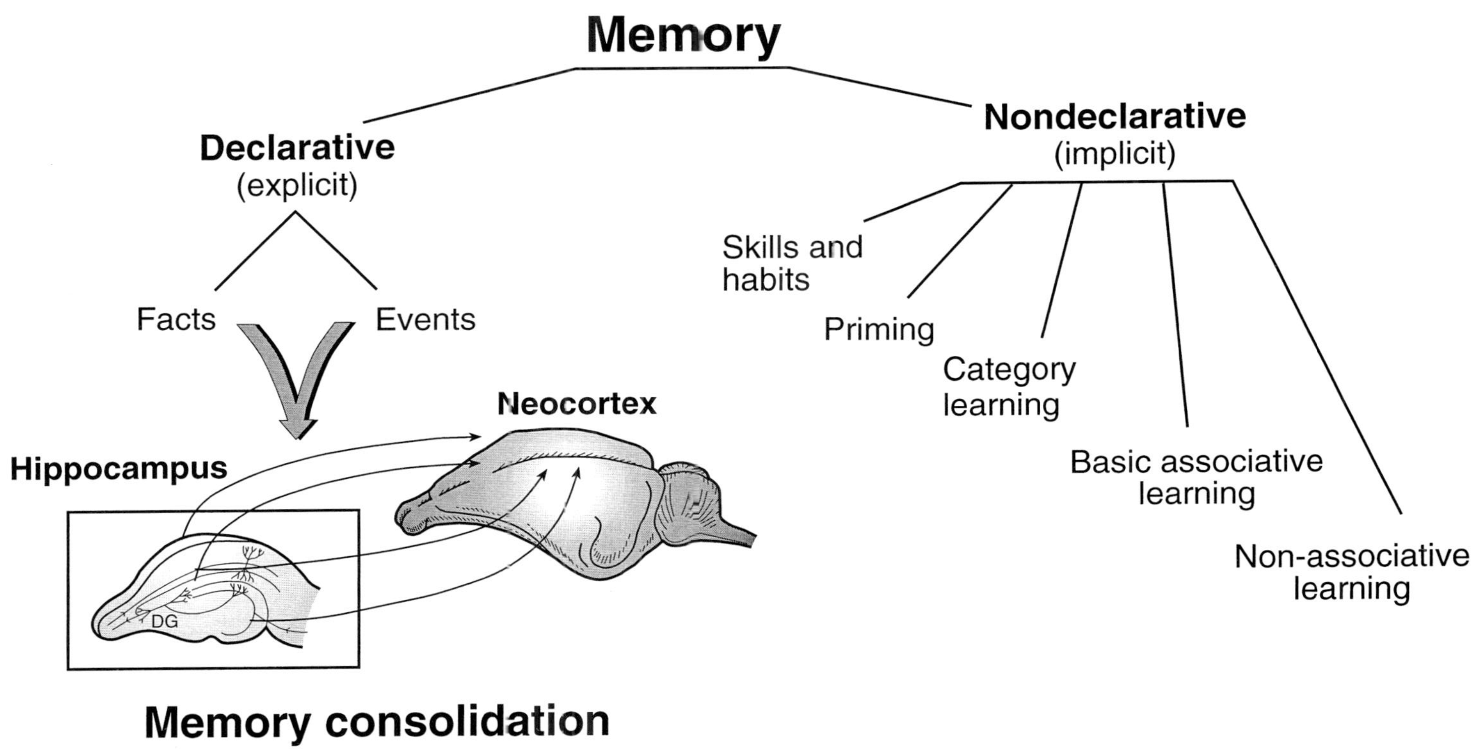

Figure 3.1. Multiple memory systems and the network scenario for memory consolidation. Memory consolidation is viewed as a property of the declarative (explicit) memory system, and depends on the gradual transfer of learned information from the hippocampus to neocortex.

3. Memory consolidation is a particular property of the declarative (explicit) memory system.
"These ideas about the significance of retrograde amnesia and reorganization of memory over time are ideas specifically about declarative memory" (Squire, 1992, p. 222).

The systems concept of memory consolidation has been able to promote an understanding of a large body of facts about the organization of human memory and its impairment in amnesia. It explains the amnestic effects of MTL lesions in humans and animals, and effectively deals with the issue of multiple memory systems in the brain. However, a number of major questions remain unresolved by this model. What are the mechanisms of relational and configural memory storage in the brains of non-mammalian species which do not possess MTL and neocortical structures? To what degree do different memory systems share common cellular and molecular components? What are the mechanisms of memory formation in non-declarative memory systems, which store learned information for a comparable and often life-long periods of time? These problems are much more efficiently tackled by molecular theories of memory consolidation.

4. MEMORY CONSOLIDATION: A MOLECULAR GENETICS SCENARIO

Is there a fundamental set of molecular processes that underlies storage of different forms of information in the nervous system? This question was first addressed in the 1960s, when pioneering experiments by Hyden and colleagues showed that learning produces a rapid activation of gene expression in the animal brain, as measured by increased incorporation of radioactive precursors into RNA and proteins (Hyden and Egyhazi, 1962, 1964; Hyden and Lange, 1968). It was also discovered that inhibitors of RNA and protein synthesis disrupted long-term but not short-term memory when injected within the restricted time period of one to two hours after learning (Dingman and Sporn, 1961; Flexner *et al.*, 1963; Agranoff, 1968; Barondes and Cohen, 1968). The narrow period of action of the inhibitors of macromolecular synthesis overlapped with the critical period when formation of memory could be disrupted by electroconvulsive shock (Agranoff, 1972). It was therefore suggested that both treatments impair the same fundamental process of long-term memory consolidation, which is thus based on *de novo* gene expression in the nervous system (Kandel *et al.*, 1987).

What makes this phenomenon particularly interesting is that the dependence of long-term memory on a "time window" of protein synthesis has been reported for such different tasks as habituation and sensitization, instrumental and classical conditioning, spatial and navigational learning, single and multiple trial learning, tasks with negative and positive reinforcement and models of sensory learning (for reviews see Barraco and Stettner, 1976; Davis and Squire, 1984). Impairment of long-term memory by RNA and protein synthesis inhibitors was observed in a variety of species including insects, molluscs, fish, birds and mammals (Agranoff, 1968; Barraco and Stettner, 1976; Davis and Squire, 1984; Montarolo *et al.*, 1986). This suggests that gene expression is a phylogenetically conserved requirement for long-term information storage in the nervous system.

It is therefore implied that the transition from the short-term to long-term memory involves a switch from information storage mechanisms which are protein synthesis-independent to those which are protein synthesis-dependent, in the same cell or even at the same synapses (Bailey *et al.*, 1996; Yin and Tully, 1996).

Other important claims of the molecular theory of memory consolidation are that this process is universal for different forms of non-declarative and declarative memory, does not depend on mammalian neuroanatomy, and is accomplished within a few hours after learning (Kandel *et al.*, 1995; Abel *et al.*, 1995).

4.1. Molecular Genetics Models of Memory Consolidation

A number of proposals were developed in the 1960s to explain the dependence of long-term memory on protein synthesis. The majority of them postulated a single process of learning-induced gene expression, responsible for the synthesis of several classes of proteins, including ribosomal proteins, synaptic proteins, structural proteins of axonal endings, and enzymes for the synthesis of membrane lipids (Gaito, 1967; Agranoff, 1968). A different and particularly interesting model was developed by Glassman (Glassman, 1969). Glassman suggested that "among the chemical events that lead to the consolidation of long-term memory, one might postulate the following sequence: protein-1—RNA—protein-2." His hypothesis about the formation of protein-1 was based on the data that the protein synthesis inhibitor puromycin simultaneously prevented changes in RNA synthesis and caused memory deficits in goldfish (Shashoua, 1968). Glassman proposed that protein-1, synthesized during learning "is an activator of specific genes which code for the RNA in the next step. This RNA codes for protein-2 which may be involved in consolidation of long-term memory, by rendering permanent the synaptic associations between neurones that were developed during short-term learning" (Glassman, 1969, p. 636).

Glassman's two-stage consolidation model received additional support from the research of Matthies and colleagues (see Matthies, 1989). This group discovered two waves of protein synthesis in the rat hippocampus after brightness discrimination training. The first wave started immediately after training, while the second was observed 6–8 hours later (Popov *et al.*, 1976). The short-acting protein synthesis inhibitor, anisomycin, disrupted long-term memory when injected into the hippocampus around the time of training, and 4–6 hours later, but not in the period between the two waves of protein synthesis (Grecksch and Matties, 1980). Based on these findings, Matthies suggested that the two phases of enhanced protein synthesis after learning represent qualitatively different consecutive stages in long-term memory formation, the first being regulatory proteins, and the second being the effector glycoproteins (Matthies, 1979, 1989).

The hypothesis about the molecular cascade in memory consolidation was further developed in a model proposed by Kandel and colleagues (Goelet *et al.*, 1986; Kandel *et al.*, 1987). According to this model, a common extracellular signal initiates separate intracellular memory processes. Short-term memory, which lasts from minutes to hours, is based on covalent modification of pre-existing proteins. For intermediate memory, which covers several hours, these modifications are prolonged by protein phosphorylation. Acquisition of long-term memory, lasting more than one day, is dependent on the induction of new genes through second messengers and constitutive transcription regulators. These regulators act by activating early effector and early regulatory genes. Early effector genes are responsible for the synthesis of proteins which retain memory for days. Memory lasting weeks and months is maintained by late effector genes which are switched on by early regulatory genes (Abel *et al.,* 1995, 1997; Bailey *et al.*, 1996).

4.2. Immediate Early Genes and Memory Consolidation

Goelet *et al.* (1986) made a specific suggestion about the nature of regulatory genes involved in long-term memory consolidation. According to their proposal, these functions may be played by a particular class of genes known as "competence" or "immediate early genes" (IEGs). Many IEGs were initially identified as proto-oncogenes, with genes *c-fos,* and *c-jun* being the best known members of this class (Curran and Morgan, 1987; Greenberg and Ziff, 1984). In normal cells, IEGs are rapidly and transiently induced by various extracellular signals including hormones and growth factors (Greenberg and Ziff, 1984; Lau and Nathans, 1985). Stimulus-induced transcription of IEGs is not prevented by protein synthesis inhibitors, indicating that all components of the cascade for signal transduction from the membrane to cell nucleus are already present in the cell before the stimulus. Besides *c-fos* and *c-jun* a number of other IEGs have been identified in recent years, including *c-myc, N-myc, L-myc, c-myb, fos-B, jun-B, jun-D, fra-1, fra-2, ets-1, ets-2, krox-20, zif/268, NGFI-B, mKR2, TIS1, TIS7, TIS8* (Sheng and Greenberg, 1990; Struhl, 1991). In total, about several hundred immediate early genes have been cloned, though only few of them have been studied in detail (Sheng and Greenberg, 1990). Many IEGs are known to encode nuclear proteins that act as transcription factors. By analogy with viral systems, the genes that are under control of immediate early genes are called "effector genes" or "late genes" (LGs) (Curran and Morgan, 1987; He and Rosenfeld, 1991).

The cascade of IEGs—LGs was initially shown to be implicated in the processes of cellular proliferation and differentiation. It is this particular function of *c-fos* and *c-myc* that was used by Kandel and colleagues to suggest the role of these genes in learning. Similar ideas were also developed by Berridge (1986).

At approximately the same time, we found that some of the proto-oncogenes including *c-fos*, *c-myc* and *c-myb* are expressed at a high level in embryonic rat brain, and disappear later after birth. This led us to test the hypothesis that these genes can be re-induced in the adult brain during learning. Initially we studied the dynamics of *c-fos* and *c-myc* mRNA levels in the cerebral cortex, hippocampus and cerebellum after active avoidance conditioning in rats. It was found that *c-fos* but not *c-myc* is strongly induced in all three brain structures 30–60 min after training (Maleeva *et al.*, 1989). Additionally, a several-fold induction of *c-fos* and *c-jun* mRNA was detected in the neocortex of mice 15 min after single-trial passive avoidance learning (Anokhin and Ryabinin, 1993). Similar *c-fos* and *c-jun* activation was seen in the chick brain after one-trial passive avoidance training (Anokhin *et al.*, 1991). Parallel experiments by other groups have demonstrated the phenomenon of IEG induction in the rat brain after such various tasks as brightness discrimination training, learning of sexual behaviour, olfactory discrimination learning and odour recognition learning, taste aversion learning and learning new motor skills in appetitive task (Tischmeyer *et al.*, 1990; Nikolaev *et al.*, 1992a,b; Baily *et al.*, 1992; Kaczmarek, 1993; Brennan *et al.*, 1994; Calamandrei and Keverne, 1994; Beck and Fibiger, 1995). In chicks, *c-fos* was induced not only during passive avoidance learning, but also during imprinting (McCabe and Horn, 1994). In songbirds, the homologue of the mammalian zif/268—ZENK is induced in auditory centres of the telencephalon when birds hear songs of their species (Mello *et al.*, 1992).

One possible interpretation of these data is that IEGs are induced in the nerve cells by behavioural stress, non-specific arousal, animal motor activity or just cellular depolarization. However, training rats and chicks in an appetitive task also produced large *c-jun* and *c-fos* mRNA induction (Maleeva *et al.*, 1990, 1991; Anokhin and Rose, 1991). Mice overtrained for 10 days in the active avoidance task demonstrated negligible IEG mRNA accumulation in the hippocampus and neocortex during testing sessions (Anokhin and Ryabinin, 1991). Similarly, low levels of *c-fos* and *c-jun* activation were observed in the forebrain of chicks which were overtrained in an appetitive visual discrimination task (Anokhin and Rose, 1991). Enhancement of ZENK expression was reported to occur when canaries were trained to associate song with a mild shock (Jarvis *et al.*, 1995). This effect was stronger than the ZENK activation produced by song alone or unpaired song presentation. During imprinting in chicks, the degree of *c-fos* induction in the brain area known to be critically involved in imprinting was correlated with the preference scores of individual birds (McCabe and Horn, 1994).

These findings suggest that IEGs activation during learning is not due simply to arousal, stress or motor activity alone (Anokhin and Sudakov, 1993). However, the role of IEGs in memory consolidation can be tested directly only by selective suppression of learning-induced IEGs expression in the brain. By employing such an approach we have recently found that antisense oligonucleotides directed against *c-fos* mRNA disrupt memory formation in a passive avoidance task in chicks, when injected into the areas of elevated *c-fos* expression during learning (Mileusnic *et al.*, 1996). This effect was specific for the long-term but not short-term memory, and was not seen with injections of the control scrambled oligonucleotides, which did not influence stimulus-induced *c-fos* expression (Mileusnic *et al.*, 1996). Suppression of *c-fos* induction by the administration of antisense oligonucleotides in the rat brain was also shown to impair retention of conditioned taste aversion (Lamprecht and Dudai, 1996) and a brightness discrimination reaction (Grimm *et al.*, 1997).

Interestingly, the IEG belonging to a family of CCAAT enhancer binding proteins (C/EBPs), was also cloned in marine mollusc *Aplysia* (Alberini *et al.*, 1994). Expression of *Aplysia* C/EBP mRNA was rapidly induced in sensory neurones by stimuli known to produce long-term facilitation—a behaviourally relevant form of synaptic plasticity (Alberini *et al.*, 1994). Microinjections of ApC/EBP antisense RNA or an antibody to ApC/EBP blocked long-term facilitation without affecting short-term facilitation (Alberini *et al.*, 1994).

Taken together these results strongly suggest that some of the IEGs, *c-fos* being one of them, are activated during the consolidation phase of long-term memory formation, and are able to act as a critical switch for the conversion of short-term to long-term memory.

4.3. CREB and the Molecular Cascade Upstream to Immediate Early Genes

How are learning stimuli translated into IEG activation in nerve cells? Recent studies in *Aplysia, Drosophila* and mice have demonstrated that a particular constitutive transcription factor,CREB, might play a decisive role in this process (Kaang *et al.*, 1993; Bourtchuladze *et al.*, 1994; Yin *et al.*, 1994).

CREB (cAMP-responsive element binding protein) belongs to the ATF family of transcription factors. These leucine-zipper proteins bind to DNA sequences called cAMP response elements (CRE), located in the upstream regulatory regions of many genes, *c-fos* being just one of them (Sheng and Greenberg, 1990). In order for CREB to become active it has to be phosphorylated at a specific amino acid, Ser-133, by a catalytic subunit of protein kinase A (PKA). Catalytic subunits of PKA are translocated into the cell nucleus after they are released from a tetrameric complex by cAMP that is generated in response to cell stimulation. This cascade mediates the effect of extracellular stimuli on expression of a variety of cAMP responsive genes in the nerve cells.

In *Aplysia*, treatments which induce long-term facilitation lead to activation of a reporter gene containing a CREB-binding site in its promoter (Kaang *et al.*, 1993). The long-term but not the short-term form of this synaptic plasticity was blocked by injecting, into the neurones involved, the oligonucleotides which contained CRE sites and thus prevented CREB-mediated transcriptional activation (Dash *et al.*, 1990).

In *Drosophila,* memory formation was examined using transgenic flies that carried a heat-shock-inducible dominant negative inhibitor of CREB (Yin *et al.*, 1994). In these flies, CREB function could be blocked at selected times before or after training. Blockade of CREB function during training specifically and completely abolished the formation of protein synthesis-dependent long-term memory, leaving intact protein synthesis-independent short-term memory.

Mice in which the CREB gene was inactivated by targeted mutation with the help of the homologous recombination technique, had impaired long-term memory in classical and contextual conditioning tasks (Bourtchuladze *et al.*, 1994). Mice were tested in a fear-conditioning paradigm, in which a conditioned stimulus (tone) and contextual stimuli (cage) were associated with an electric footshock. Normal mice could remember both the context and conditioned stimulus for many days after training, as assayed by a fearful freezing response to these stimuli. The CREB knockout mice showed normal contextual memory 30 min after training, but 30 min later started to lose it, showing a dramatic memory deficit by 24 hours after training. The freezing response to the conditioned stimulus was also impaired in these mice starting 2 hours after training.

In a recent study, Guzovski and McGaugh (1997) infused antisense oligonucleotides against CREB mRNA into the dorsal hippocampus of rats before training them in a water maze. Task acquisition and memory up to 4 hours did not differ in these animals from the control rats, while long-term memory tested at 48 hours after training was significantly impaired.

These results strongly suggest that the switch from memory which is protein synthesis-independent to that which is protein synthesis-dependent may be triggered in many species, and in many learning tasks by a phylogenetically conserved cAMP-dependent pathway of CREB-regulated transcription (Yin and Tully, 1996).

The data from the molecular genetics approaches to learning and memory allow us to make further developments of the molecular model of memory consolidation (Figure 3.2; see also Bailey *et al.*, 1996; Abel *et al.,* 1997). According to accumulating evidence, activation of the cAMP pathway may be a necessary step in the initiation of learning-related long-term cellular changes. cAMP acts through binding to the regulatory subunit of PKA

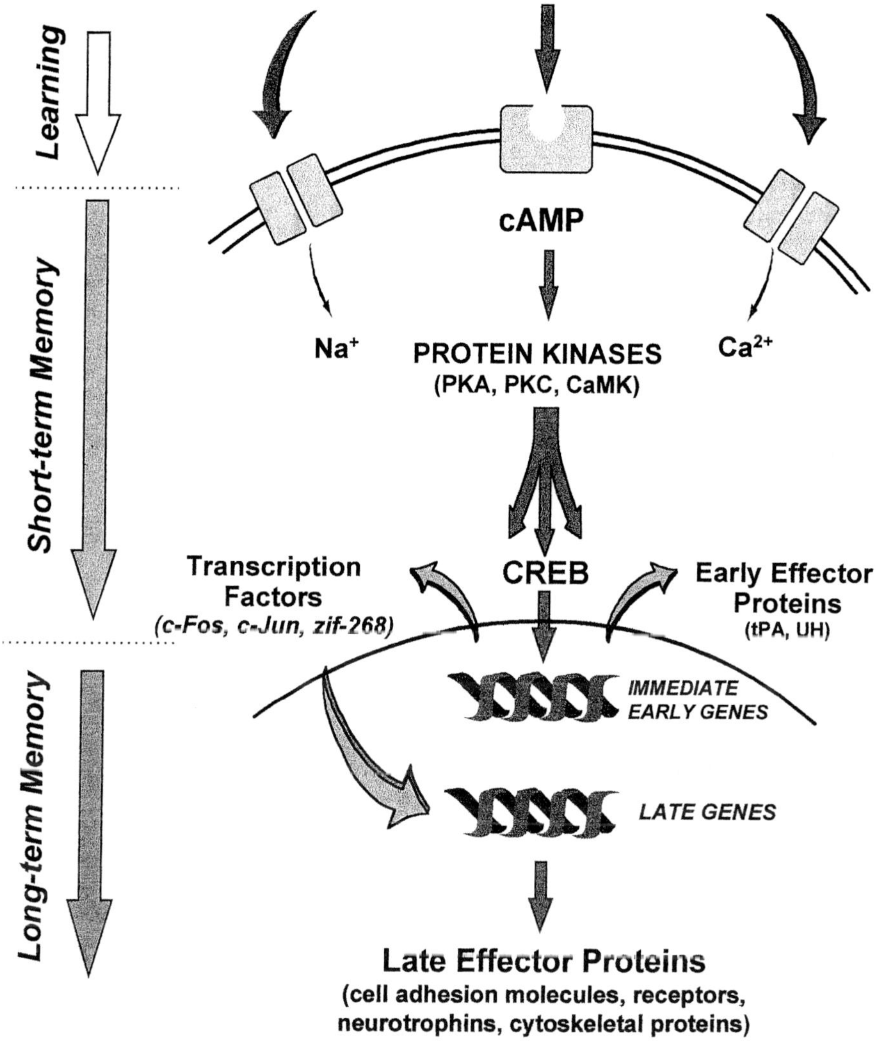

Figure 3.2. A molecular scenario for the transition from short-term to long-term memory. Consolidation of long-term memory depends on the activation of gene expression through the cascade of IEGs and LGs.

which releases its catalytic subunit. The catalytic subunit translocates to the cell nucleus and phosphorylates the constitutive transcription factor CREB, which in turn activates a family of IEGs carrying CRE elements in their regulatory regions. Some of these IEGs may encode effector proteins like tissue plasminogen activator (Qian *et al.*, 1993) or ubiquitin C-terminal hydrolase (Hegde *et al.*, 1997), which can maintain activity of the catalytic subunit of PKA by cleaving its regulatory subunits. Another group of IEGs encode transcription factors, like *c-fos* and *c-jun*. These proteins can turn on the expression of late genes, which presumably encode various structural proteins and molecules necessary for the initiation of synaptic growth. The expression of late genes constitutes the second protein synthesis-dependent time-window in memory consolidation, revealed by biochemical and behavioural-pharmacological experiments (Rose, 1995, 1996).

4.4. Main Features of Molecular Models

Like the neural systems approach, the molecular genetics approach is based on a set of fundamental assumptions about the nature of the consolidation process. However these assumptions differ from those used by systems theories. The three main ones are:

1. Memory consolidation is a cellular level phenomenon. It takes place in the same cells and synapses which were involved in the short-term memory.

"Long-term events might occur in the same cells in which short-term events occur by extending the transduction pathway into the nucleus." (Yin and Tully, 1996, p. 265).

2. Memory consolidation is based on a critical period of gene expression in the first hours after learning.

"Consolidation period is a critical period during which genes are induced that encode proteins essential for stable long-term memory" (Abel *et al.*, 1997, p. 623–624). Some of the molecular models restrict this period only to first 1–2 hours after learning (Goelet *et al.*, 1986) while others include a second wave of gene expression 4–8 hours after learning the description of the memory consolidation process (Rose, 1995).

3. The process of memory consolidation is common to declarative (explicit) and non-declarative (implicit) memory systems.

"...Even though implicit and explicit forms of learning use different mechanisms for short-term memory storage, both forms of learning seem to share a restricted number of mechanisms for long-term memory storage." (Kandel *et al.*, 1995, p. 685).

Thus, molecular genetics theories define long-term memory through its dependence on new gene expression in the nervous system. Furthermore, memory consolidation, i.e. the transition from short-term to long-term memory is believed to occur much earlier than is allowed by the network models of long-term memory formation.

5. MEMORY CONSOLIDATION: TOWARDS A SYNTHETIC MODEL

I hope that the above comparison of systems and molecular genetics approaches makes it clear that the two disciplines describe profoundly different aspects of memory formation.

Long-term memory and memory consolidation are defined within the two approaches in apparently non-overlapping ways (Table 3.1). Network models emphasise structural reorganization of memory storage sites, while molecular theories view consolidation as a transition from protein synthesis-independent to protein synthesis-dependent mechanisms at the same synapses in the same neurones. These differences are best illustrated by the role attributed to the hippocampus in the two models.

The systems approach suggests that hippocampus is involved in "time-limited" (Thompson and Kim, 1996) or short-term memory storage (for the explicit exposition of this idea, see figure 29.8 in Purves *et al.,* 1997). The hippocampus does not store long-term memories, and no consolidation occurs in hippocampal neurones. It serves only the function of a binding structure that contributes to the consolidation of memory circuits somewhere out of the MTL system, presumably in the neocortex (Alvarez and Squire,

Table 3.1. Main features of the network and molecular scenarios of memory consolidation

Basic features	*Network scenario*	*Molecular scenario*
Definition of the long-term memory	Through dependence on the integrity of the MTL system	Through dependence on protein synthesis
Relations between short-term and long-term memory	Located in different neurons and synapses	Located in the same neurons and synapses
Key feature of memory consolidation	Structural rearrangement	Gene expression
Duration of memory consolidation	Days to years	Minutes to hours
Dependence on neuronal activity	Activity-dependent	Activity-independent

1994; Thompson and Kim, 1996). The gradual development of long-term memory occurs through the strengthening of connections between neocortical cells, possibly requires the reorganization of memory representations during sleep (Winson, 1985) and thus needs lengthy periods of time. Therefore, actual long-term memory is not completely consolidated until days and even weeks after the original training (Squire, 1992).

Molecular models emphasise that protein synthesis-dependent synaptic stabilization occurs in hippocampal neurones (Bailey *et al.*, 1996). This makes the hippocampus by definition a long-term memory storage structure, and memory consolidation is thought to take place in the hippocampus itself (Abel *et al.*, 1997). The hippocampal-based long-term memory is thought to be expressed within the first hours after learning as indicated, for example, by the development of memory deficit in hippocampal-dependent tasks in CREB mutant mice (Bourtchuladze *et al.*, 1994; Bailey, 1996) and by the dynamics of RNA and protein synthesis dependent LTP expression in hippocampal neurones (Nguen and Kandel, 1996).

Further differences between molecular and systems neuroscience models lie in the domains of their suitability for different memory forms, the time required for consolidation and the species being covered (Figure 3.3a).

This analysis demonstrates that the network and molecular genetics models are dealing with two realms of memory formation which can not be directly translated into each other. Substantial experimental evidence indicates that both models describe processes which are crucial for the establishment of long-lasting memory. However neither of the two models gives a complete and satisfactory picture of the consolidation process. Network models do not permit exploration of the issue of explicit forms of memory beyond species with the MTL system. The present neural systems approach also does not address the question of structural reorganization of memory substrates in non-declarative forms of learning and the role of this process in the consolidation of implicit memory. On the other hand, the molecular genetics approach is only starting to tackle the mechanisms and functional significance of multiple waves of learning-driven gene expression at late times after learning. It also does not explore molecular mechanisms underlying gradual reorganization of memory storage sites—the subject central to the systems approach to memory consolidation.

It is therefore clear that a new synthetic model is required to accommodate findings from both systems and molecular levels (Figure 3.3b). This unified model has to account for multiple memory systems in the brain, and for multiple molecular phases of synaptic

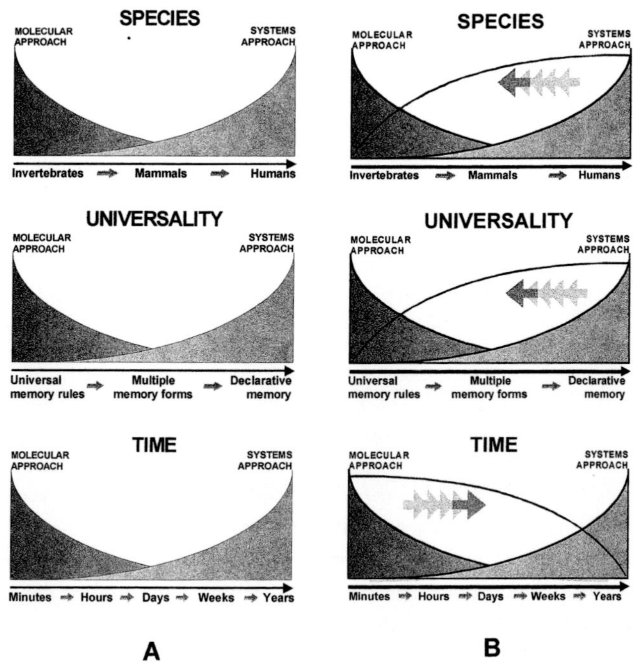

Figure 3.3. Domains for comparison of molecular and systems approaches to memory consolidation. **A.** Current studies of systems mechanisms of memory consolidation are concentrated mainly in humans and primates while molecular genetic studies prevail in species with a simpler organization of the nervous system. The systems neuroscience approach views consolidation as a process characteristic of declarative memory and taking days to years. The molecular approach attributes the consolidation process to multiple forms of memory and restricts its duration to minutes and hours. **B.** Shifts in the problem situation required for the synthetic model of memory consolidation. Network models of memory consolidation will have to cover the structural redistribution of memory storage in non-declarative memory forms and in species without the MTL system. The molecular genetic approach will have to explore the molecular bases of information transfer between temporary and permanent storage sites that occurs from days to years after the initial experience.

modifications in neurones. At the systems level, the synthetic picture should explain the memory consolidation process as a parallel establishment of a number of representations subserved by different memory systems. On the molecular side, the new model should be able to describe how each memory system goes through its own dynamics of consolidation phases, based on the processes of regulatory- and effector-gene expression in the pre- and postsynaptic neurones. A further contribution to regulation of these molecular processes is made by spontaneous endogenous neural activity within each memory system, multiple revisions of representation by recall, and reminder stimuli and interaction of this memory system with the other memory systems. The emerging picture is thus going to differ considerably from the earlier versions of single-memory-trace and single-gene-expression-phase models of memory consolidation. This simplistic interpretation should give way to "multiple draft" models, in which multiple parallel memory consolidation processes are carried out by parallel memory systems like "...all varieties of thought or mental activity—are accomplished in the brain by parallel, multi-track processes of interpretation and elaboration of sensory inputs" (Dennet, 1991, p. 111).

REFERENCES

Abel, T., Alberini, C., Ghirardi, M., Huang, Y.Y., Nguen, P. and Kandel, E.R. (1995) Steps toward a molecular definition of memory consolidation. In: D.L. Schacter, J.T., Coyle, G.D., Fischbach, M.M., Mesulam and L.E. Sullivan (eds). *Memory distortion: How minds, brains and societies reconstruct the past,* Cambridge: Harward University Press, pp. 298–328.

Abel, T., Nguyen, P.V., Barad, M., Deuel, T.A., Kandel, E.R. and Bourtchouladze, R. (1997) Genetic demonstration of a role for PKA in the late phase of LTP and in hippocampus-based long-term memory. *Cell*, **88**, 615–626.

Agranoff, B.W. (1968) Agents that block memory. In: G.C. Quarton and T. Melnechick (eds). *The neurosciences: a study program.* New York: Rockefeller Press, pp. 756–764.

Agranoff, B.W. (1972) *The chemistry of mood, motivation and memory.* New York: Plenum Press.

Alberini, C., Ghirardi, M., Metz, R. and Kandel, E.R. (1994) C/EBP is an immediate-early gene required for the consolidation of long-term facilitation in Aplysia. *Cell*, **76**, 1099–1114.

Alvarez, P. and Squire, L.R. (1994) Memory consolidation and the medial temporal lobe: A simple network model. *Proceedings of the National Academy of Sciences, U.S.A.*, **91**, 7041–7045.

Anokhin, K.V., Mileusnic, R., Shamakina, I.Y. and Rose, S.P.R. (1991) Effects of early experience on c-fos gene expression in the chick forebrain. *Brain Research*, **544**, 101–107.

Anokhin, K.V., and Rose, S.P.R. (1991) Learning-induced increase of immediate early gene messenger RNA in the chick forebrain. *European Journal of Neuroscience*, **3**, 162–167.

Anokhin, K.V. and Ryabinin, A.E. (1991) C-fos gene expression in the mouse brain after active avoidance learning. *Abstracts of the 14th Annual Meeting of the European Neuroscience Association.* Cambridge, 296.

Anokhin, K.V. and Ryabinin, A.E. (1993) Expression of c-fos and c-jun genes in the neocortex and hippocampus of mice after passive avoidance learning. *International Journal of Memory*, **1**, 67–70.

Anokhin, K.V., and Sudakov, K.V. (1993) Systems organization of behaviour: novelty as a factor that determines gene expression in the brain during learning. *Uspekhi Fisiologich. Nauk* (*Advances in Physiological Sciences*), **24,** 53–70.

Atkinson, R.C. and Shiffrin, R.M. (1968) Human memory: A proposed system and its control processes. In: K.W. Spence, J.T. Spence (eds) *The psychology of Learning and Motivation: Advances in Research and Theory. Vol.2*, New York: Academic Press, pp. 89–195.

Bailey, C.H., Bartsch, D. and Kandel, E.R. (1996) Toward a molecular definition of long-term memory storage. *Proceedings of the National Academy of Sciences, U.S.A.*, **93**, 13445–13452.

Baily, M., Nikolaev, E., Beck, J. and Kaczmarek, L. (1992) Delayed c-fos expression in sensory cortex following sexual learning in male rats. *Molecular Brain Research*, **544**, 101–107.

Barondes, S.H. and Cohen, H.D. (1968) Memory impairment after subcutaneous injection of acetoxycycloheximide. *Science, New York,* **160**, 556–557.

Barraco, R.A. and Stettner, L.J. (1976) Antibiotics and memory. *Psychological Bulletin*, **83**, 242–302.

Beck, C.H.M. and Fibiger, H.C. (1995) Conditioned fear-induced changes in behavior and in the expression of the immediate early gene c-fos: with and without diazepam pretreatment. *Journal of Neuroscience*, **15**, 709–720.

Bergson, H. (1911) *Matter and Memory*. Allen and Unwin, Winchester, MA.

Berridge, M. (1986) Neurobiology: second messenger dualism in neuromodulation and memory. *Nature, London,* **323**, 294–295.

Bourtchuladze, R., Frenguelli, B., Blendy, J., Cioffi. D., Schutz, G. and Silva, A.J. (1994) Deficient long-term memory in mice with a targeted mutation of the cAMP-responsive element-binding protein. *Cell*, **79**, 59–68.

Brennan, P.A., Hancock, D. and Keverne, E.B. (1994) The expression of the immediate-early genes c-fos, erg-1 and c-jun in the accessory olfactory bulb during the formation of an olfactory memory in mice. *Neuroscience,* **49**, 277–284.

Burnham, W.H. (1903) Retroactive amnesia: Illustrative cases and a tentative explanation. *American Journal of Psychology*, **14**, 382–396.

Butters, N. and Delis, D.C. (1995) Clinical assessment of memory disorders in amnesia and dementia. *Annual Review of Psychology*, **46**, 493–523.

Calamandrei, G. and Keverne, E.B. (1994) Differential expression of Fos protein in the brain of female mice dependent on pup sensory cues and maternal experience. *Behavioral Neuroscience,* **108**, 113–120.

Claparède, E. (1911) Recognition et moitié. *Archives of Psychology, Geneva*, **11**, 79–90.

Corkin, S. (1984) Lasting consequences of bilateral medial temporal lobectomy: Clinical course and experimental findings in H.M. *Seminars in Neurology*, **4**, 249–259.

Curran, T. and Morgan, J.I. (1987) Memories of fos. *BioEssays*, **7**, 255–258.

Dash, P.K., Hochner, B. and Kandel, E.R. (1990) Injection of the cAMP-responsive element into the nucleus of Aplysia sensory neurons blocks long-term facilitation. *Nature, London*, **345**, 718–721.

Davis, H.P. and Squire, L.R. (1984) Protein synthesis and memory: A review. *Psychological Bulletin*, **96**, 518–559.

Dennet, D.C. (1991) *Consciousness Explained.* Boston: Little Brown and Company.

DeZazzo, J. and Tully, T. (1995) Dissection of memory formation: from behavioral pharmacology to molecular genetics. *Trends in Neuroscience*, **18**, 212–218.

Dingman, W. and Sporn, M.B. (1961) The incorporation of 8-azaguanine into rat brain RNA and its effect on maze-learning by the rat: an inquiry into the biochemical bases of memory. *Journal of Psychiatric Research,* **1**, 1–14.

Duncan, C.P. (1949) The retroactive effect of electroshock on learning. *Journal of Comparative and Physiological Psychology*, **42**, 32–44.

Ebbinghaus, H. (1885) *Memory: A contribution to experimental psychology.* Reprinted 1963, Dover, New York.

Eichenbaum, H. (1992) The hippocampal system and declarative memory in animals. *Journal of Cognitive Neuroscience*, **4**, 217–231.

Flexner, J.B., Flexner, L.B. and Stellar, E. (1963) Memory in mice as affected by intracerebral puromycin. *Science, New York,* **141**, 57–59.

Gaito, J. (1967) Neurochemical approaches to learning. In: D.B. Lindsley and A.A. Lumsdaine (eds) *Brain Function and Learning*, University of California Press, pp. 1–49.

Glassman, E. (1969) The biochemistry of learning: An evaluation of the role of RNA and protein. *Annual Reviews of Biochemistry* **38**, 605–646.

Goelet, P., Castelluci, V.F., Schacher, S. and Kandel, E.R. (1986) The long and short of long-term memory: a molecular framework. *Nature, London,* **322**, 419–423.

Grecksch, G. and Matthies, H. (1980) Two sensitive periods for amnesic effect of anisomycin. *Pharmacology, Biochemistry and Behavior,* **12**, 663–665.

Greenberg, M.E. and Ziff, E.V. (1984) Stimulation of 3T3 cells induces transcription of the c-fos oncogene. *Nature, London,* **311**, 433.

Grimm, R., Schicknick, H., Riede, I., Gundelfinger, E.D., Herdegen, T., Zuschratter, W. and Tischmeyer, W. (1997) Suppression of *c-fos* induction in rat brain impairs retention of a brightness discrimination reaction. *Learning and Memory,* **3**, 402–413.

Guzowski, J. and McGaugh, J.L. (1997) Antisense oligonucleotide-mediated disruption of hippocampal cAMP response element binding protein levels impairs consolidation of memory for water maze training. *Proceedings of the National Academy of Sciences, U.S.A.*, **94**, 2693–2698.

Hacking, I. (1995) *Rewriting the Soul: Multiple personality and the Sciences of Memory.* Princeton: Princeton University Press.

He, X. and Rosenfeld, M.G. (1991) Mechanisms of complex transcriptional regulation: implications for brain development. *Neuron,* **7**, 183–196.

Hebb, D.O. (1949) *The Organization of Behavior*. New York: John Wiley.

Hegde, A.N., Inokuchi, K., Pei, W., Casadio, A., Ghirardi, M., Chain, D.G., Martin, K.C., Kandel, E.R. and Schwartz, J.H. (1997) Ubiquitin C-terminal hydrolase is an immediate-early gene essential for long-term facilitation in Aplysia. *Cell*, **89**, 115–126.

Hilts, P. (1995) *Memory's ghost: The strange tale of Mr. M. and the nature of memory.* New York: Simon and Schuster.

Hyden, H. and Egyhazi, E. (1962) Nuclear RNA changes during a learning experiment in rats. *Proceedings of the National Academy of Sciences, U.S.A.*, **48**, 1366–1373.

Hyden, H. and Egyhazy, E. (1964) Changes in RNA content and base composition in cortical neurons of rats in a learning experiment involving transfer of handedness. *Proceedings of the National Academy of Sciences, U.S.A.*, **52**, 1030–1035.

Hyden, H. and,Lange, P.W. (1968) Protein synthesis in hippocampal pyramidal cells of rats during a behavioral test. *Science, New York*, **159**, 1370–1373.

James, W. (1890) *The Principles of Psychology*, 2 Vols., Holt, New York.

Jarvis, E.D., Mello, C.V. and Nottebohm, F. (1995) Associative learning and stimulus novelty influence the song-induced expression of an immediate early gene in the canary forebrain. *Learning and Memory*, **2**, 62–80.

Kaang, B.K., Kandel, E.R. and Grant, S.G. (1993) Activation of cAMP-responsive genes by stimuli that produce long-term facilitation in Aplysia sensory neurons. *Neuron*, **10**, 427–435.

Kaczmarek, L. (1993) Molecular biology of vertebrate learning: is c-fos a new beginning? *Journal of Neuroscience Research* **34**, 377–381.

Kandel, E.R., Schacher, S., Castellucci, V.F. and Goelet P. (1987) The long and short of memory in Aplysia: a molecular perspective. *Fidia Research Foundation Neuroscience Award Lectures*, Padova: Liviana Press.

Kandel, E.R., Schwartz, J.H. and Jessell, T.M. (1995) *Essentials of Neural Science and Behavior,* Appleton and Lange, Norwalk.

Kim, J.J. and Fanselow, M.S. (1992) Modality-specific retrograde amnesia of fear. *Science, New York*, **256**, 675–677.

Korsakoff, S.S. (1889) Étude medico-psychologique sur une forme des maladies de la memoire. *Revue Philosophique*, **11**, 501–530.

Lamprecht, R. and Dudai, Y. (1996) Transient expression of cFos in rat amygdala during training is required for encoding conditioned taste aversion memory. *Learning and Memory.*, **3**, 31–41.

Lau, L.F. and Nathans D. (1987) Expression of a set of growth-related immediate early genes in BALB/c 3T3 cells. Coordinate regulation with c-fos and c-myc. *Proceedings of the National Academy of Sciences, U.S.A.*, **84**, 1182–1186.

Maleeva, N.E., Ivolgina, G.V., Anokhin, K.V. and Limborskaya, S.A. (1989) Analysis of expression of *c-fos* protooncogene in the rat cerebral cortex during learning. *Genetica,* **25**, 1119–1121.

Maleeva, N.E., Bikbulatova, L.S., Ivolgina G.L., Anokhin, K.V., Limborskaya, S.A. and Kruglikov, R.I. (1990) Activation of proto-oncogene *c-fos* in different structures of the rat brain during learning and pseudoconditioning. *Doklady Academii Nauk SSSR, (Proc. USSR Acad. Sci.)* **314**, 762–763.

Maleeva, N.E., Bikbulatova, L.S., Ivolgina, G.L., Anokhin, K.V., Limborskaya, S.A. and Kruglikov, R.I. (1991) Participation of protooncogene *c-fos* in cells of different brain structures in learning and memory processes. *Biologicheskie Membrany,* **8**, 1179–1180.

Marr, D. (1971) Simple memory: a theory for archicortex. *Philosophical Transactions of the Royal Society, London, B* **262**, 23–81.

Matthies, H. (1979) Biochemical, electrophysiological, and morphological correlates of brightness discrimination in rats. In: M.A. Brazier (ed.) *Brain Mechanisms in Memory and Learning*. New York: Raven Press, pp. 215–239.

Matthies, H. (1989) In search of cellular mechanisms of memory. *Progress in Neurobiology*, **32**, 277–349.

Mayford, M., Abel, T. and Kandel, E.R. (1995) Transgenic approaches to cognition. *Current Opinion in Neurobiology*, **5**, 141–148.

McCabe, B.J. and Horn, G. (1994) Learning-related changes in Fos-like immunoreactivity in the chick forebrain after imprinting. *Proceedings of the National Academy of Sciences, U.S.A.*, **91**, 11417–11421.

McGaugh, J.L. and Herz, M.J. (1972) *Memory Consolidation*. Albion, San Francisco.

Mello, C.V., Vicario, D.S. and Clayton, D.F. (1992) Song presentation induce gene expression in the songbird forebrain. *Proceedings of the National Academy of Sciences, U.S.A.*, **89**, 6818–6822.

Milcusnic, R., Anokhin, K. and Rose, S.P.R. (1996) Antisense oligonucleotides to c-fos are amnestic for passive avoidance in chicks. *NeuroReport,* **7**, 1269–1272.

Milner, B., Corkin, S. and Teuber, H.L. (1968) Further analysis of the hippocampal amnesic syndrome: Fourteen year follow-up study of H.M. *Neuropsychologia*, **6**, 215–234.

Mishkin, M. (1978) Memory in monkeys severely impaired by combined but not separate removal of amygdala and hippocampus. *Nature, London*, **273**, 297–298.

Mishkin, M. (1982) A memory system in the monkey. *Philosophical Transactions of the Royal Society, London, B* **298**, 85–95.

Montarolo, P.G., Goelet, P., Castellucci, V.F., Morgan, J., Kandel, E.R. and Schacher, S. (1986) A critical period for macromolecular synthesis in long-term heterosynaptic facilitation in Aplysia. *Science, New York*, **234**, 1249–1254.

Müller, G.E. and Pilzecker, A. (1900) Experimentelle Beiträge zur Lehre vom Gedächtnis. *Zeitschrift für Psychologie (supplement No. 1)* **1**, 1–300.

Nyugen, P.V. and Kandel, E.R., (1996) A macromolecular synthesis dependent late phase of long-term potentiation requiring cAMP in the medial perforant pathway of rat hippocampal slices, *Journal of Neuroscience,* **16**, 3189–3198.

Nikolaev, E., Kaminska, B., Tischmeyer W., Matthies, H. and Kaszmarek, L. (1992a) Induction of expression of genes encoding transcription factors in the rat brain elicited by behavioural training. *Brain Research Bulletin,* **28,** 479–484.

Nikolaev, E., Werka, T. and Kaczmarek, L., (1992b) C-fos protoncogene expression in rat brain after long-term training of two-way active avoidance reaction. *Behavioral Brain Research,* **48**, 91–94.

Nyberg, L., McIntosh, A.R., Cabeza, R., Nilsson, L.G., Houle, S., Habib, R. and Tulving, E. (1996) Network analysis of positron emission tomography regional cerebral blood flow data: ensemble inhibition during episodic memory retrieval. *Journal of Neuroscience,* **16**, 3753–3759.

Popov, N., Pohle, W., Ruthrich, H.L., Schulzeck, S. and Matthies, H. (1976) Time course and disposition of fucose radioactivity in rat hippocampus. A biochemical and microauto-radiographic study. *Brain Research,* **101**, 283–293.

Purves, D., Augustine, G.J., Fitzpatrick, D., Katz, L.Z., LaMantia, A.S. and McNamara, J.O. (1997) *Neuroscience*, Sunderland, Massachusetts: Sinauer Associates.

Qian, Z., Golbert, M. Colicos, M.A., Kandel, E.R. and Kuhl, D. (1993) Tissue-plasminogen activator as an immediate-early gene during seizure, kindling and long-term potentiation. *Nature, London*, **361**, 453–457.
Ribot, T. (1882) *Diseases of memory*. Appleton-Century-Crofts, New York.
Rose, S.P.R. (1991a). Memory—the brain's Rosetta stone? *Concepts in Neuroscience*, **2**, 43–64.
Rose, S.P.R. (1995) Cell-adhesion molecules, glucocorticoids and long-term-memory formation, *Trends in Neuroscience*, **18**, 502–506.
Rose, S.P.R. (1996) Cell adhesion molecules and the transition from short- to long-term memory. *Journal de Physiology (Paris)*, **90**, 387–391.
Schacter, D. and Tulving, E. (1994). *Memory systems*. MIT Press, Cambridge, MA.
Schacter, D.L., Alpert, N.M., Savage, C.R., Rauch, S.L. and Albert, M.S. (1996) Conscious recollection and the human hippocampal formation: Evidence from positron emission tomography. *Proceedings of the National Academy of Sciences, U.S.A.*, **93**, 321–325.
Scoville, W.B. and Milner, B. (1957) Loss of recent memory after bilateral hippocampal lesions. *Journal of Neurology, Neurosurgery and Psychiatry*, **20**, 11–21.
Shashoua, V.E. (1968) RNA changes in goldfish brain during learning. *Nature, London*, **217**, 238–240.
Sheng, M., and Greenberg, M.E. (1990) The regulation and function of c-fos and other immediate early genes in the nervous system. *Neuron*, **4**, 477–485.
Squire, L.R., Slater, P.C. and Chace, P.M. (1975) Retrograde amnesia: Temporal gradient in very long-term memory following electroconvulsive therapy. *Science, New York*, **187**, 77–79.
Squire, L.R. and Spanis, C.W. (1984) Long gradient of retrograde amnesia in mice: continuity with the findings in humans. *Behavioral Neuroscience*, **98**, 345–348.
Squire, L.R. (1986) Mechanisms of memory. *Science, New York*, **232**, 1612–1619.
Squire, L.R. (1992) Memory and the hippocampus: A synthesis from findings with rats, monkeys and humans. *Psychological Review*, **99**, 195–231.
Squire, L.R., Ojemann, J.G., Miezin, F.M., Petersen, S.E., Videen, T.O. and Raichle, M.E. (1992) Activation of the hippocampus in normal humans: A functional anatomical study of memory. *Proceedings of the National Academy of Sciences, U.S.A.*, **89**, 1837–1841.
Squire, L.R., Knowlton, B. and Musen, G. (1993) The structure and organization of memory. *Annual Review of Psychology*, **44**, 453–495.
Squire, L.R. and Zola, S.M. (1996). Structure and function of declarative and nondeclarative memory systems. *Proceedings of the National Academy of Sciences, U.S.A.*, **93**, 13515–13522.
Struhl, K. (1991) Mechanisms for diversity in gene expression patterns. *Neuron*, **7**, 177–181.
Teyler, T.J. and DiScenna, P. (1986) The hippocampal memory indexing theory. *Behavioral Neuroscience*, **100**, 147–154.
Tischmeyer, W., Kaczmarek, L., Strauss, R., Jork, R. and Matthies, H. (1990) Accumulation of c-fos mRNA in rat hippocampus after acquisition of a brightness discrimination. *Behavioral and Neural Biology*, **54**, 165–171.
Thompson, R.F. and Kim, J.J. (1996) Memory systems in the brain and localization of memory. *Proceedings of the National Academy of Sciences, U.S.A.*, **93**, 13438–13444.
Tully, T, Preat, T., Boynton, S.C. and Del Veccio, M. (1994) Genetic dissection of consolidated memory in Drosophila. *Cell*, **79**, 35–47.
Tully, T. (1997) Regulation of gene expression and its role in long-term memory and synaptic plasticity. *Proceedings of the National Academy of Sciences, U.S.A.*, **94**, 4239–4241.
Tulving, E. (1985) How many memory systems are there? *American Psychologist*, **40**, 385–398.
Weingartner, E. and Parker, E. (eds) (1984) *Memory Consolidation*. Lawrence Erlbaum Associates, Hillsdale.
Willingham, D.B. (1997). Systems of memory in the human brain. *Neuron*, **18**, 5–8.
Winson, J. (1985) *Brain and psyche: The biology of the unconscious*. New York: Doubleday, Anchor Press.
Yin, J.C., Wallach, J.S., Del Vecchio, M., Wilder, E.L., Zhou, H., Quinn, W.G. and Tully, T. (1994) Induction of a dominant negative CREB transgene specifically blocks long-term memory in Drosophila. *Cell*, **79**, 49–58.
Yin, J.C. and Tully, T. (1996) CREB and the formation of long-term memory. *Current Opinion in Neurobiology*, **6**, 264–268.
Zola-Morgan, S. and Squire, L.R. (1990) The primate hippocampal formation: Evidence for a time-limited role in memory storage. *Science, New York*, **250**, 288–290.

4 Informational Synthesis in Crucial Cortical Areas, as the Brain Basis of Subjective Experience

A.M. Ivanitsky

Institute of Higher Nervous Activity and Neurophysiology, Russian Academy of Sciences, Moscow, Russia
e-mail: alivanit@aha.ru

The main hypothesis developed in this paper is that the events of subjective experience emerge as a result of informational synthesis in cortical areas crucial for this mental function. Three kinds of information participate in the process of synthesis: sensory information (which signals environmental events), information retrieved from memory, and information arriving from motivational centres. This concept is based on studies of the brain mechanisms of perception and thought. Sensations are shown to arise as a result of synthesis of data describing the physical parameters and significance of the stimulus, this being achieved by neurones in the projection cortex. The mechanism of this synthesis is the circular movement of nerve impulses from the projection areas, to the associative cortex, then to the hippocampus and the hypothalamic motivation centres, with subsequent return of excitation to the projection cortex. It is also demonstrated that the process of thinking involves convergence of cortical connections upon definite centres named "interaction foci". The topography of the interaction foci is specific for particular thinking operations. Thus, in imaginative thinking, the foci are located in temporo-parietal cortex, while abstract-verbal thinking involves foci in the frontal cortex. It is suggested that information coming to foci via cortical connections is compared and synthesized in the interaction foci, and this provides the basis for decision-making. The final part of the paper addresses the functional importance of mental phenomena and their possible effect on brain processes.

KEYWORDS: brain-mind problem, brain mapping, sensation, thinking

1. INTRODUCTION

The origin and the functional value of subjective experience is one of the mysteries of the human brain. The question is whether subjective experiences are needed just to supply our life with any personal value (as expressed in the words of A.S. Pushkin's "Elegy": "I want to live, to suffer, and to think"), or whether they represent a necessary component of brain functions, and are behaviourally important.

It is evident now that these questions cannot be solved purely by deductive reasoning and philosophical analysis. The route to finding the answer lies in studies of brain functions using objective methods, and in comparing brain processes with subjective experience. The history of scientific discovery provides evidence that new advances are made on the basis of relatively simple concepts, as in the case of transmission of inherited information by triplets of the purine and pyrimidine bases. The brain is an organ designed for information processing, and it is logical to suppose that the development of mental function is linked with the organization of its information-handling processes, according to certain defined principles.

Recent years have seen ever-increasing recognition of the concept that subjective experiences result from comparison of previously existing information with new information (reflecting changes in the external or internal milieux). This idea in itself is not new. David Hume (1739/1969) proposed that the feeling of "self", which is the most important element of subjective experience, arises as a result of the "movement" of perceptions along past events.

2. INFORMATIONAL SYNTHESIS AS THE BASIS OF SENSATION

The idea of informational synthesis as the brain basis for the origin of subjective experience was first put forward by us in 1976 (Ivanitsky, 1976; Ivanitsky and Strelets, 1977; Ivanitsky *et al.*, 1984), and was based on studies of the physiological mechanisms of sensation, which is among the most elementary of mental phenomena. Psychologists have known since the 1920s that sensation arises rather slowly, some 100 msec after stimulus delivery, which is significantly later than the time at which the sensory impulses arrive in the cortex. The aim of the study was to understand what is going on during this period, and to determine which stage of brain processing corresponds to generation of a subjective image. Studies were carried out in which simultaneous measures of objective parameters of brain activity (evoked potentials, EP) and quantitative measures of perception were recorded in the same experiment. Quantitative measures of perception were obtained, using methods defined by signal detection theory (Swets *et al.*, 1961), which describes perceptual processes as the result of the interaction of two independent variables: the sensory sensitivity index d', and the decision criterion index (which depends on motivational factors). The subject had to distinguish stimuli of different strength, and press the button with the right or the left hand, according to the intensity of the perceived sensation. Afterwards, the coefficients of correlation were calculated between the physiological and psychological measures, namely between the amplitude of each EP wave and psychophysical indices. The studies were carried out in the somatosensory (Ivanitsky and Strelets, 1976) and visual (Ivanitsky and Matveeva, 1976) modalities, and essentially similar results were obtained in both cases.

It was shown that the amplitude of the early waves of the evoked potential (EP) revealed statistically significant correlation with the d' index, and those of the late ones with the decision criterion. The intermediate waves, with a latency of 140 msec for the somatosensory and 180 msec for the visual modality, correlated with both of these perceptual factors, this double correlation being revealed only for EP waves in the projection area (Figure 4.1). The amplitude of these waves was thus determined by the sensory features of the stimulus, as well as by its significance. Based on the data on the origin of evoked potential waves, we proposed a mechanism which accounts for this double correlation.

This mechanism is based on the idea of circular movement of nerve impulses, with the "central station" lying in the cortical projection area. For visual stimuli, impulses went initially from the occipital to the temporal cortex (which also plays a great role in stimu-

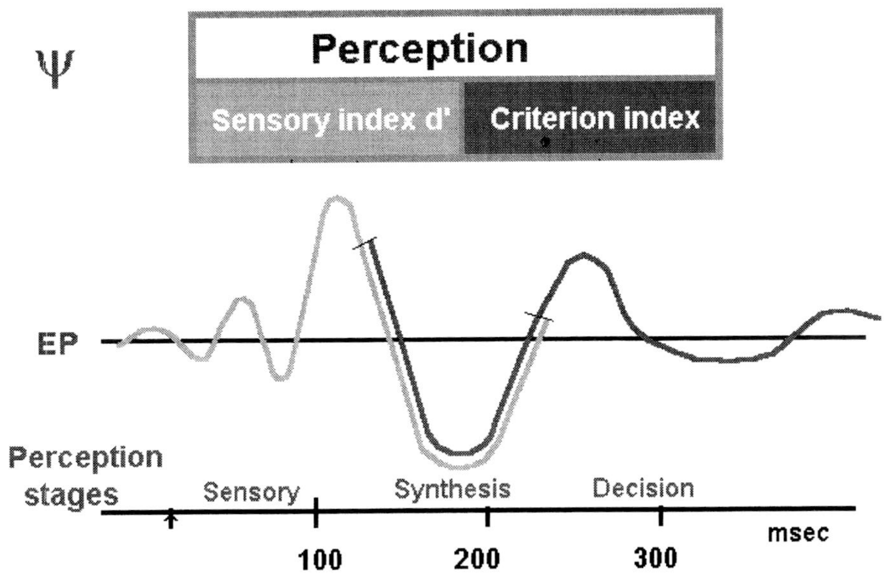

Figure 4.1. The scheme of correlations between two psychophysical indices and the amplitude of EP waves. The early EP components correlate with sensory d' index, the late ones with the criterion index, while the intermediate wave correlates with both psychophysical indices, reflecting the process of informational synthesis.

lus recognition), while for cutaneous stimuli they went from primary to secondary and tertiary areas of somatosensory cortex. After that, nerve impulses entered the limbico-hippocampal complex, and then subcortical centres for emotion and motivation. Up to this point the progressive movement of excitation corresponded completely to the scheme of a reflex arc. However, the process continued beyond this scheme: A further step was involved, which converted the arc into a loop, this step consisting of re-entry of nerve impulses into the cortex, including its projection areas, via the system of diffuse projections. This step thus represented feedback connections from the executive to the afferent centres. Due to this re-entry or return of the excitation, the nerve impulses, coming from the motivational centres, became superimposed on traces of the sensory excitation on the projection cortex neurones. At this stage (or some earlier stage) the frontal cortex also joined the process. This was revealed in the synchronization of EPs in the projection and frontal areas at time intervals from 100 to 200 msec after the stimulus (Ivanitsky and Strelets, 1979).

It has been suggested that these intermediate EP components reflect the synthesis within cortical neurones of two kinds of information about the stimulus: current information regarding the physical characteristics of the stimulus, and data retrieved from memory, on stimulus meaning.

The most interesting point, however, is that the peak latency of these EP waves coincided precisely with the time at which sensation is perceived, as measured previously in psychological experiments (Froehlich, 1929; Pieron, 1960; Boiko, 1964). It thus appears, that the synthesis of the two kinds of information about the stimulus—that which is current and that which is extracted form memory—is the key mechanism underlying

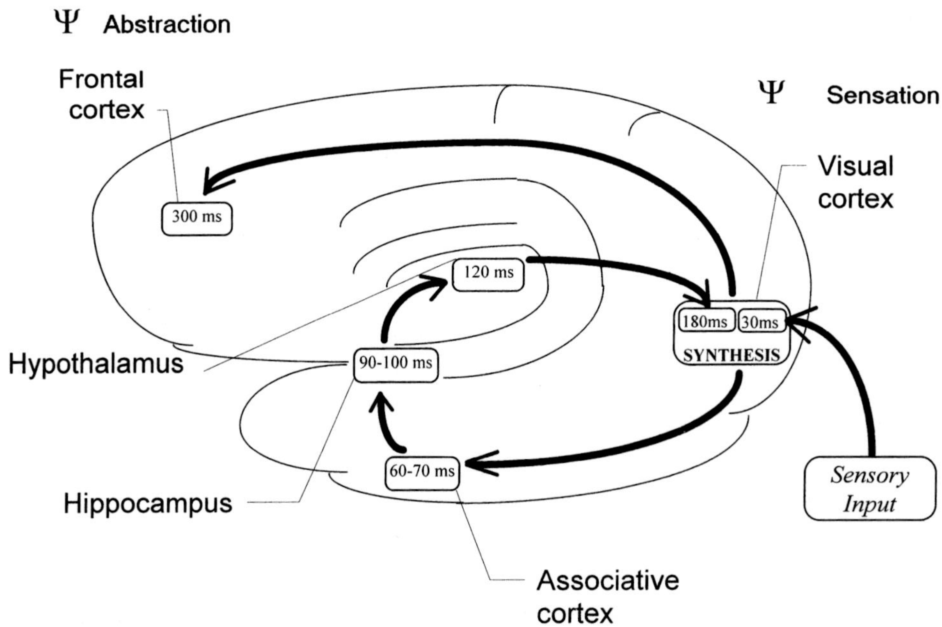

Figure 4.2. The circular movement of excitation producing the mental event of sensation. The main part of this process is the synthesis of information relating to the physical and signal properties of the stimulus, occurring in neurones of the projection cortex.

sensation, as a phenomenon described now at the physiological rather than psychic level (Figure 4.2). This represented another step towards overcoming the barrier between two levels of organization of brain processes, one of them not accompanied and the other one accompanied by subjective feelings. In terms of this concept the sequential acquisition of information from receptors leads to repetitive circulation of excitation around this loop, such that signals from the external and internal milieux are constantly being compared, resulting in mental monitoring of the ongoing changes. This process occurs with a quantized interval of some 100–150 msec, which represents the shortest duration for sensation (Goldburt and Makarov, 1971; Blumenthal, 1977).

Until recently it has not been possible to obtain detailed confirmation of this hypothesis, including the suggestion of circular excitation moving through a series of structures in the human brain, because of the ethical limitations. However, it is now appropriate to re-address this hypothesis, since, in recent years, data have appeared in the literature, directly or indirectly supporting our views—both the idea of the excitation loop, and the hypothesis that this mechanism is important for producing subjective phenomena.

The studies by Mishkin (1993) are important as the first item providing confirmation. The author studied the process of formation of memory traces in monkeys. Mishkin found that, in response to stimulus presentation, nerve impulses passed from the projection cortex to the rostral temporal-insular area, and from there the projections entered the medial temporal zone, represented by the rhinal cortex. After that the excitation went to the medial diencephalic structures, and then returned to the cortex, i.e., its medial-prefrontal areas. The excitation of these areas activated the basal cholinergic system innervating the entire brain surface, which provided the final stage of informational processing. The scheme, offered by M. Mishkin, contains more details than ours, due to the possibility of direct recording from subcortical brain structures in monkeys. It is, however, evident that the principal points of Mishkin's scheme are similar to those of our "excitation loop" hypothesis. M. Mishkin in his work does not discuss the mental aspects of this process, which is understandable, bearing in mind that the studies were carried out in monkeys.

3. RE-ENTRY OF EXCITATION AS THE BRAIN BASIS OF MENTAL FUNCTIONS

The idea about the re-entry of excitation to nervous structures, as the fundamental mechanism of subjective experience has also been suggested in recent years (independently of us) by several other authors. In most cases these authors based their hypotheses on general theoretical concepts but not on direct experimental data, i.e. the comparison of physiological and psychological data, obtained in one and the same experiment, which was our own approach. The re-entry concept has been developed to the greatest extent in the works of Edelman (Edelman and Mountcastle, 1978; Edelman, 1989), and his theory of consciousness— based on the re-entry mechanism—has become quite well known. Edelman proposes that the basis of subjective phenomena is the repetitive entry of excitation to the same neuronal groups, after additional informational processing in other groups, and that these feedback connections can be formed both between anatomically neighbouring and between distant structures. This repetitive entrance (re-entering) makes it possible to compare pre-existing data with changes occurring during one re-entry cycle. The constant monitoring of these changes due to repeated re-entry, according to this author, underlies the continuity of mental events.

A similar concept of consciousness is developed also by Sergin (1994; see also this volume). The author proposes that subjective feelings emerge as a result of cyclic circulation of excitation, which forms the phenomenon of "internal vision", this being the essence of consciousness. The hypothesis that re-entry of excitation to the primary cortex is the mechanism by which visual sensation arises was also proposed by Stoerig and Brandt (1993). These authors believe that these backward projections are less differentiated and more diffuse, providing information to different links of the visual system, and thus promoting their integration.

Studies in monkeys (Cauller and Kulics, 1991) showed that the N1 component of the EP, with a latency of 50 msec, reflected the return of excitation to the primary cortex from the secondary fields, which the authors believed to represent the mechanism underlying "conscious" tactile sensation. The latter is proved by the fact, that this wave disappears during sleep and anaesthesia, and correlates with the post-stimulus behaviour of monkeys trained to distinguish different stimuli. A similar scheme for brain organization of consciousness was suggested by Desmedt and Tomberg (1995). In this scheme conscious phenomena are based on re-entry of excitation from the dorsolateral areas of the prefrontal cortex to the areas at which sensory signals are initially projected. This is accompanied by synchronization of biopotentials at a frequency of 40 Hz.

Gray (1995) proposed a hypothesis of consciousness which, from the conceptual and neurophysiological points of view, was rather well developed. Gray suggested that the content of consciousness is determined by the activity of a comparator in the subiculum (a part of the hippocampus), together with backward connections from this comparator to the set of neurones in the cortical perceptual system. This set of neurones also supplies activity which enters the given comparator, after taking into account the results of the ongoing process of comparison. The idea that limbic structures have an important role in the genesis of subjective experience is in good agreement with the data of Mishkin *et al.* (1991) showing that these structures are closely connected with explicit memory, recognition and memory recall. However, the hippocampus clearly does not play such a major role in the higher nervous functions. That is why we would prefer a hypothesis in which the leading role in mental activity is played by informational synthesis in the neocortex, especially as this is in good agreement with the data provided by various investigators, as cited above. The data given above are summarized in the following table.

An important element in assessing the re-entry hypotheses is the correspondence between the time scale of the cerebral processes suggested by these hypotheses and the times at which events are experienced at the subjective level. Edelman noted that one excitation cycle took up to 150 msec. Adding to this the time needed for sensory impulses to reach the cortex gives a total time quite close to the time delay found in our experiments. Gray pointed out that consciousness is quantized predominantly by processes associated with the frequency of the theta rhythm, which gives a time of 1000: 6 = 167 msec. Simonov (1979) mentioned some time ago the importance of the theta frequency in this context. Desmedt and Tomberg, in the work cited above, noted that the 40-Hz synchronization process developed over a time period of 100 msec after the appearance of a potential in the primary cortex, and before the start of the P300 wave,

Table 4.1. Re-entry and informational synthesis as the brain basis of mind

*1. **A. Ivanitsky, 1976, 1984***
Informational synthesis provided by excitation re-entry is the brain basis of sensation
A. Ivanitsky, 1990
Interaction foci as cortical formations providing informational synthesis in thought
*2. **G. Edelman, 1978, 1989***
Re-entering process as the brain basis of consciousness
*3. **L. Caullier, A. Kulics, 1991***
Backward projections to S1 cortex signal touch sensation in monkeys
*4. **P. Stoerig, S. Brandt, 1993***
Nerve impulses returning to the primary visual cortex underlies visual feeling
*5. **V. Sergin, 1994***
Re-entering as the basis of "Inner Vision" phenomenon
*6. **J. Desmedt, K. Tomberg, 1995***
The re-entry of nerve impulses from prefrontal cortex back to primary somatosensory area is the basis of conscious touch sensation
*7. **J. Gray, 1995***
The comparison of information in subiculum area of hippocampus determines the content of consciousness

i.e., within the 100–200 msec time period. Shevelev (1997) considers that information processing within the visual cortex takes about 200 msec.

The following experimental observations are also of interest. Libet *et al.* (1967) made intraoperative recordings of evoked potentials from the cortical surface, arising in response to electrical stimulation of the skin, and found that weak subthreshold stimuli produced only the early wave of the response in the cortex, with latencies of up to 100 msec. Stronger stimulation produced additional late oscillations in EP, with a latency of 150 msec, and this was accompanied by the appearance of subjective experience. This latency was virtually the same as the latency of the EP waves recorded in our studies, which had the double correlation with both perceptual indexes. Libet carried out additional studies in which the somatosensory cortex was stimulated directly, and concluded that subjective events are delayed with respect to cortical events, by a period of 500 msec. However, we believe this time period to be excessive, and that direct stimulation of the cortex may disrupt the finer neuronal processes underlying sensation.

Baziyan and Lyubimov (1990) studied the suppression of visual images due to eye movements, and found that the early components of visual EPs did not change, while the later waves, with latencies of greater than 100 msec, were reduced in amplitude. Czigler and Csibra (1992) showed that when series of visual stimuli were presented, in which some stimuli differed from most of the others, presentation of the "oddball" stimuli was followed by the appearance of a negative oscillation in the posterior areas of the hemispheres, some 140–180 msec after stimulus presentation. The authors associated this oscillation with the visual version of the phenomenon of "processing negativity" (Naatanen, 1982) which in part reflects the processes of selective attention and the comparison of current stimulus properties with information held in memory. We believe that all these data are in good agreement with the results of our own studies.

In humans, a sensation which has arisen is recognized and categorized at a later stage: These processes occur not in the primary projection areas but in the frontal areas of the cortex. In our studies (reviewed in section 2, above), this was regarded as the third stage of perception, characterized by a correlation between the decision criterion and the late EP waves, including the P300 wave. Verbal functions are usually involved in this process. This was demonstrated by Salmelin *et al.* (1994) who recorded brain magnetic fields while subjects considered a variety of pictures. Even when the subject was not required to name the item represented, responses were also seen in the verbal area of the left hemisphere. This occurred some 400 msec after presentation of the stimulus, i.e. 200 msec after the image was perceived, which in these experiments was defined as the appearance of responses in the visual and temporo-parietal areas of the cortex, with a latency of about 200 msec.

Another study (Thorpe *et al.*, 1996) on the EP during the recognition of noisy pictures, which either contained or did not contain an animal image, showed that differences in the pattern of EPs started from 150 msec after picture presentation, this being interpretable as the onset of the process of recognition of the experienced sensation. Baars (1993) analyzed the psychological literature and came to the conclusion that images arise within the first 200 msec of stimulus presentation, with categorization occurring subsequently, at 200–500 msec.

This leads to the suggestion that during the time of one "quantum" of subjective experience, the human mind is at the preverbal stage. The time course thus shows that more complex mental processes do not displace simpler processes, but are "superimposed" upon them. This topic is discussed in more detail below.

Finally, there are another two studies which provide indirect support for our hypothesis. These studies demonstrate that, contrary to the widely held view, excitation of neurones in the primary cortex is not sufficient for producing sensation, despite its being a necessary element for sensation. Crick and Koch (1995) suggested that visual sensation requires the co-ordinated functioning of the visual cortex and the hippocampus, as well as forward connections to the frontal cortex. The presence of such forward connections has been demonstrated for fields V4 and MT (and possibly also for V2 and V3), but not for field V1. Stoerig and Brandt found all these parts of the visual system are included amongst the destinations of the reverse projections.

Crick and Koch used data obtained in a number of psychophysical experiments to support their views. Thus, it was established that presentation of visual stimuli consisting of gratings of different frequencies resulted in masking of the lower-frequency grating by the higher-frequency grating, although this higher-frequency grating was still perceived as a gray background. This was explained by the suggestion that masking occurred at the level of high-resolution neurones in the primary cortex, which themselves do not produce sensation. It is also known that images on the retinas of the two eyes and their corresponding projections upon the primary cortex are not identical, which is the basis of binocular vision. Nonetheless, the subject sees a single image, which is generated at a stage later than that occurring in field V1.

Damyanovich (1996) studied patients who had lost skin sensitivity after cerebral insults to the internal capsule, and found that somatosensory EPs could be recorded in the projection cortex in some of these patients, starting with the early components (though the early N19 wave, which Ivanitsky and Strelets [1976, 1977] found to be highly correlated with 'd' was absent). This can be considered as the direct evidence that the arrival of nerve impulses in the primary cortex is not sufficient for producing a sensation.

The importance of these studies is that they show that mental processes are not only consequent upon—not simply "following up" the physiological processes—but also need a specific organization of (and interaction between) brain structures. The same conclusion was made also by Stoerig and Brandt (1993) based on the fact that a variety of primary visual cortex lesions involved alterations in visual function which were termed blindsight, as distinguished from blindness. A patient reported that he saw nothing, even though he could respond to stimuli of particular positions and colours. These points indicate that the question of the origin of mental function is not merely theoretical, but that its solution is needed at a practical level for understanding the features of neurological syndromes, and even more for diseases of mental origin.

In summary, the experimental data, although obtained by a variety of different methods, are quite unambiguous; hypotheses of the cerebral basis of mental function based on these data are all quite similar, despite having been proposed by scientists belonging to different schools. This generates optimism, and suggests that we are now approaching a good understanding of the key mechanism underlying subjective experience: Mental function is based on comparison and synthesis of available information with data retrieved from memory, and this occurs by the mechanism of re-entry of excitation into the area of primary projection. The key step in creating sensations is such synthesis of information in the projection cortex, with the result that the sensation is based on the physical characteristics of the stimulus, coloured by its "feeling." An important point here is that the meaning of the stimulus, despite being involved in forming the sensation, is included at this stage of perception in an unclear, implicit, form. Awareness of this meaning occurs at a later stage, when the frontal cortex becomes involved. It is also of note that the concept of sensation production by means of synthesis of the various properties of a stimulus is also close to the ideas of Anokhin (1978), who suggested that mental function is a generalization of all existing information, which thus acquires a role as an important determinant of behaviour.

4. THE MECHANISMS OF THOUGHT: INTERACTION FOCI AS DYNAMIC FORMATIONS IN THE CORTEX PROVIDING INFORMATIONAL SYNTHESIS

We subsequently developed the hypothesis of information synthesis as applied to the cerebral mechanism of thought operations. These studies were carried out mainly using brain potential mapping methods developed in our laboratory in the late 1980s. This technique was named *intracortical interaction mapping (IIM)* (Ivanitsky, 1990, 1993a). The method we developed was based on the fundamental concept that synchronization of potentials facilitates the establishment of connections between brain structures (Rusinov, 1969; Livanov, 1972). This concept arises from some ideas of classical Russian neurophysiology, and was later confirmed in a number of studies including those using mathematical modeling of neural processes by C. von Malsburg (1981; see also Abarbanel *et al.*, 1996).

The method of intracortical interaction mapping is based on three theoretical premises:

(1) Groups of cortical neurones are functionally specialized, and each makes its own contribution to information processing. Specialization is determined both by the cortical topography of a group, especially in projection regions, and by properties acquired due to learning processes during ontogenesis. These ideas are close to Edelman's neuronal group selection theory (Edelman and Mountcastle, 1978). Theoretical and experimental data show that mental functions emerge as a result of these integrations of neuronal groups into united systems.

(2) Cells within groups are connected by a system of forward and reverse connections such that the group acquires the properties of a neuronal oscillator (Madler *et al.*, 1991; Kasanovich and Borisyuk, 1994; Kiseleva *et al.*, 1994; Başar and Schürmann, 1996), which has its own characteristic discharge frequency. The frequency is generally lower than the discharge frequencies of individual neurones, and approximates to the frequencies of the EEG. Some cortical neurones also have pacemaker properties. Analysis of EEG and EP spectra, using methods such as Fast Fourier Transformation, can separate the activities of the major cortical oscillators, which are detected as components within the frequency spectrum.

(3) Coincidence of the frequency characteristics of different neuronal oscillators promotes the formation of functional connections between them. This is because in this case signals from one neuronal group all reach another group at the same phase of its excitation cycle. When this phase is the exaltation phase, the excitation thresholds of neurones

in the second group are at a minimal level, which facilitates their involvement in concerted activity with the first group. In the refractory phase, the connection acquires a latent, inhibitory nature. Transition from one type of connection to the other may be quite rapid, taking a few cycles over a fairly short period of time, this time period being compatible with the rate at which mental processes occur.

Consideration of these points suggests the conclusion that precise coincidence of frequency peaks in the bioelectrical activity spectra of different regions of the cortex is evidence that these regions contain groups of neurones which are functionally connected. Stressing the agreement of spectral frequency characteristics, this method is thus insensitive to phase changes, and detects both the excitatory and inhibitory interactions between cortical fields. This is the major difference between this method and the more widely used coherence method, which detects connections between points of the cortex (identified on the basis of synchronized activity) only when the phase difference does not change during the relevant period of time. If the phase difference does change, the coherence method fails to detect the connection which, according to Bullock *et al.* (1995) results in errors.

The ideas behind the concept of intracortical interaction mapping were realized in appropriate computer programs written by G. Ivanitsky and O. Kashevarova. These programs used the following algorithms: (1) fast Fourier transformation for segments of EEG or EP recordings; (2) spectral windowing (smoothing) as needed; (3) selection of the major spectral components from each of the major EEG bands. The criteria for selection were determined by the investigator, using a comparison of the component with the mean energy level of the spectra, etc.; (4) the search for components in a given EEG sample coinciding to a defined level of precision (usually one spectral quantum) with components in the spectra of all the other derivations; (5) the number of coinciding components is calculated for each lead and each EEG band. This number is normalized with respect to the number of leads minus one; (6) construction of brain maps. Two types of map have been developed: an interpolation map, which identifies connection centres, and an "arrow" map, which shows connections between different cortical regions. (Step 5 is excluded for this second type of map.)

The method was tested in a rather simple task when the subject was asked to move rhythmically the finger either of his left or right hand. His EEG was recorded and then an interaction map was built. The system of connections in contralateral cortical zones was seen in this map (Figure 4.3). It is remarkable that in the right hemisphere (while moving the left hand finger) the connections were the built in alpha frequency band, and in left hemisphere (right hand finger movements), in beta band frequencies. Another difference was that in the right hemisphere the bands converged to the parietal areas, and in left hemisphere to the frontal areas. These features could be caused by the higher control of the frontal zone over the movements of the dominant hand. Another verification of the method was obtained while studying cortical connectivity during mental image construction (in experiments to be described later). During the stage at which the image was created the cortical map included two connections centres. The main one was in the occipital cortical area and the secondary one in the temporal zone. In a number of studies (Glezer, 1985; Lamb *et al.*, 1989) it was shown that the temporal cortex is involved in visual image recognition. These authors however obtained their data in experiments on cortical fields with operative damage in animals, or in clinical studies in patients with

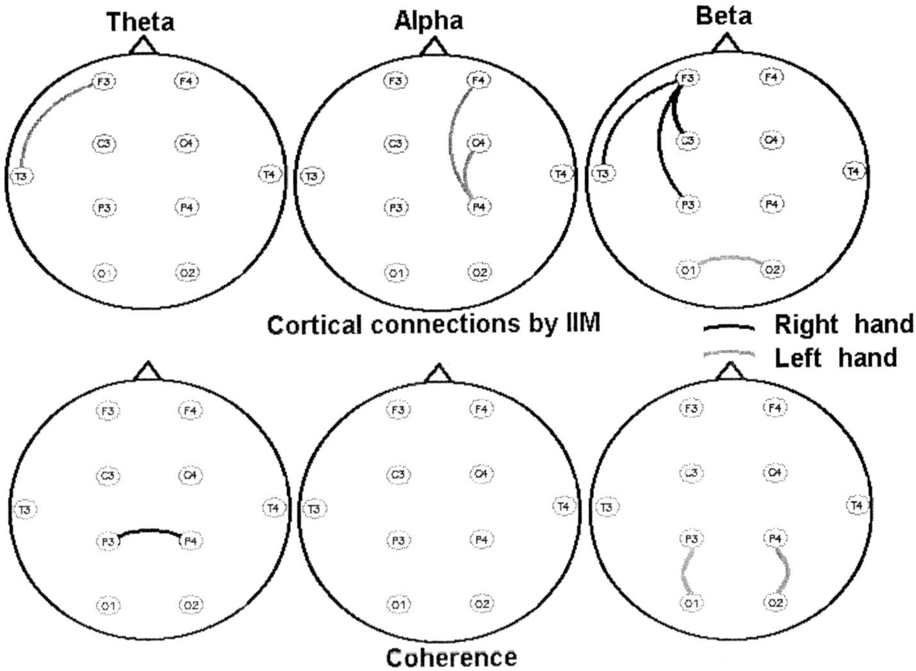

Figure 4.3. Above:- The cortical interaction maps ("arrows" version) in repetitive movements of the left and right hand fingers. Connections in theta, alpha and beta bands frequencies are given. Below:- results of coherence analysis in the same tasks.

stroke, with lesions located in the temporal region. With our method we could testify to the involvement of the temporal zone in visual image processing using non-invasive and rather simple procedures (Figure 4.4).

The intracortical interaction method was used for studying cortical connections during different types of thought operations. Tasks involving imagination, spatial and abstract-verbal thinking were presented to subjects using a monitor screen. In the imaginative thinking task, subjects had to recognize emotions on photographs of faces, where the actor expressed one of four basic emotions: joy, fear, anger, and grief, as well as mixed states. In the spatial task subjects had to compare two geometrical figures, to determine whether they were identical or mirror-symmetrical. The verbal task consisted of solving anagrams or selecting the odd one out of four words, where the odd word was of a different semantic category.

The EEG was recorded from ten electrodes positioned according to the 10/20 scheme in occipital, parietal, temporal, central and frontal electrodes in left and right hemispheres. Brain signals were amplified over a frequency band from 0.5 to 70 Hz. The vertical and horizontal EOG was also recorded, and a special program was used to exclude EEG deterioration caused by eye movement artefacts. Then the cortical interaction maps were built using the procedure as described above.

These studies showed that a simple and fairly symmetrical pattern of connections characteristic of the resting state underwent alteration when mental activity commenced. The

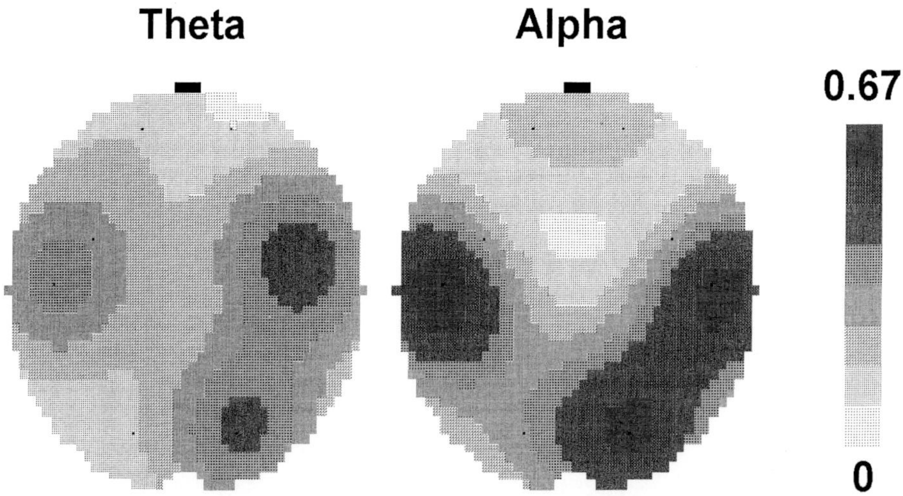

Figure 4.4. The intercortical interaction map (interpolation version) in theta (left) and alpha (right) bands frequencies in the situation when the subject searched for a visual mental image to be built from a defined number of simple elements. The centres of connections are seen in occipital and temporal cortical zones.

connections started to converge on defined cortical regions, forming "nodes" or "connection centres," which were named interaction foci. The topography of interaction foci was specific for different types of thought operation. Thus, in the case of imaginative thinking, foci were located predominantly in the temporo-parietal areas, while in abstract-verbal thinking, foci were located in the frontal regions of the cortex. Spatial tasks, including elements of both types of thinking, involved formation of foci initially in the posterior and subsequently in the anterior regions of the cortex (Ivanitsky, 1993b; Ivanitsky and Ilyuchenok, 1992; Nikolaev, 1994; Nikolaev *et al.,* 1996; Sidorova and Kostyunina, 1991) (Figures 4.5, 4.6).

Generalization of these results indicated the existence of two cognitive systems in the brain. The first of these, associated with the temporo-parietal regions of the cortex, is responsible for imaginative thought processes. The second system, for abstract-verbal thinking, is located in the frontal regions of the cortex. The systems are thus located along the anteroposterior axis of the brain, and their separation occurred earlier in evolution than specialization of the hemispheres. An additional difference between the systems, other than their function and topography, is that imaginative thinking is predominantly intuitive and implicit, and occurs without clear consciousness of the sequence of thought operations, while abstract-verbal thinking, on the other hand, is rational and explicit, since the thought processes are experienced as being controlled by the subject. This item will be developed in details later in this paper. It is noteworthy that activation of the "other" cognitive system, as indicated by appearance of foci in the frontal cortex during imaginative thinking tasks, and in the temporo-parietal cortex during verbal tasks, resulted in failure to make an appropriate decision.

Verbal transformation of the functions of the hemispheres, which occurred with the appearance of speech, is "superimposed" on the functional characteristics of the cognitive

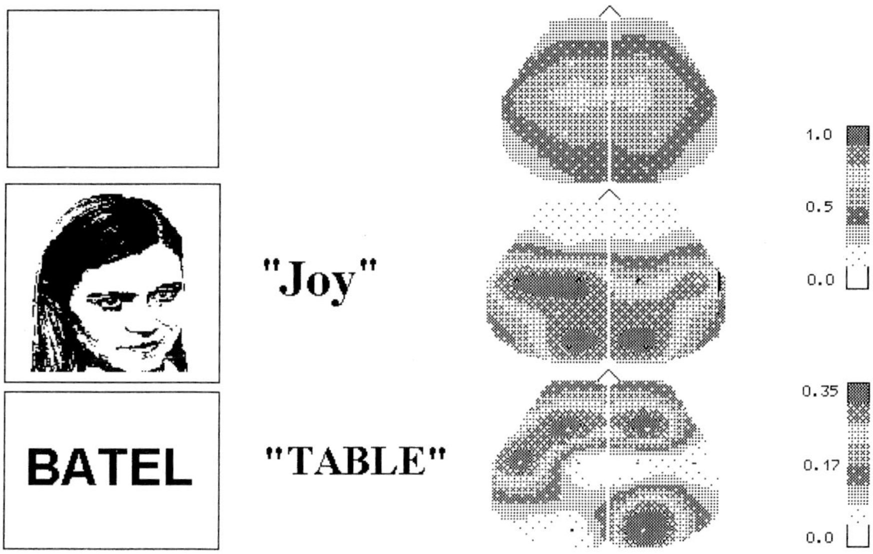

Figure 4.5. The cortical interaction maps at rest (no task is delivered, above), in imaginative thinking (recognition of emotions on face photos, middle row), and verbal thinking (anagram solution, below) . From left to right: the task example, the subject's verbal response, and the interaction map in alpha band frequencies. Average of a group of ten subjects. The last two seconds before the verbal response, detected from EMGs of the mouth muscle, were analyzed. The scales show the normalized connection numbers.

systems. Studies in which subjects were given the task of mentally constructing a visual image from a limited set of simple elements showed that in those subjects who invented rather realistic pictures the foci appeared mainly in their right hemispheric regions, while subjects who constructed abstract images had foci predominantly in the left hemishere. However, in both cases, on fulfilment of the first stage of this task—the search for the visual image to be constructed—foci were present in the occipital and temporal areas of the hemispheres (recognition zones—see Figure. 4.4), while at the stage at which the image was constructed from the set of simple elements (angles and inclined lines), foci were present in the frontal cortex (Ivanitsky *et al.*, 1990).

It is important that reaching the solution for all types of task, even when no verbal response was needed, was accompanied (and perhaps determined) by the functional involvement of the speech area of the left temporal lobe.

These concepts about the existence of two cognitive cerebral systems is in good agreement with data from other authors, particularly those of Posner *et al.* (1988) and Posner and Rothbart (1994), who provided detailed descriptions of two "neuronal networks" located in the temporo-parietal and frontal areas of the hemispheres. These authors carried out studies using positron emission tomography: This method, with its high spatial resolution and generation of three-dimensional images, allowed these investigations to make a invaluable contribution to understanding the cerebral basis of mental function. However, studies of the primary effects of neuronal excitation, based on measurement of brain potentials or magnetic fields, also have a number of advantages. Apart from their high temporal resolution, these methods allow the investigator

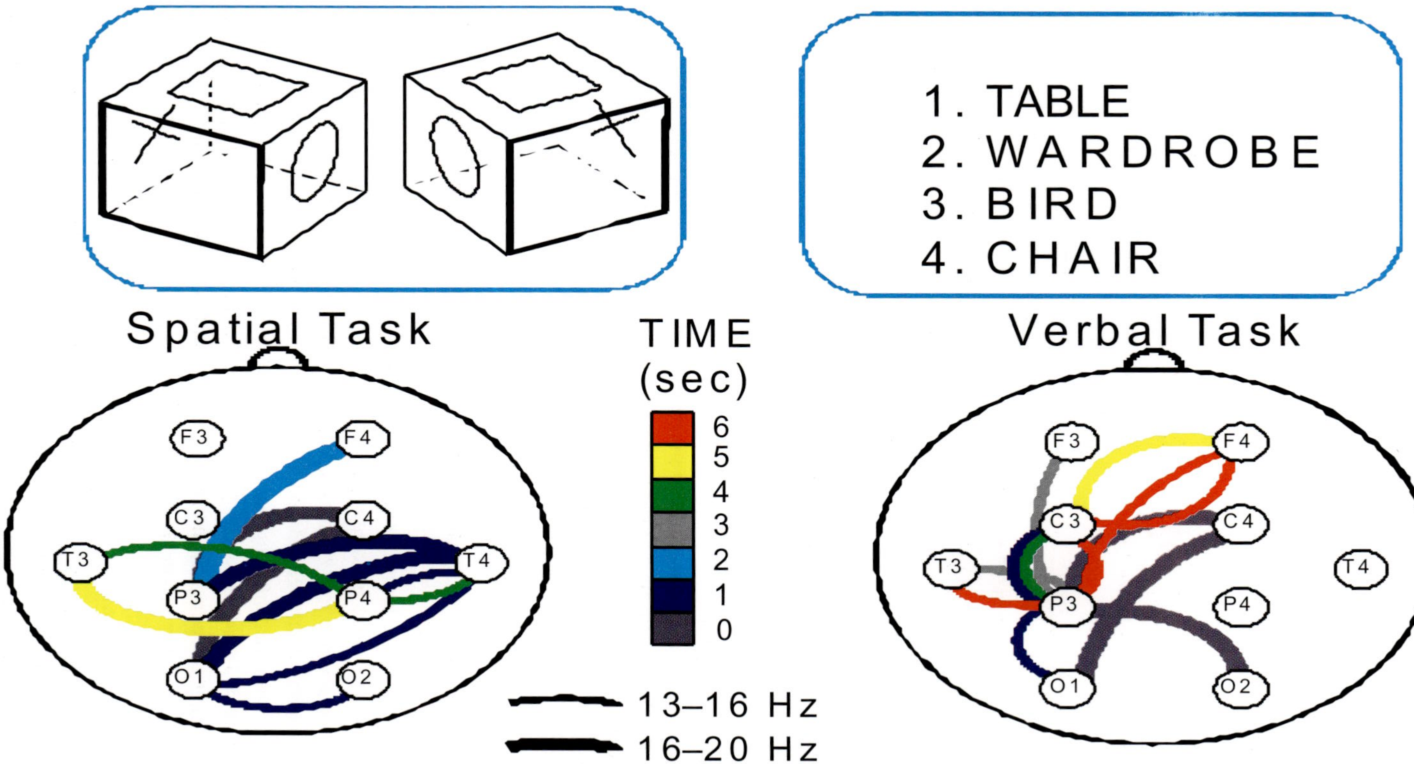

Figure 4.6. Cortical connections ("arrows") in two thinking tasks in beta-band frequencies (13–20 Hz). The spatial task was comparison of geometrical figures, the verbal task included the search for one in four words related to another semantic category. Only statistically significant connections in comparison with visuo-motor control are shown, for a group of 43 subjects. The color of connections (according to the scale below) indicates the time of appearance of connections during the process of solving the task. The connections thickness designates in which of two beta sub-bands connection was formed.

to address the questions not only of "where" but also of "how" brain information processing occurs. Studies of the patterns of neuronal connections can also produce descriptions of the organizational principles of neuronal processes underlying mental function.

As already discussed, one of the principles of organization of cortical connections for thought consists of convergence of these connections upon defined centres, i.e., interaction foci. Connections arriving at a focus operate at different frequencies: This is how foci are formed, since connections operating at a single frequency would form a homogeneous network without centres. It can be suggested that each connection carries its own particular information to the centre from a defined region of the cortex, or from a particular subcortical structure. In the focus, this information can be compared and recombined in certain ways. The major function of an interaction focus is thus to synthesize information, i.e., to carry out a process similar to that which is seen in the projection cortex during generation of a sensation. The main difference is that the role of the sensory signal can, in the present case, be carried out by information stored in working memory (for example, about the conditions of a soluble task), and the leading role in the processes of information synthesis is played not by the primary projection cortex, as in generation of sensation, but by the associative cortex. Within the focus, information held in working memory is compared with information retrieved from long-term memory and signals arriving from the motivational centres. It has been proposed that the result of these comparisons made in the focus is the eventual function of the thought process, i.e. decision-making. Subjectively, this is perceived as the process of thinking and finding an answer. However, there are some differences in experiencing these functions when they occur during the operation of imaginative and abstract cognitive systems.

From the physiological point of view, the focus thus performs functions analogous to those of a command neurone in lower animals (Kupferman and Weis, 1978; Sokolov, 1979). However, the complexity and varied nature of the incoming information in humans and higher animals requires a correspondingly more complex structure. This structure hypothetically consists of groups of neurones with different frequency characteristics, each tuned to a peripheral group of neurones with an identical frequency. The nature of these connections must be two-way, i.e., both forward and reverse: If two groups of neurones have the same frequency, they must be equally able to detect and transmit information from the connected group, depending on the phase-ratio of their oscillations. The single loop involved in the perception of sensation is thus replaced by a system of loops connected together at one centre.

Within a focus, groups of neurones must be joined by connections formed in a different manner: Since these neurones work at different frequencies, the principle of the equal excitation cycle cannot be applied here. These connections must apparently be fixed (hard-wired), this being determined by structural changes in synapses, which are efficient at *any* phase in the neurone or neurone oscillator excitation cycle, except for the absolute refractory phase (Figure 4.7). The concept of mental function occurring by means of a combination of hard-wired and labile connections was first proposed by Bechtereva (1980).

The hypothesis of interaction foci and their functional role is in good agreement with data obtained by Damasio (1994), whose functional magnetic resonance studies led to the conclusion that the active areas of the brain, which were detected when subjects carried

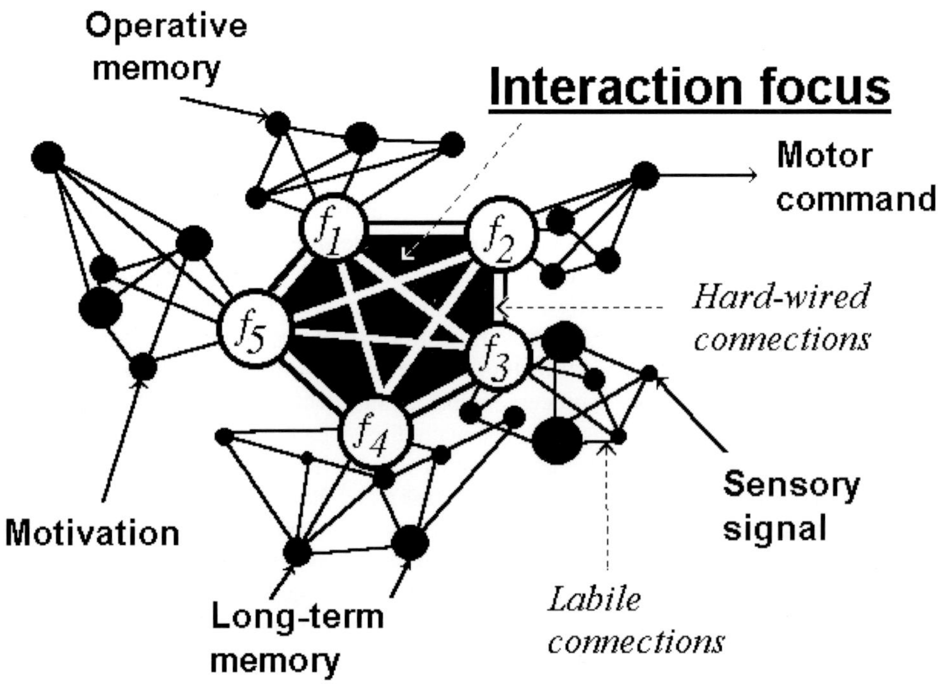

Figure 4.7. A hypothetical scheme of an interaction focus. The focus consists of groups of nerve cells distinguished by their different frequency parameters (*f1–f5*), and connected with groups of peripheral neurones by labile connections based on their identical frequency characteristics. Groups within the focus are linked by hard-wired connections based on structural changes in synapses. This structure forms the focus which synthesizes information circulating in different networks, resulting in decision-making.

out a variety of psychological tests, were merely areas on which different types of information converged. The term "focus" was used in a similar sense by Gevins *et al.* (1994).

In summary, the data obtained from studies of the mechanisms of perception and thought can be unified by the single principle of information synthesis as the cerebral basis underlying the genesis of a new function, i.e., subjective experience.

The neural network model used in the interaction focus concept, in which the network consists of neurones of different levels of lability and is also constructed on the hierarchical principle, has a number of advantages as compared with a uniform, "isolabile" neural network. The most important of these is the high information capacity of such networks, which overcomes one of the major difficulties associated with hypotheses suggesting that mental states are encoded by homogeneous neural networks. The concept of interaction foci is also in good agreement with ideas of Prigogine and Stengers (1984), who proposed so-called dissipative structures, which arise from chaos based on the self-organization principle. Interaction foci can arise during the learning of a defined habit by self-organization of neuronal groups with different frequency properties, and represent a cortical nucleus corresponding to a particular mental function. A system of labile connections is formed around this fixed nucleus; the overall system determines the qualitative individuality and uniqueness of the mental state being experienced.

5. THE PROBLEM OF THE "SELF"

An important component of internal experience is the sensation of "self" as the subject of perception and action. Thus, a discussion of the cerebral basis of mental function cannot ignore this perhaps very complex question. Swensson (1994) suggested that any theory attempting to describe the cerebral mechanisms of mental function must include an explanation for the "first-person viewpoint" phenomenon. The main difficulty with this question is that traditional physiological methods of seeking the cerebral localization of a given function are not applicable here, because of a logical contradiction known as the homunculus regression paradox (the homunculus being a hypothetical structure integrating the "self"). The paradox is as follows. If the homunculus is supposed to be located in a defined part of the brain, an explanation of how it is integrated with its surroundings would be needed. This would require the homunculus to contain its own "sensory" systems, along with motivational structures, etc., almost to the level of needing an entire brain. Continuing this reasoning, the homunculus would need to contain another such entity, and so on *ad infinitum* (Crick, 1979).

According to the concept proposed here, the feeling of "self" arises in the brain as a result of reviewing long-term memory contents during the process of comparing two or more information streams. In the case of sensation, this comparison is between information from the external milieu and and that from memory, and, in the case of thought, the comparison is between working and long-term memory. The association of "self" with memory is evident: The "self" is none other than the totality of recollections of one's own impressions, thoughts, past actions, and the responses of others to these actions. A. Tennyson wrote: "I am a part of all that I have met" ("Ulysses"). We believe that this approach to the question may provide a way around the difficulties.

The concept of "self", as a brain-wide dynamic system stored in memory, eliminates the question of where the "self" is located. On the other hand, the "homunculus regression" problem can be solved by supposing that a memory is transferred to the level of mental experience in response to an external signal (or by a comparison with operative memory). The homunculus thus uses sensory and other systems of the brain, so that it is non-regressive. Thus the homunculus is everywhere in the brain, but it is impossible to find him out until he discovers himself coming for his rendezvous with the external signal. It is noteworthy that the brain circuitry underlying mental phenomena is designed in such a way that, from one side, the external signal could not be perceived without recollection of memory traces and, from the other side, the memory traces to be retrieved and experienced need an external trigger. The same is correct for the interrelation between the working and long-term memory.

These ideas are thus close to those of Hume (1739/1969), as discussed at the beginning of this article. However, Hume saw one difficulty in his constructs, which he felt to be insuperable. Hume felt that perceptions—the only things we actually perceive—are too transient and unconnected, and were thus insufficient to maintain the major properties of "self", namely constancy and continuity, in other words, that main feature of "self" which Hume identified as "personal identity." We believe that data indicating that sensation does in fact include memory allow this contradiction to be overcome.

It is interesting to follow up how the concept of "self" is modified in the subjective sphere in relation to the location of cortical areas where information synthesis occurs. In perception, the external world is presented to the subject, and has the appearance of being independent of him. In imaginative thinking, the subject seeks a solution, which comes in an apparently spontaneous manner, by awareness and recognition. Finally, in abstract thought, the feeling of "self" is experienced as a factor controlling a directed search for a solution, and guiding the sequence of thought operations. These differences are apparently caused by the fact that the function of the evolutionarily-later parts of the cortex (the frontal cortex as compared with the temporo-parietal cortex and, to a greater extent the projection areas) is experienced as a more active and consciously controlled type of function. Apart from the evolutionary factor, these differences would also appear to be determined by the general principle of construction of the central nervous system with its posterior perceiving and anterior executive areas.

The data on the exclusive role of the frontal area in experiencing volitional self-control functions are in good agreement with the concept of the executive attention network, developed by Posner and Rothbart (1994). This network is located in the anterior cingulate zone, controls other attentional networks (such as the visual orienting system) and supervises the operations of working memory. Posner and Rothbart suggest that the executive attention network is responsible for a subject's conscious awareness. It is of interest that the activation of the executive system occurs at 150 msec after the stimulus presentation, this value coinciding rather closely with the time of sensation, as determined in our experiments. It was also found that at this time interval the cortical connections between projectional and frontal areas are established (Ivanitsky and Strelets, 1979).

Considering the great importance of the executive attention network concept, we think, however, that no single cortical structure is responsible for conscious experience. The brain mechanism underlying mental events is presumably based on some universal principles of brain informational processing, and the circuit for re-entry, providing comparison and synthesis of information, provides much evidence that it serves as an essential feature for this mechanism. Posner and Rothbart (1994) also consider that re-entry is one of the basic design principles of the working brain.

6. THE SUBJECTIVE EXPERIENCE OF VERBAL FUNCTIONS

Speech constitutes a large and important part of our conscious experience and plays a major role in the phenomenon of human consciousness. Simonov (1981, 1993) defined consciousness as knowledge which can be transferred to other people in an abstract form. He supposed that consciousness arises during this intercourse, and is thus communicative in nature. The idea that consciousness is social in nature has been taken up by a number of other authors (Hesslow, 1994; Frith, 1995).

The discussion thus far has, for a number of reasons, been concerned mainly with the mechanisms of preverbal mental functions. Firstly, we supposed that the search for the cerebral bases of mental function should begin with its simpler forms, especially since these simpler forms in humans have undoubtedly retained their importance. Speech, like other forms of abstraction, cannot substitute for direct sensory perception, for example of the colour blue. As discussed above, the times of appearance of sensations and their recognition are separated by an interval of about 100–200 msec. This sequence of events can be retained in more complex mental functions. For example, Einstein wrote that his theoretical ideas initially appeared as unclear images and only then appeared in their completed form. In our studies of thinking, connections with speech centres generally arose at the later stages of consideration of a task, before the decision-making point.

However, the importance of mechanisms responsible for preverbal forms of mental function, i.e., the re-entry of excitation, is obviously not limited to this item. These deep mechanisms would appear to be quite universal, and produce, with some further elaboration, the subjective experience of speech functions—hearing and perception of another person's words, as well as the perception of one's own inner speech. A number of investigators have made attempts to explain the mechanisms of inner speech. According to one such hypothesis, inner speech is based on proprioceptive sensations resulting from small, involuntary contractions of the articulatory muscles during verbal thought. However, this hypothesis has been refuted, since administration of large doses of curare-like agents to volunteers fully blocked muscle contractions, but had no effect on the ability to think and to use inner speech (Smith *et al.*, 1947; Weisberg, 1980). Further refutation of this hypothesis, which is based on the idea that the mere arrival of sensory impulses in the cortex is sufficient for the appearance of sensation (which is now known not to be the case) is as follows: Transmission of signals to muscles, muscle contraction, and reverse transmission of sensory signals to the cortex would need at least 300–500 msec. This would produce a significant discrepancy between the time at which the thought occurred and its subjective perception, which would make sequential inner speech impossible, and also make these perceptions themselves unnecessary, because they would only follow thoughts and not underlie them. The mechanisms of mental experience and inner speech must therefore be intracerebral, and must be based on a single integrated system of connections between associative zones of the cortex and the speech areas.

To understand the actual mechanism of such integration, the fact that interaction foci appear in the left temporal zone at the final phase of thought is of importance. It provides evidence that sensory-verbal areas are involved in decision making, and that the information synthesis mechanism participates in this process. Important data were also reported by Posner and Rothbart (1994), who studied the anatomy and timing of cortical activation, in a task requiring discrimination of complicated visual signals, such as verbal ones. These authors showed that the process was initiated by activation of the visual cortex to compute some features of the visual image, then the frontal zones were involved in semantic analysis, and finally secondary activation occurred of the same visual area which initially analyzed visual features. The time interval between these two activations of the visual field was 150 msec (we meet this value again and again—it is indeed the time for converting the brain events into mental ones due to the re-entry cycle).

One may conclude, on the basis of these data, that, in perceiving auditory verbal stimuli, as well as in inner speech, the words "hearing" or "perception of inner sounds" are defined by the return of excitation from the frontal to the auditory cortex, and that a similar closed circuit is involved for visual signals, where the re-entry to visual fields defines the word "seeing". These ideas are also compatible to Edelman's (1989) suggestion that speech-associated "higher-order consciousness" is based on the same principle of re-entry of excitation into fields of the frontal, parietal, and temporal areas of the cortex, which are responsible for particular functions, the speech centres expressing the incoming information in the form of the appropriate phonemes.

7. THE FUNCTIONAL ROLE OF MENTAL EXPERIENCE

Finally, there is one further question, which was mentioned at the beginning of this article—that of the functional significance of subjective experiences, and their role in behaviour. Mental functions, as a result of information synthesis, contain an integrated assessment of a situation which can be used for efficient determination of a behavioural response. The elements of this generalization are apparent even in the simplest mental functions, such as sensation. In thinking, information synthesis includes not only the integration, but also the recombination of previously existing information: This is the basis of decision-making. This applies both to the perceptual decisions (recognition of the stimulus), and, to a greater extent, to decisions with regard to an action.

The evolutionary appearance of speech and its associated human consciousness produced fundamental changes in the abilities of the brain. Encoding the world as internal experiences, in the form of abstract symbols, makes this world of experiences, with its thoughts and feelings, available to other people, thus creating a common spiritual space which permits communication and accumulation of knowledge. Because of this, each new generation of people does not live in the same way as the previous generation—this is a sharp contrast with the lives of animals, whose lifestyle remains constant for thousands of years. Biological evolution, with its rules of survival, is thus replaced by evolution (and revolution) occurring in people's minds.

A more difficult question is that of the role of mental phenomena as factors affecting on-going cerebral processes, or even controlling these processes. It can, of course, be suggested that a stable systems of connections, responsible for the processes of information synthesis and underlying mental functions, form an entity which directs the movement of neural processes along learned systems of connections. However, this is only an apparent answer to the question, and leaves the role of the mental principle itself unclear: Integration centres, in the form of interaction foci or other structures, are complex but are nonetheless physiological structures, so the discussion becomes one of the effects of physiologically more complex structures upon physiologically simpler structures, rather than one of the effects of mental function on the brain.

The question does, however, appear to find an answer, in that mental functions arising from information synthesis have a new quality as compared with purely physiological processes. This new quality in turn follows a different developmental logic, i.e., a sequence of events which follows a different set of rules, which are of a higher order. Thus, a chain of thoughts is determined by their internal content, and develops according to the rules of logic and deductive reasoning (let us take, for example, the syllogism, e.g. 'All men are mortal; Socrates is a man: therefore Socrates is mortal').

The idea that mental events apply a different type of logic to the physiological processes on which they are based has been suggested by a number of authors (Stoerig and Brandt, 1993; Sperry, 1994). In essence, this is the only way to explain why mental phenomena are needed for organizing complex behaviour and preventing them from being mere "epiphenomena." These ideas are not ordinary ones, but, in essence, this is the only possibility of finding the way out from the closed ring, and to understand the sense of mental phenomena, saving them from the fate of "epiphenomena". In this quality, psychic experiences could not arise and be preserved in evolution, as nature abhors not only "a vacuum" but also "uselessness".

However, the sequential appearance of ideas about the special logic of mental functioning and its pre-eminence over physiological functioning can be supplemented with the following step, expressed as the question of whether the concept of "free will" acquires a real, rather than a symbolic sense. The internal logic of mental events is such that it has the ability to select a behavioural act on the basis of a subjective (but valid) assessment of the importance of one or another factor or motive for behaviour. Recognition of the overall inhomogeneity of these assessments also allows alternative solutions to be selected.

Such a seemingly-fantastic concept should not be rejected (nor accepted) without detailed consideration. The complexity of this problem requires an extraordinary hypothesis ("crazy" in the words of N. Bohr).

For example, the argument that this approach contradicts the principle of determinism is rejected in the sense that it merely retreats from the generally accepted "ascending determinism" principle, which states that the whole is completely determined by the sum of its parts. However, if this type of determinism is regarded as the only type possible, then all the phenomena of nature would have to be regarded as predetermined, starting from the moment of the initial big bang which created the universe. The world is in fact more complex. New forms of organization, arising during the process of development, confer new properties on matter as a whole, and these affect the behaviour of its parts. The principle of determinism is not refuted by this approach, but is merely replaced by the concept of two-dimensional determinism—both ascending (bottom-up) and descending (top-down). (The "mystical" aspect of descending determinism disappears immediately, on consideration of Sperry's example of a wheel rolling down a mountain (genuinely top-down) and pulling down the molecules of which the wheel is made).

These ideas are of great importance in terms of the question of the relationship between mind and brain, which is one of the most complex questions of contemporary science. The hypothesis presented here is an attempt to explain the nature of mental function as the consequence of a defined organization of cerebral processes. The effect of this organization is that information synthesis takes place in a unified centre, encoded in cerebral processes, with the result that these processes acquire a new quality and a developmental logic.

The present work is thus devoted to the question concerning where, how and why mental events arise on the base of brain work. During movement along this chain of questions, the answers come to be more and more complicated and hypothetical. In the end, the question, why this all takes place, i.e. why informational synthesis leads to the subjectively experienced events of the colour and sound, happiness and sadness, the feeling of our own thoughts and will, can probably never be answered. That is why the only possible answer, that "nature is organized in this manner", will not calm our brain, that being the manner of our brain's construction.

ACKNOWLEDGMENTS

The study was partly supported by: 1) Grant N99-04-48229 of Russian Foundation for Basic Research; 2) Grant N99-06-0059 of Russian Foundation for Humanities; 3) Grant N97-38 ESSI by James S. McDonnell Foundation.

REFERENCES

Anokhin, P.K. (1978) Psychic form of the reflection of the reality. In: *Selected Works. (Philosophical Aspects of the Functional System Theory).* Moscow: Nauka (in Russian), pp. 338–360.

Abarbanel, G.D.J., Rabinovich, M.I., Selverston, A., Bazhenov, M.V., Huerta, P., Sustchik, M.M. and Rubchinsky, L.L. (1996) Synchronization in neuronal ensembles. *Uspekhi fizicheskikh nauk*, **166**, 363–390 (in Russian).

Baars, B. (1993) *Cognitive Theory of Consciousness*. New York: Cambridge University Press.

Başar, E. and Schurmann, M. (1996) Alpha rhythm in the brain: Functional correlates. *News in Physiological Sciences*, **11**, 90–96.

Baziyan, B.H. and Luybimov, N.N. (1990) Evoked potentials at the glance fixation and the saccadic movements of the human eyes. *Fiziologia Cheloveka,* **16**, 28–35 (in Russian).

Bechtereva, N.P. (1980) *The Healthy and Sick Human Brain.* Leningrad: Nauka (in Russian).

Blumenthal, A.L. (1977) *The Process of Cognition*. New York: Engelwood Cliffs.

Boiko, E.I. (1964) *Human Reaction Time.* Moscow: Medicina (in Russian).

Bullock, T.H., McClune, M.C., Achimowicz, J.Z., Irogui-Madoz, V.J. and Druckrow, R.B. (1995) Temporal fluctuations in coherence of brain waves. *Proceedings of the National Academy of Sciences, U.S.A.*, **92**, 11568–11572.

Cauller, L.J. and Kulics, A.T. (1991) The neural basis of the behaviorally relevant N1 component of the somatosensory-evoked potential in S1 cortex of awake monkeys: evidence that backward cortical projections signal touch sensation. *Experimental Brain Research*, **724**, 607–619.

Crick, F.H. (1979) Thinking about the brain. In: *The Brain, Scientific American*, **240**, 181–188.

Crick, F. and Koch, C. (1995) Are we aware of neuronal activity in primary visual cortex? *Nature, London,* **375**, 121–123.

Czigler, I. and Csibra, G. (1992) Event-related potentials and identification of deviant visual stimuli. *Psychophysiology*, **29**, 471–485.

Damasio, A. (1994) *Descartes' Error: Emotion, Reason and the Human Brain.* Putham: Grosset.

Damyanovich, E. (1996) The functional organization of the somatosensory behavior in norm and while changed reactivity of the human brain. *Synopsis of thesis for Candidate of Biological Science.* Institute of Higher Nervous Activity and Neurophysiology, Moscow, Russia (in Russian).

Desmedt, J. and Tomberg, C. (1995) Neurophysiology of preconscious and conscious mechanisms of the human brain. In: *Abstracts of the Xth International Congress of Electromyography and Clinical Neurophysiology,* Kyoto, Japan, October 15–19, S4.

Edelman, G.M. (1989) *The Remembered Present. A Biological Theory of Consciousness.* New York: Basic Books.

Edelman, G. and Mountcastle, V. (1978) *Mindful Brain. Cortical Organization and the Group Selection. Theory of Human Brain Function.* Cambridge, Mass., and London, England: The MIT Press.

Frith, C. (1995) Consciousness is for other people. *Behavioral and Brain Sciences,* **18**, 682–683.

Froehlich, F.W. (1929) *Die Empfindungszeit: Ein Beitrag zur Lehre von der Zeit-Raum und Bewegungsempfindung.* W. Fischer Verlag, Jena.

Gevins, A., Cutillo, B., Desmond, J., Ward, M., Barbero, N. and Laxer, K. (1994) Subdural grid recordings of distributed neocortical networks involved with somatosensory discrimination. *Electroencephalography and Clinical Neurophysiology,* **92**, 282–290.

Glezer V.D. (1985) *Vision and Thinking*. Leningrad: Nauka (in Russian).

Goldburt, S.N. and Makarov, P.O. (1971) The measurement of the reaction time to the appearance of the short sensory (auditory) stimuli for the measurement of the duration of the sensation. *Doklady Academii.de Nauk SSSR*, **198**, 1237–1238 (in Russian).

Gray, J.A. (1995) The contents of consciousness: A neuropsychological conjecture. *Behavioral and Brain Sciences*, **18**, 659–676.

Hesslow, G. (1994) Will neuroscience explain consciousness? *Journal of Theoretical Biology,* **171**, 29–40.

Hume, D. (1739/1969) *The Treatise on the Human Nature*. E.C. Mossner (ed., and introduction) Harmondsworth, Middlesex, Penguin Books.

Ivanitsky, A.M. (1976) *Brain Mechanisms of the Signal Evaluation.* Moscow: Medicina (in Russian).
Ivanitsky, A.M. (1990) The consciousness and reflex. *Zhurnal Vysshey Nervnoy Dejatelnosty,* **40**, 1058–1062 (in Russian).
Ivanitsky, A.M. (1993a) Consciousness: criteria and possible mechanisms. *International Journal of Psychophysiology,* **14**, 179–187.
Ivanitsky, A.M. (1993b) Interaction foci, informational synthesis and mental activity. *Zhurnal Vysshey Nervnoy Dejatelnosty,* **43**, 213–227 (in Russian, translated in *Neuroscience and Behavioral Physiology,* 1994, **24**, 239–246).
Ivanitsky, A.M. and Ilyuchenok, I.R. (1992) Brain biopotentials mapping at verbal task solution. *Zhurnal Vysshey Nervnoy Dejatelnosty,* **42**, 625–635 (in Russian).
Ivanitsky, A.M. and Matveeva, L.V. (1976) The relationship between the evoked potentials parameters and the sensory-perceptive process structure. *Fiziologia Cheloveka,* **2**, 386–399 (in Russian).
Ivanitsky, A.M., Podkletnova, I.M. and Taratynova, G.V. (1990) The study of the intracortical interaction dynamics in thinking process. *Zhurnal Vysshey Nervnoy Dejatelnosty,* **40**, 230–399 (in Russian).
Ivanitsky, A.M. and Strelets, V.B. (1976) Evoked potential and the psychophysical characteristics of perception. *Zhurnal Vysshey Nervnoy Dejatelnosty,* **26**, 793–801 (in Russian).
Ivanitsky, A.M. and Strelets, V.B. (1977) Brain evoked potentials and some mechanisms of perception. *Electroencephalography and Clinical Neurophysiology,* **43**, 397–403.
Ivanitsky, A.M. and Strelets, V.B. (1979) The functional connections between different regions of the cerebral cortex at external stimulus perception. *Zhurnal Vysshey Nervnoy Dejatelnosty,* **29**, 1071–1074 (in Russian).
Ivanitsky, A.M., Strelets, V.B. and Korsakov, I.A. (1984) *Brain Informational Processing and Mental Activity.* Moscow: Nauka (in Russian).
Kasanovich, Ya.B. and Borisyuk, R.M. (1994) The synchronization in neuronal network of the phase oscillators with the central element. *Matematicheskoe Modelirovanie)* **6**, 45–60 (in Russian).
Kiseleva, I.V., Medvedev, A.V. and Frolov, A.A. (1994) The analysis of the statistical characteristics of the brain potentials. *Zhurnal Vysshey Nervnoy Dejatelnosty,* **39**, 783–788 (in Russian).
Kupferman, I., Weiss, K.R. (1978) The command neuron concept. *Behavioral and Brain Sciences,* **1**, 3–8.
Lamb, M.R., Robertson, L.C. and Knight, R.T. (1989) Attention and the interference in the processing of global and local information: effects of unilateral temporal-parietal lesion. *Neuropsychologia,* **27**, 471–483.
Libet, B., Alberts, W.W., Wright, E.W.E., Jr. and Feinstein, B. (1967) Responses of human somatosensory cortex to stimuli for conscious sensation. *Science, New York,* **158**, 1597–1600.
Livanov, M.N. (1972) *The Spatial Organization of the Brain Processes.* Moscow: Nauka (in Russian).
Madler, C., Schwender, D. and Pöppel, E. (1991) Neuronal oscillators in auditory evoked potentials. *International Journal of Psychophysiology,* **11**, 55.
Malsburg, C., v.d. (1981) The correlation theory of brain function. *Internal Report 81–2.* Department of Neurobiology Max Plank Institute for Biophysical Chemistry.
Mishkin, M. (1993) What is recognition memory and what neural circuits are involved? In *Abstracts of XXXII Congress of the International Union of Physiological Sciences, Aug. 1–6, 1993, Sunday, Glasgow,* pp. 42–43.
Mishkin, M., Horn, G., White, N. and Schacter, D. (1991) Cerebral memory systems. In *Third IBRO Congress of Neuroscience, August 4–9, 1991, Montreal, Canada. Abstracts.* S 19, p. 4.
Naatanen, R. (1982) Processing negativity, evoked potential reflection of selective attention. *Psychological Bulletin,* **92**, 605–640.
Nikolaev, A.R. (1994) Investigation of the stages of the mental rotation of complex figures with the intracortical interaction mapping technique. *Zhurnal Vysshey Nervnoy Dejatelnosty,* **44**, 441–447 (in Russian, translated in *Neuroscience and Behavioral Physiology.*, **25**, 228–233).
Nikolaev, A.R., Anokhin, A.P., Ivanitsky, G.A., Kashevarova, O.D. and Ivanitsky, A.M. (1996) The spectral EEG reconstructions and the cortical connections organization in spatial and verbal thinking. *Zhurnal Vysshey Nervnoy Dejatelnosty,* **46**, 831–848 (in Russian).
Pieron, H. (1960) *La Sensation.* Paris: Presse Université. France.
Posner, M.I., Petersen, S.E., Fox, P.T. and Raichle, M.E. (1988) Localization of cognitive operations in the human brain. *Science, New York,* **240**, 1627–1631.
Posner, M.I. and Rothbart, M.K. (1994) Constructing neuronal theories of mind. In: C. Koch and J. Davis (eds) *Large-Scale Neuronal Theories of the Brain,* Cambridge, Mass.: MIT Press, pp. 183–199.
Prigogine, I. and Stengers, I. (1984) *Order out of Chaos. Man's New Dialog with Nature.* London: Heineman.
Rusinov, V.S. (1969) *Dominant. Electrophysiological Study.* Moscow: Medicina (in Russian).
Salmelin, R., Hari, R., Lounasman, O.V. and Sams, M. (1994) Dynamics of brain activation during picture naming. *Nature, London,* **368**, 463–465.
Sergin, V. Ya. (1994) The consciousness as the system of inner vision. *Zhurnal Vysshey Nervnoy Dejatelnosty,* **44**, 627–639 (in Russian).

Shevelev, I.A. (1997) Temporal signal processing in the visual cortex. *Fiziologia Cheloveka*, **23**, 68–79 (in Russian, translated in *Human physiology*, **23**, 186–196).
Sidorova, O.A. and Kostyunina, M.B. (1991) The participation of cortical areas of the brain in processes of the perception and reproduction of emotional states of man. *Zhurnal Vysshey Nervnoy Dejatelnosty,* **41**, 1094–1101 (in Russian, translated in *Neuroscience and Behavioral Physiology*, 1993, **23**, 135–141).
Simonov, P.V. (1979) Memory, emotions and dominant. In T. Oniani (ed.) *Gagrskiye Besedy,* **7**. *Neurophysiological Basis of Memory,* Tbilisi: Mezniereba, pp. 358–377 (in Russian).
Simonov, P.V. (1981) *Emotional Brain.* Moscow: Nauka (in Russian).
Simonov, P.V. (1993) *The Creative Brain. The Neurobiological Basis of Creation.* Moscow: Nauka (in Russian).
Smith, S.M., Brown, H.O., Toman, J.E.P. and Goodman, I.S. (1947) Lack of cerebral effects of D-tubocurarine. *Anesthesiology,* **8**, 1–14.
Sokolov, E.N. (1979) The conceptual reflectory arc. In: T. Oniani. (ed.) *Neurophysiological Basis of Memory, Gagrskiye Besedy,* **7**. Tbilisi: Mezniereba, pp. 104–117 (in Russian).
Sperry, R.W. (1994) The perspectives of the mentalist revolution. The appearance of the new scientific philosophy. In *Brain and Mind.* Moscow: Nauka, p. 20–44 (in Russian).
Stoerig, P. and Brandt, S. (1993) The visual system and levels of perception: properties of neuromental organization. *Theoretetical Medicine*, **14**, 117–135.
Swensson, G. (1994) Reflections on the problem of identifying mind and brain. *Journal of Theoretical Biology*, **171**, 93–100.
Swets, Y., Tanner, W. and Birdsall, T. (1961) Decision process in perception. *Psychological Reviews*, **68**, 301–340.
Thorpe, S., Fize, D. and Marlot, C. (1996) Speed of processing in the human visual system. *Nature, London*, **381**, 520–522.
Weisberg, R.W. (1980) *Memory, Thought and Behavior.* New York, Oxford: Oxford University Press.

5 Nature of Sensory Awareness: The Hypothesis of Self-identification

V.Ya. Sergin

Neuroinformatics Laboratory, Russian Academy of Sciences Far East Division, 9 Piyp Ave, Petropavlovsk-Kamchatsky, Russia
e-mail: volcan@svyaz.kamchatka.su

What kind of coordinated neuronal activity of the brain is likely to produce the mentally experienced phenomenon of awareness? There is no conclusive answer to this question, though it is a key to the understanding of any form of consciousness. This paper introduces the notion of a special kind of coordinated neuronal activity, which achieves the process of *signal self-identification.* ("auto-identification"). The process of self-identification consists of the relay of a specific pattern of excitation produced by a stimulus in one or several cortical areas, back to neurones of these cortical areas through massively parallel feedback. The coinciding (identical) patterns of excitation produced by the stimulus and by relay through back-projections add together on the same neuronal structures, thus making them fire vigorously. This cyclic process accentuates the specificity and enhances the mapping of the stimulus in terms of signal intensity, thus providing the best conditions for stimulus categorisation by distributed long-term memory. The result of categorisation, a symbol or image, is expressed physiologically by a pattern of neuronal activity, which is also included in the cycle of self-identification, thus providing for mapping of the subjective meaning of the sensory features of the stimulus. Such mapping of the stimulus means that the process of perception passes from the physiological (objective) to the mental (subjective) level.

KEYWORDS: awareness, consciousness, vision, sensation, self-identification

1. INTRODUCTION

How do humans become aware of anything, such as a flash of light, a scent or a pain? Despite abundant experimentation on conscious perception, it is still unclear what physiological mechanisms may produce the mental phenomenon of awareness. There have been rather few attempts to resolve this problem, although it is the key to understanding any form of conscious cerebral activity.

Existing conceptual works give prominence to the idea of a critical role for feedback (signal re-entry) in mechanisms producing the phenomenon of awareness. The functional role envisaged for feedback is different, depending on the proposed mechanism of awareness. There are concepts that awareness arises as a result of synthesis of sensory information and information stored in memory (Ivanitsky, 1976; Ivanitsky *et al.*, 1984), current associative recall—the "remembered present" (Edelman, 1978, 1989), identification of sensory input with the contents of sensory memory (Sergin, 1992, 1994a,b,c), as a result of self-referent processes (Harth, 1995), comparison between the forecast and the reality in sensory inputs (Gray, 1995), "adaptive resonance" of the expected and the actual input patterns (Grossberg, 1995), and other processes carried on through feedback. There are also ideas of the importance of intensely synchronous neuronal firing, the dominating role of the neocortex, and the importance of multilevel explicit symbolic interpretation of sensory data for awareness (Crick, 1984; Crick and Koch 1990). The idea that intensely synchronous firing of neurones, which forms fields of high cortical activity, plays an important role in the generation of the phenomena of consciousness, has deep roots in physiology (Pavlov, 1951; Livanov, 1972; Simonov 1990) and is confirmed by the latest experimental data (Sviderskaya *et al.*, 1993; Llinas and Ribary, 1993; and others).

These notions have a substantial experimental basis, and the aforesaid processes may indeed be involved in the mechanisms of awareness. However, it is undeniable that the processes of information synthesis, associative recall, data comparison, stimulus feature binding or synchronous neuronal firing occur in many other types of coordinated neuronal activity underlying behaviour and cognition. These processes look insufficiently specific to explain the unique phenomenon of awareness. The key, and the most mysterious aspect of the problem remains unclear: How does coordinated neuronal cerebral activity (a physiological process) produce the phenomenon of awareness (a mental process)?

There is another question which is important in terms of experimentation. The functioning of re-entrant systems has a cyclic character and involves large areas in the brain. How then do these cyclic processes, which underlie the mechanisms of awareness, relate to the electric activity of the brain? Do they manifest themselves in low-frequency electric activity in the theta- and alpha-bands, as follows from some works (Edelman 1978, 1989; Ivanitsky, 1987, 1996; Gray, 1995); or do they correspond to high frequencies, of the beta- and gamma-bands (Crick and Koch, 1990; Sergin, 1991, 1992; Desmedt and Tomberg, 1995)?

2. HYPOTHESIS OF SELF-IDENTIFICATION

This work discusses only awareness of primary sensory stimuli. Discussion of other stages of the process of conscious perception is beyond the scope of this paper. We also avoid, wherever possible, discussing all other matters related to the problem of consciousness or attention. Our efforts are focused on the sole problem of revealing the physiological mechanisms producing the mental phenomenon of awareness. The term "awareness" is not defined, in the hope that the context of the paper will make the reader's intuitive understanding easier.

As is well known, in the process of perception, a stimulus produces a specific distribution of neuronal activity in one or more areas of sensory cortex. One can assume that a specific pattern of excitation in output neurones is relayed through massively parallel feedback, returing again to neurones of the same cortical areas. The coinciding (identical) patterns of excitation produced by the stimulus itself, and by relay through back projections, are added together on the same neuronal structures, thus inducing firing in an increasing number of neurones, and enhancing the intensity of excitation in these structures. Such a cyclic positive feedback process produces an "explosion" in the intensity of the specific pattern of excitation. At the same time, background excitation in neuronal structures not activated by the stimulus is of low intensity, and its frequency and phase distribution are random. Their uncoordinated interaction cannot induce a rapid increase in background noise. Moreover, intense excitation, which is produced by the stimulus, may cause dynamic inhibition in surrounding neuronal structures (Mountcastle 1978). Therefore, the specific pattern of excitation acquires high intensity and contrast. The specificity of spatial excitation in the cortex accentuates specific features of the stimulus, thus providing adequate mapping of it.

A mechanism of enhancement of the prominence of specific signals (through amplification of their intensity) and memorising (based on the duration of circulation within a closed circuit) might originate in the course of evolution. This mechanism could map vital events, such as dangerous physical and chemical effects of the environment. The adaptive utility of this mechanism might make it subject to evolutionary selection and lead to an expanded set of physical and semantic characteristics of the signals to be mapped. Such evolutionary processes might ultimately form an apparatus of intensity mapping of actually significant signals.

The identification of a pattern of stimulus-produced excitation with itself by its feedback to the input is the process of *self-identification*.* This process accentuates the specificity and enhances the intensity of mapping of the stimulus, thus providing the best conditions for its categorisation by distributed long-term memory. Presumably, parallel processes of self-identification and categorisation of signals underlie the mentally experienced phenomenon of awareness.

The process of self-identification proceeds due to coincidence (in the principle features) of the feedback pattern with the pattern of cortical excitation. Such coincidence only becomes possible in the event that no change occurs in the input excitation during circulation of output excitation within the feedback circuit. Otherwise a lagging feedback pattern will not coincide with the current cortical excitation pattern, which will make intensity mapping of specific features of the stimulus (and therefore awareness) impossible.

For example, if two brief successive flashes of light of different colours follow each other, the feedback pattern produced by the first flash will overlap the pattern of cortical excitation produced by the second flash. The resulting distribution of cortical neuronal activity should then correspond to the blend of the colours of the first and the second flash. Becoming aware of these successive flashes as individual events is therefore impossible. In order to become aware of successive flashes of light of different colour, it is necessary that the duration of the flashes, or the interval between them, should exceed the duration of the cycle. In this case the process of self-identification is completed for each flash separately, which makes it possible to become aware of them. Indeed, it has been experimentally established that two successive flashes of light—red and green—each lasting 20 ms, are perceived by a subject as a single yellow flash (Crick and Koch,

---

* In clarification, the phrase "self-identification" does not refer to the personal self or "ego", but to the identification of a pattern of neuronal activity with itself. Thus, "self" in the phrase "self-identification" has implications similar to those in "self-organization", sometimes used in brain research. A phrase which is more or less equivalent is "autoidentification" (ed)

1992). A longer flash—up to 60–70 ms each—results in the successive perception of red and green.

Therefore, in order for one to become aware of a random sequence of signals, it is necessary that their duration (or the intervals that separate them) should be longer than the duration of the cycle. In the case of a temporally continuous signal, it is necessary that its change or displacement should not exceed a liminal value during the cycle, and this makes it possible to realize the process of self-identification. It is unlikely that this requirement contradicts well-known facts. For example, one can easily follow the motion of a luminescent spot, but if the velocity of the motion exceeds a certain liminal value, the subject sees only a luminescent line.

The temporal criterion for realization of the process of awareness makes the hypothesis experimentally verifiable and gives certain hints of its anatomical basis. The minimum duration of circulation may be provided by vertical neuronal chains with feedback, distributed in the projection and associative cortical areas and in certain subcortical structures (Figure 5.1). Spatially distributed vertical columns (Mountcastle, 1978) appear to be the likely candidates for such role. Neuronal cortical chains connected with cells of subcortical structures, for instance, thalamo-cortical circuits (Steriade *et al.*, 1993) also look attractive.

In the case of a spatial stimulus, for instance a visual image, simultaneous identification of all its spatial characteristics is a necessary condition for awareness. Simultaneous identification requires synchronous circulation of signals in spatially distributed neuronal structures of the respective cortical areas. Synchronous circulation of signals in projection and associative cortical areas responding to a multimodal stimulus is a condition for integrated awareness. Therefore, the spatial criterion for realization of the

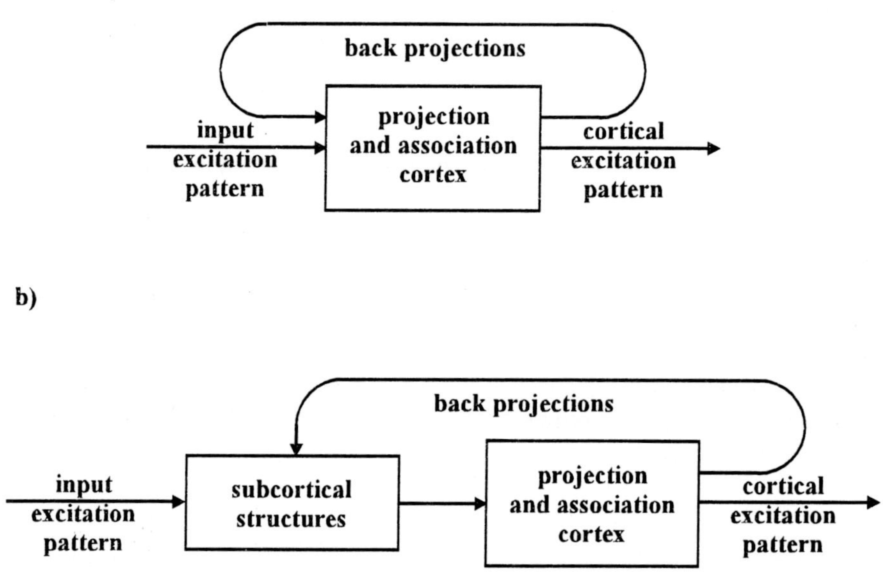

Figure 5.1. Diagram of the process of self-identification (plausible variants)

process of awareness follows from the hypothesis of self-identification: Circulation of signals in different cortical areas responding to a stimulus should be synchronous (i.e. concurrent).

Self-identification takes place within the duration of circulation of a signal in a closed circuit and its result is a single event, namely signal awareness. Therefore, the mechanism of self-identification produces discrete events at discrete intervals equal to the duration of the cycle. In such a mechanism of awareness, cycle duration is the shortest discriminable period of time. Successive signals falling within one cycle should therefore be mentally perceived as simultaneous. Signals falling into different cycles should be perceived as successive.

An *a priori* estimate of the likely duration of the cycle of self-identification is possible, based on the requirements of an organism's adaptation to the conditions of its environment. It follows from the postulated mechanism of self-identification that any signal should first circulate through at least one cycle so that the brain could be aware of it. The duration of a cycle is the period of a continuous process of quantisation. Then the duration of the cycle should be shorter than the typical temporal intervals in environmental changes, such as are vital for the organism. Otherwise important changes will occur in the environment, but awareness will not catch up with events. Vital events, such as an animal's movements (those of a predator or prey) including for instance chasing, jumping, striking with a paw, and, likewise, the timing of response delay, have frequency spectrums the periods of whose high-frequency components are equal to approximately 0.1 sec. (The operator's minimum response delay in laboratory conditions is 180 msec to respond to a visual signal, and 140 msec to respond to an auditory signal). The measured spectrums of turbulence near ground level (involving vortices and showers, movements of branches of trees and bushes and grass stalks) have an abrupt fall at approximately 0.1 sec. Representation of a continuous signal by discrete sampling requires at least two readings per period of the highest-frequency components of the continuous signal frequency spectrum. The period of quantisation should thus be less than or approximately equal to 50 msec, which constitutes the theoretic estimate of the cycle duration (Sergin, 1991, 1992).

If the postulated mechanism of signal self-identification does exist, the theoretically predicted characteristics of the process of awareness should comply with experimental data.

1. Successive events, which occur within one cycle of self-identification, should be perceived as simultaneous. If, for example, two successive signals fall within one cycle of self-identification, they should fuse into a single signal. It has indeed long been established in experimental psychology that a rapid succession of a faint and a strong signal is perceived as blended. The first signal is believed to be masked by the second. This phenomenon, observed in the visual, auditory and tactile modalities, is referred to as "backward masking". The self-identification model agrees with these experimental data, and explains the mechanism of "masking," which consists of fusion of the signals in accordance with their weights.

Differently timed components of a spatial image falling within one cycle of self-identification should fuse into a single image. Then, if an image (for example a geometric figure or a printed word) is split into two complementary spatial components, neither of which mean anything if taken separately, and the two are presented one after the other within a sufficiently short period of time, they should be perceived as a single image. A longer interval between presentations of the two components places them in different cycles of self-identification. In this case, the components will not fuse into the original image, and its perception is impossible. Numerous tachistoscopic experiments established in the 1970s that successive presentations of complementary components within a short period of time does indeed lead to recognition of an image. An interval between presentations of the components in excess of 100 ms makes recognition of an image impossible (see Hoffmann, 1982).

The mental-level indivisibility of successive events falling within one cycle of self-identification is compatible with experimental data of another kind. Hylan (1903) established as long ago as early this century that six consecutively exhibited letters seem simultaneous if they fall within an interval of approximately 80 ms. Research on this phenomenon in the decades that followed led to the establishment in psychology of the notion of the "perceptual moment", which is the longest interval of time within which successive perceptual events are perceived as simultaneous. The "perceptual moment" happens therefore to be the duration of one cycle of self-identification. Examination of such phenomena as flicker fusion, apparent motion and other phenomena, which reflect the temporal structure of perceptual processes, reveals their full compatibility with the mechanism of self-identification of signals (Sergin, 1994a).

2. *If sensory signals are processed serially in a cyclical fashion in the mechanism of awareness, then signals which fall into different cycles should be perceived as successive.* Then, the minimum period of time needed for discrimination between successive signals is equal to the duration of the cycle. Theoretically, there should thus be a temporal threshold for discrimination between successive events, the value of which coincides with the duration of the cycle. The temporal threshold should not depend much on the modality of signals since it is produced by a mechanism of the same type at the cortical level, regardless of modality.

The threshold for discrimination between successive stimuli was established experimentally as long ago as the 1960s; it happens to be approximately the same for the auditory, visual and tactile modalities and for alternating stimuli of different modalities, and was about 60 ms (Hirsh and Sherrick, 1961; Kristofferson, 1967; Efron, 1973), which is close to the theoretical estimate of cycle duration. If the threshold for discrimination between successive events is produced by the duration of synchronous circulation of excitation in neuronal structures, training should change it like any other physiological process. "The striking effect of learning" (Efron, 1973) has indeed been discovered. In trained subjects, the threshold for discrimination between successive stimuli is as low as 15–20 ms in the auditory, visual, tactile and alternating modalities (Hirsh and Sherrick, 1961). The approximate equality of the thresholds for discrimination in different modalities, and the equal changes in them as a result of training, despite the fundamental anatomical and physiological differences between the respective perceptual organs, provides evidence for the existence of a universal signal-processing mechanism irrespective of modality. Both qualitative and quantitative characteristics of the temporal threshold for signal discrimination thus agree well with theoretical predictions.

Although the duration of the cycle limits the temporal resolution of perceptual events in the process of awareness, signals should keep their subliminal (nonconscious) temporal structure. They should, in particular, keep the temporal sequence of the signal components, although one cannot be aware of the sequence. The conservation of nonconscious information of the temporal sequence of signal components should be revealed in the ability of the subject to establish the similarity or difference between stimuli consisting of identical components in different orders. Efron (1973) showed experimentally that unconscious information of the order of two successive microsignals constituting a short stimulus did remain intact. In his experiments, auditory stimuli lasting several tens of milliseconds each, consisted of two shorter signals (microsignals) of different sound frequencies. The stimuli differed in the order of microsignals of different frequencies. Subjects were unable to report explicitly the order of microsignals in each stimulus, but could nevertheless discriminate between stimuli with different sequences of microsignals. Experiments in the visual and tactile modalities also showed the conservation of nonconscious information about the order of signal components within a stimulus. Similar result were obtained for stimuli consisting of three components (Efron, 1973).

3. FREQUENCY AND SPATIAL CHARACTERISTICS

The hypothesis of self-identification has consequences related to temporal characteristics of signal awareness, which can be formulated precisely, and which agree well with experimental data on the temporal structure of the process of awareness. Examination of extensive psychological and psychophysical data related to phenomena such as the perceptual moment, temporal thresholds, temporal summation, backward masking, flicker fusion etc, provides an estimate for the duration of the self-identification cycle of an order of several tens of milliseconds, which may vary between 10 and 100 msec (Sergin 1992, 1994a,c). We may therefore estimate the respective frequency band of the cyclic processes at values between 10 Hz and 100 Hz.

If the mechanism of self-identification underlies signal awareness, the frequency of its operation should be related to the frequency of events perceived, which the organism needs for adaptation to changing environmental conditions. The mechanism of self-identification should respond to an increase in the inflow of information I through any perceptual channel with an increase in the cycle frequency f_c, in order to promptly include it in the process of awareness. Therefore, $f_c \sim I$. This relation has two asymptotes. If input information is too great, the frequency reaches a value which it cannot physiologically exceed. No further increase in information input can then raise the cycle frequency. If input information flow is too small, the frequency falls to its lowest possible value, which corresponds to the state of relaxation. An increasing function with two asymptotes, one at approximately 10 Hz and the other at approximately 100 Hz, may therefore represent the dependence of the cycle frequency on input information flow (Figure 5.2).

The frequency characteristics of the mechanism of self-identification makes it possible to predict certain experimentally verifiable properties of the process of perception. If conscious perception is a discrete successive process, the human ability to determine the duration of short temporal intervals should be limited by an error equal to the value of the corresponding "quantum" of time (duration of the cycle). If, in a psychophysical problem of time estimation, the only variable is their duration D, then information inflow $I \sim 1/D$. Let τ_c be the duration of the cycle. If $fc = 1/\tau_c$ then, since $f_c \sim I$ we find that $\tau_c \sim D$.

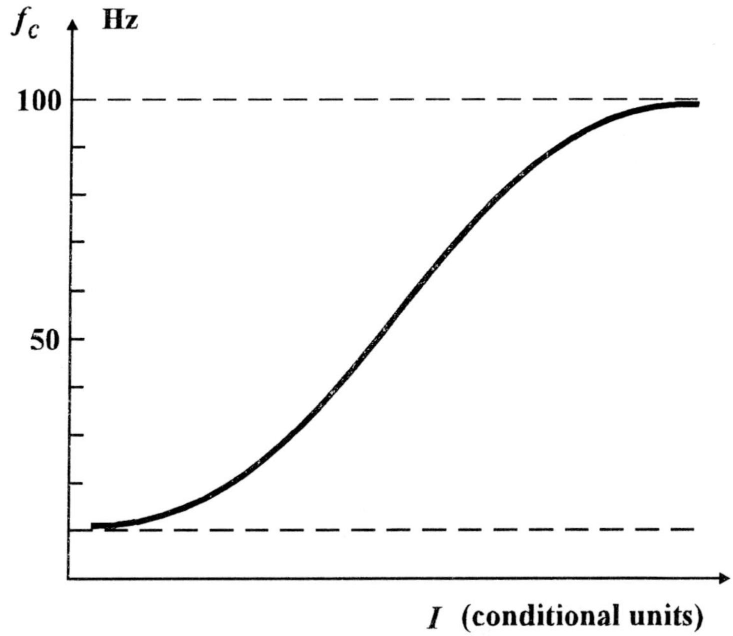

Figure 5.2. The dependence of the self-identification cycle frequency *fc* on information flow *I*

That is, the self-identification cycle duration should increase as the duration of the estimated interval increases, and should decrease as the duration of estimated intervals decreases. The minimum error of interval estimations should be approximately 10 ms. As the duration of estimated intervals increases, the duration of cycles increases too, which may increase the error to 100 ms. The relative error in the estimate of interval duration in the linear range should remain approximately constant. That is, $\delta D/D = constant$.

These theoretical predictions agree well with the results of experimental research of Kristofferson (1967, 1980, 1984), who discovered the effect of quantisation of the subjective estimate of the duration of temporal intervals. In these works, it was established experimentally that the value of the time quantum is a function of the duration of the estimated intervals. Doubling or halving the duration of the estimated intervals a given number of times doubles or halves the value of the quantum the same number of times. As the duration of estimated intervals changes from 100 ms to 800 ms, the value of the quantum changes from 12 ms to 100 ms. That is, the value of the time quantum changes quite like the period of the cycle of self-identification is supposed to do. Kristofferson (1984) arrived at the conclusion that the quantisation of subjective estimates of interval duration is caused by a periodic process which provides for internal timing.

Synchronous cyclical processes of self-identification take place in spatially distributed cortical neuronal structures of the brain, and should constitute an important part of cerebral electric activity. Therefore, temporal and frequency characteristics of the cyclical processes, which were established on the basis of psychological and psychophysical data, should fit independent experimental data of electro- and magnetoencephalography. Indeed, in the state of wakefulness and under intense sensory stimulation, high-frequency beta- and gamma-band oscillations (14–30 Hz and 30–100 Hz, respectively) dominate cortical electric activity. In a state of relaxation and in the absence of external stimuli, with the eyes closed, cerebral electric activity shifts into a low-frequency mode dominated by alpha-rhythm (8–13 Hz). That is, cortical electrical activity reveals, overall, a dependence of the oscillation frequency of EEG potentials on the rate of input information flow. This agrees, in general terms, with the frequency characteristics built on the basis of psychophysical data (Figure 5.2).

The process of self-identification, which proceeds by way of simultaneous circulation of signals in homogeneous neuronal structures in a limited cortical area, should produce in that area a field of spatially coherent oscillations. A stimulus containing different components, for example, boundaries, colour and motion, may simultaneously produce several fields of coherent oscillations, with frequencies of their own, in different projection and associative cortical areas. In this case, rapidly changing features may be self-identified, due to high frequencies of circulation, while less mobile and stable characteristics are self-identified by low frequencies. The spectrum of cortical electric activity present at any time may therefore contain many fields of spatially coherent oscillations differing in frequency and topographic distribution, which provide for integrated and synchronous mapping of a changing stimulus. Possible ways in which integration of the features of a stimulus can be mapped by simultaneous oscillations in neuronal activity of different frequency, are discussed in the works of Damasio (1989), and Borisyuk *et al.* (1994).

Recent investigation in human cerebral electric and magnetic activity, where computerised analysis of detailed spatio-temporal structure of processes has played a central role, has indeed discovered fields of spatially coherent oscillations (Sviderskaya and Shlitner, 1990; Llinas and Ribary, 1993, etc.). One work (Sviderskaya and Shlitner, 1990) describes electrophysiological experiments where potentials were measured by 48 macro-electrodes placed on the subject's head to form a grid of 8 arcs containing 6 electrodes each, spaced at regular intervals between the frontal and occipital poles. The EEG measurements were processed by specially designed software, which made it possible to estimate the cross-correlation ratio and frequency-specific coherence and phase characteristics of potentials. The result of the computer analysis was the establishment of numerous very distinct fields of spatially coherent oscillations in the cortical electrical activity. These fields of coherent oscillations of various frequencies are characterised by various topographic distributions. Another work (Sviderskaya *et al.*, 1993) shows that fields of spatially coherent oscillations of various frequencies have areas of mutual overlap, which occur in areas of high local synchrony of potentials (determined from cross-correlation coefficients). A linear dependence has been discovered between the spatial synchrony of potentials and the number of narrow-band coherent oscillations in such a cortical area. It was noted that the intensity of local activation in a certain cortical area is higher, the more fields of spatially coherent oscillations of various frequencies occur within its boundaries.

There exist numerous data on fields of coherent oscillations in the cerebral cortex obtained in animal experiments. For example, Freeman (1992) found, in an extensive program of research on olfactory processes, spatially coherent oscillations of electric activity in the olfactory bulb and olfactory cortex in the band between 20 Hz and 90 Hz.

It turned out that each smell is identified by a certain spatial distribution of amplitude values of coherent oscillations in the olfactory bulb, so that exposure to different smells produces different coherent excitation patterns (Freeman, 1991).

4. THE SYSTEM OF AWARENESS

Coding and processing of specific features of stimuli are related to the functioning of a specific system of cerebral activation. However, cyclical processes of self-identification cannot be activated only by a specific system, since awareness would then be confined to the field of stimulation, which is not the case. Llinas and Ribary (1993), proceeding from observations derived from different sources, arrive at the conclusion that the specific system provides content, and the non-specific system provides temporal binding of the contents into an integrated cognitive experience (awareness). The activation of processes of self-identification by the non-specific reticular-thalamic system may provide connection between processes of awareness and internal motivation, which gives perception some freedom from the stimulation field. Moreover, the non-specific system may simultaneously activate processes of self-identification of signals of different modalities (submodalities) and different levels (sensory and cognitive) through binding them (by pattern simultaneity) into one integral awareness. Since processes of self-identification are expressed in simultaneous fields of spatially coherent potential oscillations at different frequencies, their triggering from the state of relaxation should appear as "desynchronisation" in the EEG. One experimental study (Sviderskaya *et al.,* 1993) shows that the phenomenon of "desynchronisation" in slow high-amplitude EEG rhythms actually consists of its being replaced by a multitude of fields of spatially coherent oscillations of different frequencies.

Processes of self-identification are a means used to map stimuli of actual significance. Therefore they should be triggered (or not triggered) after the significance of a stimulus has been determined. That is, the theoretically necessary sequence involved in the process of conscious perception should first include a non-conscious evaluation of the actual significance of the stimulus, and then awareness. Then, in response to the stimulus, synchronous high-frequency oscillations in cortical potentials should arise with a delay from the time of arrival of the signal at the projection area, or not arise at all. It has been established in psychophysical experiments involving simultaneous registration of waves of induced potential that the significance of a stimulus is indeed evaluated before the sensation, in the first 100 or 150 ms, still at the non-conscious level (Ivanitsky, 1987; Kostandov, 1988). According to data of Gray (1994), the beginning of synchronisation of high-frequency oscillations is not related to the beginning of the effect of the stimulus and may be delayed for 50 to 100 ms. Desmedt and Tomberg (1995) state that synchronisation of oscillations at 40 Hz develops within 100 ms after potential arises in the primary cortex. According to data of Bressler (1995), synchronisation of high-frequency oscillation is delayed for 80 to 100 ms.

Self-identification of features of a stimulus produces coherent neuronal activity, thus forming a specific pattern and raising the signal/noise ratio for a very brief period of time. This provides the best conditions for *categorisation* of the pattern by distributed long-term memory (as described, for example, in the works of Grossberg [1988, 1995,

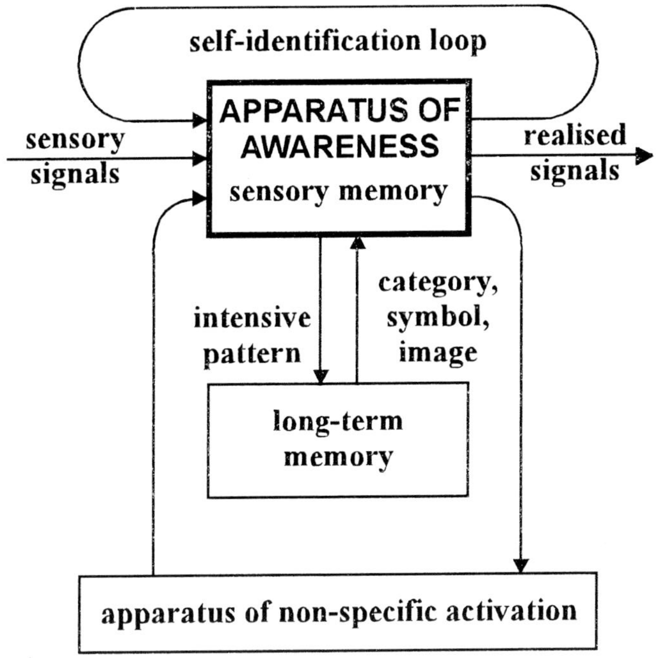

Figure 5.3. Block diagram of the system of sensory self awareness

etc]). The result of categorisation—a symbol* or "image"—expresses the subjective sense of sensory features of the stimulus. The categorical mapping of symbols is a response of memory to input excitation and is physiologically expressed in a pattern of cortical neuronal activity. This pattern is also involved in the cycle of self-identification, thus providing intensity mapping of the subjective sense of the stimulus (Figure 5.3). This event should correlate with the moment of stimulus awareness and it should be physiologically expressed in synchronisation of oscillations of potential in extensive cortical areas. The symbolic mapping of the stimulus means the transition of the process of perception from the physiological (objective) level to the mental (subjective) level.

The spatial distribution of long-term memory means that the memory of specific features of a stimulus, such as its boundaries, motion or colour, may be located in respective primary or specialised cortical areas, and the memory of binding the features into objects

* Sergin writes (in clarification) "A symbol may be a simplified image which stands in place of the real thing (such as a yellow disc may be a solar symbol) or a token which represents something. A symbol may consist of simpler symbols of one or several submodalities, therefore symbolic representation may be a multi-level representation (e.g. wood, trees and glades, branches, leaves and grass and other details). Symbols are products of human intelligence and symbolic representation is therefore originally subjective. One stimulus field (e.g. a Rorschach test inkblot) may produce different symbolic interpretations in different subjects. Pictograms, hieroglyphs, numbers, alphabetic characters, etc. may make particular cases of symbols. Symbolic representations of such kind are also subjective, as they are determined by the subject's culture."

or images, is located in the associative cortical areas of different levels (Damasio, 1989). Therefore simple unimodal stimuli may be categorised in the posterior areas of the cortex, while more complex stimuli, will involve intermediate sensory areas. Complex scenes, multimodal events or their temporal sequences may be categorised and subjectively represented in the frontal associative areas of the brain, as well as the temporal area of the cortex and the hippocampal system. Then, in response to input excitation, memory produces a pattern of neuronal activity in the same cortical areas as are involved in the perception of the stimulus. If the memory response corresponds at least approximately to the stimulus, memory produces a pattern close to the pattern of sensory excitation (similar in principal features). Identical components of these patterns add together in the same neuronal structures, accentuating the most significant features of the stimulus. The resulting pattern of neuronal activity is again subjected to categorisation, which produces a new pattern of subjective representation. This cyclical process ends in a pattern of sensory excitation becoming approximated by the best version of symbolic interpretation which the subject's memory has at its disposal.

Mapping of the stored data of memory by a neuronal activity pattern in the sensory cortex represents these data in the same form as external signals. Data stored in long-term memory in an implicit form are thereby converted into an explicit form. Explicit representation of internal data allows their categorisation and symbolic interpretation in the same manner as applies to external signals . Therefore mapping of symbolic data by a neuronal activity pattern is its representation to the subject (i.e. to an integrated conscious "self") as an external world description element. As a result, the external world is represented to the subject in his own terms (or symbols, or images), which is the most specific experience of subjective perception. The ability of memory to project its contents to the sensory cortex and produce thereby a specific neuronal activity pattern may form the well-known ability of human mentality to project its subjective representations to the real world.

At whatever level awareness might take place, be it a local peripheral area of the sensory cortex or extensive frontal areas of the neocortex, an act of awareness causes activation of other portions of the brain, response of centres of emotion and motivation, and the motor and vegetative systems, thus producing experiences and actions adequate to subjective interpretation of the stimulus. As a result, the brain and the organism respond as a systemic entity. The magnitude of response and its specificity depend little on the intensity or objective contents of the stimulus and are almost entirely determined by its subjective meaning in a given situation.

In the case of a simple unimodal stimulus, mapping and categorisation proceed in the primary sensory (projection and associative) cortical areas, which should take the shortest possible period of time. Becoming aware of a flash of light or a sound may therefore take one or two cycles. Categorisation of stimuli having complex physical and semantic features requires involvement of the intermediary and higher associative cortical areas and should take more time. Becoming aware of them should therefore require many successive cycles.

It has been established in experimental psychology that in the process of perception, for instance during gaze fixation, a sensory image is processed consecutively, stage by stage, first with discernment of its general features, then becoming increasingly more detailed and specific (Hoffman, 1982). Consecutive awareness of features of a fixed image coincides well with consecutive cyclic processes of self-identification and symbolic interpretation of data. Since gaze fixation typically lasts for 100–200 ms, the process of conscious perception may include 5 to 10 cycles. A similar estimate is produced for the olfactory system. Inhalation is for the olfactory system the same as gaze fixation for the visual system, and one instance of this also last a fraction of a second. High-frequency coherent oscillations discovered by Freeman (1992) are timed with inhalation, and have the form of packets consisting of approximately ten waves each.

Self-identification is a means of enhancement of the prominence of specific features, which makes their inclusion in an integrated image possible. Consecutive selection (i.e. from cycle to cycle) and memorising of selected features is the process of selective integration. The comparison of current integrated features with data stored in long-term memory makes current categorisation possible. Every act of categorisation specifically activates the perceptual system, directing its resources to revealing quite definite sensory features. Selective integration of sensory features, current categorisation and active search for specific features makes it possible to implement continually accurate interpretation of the stimulus field, and to control the process of perception.

5. NATURE OF VISION AND SENSATION

The human faculty to see is as mysterious as the faculty of awareness. Neither telescope, nor video camera, nor video robot see, though they can map and process visual information. Humans and animals see, however. Visual awareness and vision are empirically indistinguishable and constitute the same mental phenomenon. Visual awareness is of course something more than intensity (brightness) mapping: It includes understanding (by way of interpretation of the mapping). Seeing also includes understanding. The Russian word "videt'" ("see") is often used in the meaning of "understand". The English word "see" means both "see" and "understand". At the dawn of mathematics in ancient Greece, a relevant geometrical construction was deemed the proof of a statement (theorem): "Now see."

Awareness is a form of secondary processing of pre-selected of signals, the purpose of which is explicit representation of actually significant data. Operation of the system of awareness forms a selected multimodal flow of explicit knowledge, apparent, represented to the subject ("the self"). Similarly, there is a flow of information of which the subject is unaware (implicit, concealed knowledge). The universal system of explicit representation of data reaches its maximum efficiency in visual perception, thus producing the mental phenomenon of vision. Awareness imparts to the process of perception a new quality in terms of the mental dimension, allowing us not only to look and respond adequately, but also look and see; and not only listen and act, but also listen and hear.

The functioning of consciousness is always related to internal representations of images, symbols, sounds, smells, touches, etc, and is inseparable from them. No conscious cerebral activity is possible without processes of explicit internal representation of external events or internally generated signals and images of different modalities. The mechanism of self-identification makes it possible to accentuate both external and internal signals. Therefore, the physiological mechanism of review of one's own thoughts and mental images (giving them explicit representation) may be quite similar to the mechanism of awareness of sensory data. The system of awareness may thus be a universal means of extra- and introspection (Sergin 1994a,c).

Not all sensory data require accurate and detailed awareness. If some information has no actual value, the process may be interrupted at the stage of awareness of the general (qualitative) character of stimulation or the mere fact of its presence. This pre-awareness may well be the primitive sensation (qualia). That is, sensation may be the early stage of awareness, which maps only the qualitative features of the stimulus. If details of the stimuli are not mapped, they may be represented in a simple (e.g. single-parameter) form, which makes it possible to include many events simultaneously in the process of perception. The extension of the parallel flow of noticeable (accentuated) signals gives the organism important adaptive advantages, and that property of perception could be the object of evolutionary selection.

Note, that certain stimuli, which, for example, are too faint or quickly changing, are physically inaccessible to detailed awareness and may only be sensed. Data of many receptor organs (for example, of balance, temperature and pressure), and data about the internal environment of the organism, as opposed to visual and auditory information, are always mapped in a qualitative form. As a result, the vast majority of stimuli of the environment and internal world of the subject are merely sensed. But it is just this multimodal and multi-image flow of stimuli reaching pre-awareness which generates the sense of life, the sensation of being in the surrounding world.

Sensation, like awareness, is produced by intensity mapping of a stimulus. However, in the case of sensation, interpretation of the mapping is characterised by undivided, rather than differentiated categories. Clear awareness is characterised by highly differentiated categorisation and multi-level interpretation of subjective representations. Categories, symbols and images constitute one fund of knowledge common for all people and created by the cultural evolution of the mankind. This is exactly why con-sciousness is common knowledge (Simonov, 1987). In awareness, the mental processes of categorisation and symbolic interpretation prevail. Highly differentiated mapping of data makes their analysis and synthesis possible, while explicit symbolic representation makes it possible to operate on the data as external objects. As a result, it becomes possible to use knowledge, which constitutes the most important contents of conscious activity of the brain (the relevant psychophysiological mechanisms are discussed in the works of Sergin (1994a,b,c).

From these notions the differences between human and animal sensory awareness follow naturally. Self-identification as a mechanism of enhancement of the prominence of vital events could appear at the early stages of evolution, and all mammals probably have perceptual consciousness. However, animal sensory awareness has no apparatus of symbolic interpretation in so far as it is determined by cultural evolution. It is more physiological and thus differs from human. Nevertheless, this is just why consciousness of animals has one fundamental advantage: It does not replace objective sensory data with subjective interpretation based on a priori knowledge. Animal consciousness may be characterised as essential, capable of immediately seeing the essence of objects and events. As opposed to animals, symbolic description of the world dominates in the humans, and man is capable primarily of perceiving what this description includes. We see what we know and this is not always what is actually there.

6. CONCLUSION

This paper postulates a process of self-identification, which consists in the relay of a specific pattern of excitation produced by a stimulus in one or several cortical areas, back to the neurones of these areas through massively parallel feedback. The coinciding (identical) patterns of excitation produced by the stimulus and relayed through back projections add together on the same neuronal structures, thus inducing their intense firing. This cyclic process of collective neuronal activity accentuates the specificity and enhances the intensity of the mapping of the stimulus, thus providing the best conditions for its categorisation by distributed long-term memory.

The result of categorisation, a symbol or an image, expresses the subjective meaning of the sensory features of the stimulus and is mapped physiologically by a pattern of neuronal activity in the cortex. Mapping of symbolic information by a pattern of neuronal activity in the cortex is its representation to the subject (or "the self") as an element of description of the outer world. As a result, the outer world is represented to the subject, this process being the most specific experience of the mental phenomenon of sensory awareness.

The cyclical mechanism of self-identification determines the discreteness and succession of the processes of awareness. The duration of the cycle depends on the frequency of perceived events and varies between 10 ms and 100 ms. This is the minimum period of time which may be discriminated at the mental level. Examination of extensive data related to such phenomena as the perceptual moment, temporal thresholds, temporal summation, backward masking, etc, reveals full agreement of theoretically predicted and experimentally established data on the temporal structure of processes of sensory awareness. Predicted electrophysiological phenomena, such as fields of coherent potential oscillations, their frequencies and duration of occurrence agree well with the latest electro- and magnetoencephalographic data.

The compatibility of the mechanism of self-identification with independent psychological, psychophysical and electrophysiological data provides evidence of its reality. This opens the prospect of experimental research into the anatomical apparatus, physiological mechanisms and mental structure of the processes of awareness, on a new conceptual basis.

REFERENCES

Borisyuk, G.N., Borisyuk, R.M., and Kazanovich, Y.B. (1994) Modelling "pre-attention" and "attention" information processing by synchronisation of neural activity. *Radiophysics. Bulletin of Higher Education*, **37**, 933–944.

Bressler, S.U. (1995) Large-scale cortical networks and cognition. *Brain Research*, **20**, 228–304.

Crick, F. (1984) The function of the thalamic reticular complex: The research light hypothesis. *Proceedings of the National Academy of Sciences, U.S.A.*, **81**, 4586–4590.

Crick, F. and Koch, C. (1990) Some reflections on visual awareness. *Cold Spring Harbor Symposia on Quantitative Biology, Volume 55*, pp. 933–962, Cold Spring Harbor Laboratory Press.

Crick, F. and Koch, C. (1992) The Problem of Consciousness. *Scientific American*, **267**, 152–159.

Damasio, A.R. (1989) The Brain Binds Entities and Events by Multiregional Activation from Convergence Zones. *Neural Computation*, **1**, 123–132.

Desmedt, J. and Tomberg, C. (1995) Neurophysiology of preconscious mechanisms of the human brain. *Abstract of the Xth International Congress of Electromyography and Clinical Neurophysiology*. Kyoto, Japan.

Edelman, G.M. (1978) Group selection and phasic re-entrant signaling: a theory of higher brain function. In: *The Mindful Brain,* pp. 51–100 Cambridge, MIT Press.

Edelman, G.M. (1989) *The remembered present. A biological Theory of consciousness.* New York, Basic Books.

Efron, R. (1973) Conservation of temporal information by perceptual systems. *Perception and Psychophysics*, **14**, 518–530.

Freeman, W.J. (1991) The Physiology of Perception. *Scientific American,* **264**, 78–85

Freeman, W.J. (1992) Tutorial on Neurobiology: from Single Neurons to Brain Chaos. *International Journal of Bifurcation and Chaos,* **2**, 451–482.

Gray, C.M. (1994) Synchronous Oscillations in Neuronal Systems: Mechanisms and Functions. *Journal of Computational Neuroscience*, **1**, 11–38.

Gray, J.A. (1995) The contents of consciousness: A neuropsychological conjecture. *Behavioural and Brain Sciences*, **18**, 659–722.

Grossberg, S. (1988) Nonlinear neural networks: Principles, Mechanisms and Architectures. *Neural Networks*, **1**, 17–61.

Grossberg, S. (1995) The attentive brain. *American Scientist*, **83**, 438–449.

Harth, E. (1995) *The creative loop.* Addison-Wesley Publishing Company.

Hirsh, I.J. and Sherrick, C.E. (1961) Perceived order in different sense modalities. *Journal of Experimental Psychology*, **62**, 423–432.

Hoffmann, J. (1982) *Das Active Gedächtnis*. Berlin: VEB Deutscher Verlag der Wissenschaften.

Hylan, J.R. (1903) The distribution of attention. *Psychological Review*, **10**, 373–440 and 498–533.

Ivanitsky, A.M. (1976) *Brain mechanisms of signals estimating.* Moscow: Medicine Publishers. (in Russian).

Ivanitsky, A.M., Strelets, V.B. and Korsakov, M.A. (1984) *Information Processes of the Brain and Mental Activity.* Moscow: Nauka Publishers (in Russian).

Ivanitsky, A.M. (1987) Psychic activity and the organization of brain processes. *Vestnik of Academy of Medical Sciences,* **8**, 14–20 (in Russian).

Ivanitsky, A.M. (1996) Brain basis of subjective experience: information synthesis hypothesis. *Journal of Higher Nervous Activity*, **46**, 241–252 (in Russian).

Kostandov, E.A. (1988) Conscious and unconscious forms of human higher nervous activity. In: *Mechanisms of Human Brain Functioning, Part One, Human Neurophysiology,* Leningrad: Nauka Pablishing House, pp. 491–526.

Kristofferson, A.B. (1967) Attention and Psychophysical Time. *Acta Psychologica*, **27**, 93–100.

Kristofferson, A.B. (1980) A Quantal Step Function in Duration Discrimination. *Perception and Psychophysics*, **27**, 300–306.

Kristofferson, A.B. (1984) Quantal and Deterministic Timing in Human Duration Discrimination. *Annals of New York Academy of Sciences*, **423**, 3–15.

Livanov, M.N. (1972) *Spatial organisation of cerebral processes*. Moscow: Nauka Publishers. (in Russian).

Llinas, R. and Ribary, U (1993) Coherent 40-Hz oscillation characterizes dream state in humans. *Proceedings of the National Academy of Sciences, U.S.A.*, **90**, 2078–2081.

Mountcastle, V.B. (1978) An organizing principle for cerebral function: The unit module and the distributed system. In: *The Mindful Brain*, Cambridge, MIT Press, pp. 7–50.

Pavlov, I.P. (1951) *Twenty-year experience of objective studies of animal higher neural activity. Complete Works, 2nd edition*, Moscow and Leningrad, volume 1.

Sergin, V.Ya. (1991) Brain as Neurocomputer: The Macrostructure of Intelligence. In: A.V. Holden and V.I. Kryukov (eds). *Neurocomputers and Attention, Vol. 2.* Manchester: Manchester University Press, pp. 771–781.

Sergin, V.Ya. (1992) A Global Model of Human Mentality. In: R. Trapple (ed.) *Cybernetics and Systems—92, Vol. 1*. pp. 883–890, Singapore: World Scientific Publishing Co.

Sergin,V.Ya. (1994a) Consciousness as an Inner Vision System. *Journal of Higher Nervous Activity,* **44**, 627–639 (in Russian).

Sergin, V.Ya. (1994b) Consciousness as a data-processing system. *Neural Network World*, **4**, 601–608.

Sergin, V.Ya. (1994c) Mechanisms of Consciousness. In: R. Trapple (ed.) *Cybernetics and Systems—94, Vol. 2.* Singapore: World Scientific Publishing Co, pp. 1887–1894.

Simonov, P.V. (1987) *Motivated brain.* Moscow: Nauka Publishers (in Russian).

Simonov, P.V. (1990) The Light Spot of Consciousness. *Journal of Higher Nervous Activity*, **40**, 1040–1044.

Steriade, M., McCormick, D.A. and Terrence, J.S. (1993) Thalamo-cortical oscillations in the sleeping and aroused brain. *Science, New York,* **262**, 679–685.

Sviderskaya, N.E. and Shlitner, L.M. (1990) Coherent Cortical Electric Activity Structures in the Human Brain. *Journal of Higher Nervous Activity*, **16**, 12–19 (in Russian).

Sviderskaya, N.E., Korolkova, T.A. and Tishaninova, L.V. (1993) The fields of the higher activity: electrophysiological correlates. *Journal of Higher Nervous Activity*, **43**, 1080–1087 (in Russian).

6 Brain Mechanisms of Emotions

P.V. Simonov

Institute of Higher Nervous Activity and Neurophysiology, Russian Academy of Sciences, Moscow, Russia
e-mail: BALABAN@IHNA.MSK.RU

At the 23rd International Congress of Physiological Sciences (Tokyo, 1965) experimental results led us to the conclusion that emotions were determined by the actual need, and the estimated probability (possibility) of its satisfaction. A low probability of need satisfaction leads to negative emotions, which are actively minimized by the subject's behaviour. An increased probability of satisfaction, as compared to the earlier forecast, generates positive emotions, which the subject tries to maximize — that is to enhance, to prolong, or to repeat. We named our concept the Need-Informational Theory of Emotions. According to this theory, motivation, emotion and estimates of probability have different neuromorphological substrates. Activation of the frontal parts of the neocortex, through hypothalamic structures which generate motives, orients behaviour towards signals which have a high probability of being reinforced. At the same time the hippocampus is necessary for reactions to signals of low probability events, which are typical for the emotionally excited brain. By comparison of motivational excitation with available stimuli or their engrams, the amygdala selects a dominant motivation, destined to be satisfied in the first instance. In the case of classical conditioning and escape reactions, reinforcement is related to the involvement of hypothalamic neurones responding during negative emotions, while in the case of avoidance reactions, neurones related to positive emotions are involved. The role of the left and right frontal neocortex in the appearance of positive or negative emotions depends on informational (cognitive) functions.

KEYWORDS: motivation, amygdala, hippocampus, learning, cortical asymmetry

1. INTRODUCTION

William James—the author of one of the first physiological theories of emotions more than a century ago—published a paper with a most remarkable title: "What is emotion?" (James, 1884). Nevertheless, one hundred years after this question was formulated, we find in the textbook "Human Physiology" the following revelation: "Despite the fact that each of us knows what emotions are, it is impossible to give the emotional state a precise scientific definition... . At the present time there is no generally accepted scientific theory of emotions, nor any precise data concerning which centers emotions arise in, how they arise, or what their nervous substrate is" (Schmidt and Thews, 1983).

2. THE NEED-INFORMATION THEORY OF EMOTIONS

At the 23rd International Congress of Physiological Sciences (Tokyo, 1965), the results of psychophysiological experiments brought me to the conclusion that human emotions were determined by some actual need and the estimated probability (possibility) of its satisfaction, on the basis of phylo- and ontogenetic experience (Simonov, 1975; Simonov, 1986). The individual makes this estimation involuntarily (sometimes unconsciously), comparing information about the means and time that are predictably necessary for satisfaction of this need, with information about the circumstances actually present. A low probability of goal achievement leads to negative emotions (fear, alarm, fury, grief, etc.) which are actively minimized by the subject's behaviour. An increased probability of satisfaction, as compared to an earlier prognosis, generates positive emotions of pleasure, joy and encouragement, which the subject tries to maximize, that is, to intensify, continue or repeat. I called this concept "The Need-Informational Theory of Emotions", in order to attach great importance to the subjects' estimation of the probability of need satisfaction in the genesis of emotions (Simonov, 1984).

In its most general form, the rule for the genesis of emotion may be presented as a structural formula:

$$E = f\{-N\,(In - Ia)\}$$

where E is emotion, its degree, quality and sign; N is the power and quality of the actual need in the broadest sense of the word. For humans, this includes not only vital needs like hunger, thirst and sex, but also diverse social and idea-related (spiritual) needs including the most complicated and lofty ones. $(In - Ia)$ is the estimated probability (possibility) of need satisfaction on the basis of phylo- and ontogenetic individual experience. In refers to information about the means and time prognostically necessary for satisfaction of the need. Ia designates information about the means and time available to the subject at a given moment. The term "information" in the equation implies information to be both a quantity and a quality, that can be determined as the change in probability of goal achievement. Inspection of this equation shows that when $Ia>In$, positive emotions are generated, and when $In>Ia$ negative emotions are produced. The word "motivation", used in the following paragraphs, also requires definition: Activation of traces (engrams) of external objects capable of satisfying a need transforms the need into a motivation.

3. THE CEREBRAL BASIS FOR ADAPTIVE FUNCTIONS OF EMOTIONS

The results of neurophysiological experiments show that needs, motivation and emotions have different morphological substrates. Thus, on stimulation of areas of self-stimulation in the lateral hypothalamus by electric currents of gradually increasing strength, behavioural reactions of rats always occur in the same sequence. Weak stimulation elicits a generalized searching behavior which is not directed to the objects in the chamber—food, water, or the lever for self-stimulation. Current increase elicits evidence of motivation—eating, drinking and gnawing behaviours. Further current increase elicits self-stimulation behaviour with related motivational effects; and at the next stage, only self-stimulation takes place (Table 6.1).

Table 6.1. Relationship between motivating and rewarding stimulation of lateral hypothalamus

Behavior	*Strength of current (mA)*
Exploratory	0.05–0.01
Eating, drinking, gnawing, etc.	0.1–0.25
Consummatory behavior and self-stimulation	0.25–0.35
Self-stimulation	0.35–0.5

In zones of self-stimulation of the lateral preoptic area and lateral hypothalamus, two classes of neurones were recorded, which specifically changed their activity either to motivational states (linked with negative emotions) or to emotionally-positive (reinforcing) states, when these were elicited by electric stimuli and by natural stimuli (for instance, change of the level of alimentary and water motivations). Neurones of the second type (reinforcing) were maximally activated during stimulation by the current eliciting self-stimulation, and they were also activated at satiation. Motivational and emotionally positive (reinforcing) behaviours oppose each other, and elicit reciprocal changes in activity of the first and second type neurons.

In experiments of our collaborators (Mikhailova and Zaichenko, 1993) classical and operant conditioned reflexes were produced in rats, in which the conditioned signal (light) correlated with emotionally negative intracerebral stimulation of the dorsolateral tegmentum. It was shown that neurones of the above two types in the lateral hypothalamus participated in different ways in the realization of these reflexes. Figure 6.1 demonstrates changes in impulse activity of motivation-related neurones (upper panel) and neurones related to positive reinforcement (lower panel) during realization of conditioned defensive reflexes: classical, escape and avoidance. Columns show the frequency of discharges during the following epochs: (1) 5 sec before the conditioned stimulus; (2) during the action of light; (3) during the combined light and current; (4) after switching off the stimuli. It can be seen that realization of classical conditioned reflexes was accompanied by suppression of spiking in the reinforcing neurones. Escape reactions were accompanied by intensification of activity in motivational neurones. For avoidance reactions, there was an increase in activity in positively reinforcing neurons, but only when they were well-elaborated, and the rat was thus not punished by electric current.

These data allow one to answer the question continuously discussed in the literature: What serves as a reinforcement in operant defensive reflexes? In the case of classical reflexes and escape reactions, the emotionally negative state of fear serves as a reinforcement. However, successful accomplishment of avoidance reactions involves the mechanisms of positive emotions in the process of making decisions about behaviour.

I have already noted above that the influence of emotions on behaviour is determined by the animal's attitude to its emotional state, and is dominated by the principle of maximization of positive emotions and minimization of the negative ones. This principle is accomplished by the influence of hypothalamic structures representing motivational and emotional states upon neocortical areas concerned with informational (cognitive) and motor-organizing functions. This is confirmed by analysis of the spatial synchronization of electrical activity in brain structures, during self-stimulation in rats by weak constant currents (Pavlygina *et al.*, 1976). Figure 6.2 shows the percentage of cases in which

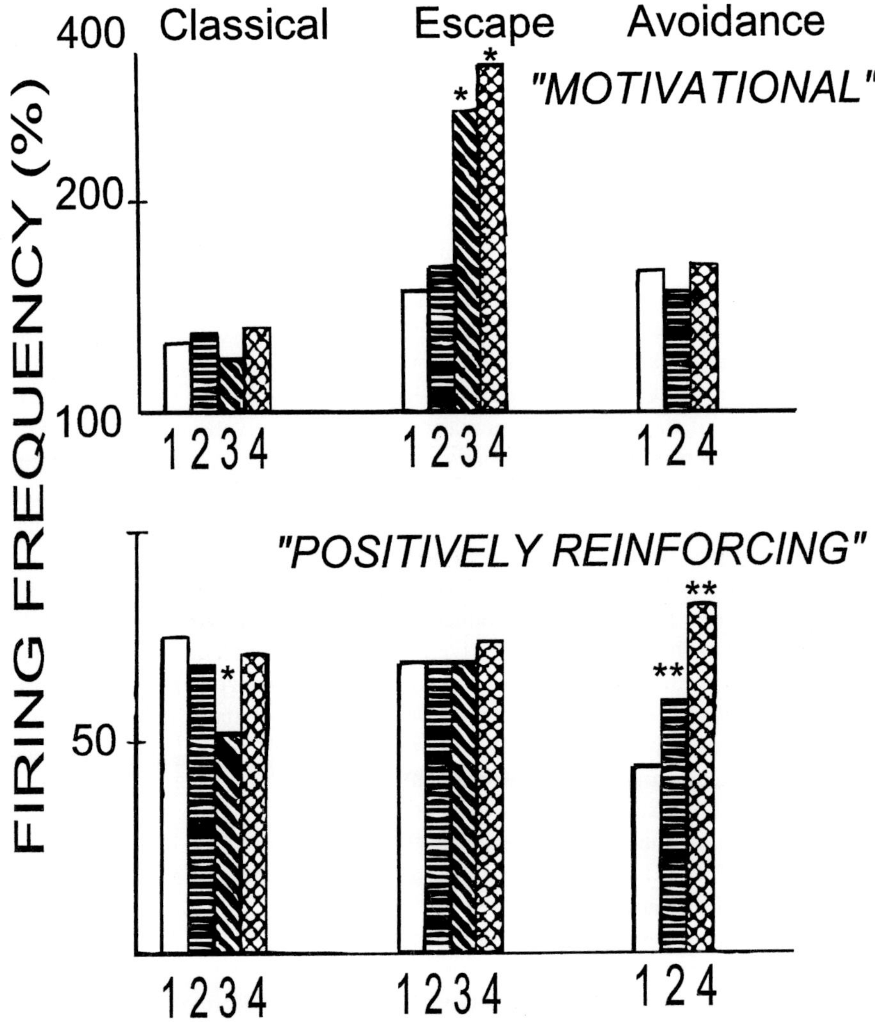

Figure 6.1. Changes in impulse activity of motivational (upper panel) and positively reinforcing (lower panel) neurones of the lateral hypothalamus during realization of conditioned defensive reflexes: classical, escape and avoidance. Columns consequently show: (1) frequency of discharge 5 sec. before the conditioned stimulus, (2) during the action of light, (3) during the combined light and current, (4) after switching off the stimuli *– p < 0.05; ** b < 0.01

significant coherence ($p < 0.05$) was observed in alpha- and theta- frequency ranges, when potentials in the following brain structures were compared: 1—the emotionally positive point of the hypothalamus *vs* the motor region of the cortex; 2—the motor cortex *vs* the visual cortex; 3—the emotionally negative point of the hypothalamus *vs* the motor region of the cortex. The analysis was carried out in the following behavioural states: I—When the animal is in a calm state; II—immediately before pressing the lever; III—while pressing the lever; IV—the period of leaving the lever; V—after leaving the lever.

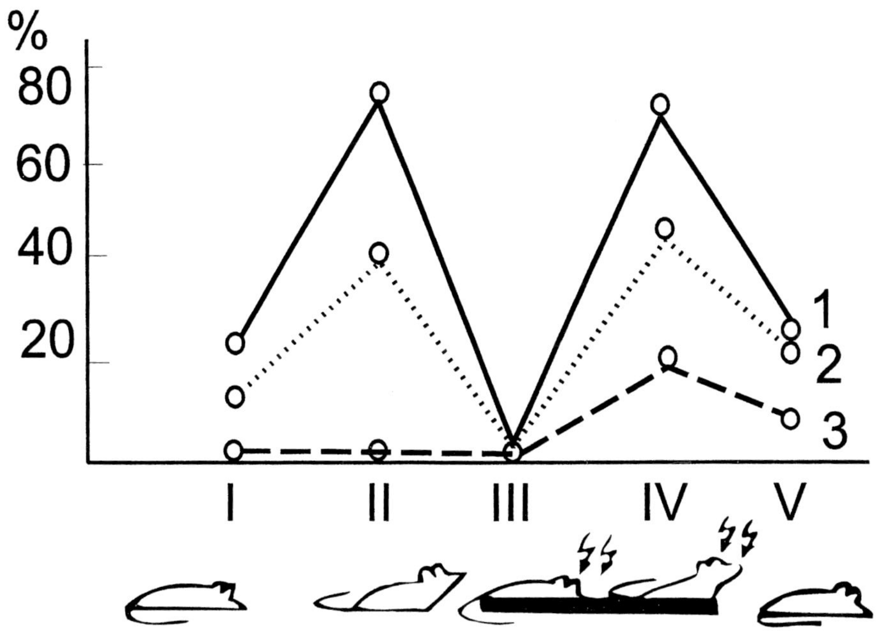

Figure 6.2. Percentage of significant changes in coherence of electrical activity in the rat brain at different steps of self-stimulation. 1—emotionally positive point of the hypothalamus—motor region of the neocorex; 2—motor neocortex—visual neocortex; 3—emotionally negative point—motor neocortex. (Electrical activity was recorded during spontaneous performance of the task, with subsequent analysis of different phases of behaviour—see text.)

Examination of the graph, shows clearly that immediately before lever pressing there is a sharp rise (by a factor of more than 3) in the percentage of cases in which there is a statistically significant coherence of electrical activity in the emotionally positive point of the hypothalamus and the motor region of the cortex.

This increase in coherence demonstrates that nervous pathways become ready for spread of excitation along three channels of conditioned connections: (i) from the initially stimulated emotionally positive point to the motor cortex; (ii) from here to the visual cortex; and (iii) also between the visual analyzer which receives the conditioned signal of future reinforcement (the sight of the lever, its location in the chamber, etc.) and the motor cortex, since it is the view of the lever which will direct the animal's movement initiated by the trace of arousal from the emotionally positive region.

While the rat is on the lever, coherence decreases, the animal receives reinforcement, and becomes totally passive. Immediately before the rat leaves the lever, there is for the first time an increase in coherence between the negative point and the motor region of the cortex: Excitation in negative structures is ready for transformation into the motor reaction of avoidance. After the animal leaves the lever, the percentage of cases in which there is a statistically significant increase in coherence returns to its original value. At this stage, only traces of the emotionally negative state can still be seen when comparing electrical activity in the negative point with the motor region of the cortex.

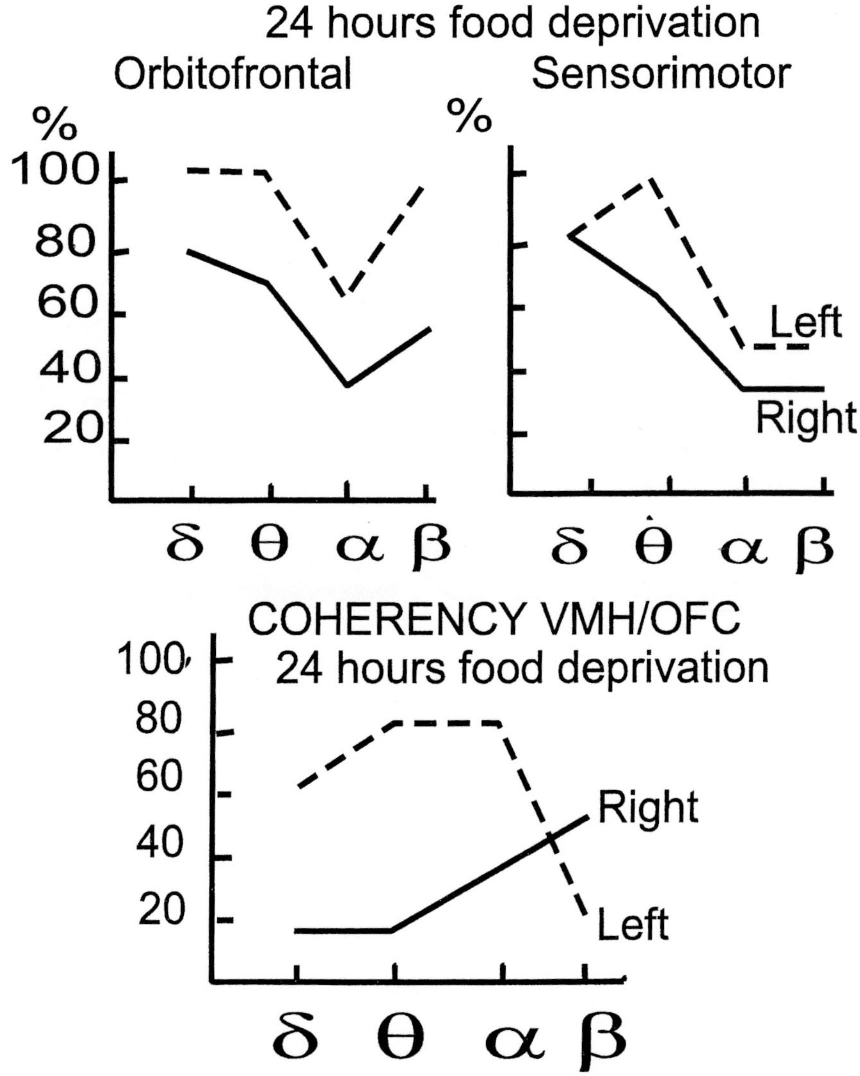

Figure 6.3. Changes in spectral content of electrical activity in orbitofrontal and sensorimotor cortical regions of the right (solid line) and left (dashed line) hemispheres after 24 hours food deprivation in rabbits. Lower panel—estimation of coherency in potentials in the ventromedial hypothalamus and orbitofrontal cortex in the right (solid line) and left (dashed line) hemispheres of the rabbit in a state of hunger after 24 hours deprivation; abscissa—frequency ranges, ordinate—mean percentage of cases of statistically significant increases in coherency.

R.A. Pavlygina and Yu.V. Lyubimova (1994a, 1994b) in their investigations have shown that motivational influences of the hypothalamus upon the neocortex are asymmetrical. Figure 6.3 (upper panels) summarizes data showing statistically significant decreases in the power of electrical activity in various frequency bands, in rabbits after 24 hours food deprivation. Results presented are from the orbitofrontal and sensorimotor cortical regions of the right (solid line) and left (dashed line) hemispheres. Asymmetry is expressed not only in the prevailing activation of the left hemisphere, but also in the intensification of interactions within the left hemisphere: The lower panel of Figure 6.3 demonstrates statistically significant increases in estimates of coherency, comparing potentials in the ventromedial hypothalamus and orbitofrontal cortex in the right (solid line) and left (dashed line) hemispheres in rabbits in a hungry state after 24 hours deprivation.

Interhemispheric asymmetry during natural alimentary motivation is also revealed by recording impulse trains of single neurones in the visual and sensorimotor regions of the neocortex of rabbits, and by investigation of interaction of these neurones (Pavlova, 1995). In Figure 6.4 such neuronal data are presented for the left (1) and right (2) hemispheres of the neocortex in hungry (open columns) and satiated (shaded columns) rabbits. In A, the percentage of pairs of neurones working correlatively is shown. In B the probability of peaks (positive values) and gaps (negative values) appearing in the corresponding bins of crosscorrelation histograms is shown (abscissa—time in ms; ordinate—probability; N—number of crosscorrelation histograms). In the inset the location of recording electrodes is shown. Judging by the neuronal impulse activity, the cortex of the left hemisphere shows more correlated activity in hungry rabbits while the right hemispheric cortex does so in satiated ones (N values in Figure 6.4B). The most pronounced differences were observed in neurones of the frontal regions, while less pronounced differences were seen in sensorimotor neurones.

All recent data suggest that the hypothalamus is the key structure for realization of the most ancient *reinforcing function* of emotions, which serves to solve the universal behav-

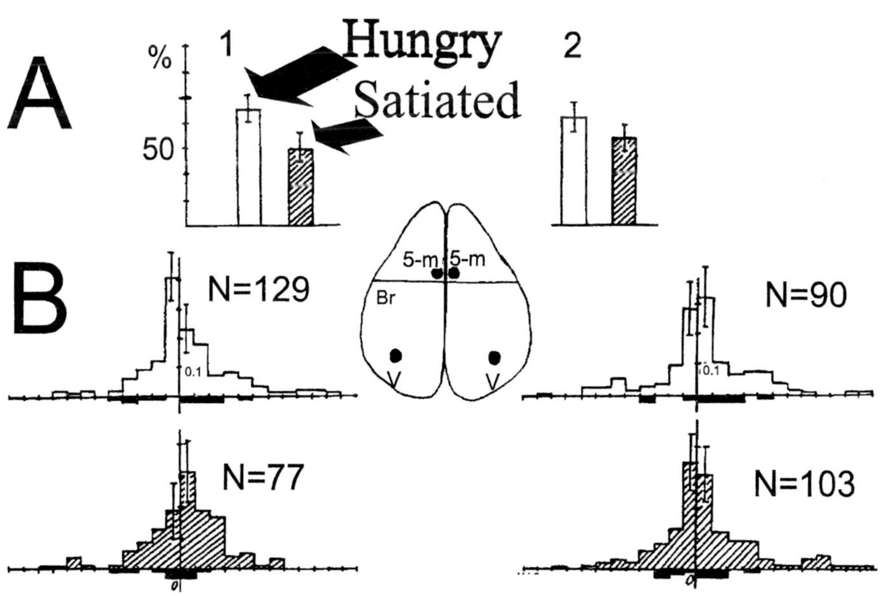

Figure 6.4. Neocortical neurons with correlating activity in hungry and satiated rabbits. Bin width (lower panels) in msec.

ioural task of maximization-minimization of a given emotional state, and is expressed in behaviour as approach or avoidance. In comparison with the hypothalamus, after lesions of the amygdala, consumption of food and water, and the response to blood glucose level during food deprivation are essentially unaltered. According to M.L. Pigareva (1983), bilateral destruction of the amygdala does not prevent the development of either alimentary or defensive conditioned reflexes in rats. However, the picture changes radically in the case of competition between coexisting motivations, when it becomes essential for the animal to distinguish the dominant need, that is, the one requiring immediate satisfaction.

A good experimental model of such a situation is the elaboration of conditioned switching of heterogeneous conditioned reflexes by the method of E.A.Asratyan (Asratyan, 1965) when the same conditioned signal (tone) is reinforced in the morning by food, and in the evening by a noxious shock. As judged by the percentage of correct defensive and feeding conditioned reactions, destruction of the amygdala in rats results in a failure to achieve switching, this failure persisting for a period of forty days. Nevertheless, solution of such a behavioural task is possible when a sufficient imbalance is artificially created between competing motivations and corresponding emotions, such as between hunger and fear.

Figures in Table 6.2 show the number of rats in each group, and figures in brackets show the number of animals in which conditioned switching was elaborated in 60 days. The criteria for elaboration were that three sessions in succession should be carried out, in each of which a 100% fulfillment of either alimentary and defensive conditioned reflexes was to be achieved. Amygdalectomized rats managed to fulfill this criterion if a strong noxious stimulus was paired with 24-hours of food deprivation, or, alternatively, a weak painful stimulus was applied against a background of hunger produced by three-days of deprivation. However, with lower intensity currents and a lesser degree of food deprivation, transfer could not be shown in the amygdalectomized rats. In other words, the amygdala plays a crucial role in realizing the *function of transswitching behaviour* associated with emotions, and in choice of the motivation corresponding not only to one or another need, but to the external conditions of its satisfaction in the given context and at the relevant moment. The amygdala is involved in the process of organization of

Table 6.2. Effect of amygdala destruction on the development of conditioned reflex switching in rats

Strength of current, mA	*Duration of food deprivation*			
	One day		*Three days*	
	Groups of rats			
	Intact	*Operated*	*Intact*	*Operated*
0.4	4(0)	5(0)	8(2)	8(5)
0.6	5(2)	5(0)	8(4)	8(4)
0.8	5(3)	5(0)	5(5)	5(0)
1.0	5(4)	5(0)	5(3)	5(0)
1.2	7(5)	7(4)	6(4)	6(1)
1.4	8(3)	8(5)	6(3)	6(0)

behaviour at a comparatively late stage, when the manifest needs are already being compared with the possibility of their satisfaction, and are being transformed into the corresponding emotions.

As for the estimation of the probability of satisfying a need (probability of reinforcement, or *In* – *Ia* in the structural equation), it is made by the "informational" structures of the brain, such as the hippocampus, and the frontal regions of neocortex. The most striking defect in hippocampectomized animals turns out to be their sensitivity to situations with a low probability of conditioned signals receiving reinforcement (Figure 6.5). In experiments with different levels of probability of reinforcement, when the probability of reinforcement of alimentary conditioned reflexes is 100% or 50%, hippocampectomized rats (dashed lines) are less advanced than the intact ones (solid lines), but still manage the task. When the probability of reinforcement is only 33% to 25%, elaboration of conditioned reflexes turns out to be beyond their abilities. In experiments similar to those

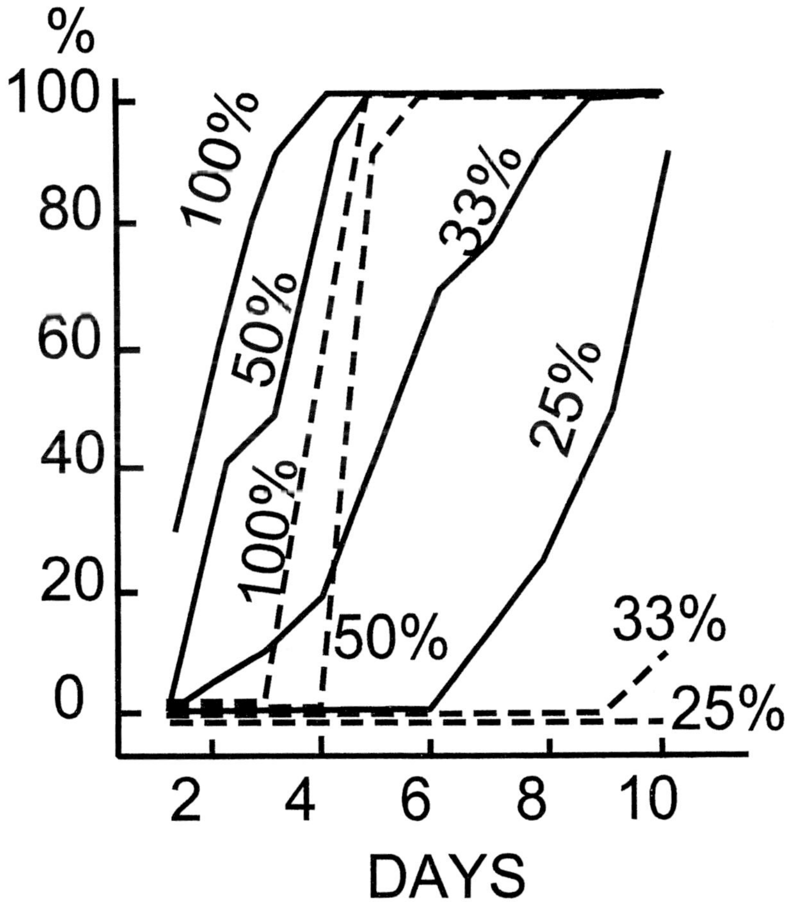

Figure 6.5. Percentage of correct CR in control and hippocampectomized rats in conditions of different probability of reinforcement.

mentioned above, with conditioned switching, the probability of reinforcement by food or tone was arranged to be high in tests in the morning and low in evening tests, while the probability of reinforcement of the same tone by noxious stimulation is reciprocal. In normal rats, ten days of training to elaborate switching of defensive and alimentary conditioned reflexes were unsuccessful. Later, the same rats after bilateral hippocampectomy ignored reinforcement of low probability, and stable conditioned switching could be formed in two weeks. Bilateral hippocampectomy not only facilitates elaboration of conditioned switching, but also eliminates signs of emotional tension in these animals, as observed by recording the heart rate.

The capacity of the hippocampus to react to signals of low-probability events allows one to consider it as a key structure for realizing the *compensatory* function of emotions, which helps to substitute for lack of information. This function is manifested not only in the hypermobilization of the vegetative shifts (increased heart rate, rise in blood pressure, release of hormones in the blood, etc.) which generally exceed the actual needs of the organism. The appearance of emotional tension is also accompanied by a switching to behaviours different from those characteristic of the calm state. In this context, it is not an accident that the first stage of formation of a conditioned reflex—the stage of generalization—was also called the "emotional stage" by the psychiatrist V. Osipov (a student of Pavlov). An emotionally aroused brain reacts to a wide range of signals presumed to be significant signals, the actual significance of which, and their correspondence or lack of correspondence to reality, will become clear only later in the process of conditioned reflex stabilization.

The behavioural, electrophysiological and neuroanatomical characteristics of the stage of generalization of the conditioned reflex display similarity to the manifestation of Ukhtomsky's dominant (Pavlygina, 1991a; Pavlygina, 1991b). The principle of the dominant was elaborated by A. Ukhtomsky based on thorough behavioural analysis of the rule of evaluation of external signals and responses to these signals. It states that certain external stimuli can create in the brain a locus of increased excitability which changes the responsiveness: Animals can respond to sound by eye-blinking if the "dominant state" has been previously created by air-puffs in the eye (Pavlygina, 1991a,b). Emotions may play a significant role in the appearance and maintenance of the dominant state in the brain.

In the process of consolidation, a conditioned reflex is accompanied by a fall in emotional tension and a simultaneous shift from a dominant (generalized) reaction to strictly selective reactions to the conditioned signal, and then the re-emergence of emotions leads to secondary generalization. Thus, the growth of emotional tension widens the variety of engrams retrievable from memory, and on the other hand, it lowers the criteria for "decision-making" when these engrams are compared with the present stimuli. The greater the anxiety, the more often the subject responds to a neutral stimulus as if it were an aversive one.

The hypothetical dominant reaction is advantageous only in conditions of pragmatic uncertainty. Information deficit is present early in conditioning, and thus there is a low estimated probability of need fulfillment (i.e. $In - Ia$ in the structural equation is low). This information deficit is substituted later by exploratory behavior, perfection of skills and the mobilization of engrams held in the memory. The compensatory significance of negative emotions consists of their *substitutional* role. As regards positive emotions, their compensatory function is realized via their effects on the need which initiates the behaviour. In a difficult situation, when there is a low probability that a goal will be achieved, even a small success (increase in probability) will generate positive emotions which strengthen the need according to a rule which follows from the stuctural formula for emotions.

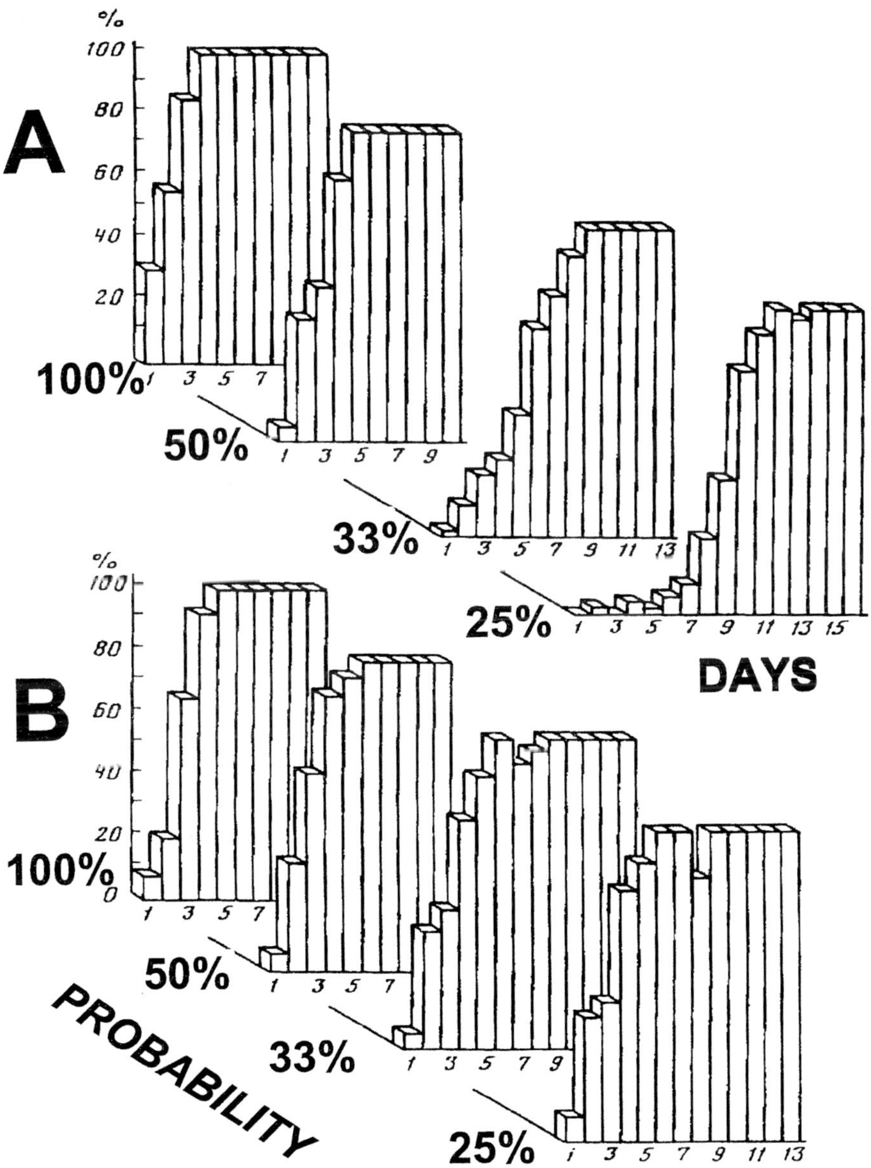

Figure 6.6. Percentage of CR in control rats (A) and after destruction of frontal cortex (B) in conditions of different reinforcement probability.

As described above, the hippocampus participates in a variety of situations with low probability events. In distinction to the hippocampus, the second "informational" brain structure—the frontal neocortex—orients behaviour towards signals of high-probability events. Figure 6.6 shows the percentage of motor alimentary conditioned reactions in intact rats in conditions of different probability of reinforcement. The abscissa shows days of tests. It can be seen that the elaboration of a conditioned reflex is retarded when the probability of reinforcement is low. Destruction of the frontal cortex eliminates the effect of a low probability of reinforcement. In other words, signals with different probabilities of being reinforced with food become equally effective after frontal cortex ablation. This result is especially interesting, since, to judge from available data, the frontal regions of the cortex in rats do not differ in their essential functions from the frontal cortex of higher vertebrates, including primates (Koeb, 1974).

4. HEMISPHERIC ASYMMETRY OF EMOTIONS

While discussing the influence of the hypothalamus on the neocortex, I have already mentioned the functional asymmetry of the left and right frontal brain regions. Analysis

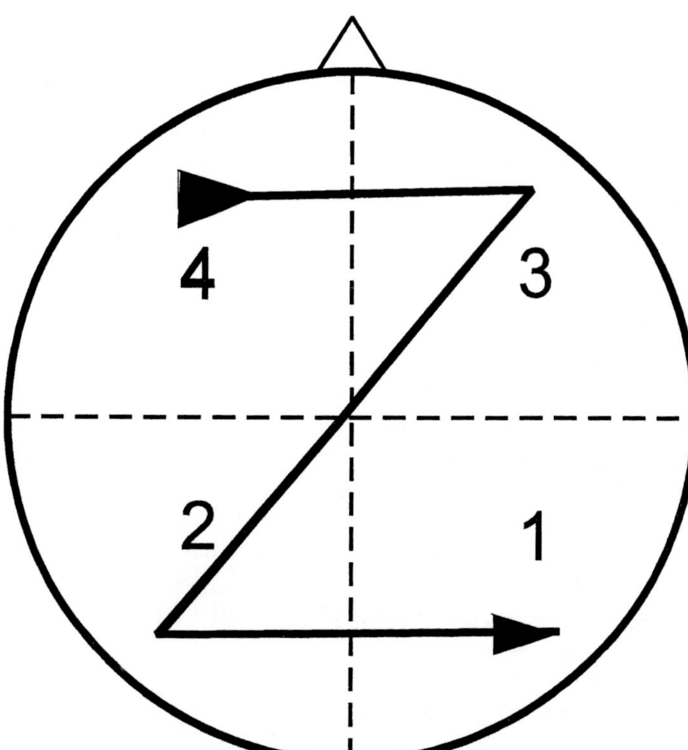

Figure 6.7. Elaboration of conditioned reflex evokes shift of maximal activation focus.

of brain electrical activity in humans, dogs and rats showed that on repeated fulfillment of the same tasks and in the process of elaboration of classical conditioned reflex, a zone of higher activation (short latency of EPs, the highest alpha-rhythm frequency, etc.) shifts from anterior parts of the left hemisphere to the anterior and then to the posterior areas of the right one (Figure 6.7). In other words novelty is a significant factor when discussing of the topic of lateralization of emotions (Simonov *et al.*, 1995).

Intensity of emotional tension, irrespective of its sign, is often related to activity of temporoparietal regions of the right hemisphere (Davidson, 1993; Heller, 1993). In contrast to connections of the left hemisphere with reticular and brainstem regions, that region has well-developed connections with diencephalic structures (Simonov, 1991). This determines the way in which emotional tension is expressed in terms of changes in vegetative functions, such as changes in the skin-galvanic reflex, heart rate, blood pressure, cortisol secretion, etc. A significant role of the temporal cortex of the right hemisphere in realizing emotional reactions is revealed in animals as well (Crowne *et al.*, 1987).

Concerning the sign (positive or negative) of emotions, Heller and Davidson suggested that it was determined by the ratio between the activities of the left (LFC) and right (RFC) regions of the frontal cortex. W. Heller (1993) presented this rule as two inequalities:

LFC > RFC—positive emotions;
RFC > LFC—negative emotions.

In the above-mentioned papers there is still no answer to the question of what determines the specificity of left and right frontal cortex in the genesis of positive and negative emotions. It would be simplistic to suggest that "centers" of corresponding emotions are localized in these two brain structures. According to the "Need-Informational Theory of Emotions", positive emotions arise when the *available* information exceeds a prognostically *necessary* information, and, conversely, negative emotions arise when that which is necessary is greater than what is available. Comparison of these inequalities with those proposed by Heller (see above) suggests that the RFC preferentially deals with pragmatic information, required for satisfaction of need, (i.e. earlier experience stored in memory), whereas the LFC processes the most recent and currently available information. Taking into account the specificity of informational (cognitive) functions, which are realized by the left and right frontal neocortex, allows one to interpret the lateralization of positive and negative emotions and the role of these brain structures in the genesis of emotional states (Simonov, 1994).

The data on functional specialization of the cerebral hemispheres in humans are of significance for understanding the neurophysiological mechanisms of creative intelligence (Simonov, 1993). In experiments by Monakhov and co-authors, subjects were shown pictures with two images, one of which was easily recognizable, and the other, disguised. Upon identification of each image, subjects were required to press a button. Electrical activity was recorded in various parts of the cerebral cortex, using a method of construction of a topographic map reflecting interactions of electrical activity in 24 recording points (Monakhov *et al.*, 1989, 1990). Maximal changes in electrical activity were observed in the left frontal cortex and in the right central region just prior to picture presentation. At the moment of identification of the first image, which usually happened two seconds after presentation of the picture, a dramatic rise in electrographic activity in the frontal cortex was observed. Seven or eight seconds prior to pressing the button (indicating identification of the second, masked image) the right frontal region became activated. This sudden peak of activity lasted for 4–5 seconds, and 1–2 seconds before pressing the button the electrograph showed a background pattern of activity. Enhanced activity in frontal, occipital and parietal lobes of the right hemisphere was observed in subjects who gave the most productive answers during solving problems that called for productive imagination.

Let me stress that we still know too little about the neurophysiological mechanisms of creative intelligence. But I might try to give a schematic representation of how various parts of the brain are involved in the realization of creative behavioural acts (one can regard this speculation as a program for further experimental studies).

In all likelihood, the nuclei of the amygdaloid complex have a significant role to play in determining a dominant motivation that initiates a search of information necessary for problem solving. The hippocampus makes it possible to expand a set of data extracted from memory and use it as material for hypothesis formation. The hypotheses thereby produced are generated and evaluated in the lobes of the right hemisphere; irrelevant versions are discarded in the process of decision making. In addition, the hippocampal neurones, the detectors of novelty, can respond both to new stimuli and to a recombination of data stored in memory.

Interaction of the amygdaloid complex, the hippocampus and the prefrontal cortex may be responsible for the phenomenon of insight, that is, sudden comprehension without use of previous experience. This also involves the caudate nucleus where neurones have been described, controlling decision making and the antecedent stages (where the final choice is delayed).

Interaction of the interior parts of the left and right cerebral hemispheres in creative persons gives rise to a "dialogue of two voices—one, imaginative, and the other—critical." The hypotheses thus conceived are selected in a conscious logical manner for subsequent empirical verification. This means that the functional asymmetry of the two cerebral hemispheres emerges as the most adequate neurobiological base of coordination between the conscious and the unconscious components of the creative process.

ACKNOWLEDGEMENT

Supported by Russian Foundation for Fundamental Research (grants No 96-04-48197 and No 96-15-97833) and Russian Humanitarian Scientific Foundation (grant No 97-06-08006).

REFERENCES

Asratyan, E.A. (1965) *Compensatory Adaptation, Reflex and the Brain Theory*. Pergamon Press, Oxford.

Crowne, D., Richardson, C. and Dawson, K. (1987) Lateralization of emotionality in right parietal cortex of the rat. *Behavioral Neuroscience*, **101**, 134–138.

Davidson, R. (1993) Parsing affective space: Perspectives from neuropsychology and psychophysiology. *Neuropsychology* **7**, 464–475.

Heller, W. (1993) Neuropsychological mechanisms of individual differences in emotion, personality and arousal. *Neuropsychology,* **7**, 476–489.

James, W. (1884) What is emotion? *Mind*, **4**, 188–205.

Koeb, B. (1974) Social behavior of rats with chronic prefrontal lesions. *Journal of Comparative and Physiological Psychology*, **87**, 466–474.

Mikhailova, N.G., Zaichenko, M.I. (1993) Hypothalamic neurones and defensive reflexes. *Zhurnal Vysshey Nervnoy Dejatelnosty*, **43**, 496–506.

Monakhov, K.K., Kulikov, M.A., Cheremushkin, E.A., Kurova, N.S., Vorobeva, T.A. (1989) A method for mapping the characteristics of the connection of electrical processes in the human and animal cerebral cortex. *Zhurnal Vysshey Nervnoy Dejatelnosty*, **39**, 1173–1176.

Monakhov, K.K., Vorobeva, T.A., Cheremushkin, E.A. (1990) The role of the extralogical form of thinking in cognitive activity. *Zhurnal Vysshey Nervnoy Dejatelnosty*, **40**, 1073–1079.

Pavlova, I.V. (1995) Interactions of rabbit's neocortical neurones in the state of hunger. *Zhurnal Vysshey Nervnoy Dejatelnosty,* **45**, 1202–1205.

Pavlygina, R.A. (1991a) The dominant and the conditioned reflex. In: *Soviet Scientific Reviews,* Section F, **5**, Part 1. Harward Academic Publishers, Cambridge, p. 37–71.

Pavlygina, R.A. (1991b) Reinforcement as discontinuation of a motivational dominant. In: *Systems Research in Physiology*, p. 51–64 Amsterdam, Gordon and Breach.

Pavlygina, R.A., Lyubimova, Yu. V. (1994a) Spectral characteristics of rabbit's brain electrical activity in the state of hunger. *Zhurnal Vysshey Nervnoy Dejatelnosty*, **44**, 57–63.

Pavlygina, R.A., Lyubimova, Yu.V. (1994b) Correlative characteristics of the brain electrical activity of rabbit in the state of hunger. *Zhurnal Vysshey Nervnoy Dejatelnosty,* **44**, 532–540.

Pavlygina, R., Trush, V., Mikhailova, N., Simonov, P. (1976) Self-stimulation by direct current as a model for studying mechanisms of motivated behavior. *Acta Neurobiologiae Experimentalis,* **36**, 725–734.

Pigareva, M.L. (1983) *Experimental neuropsychology of emotions* (in Russian). D. Sci. Thesis, Moscow.

Schmidt, R., Thews, G. (eds). (1983) *Human Physiology*. Springer-Verlag, Berlin.

Simonov, P. (1975) Parameters of action and measuring emotions. In: L. Levi (ed.) *Emotions: Their Parameters and Measurement*. Raven Press, New York, p. 421–432.

Simonov, P. (1984) The need-informational theory of emotions. *International Journal of Psychophysiology*, **1**, 277–289.

Simonov, P. (1986) *The Emotional Brain*. Plenum Press, New York—London.

Simonov, P. (1988) Interactions of brain structures in the process of behavior organization. *Neuroscience and Behavioral Physiology,* **18**, 162–169.

Simonov P V *The Motivated Brain* (1991) Gordon and Breach Science Publishers. Philadelphia. pp. 120–121.

Simonov, P.V. (1993) *Creative Brain. Neurobiological basis of creation*. (In Russian). Moscow:Nauka.

Simonov, P. (1994) Functional asymmetry of the frontal neocortex: implications for emotions. *Doklady Biological Sciences,* **338**, 416–417.

Simonov, P.V., Rusalova, M.N., Preobrazhenskaya, L.A., Vanezian, G.L. (1995) Novelty and asymmetry of brain. *Zhurnal Vysshey Nervnoy Dejatelnosty,* **45**, 13–17.

7 The Functional Significance of High-frequency Components of Brain Electrical Activity

V.N. Dumenko

Institute of Higher Nervous Activity and Neurophysiology, Russian Academy of Sciences, Moscow, Russia
balaban@ihna.msk.ru

At present, the significance of high-frequency (HF) components in brain electrical activity (EA) is discussed widely in the literature. In particular, the gamma range of the human EEG is considered to be one of the indices of cognitive activity. In this paper a review is presented of the author's own data concerning the functional assessment of HF components (both low-voltage components, and bursts of high amplitude) of the neocortical broad-band EA (1–100, 1–256 Hz) recorded in the course of instrumental conditioning in dogs. Three different techniques of EA analysis were used in the study: 1) Fast Fourier Transformation of broad-band EA. 2) An alternative method of non-harmonic expansion of EEG curves, taking their shape into account. 3) Factor Analysis of the EA spectral densities. Combination of these techniques allowed one to obtain some novel evidence indicating that HF fluctuations have a high information content. The precision of spatial localization of the HF components of EA in the cortex is much more prominent than for the traditional 1–30 Hz range. The results of factor analysis suggest functional heterogeneity of the HF band. The data obtained open new avenues for research on the phenomena of the functional mosaics, as one of fundamental mechanisms in organization of neocortical brain activity.

KEYWORDS: Neocortex, Electrical activity, High-frequency components, Methods of analysis, Alimentary instrumental conditioning, Dogs

1. INTRODUCTION

In the overwhelming majority of studies of brain electrical activity (EA) the frequency band of 1–30 Hz (from delta up to beta-3) has been studied. However, recently a number of reports have been published concerning the involvement of frequencies higher than 30 Hz (the gamma-range of the human EEG) in processes of cognitive activity (Bressler, 1990; De France and Sheer, 1988; Pulvermuller *et al.*, 1995; Ray and Cole, 1985; Sokolov, 1990, 1996). Moreover, ideas about a functional role for this range encompass not only processes of sensory integration and perception, but have even considered it as a basic condition for the formation of consciousness (Barinaga, 1990; Crick and Koch, 1990). It is necessary to emphasize that gamma-frequency activity is characterized by its low amplitude (up to 10 μV), several times below that of the dominant alpha-waves (40–50 μV). High-frequency (HF) fluctuations are commonly attributed to the combined gamma-band, regarded by many authors as a summation of noise, which masks any useful signal. The difficulties of recording the HF-components in humans through the scalp are due to the occurrence of artefacts from muscle potentials in the head. In addition, the averaging properties of the scalp make any evaluation of the localization of gamma-waves essentially more complicated.

In this context investigations of low amplitude HF-components in brain EA of animals with implanted electrodes are of interest. However, such works are comparatively rare. Technical difficulties of this approach are determined by the necessity of using sensitive broad-band amplifiers, with low noise. Physiological difficulties are connected with the need to create states in the brain in which there is an intensification of synchronization in the activity of neuronal groups in the location of the macroelectrode. These states allow one to localize low-amplitude HF fluctuations in the total EA (Dumenko, 1977, 1988).

For more than 20 years the author of the present article has been investigating the features of the neocortical EA in dogs, during the process of conditioning. The originality of the approach is determined by use of a wide-band of frequencies for analysis (1–100, 1–256 Hz).

Dogs were chosen as the experimental animals because the high-voltage slow waves, which are characteristic of the rabbit's EA (theta-rhythm: Efremova *et al.*, 1981) are usually not the dominant type of activity in the dog's neocortex. This permits the extraction of low-voltage HF components (up to 10 μV) without prior analog filtration. In our research HF-fluctuations were found in EA of the neocortex in the process of instrumental conditioning (Dumenko, 1977–1995).

Earlier work was based on the analysis of EA parameters in the course of conditioning by means of the Fast Fourier Transformation (FFT) technique (power spectra, coherence, phase shifts). From such work, it was assumed that the dynamics of HF fluctuations in interstimulus intervals, when the animals are in the state of concentrated *selective* attention in *expectation* of the conditional signal, reflects cognitive activity of animals (Dumenko, 1995).

The present review has been divided into 3 parts: HF activity of low amplitude (section 2); HF activity of high amplitude (section 3); The functional significance of HF components of the brain EA (section 4).

During the experiments each animal was implanted with 12 epidural electrodes, which were placed in various combinations over the motor, orbital, auditory, visual and parietal areas of one hemisphere, according to the atlas of Adrianov and Mering (1959). Two reference electrodes were implanted in the nasal bones of the skull. Broad-band amplifiers (1–1000 Hz) with a low level of noise (up to 2–3 μV) were used. The animals were trained to press the pedal of a feeder, with a defined effort (600–1000 g), to obtain meat reinforcement in response to a conditional signal (tone 1000 Hz). A 500 Hz tone served as a differential stimulus. This paradigm of conditioning favoured the creation in the animals of a significant emotional effort, which did not abate during the whole interstimulus interval (2–3 minutes), and was monitored using vegetative parameters (respiratory pattern, electrocardiograms). Parameters of EA during interstimulus intervals, as well as short-term (1s) EA reactions on presentation of the conditional stimuli were analyzed (Dumenko, 1985, 1992; Dumenko and Kozlov, 1989). Parameters of EA in the interstimulus intervals were evaluated by means of the FFT (power spectra, coherence, phase shifts) (Dumenko, 1985, 1995), and factor analysis of values for spectral density (Dumenko *et al.*, 1995, 1996). Moreover, we used a non-harmonic form of analysis, developed together with M.K. Kozlov (involving the expansion of the EA waves with respect to their shape) (Dumenko and Kozlov, 1993, 1995; Kozlov and Dumenko, 1990). Parameters of EA reactions to the presentation of conditional signals were evaluated with the help of the FFT technique (Dumenko and Kozlov, 1989).

2. HF-ACTIVITY OF LOW-AMPLITUDE

Noteworthy among the relatively few studies of HF-components of low voltage, are several authors who have mentioned gamma-waves (40–60 Hz) in EA of the cortex in cats and rats, but without involving learning (Boeijinga and Lopes da Silva, 1988; Boujer *et al.*, 1987; Vanderwolf and Baker, 1986). In the hippocampus of waking cats, rabbits and rats are found (in addition to the traditional theta rhythm) fluctuations at 40–50 Hz, which disappear during anaesthesia (Bragin *et al.*, 1995; Buzsaki *et al.,* 1992; Kanamori, 1985; Vanderwolf and Baker, 1986). In the opinion some of the authors, this type of EA is more sensitive to change of behaviour than is the theta rhythm.

2.1. EA In Interstimulus Intervals

2.1.1. The analysis by means of FFT

The dynamics of HF components (40–200 Hz) were studied, these being characterized by low amplitude (5–10 μV), that is, several times below that of the dominant EA fluctuations (30–40 μV) (Figure 7.1[A,B]). Because of the small amplitude of the HF activity, the main waves of EA were filtered (over the band 70–1000 Hz) with the purpose of revealing of the HF-parameters. Thus, fluctuations up to 150 Hz were observed (Figure 7.1[C]). The form of the autocorrelation function indicated the absence of periodicity (Dumenko, 1977). The dynamics of this HF-activity, in particular during presentation of extra stimuli, or during the transition from drowsiness to wakefulness (and *vice versa*) (Figure 7.1 [A,B]), including the development of nembutal sleep, testified to the physiological role of this form of activity (Dumenko, 1977, 1992).

For analysis in a band 1–100 and 1–256 Hz with a resolution 1 Hz, 4 or 6 second segments of EA were used (the sampling frequency being 512 Hz), taken in the middle of 2–3 minutes of the interstimulus interval. The power of EA fluctuations was evaluated on a logarithmic scale, which ensured that comparison of data could make strong distinctions in the power level at different frequencies. Two explicitly different states were compared—the state of quiet wakefulness before the beginning of conditioning, and active wakefulness against a background of a stable motor skill (shown in 90–95% of manifestations of conditioned reflexes), together with an absence of motor reactions to differential stimuli, or of interstimulus pressings on the pedal.

Autospectra (ASP) of the various brain areas of one animal, shown in Figure 7.2, demonstrate that maximal values correspond to the traditional 1–30 Hz band, and the minimal ones to the HF fluctuations (70–100 Hz). After elaboration of a motor skill, the power of the HF fluctuations increased, the peaks became more prominent, and the power of the traditional frequency band was lowered, to a greater or lesser extent. These

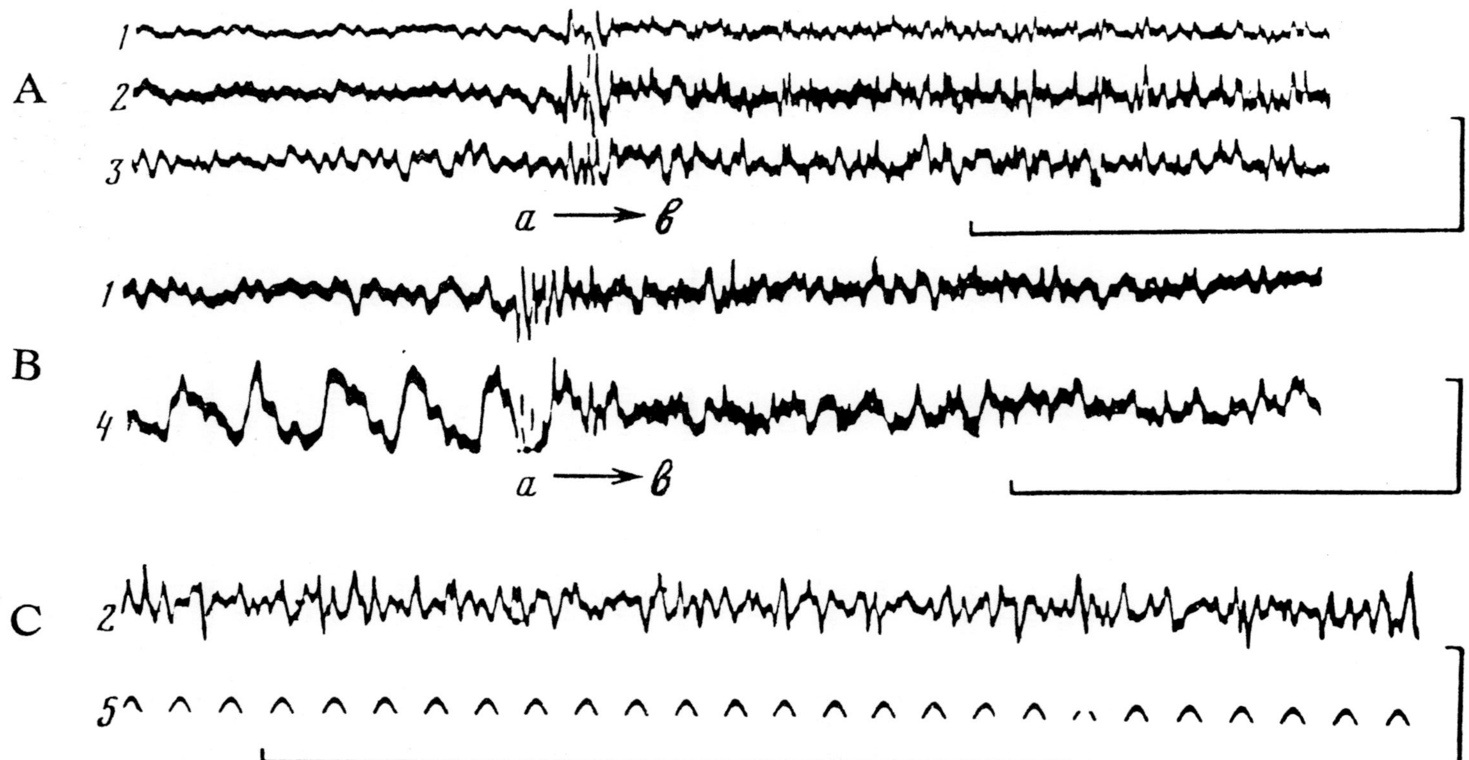

Figure 7.1. Examples (**A, B**) of the appearance of low-amplitude high-frequency activity in the time of transition from drowsiness (a) to waking (b). Areas: **1**—motor, **2**—orbital, **3**—auditory, **4**—visual, **5—**50 Hz. **C**—high-frequency components after analog filtration (band 70–1000 Hz). Time 0.5 sec, calibration—50 μV.

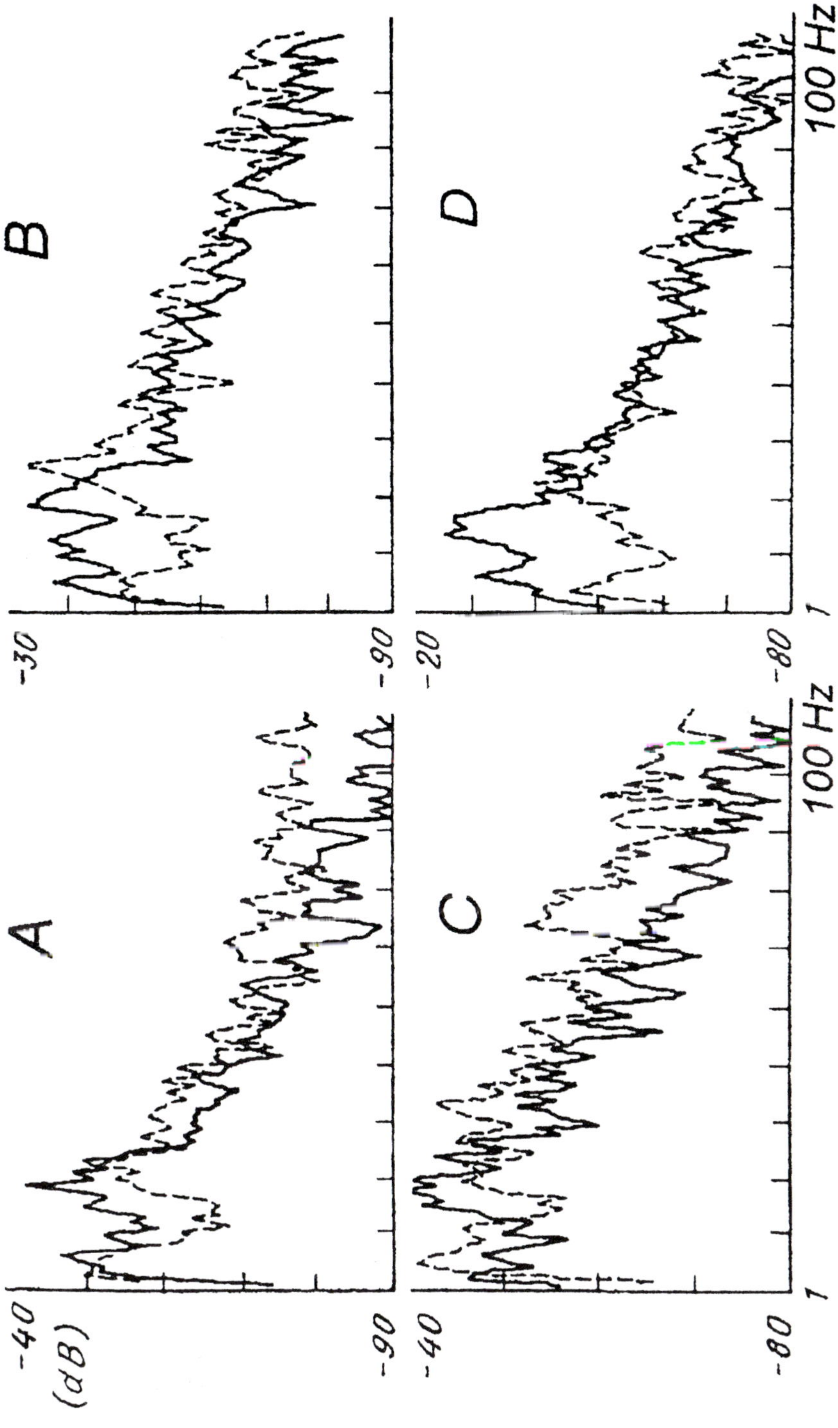

Figure 7.2. Autospectra of cortical electrical activity of various areas in a dog before (solid line) and after (dashed line) conditioning. **Abscissa:** frequency 1–100 Hz (resolution 1 Hz); **ordinate:** level of power (logarithmic scale, dB). The cortical areas: **A**—motor, **B**—orbital, **C**—auditory, **D**—visual.

Table 7.1. Distribution of minima (A) and maxima (B) of autospectra of electrical activity before the beginning (experiments 2–3) and after (experiment 33) conditioning in one of the animals

Experiments	*Sample size (n)*	*A*				*B*				
		1	*2*	*3*	*4*	*5*	*6*	*7*	*8*	*9*
2–3	84	20	***31***	18	15	12	19	***26***	19	8
33	120	–	46	***66***	8	***44***	36	27	13	–

In the Table numbers 1–9 correspond to power levels from –100 up to –20 dB (steps of 10 dB). Modes of distributions are italicized (bold).

tendencies were observed in a number of experiments, as assessed using the the minimal (Table 7.1 [A]) and maximal (Table 7.1[B]) values of ASP.

On averaging the ASP are smoothed to a large extent. Therefore, we have shifted from estimating the average values to plotting histograms of the distributions of 8 allocated levels of power, for bands in 20 Hz width from each derivation (Dumenko, 1985, 1988). The various distributions were compared on the basis of the chi-squared test (Dumenko, 1992).

During the stage of elaborating the skill, the power of HF fluctuations in a band 40–100 Hz increased by 10–20 dB ($p < 0.01$); in a number of areas the power of fluctuations in the 1–30 Hz band was reduced or stayed at its previous level (Figure 7.3). Observed changes were also evident at higher resolution, by plotting histograms with frequency-bins of 4 Hz instead of 20 Hz (Dumenko, 1995a). On broadening the band for analysis up to 256 Hz we did not generally observe any increase of the power level of HF fluctuations above 170 Hz (Dumenko, 1992).

There is a very characteristic behaviour of the animals corresponding to the dynamics of EA in the interstimulus intervals when a stable skill has been aquired. During the whole interval (2–3 minutes) animals were in a state of *selective attention* directed to feeder, *expecting* presentation of the conditional stimulus. This state was so dominantly expressed, that it was not usually eliminated by the presentation of external stimuli (Dumenko, 1992).

Histograms for distributions of values of coherence and phase shift between potentials were plotted for EA from pairs of areas (Figure 7.4). At the stage of attainment of a stable skill high levels of coherence (0.7 and above) occurred. The larger share of this measure was due to the HF-components, within the frequency range 40–100 Hz (when analysing the 1–100 Hz band; Figure 7.5), and within the frequency range of 40–170 Hz (when analyzing the 1–200 Hz band), although their powers were several times lower than the power of the dominant fluctuations in the 1–30 Hz band. In parallel with increases of high coherence we observed a significant lowering for the values of phase shift between potentials of the areas compared (Figure 7.4 [A,B]). This was displayed as an increased share of the small phase shifts (±15°) (Dumenko, 1992).

The completion of the learning process was accompanied by a significant ($p < 0.05$) decrease in coefficients of variation of the spectra power and of coherence. Reduction of variability of these parameters probably reflects the formation of a steadier system of the interregional relations (Dumenko, 1992).

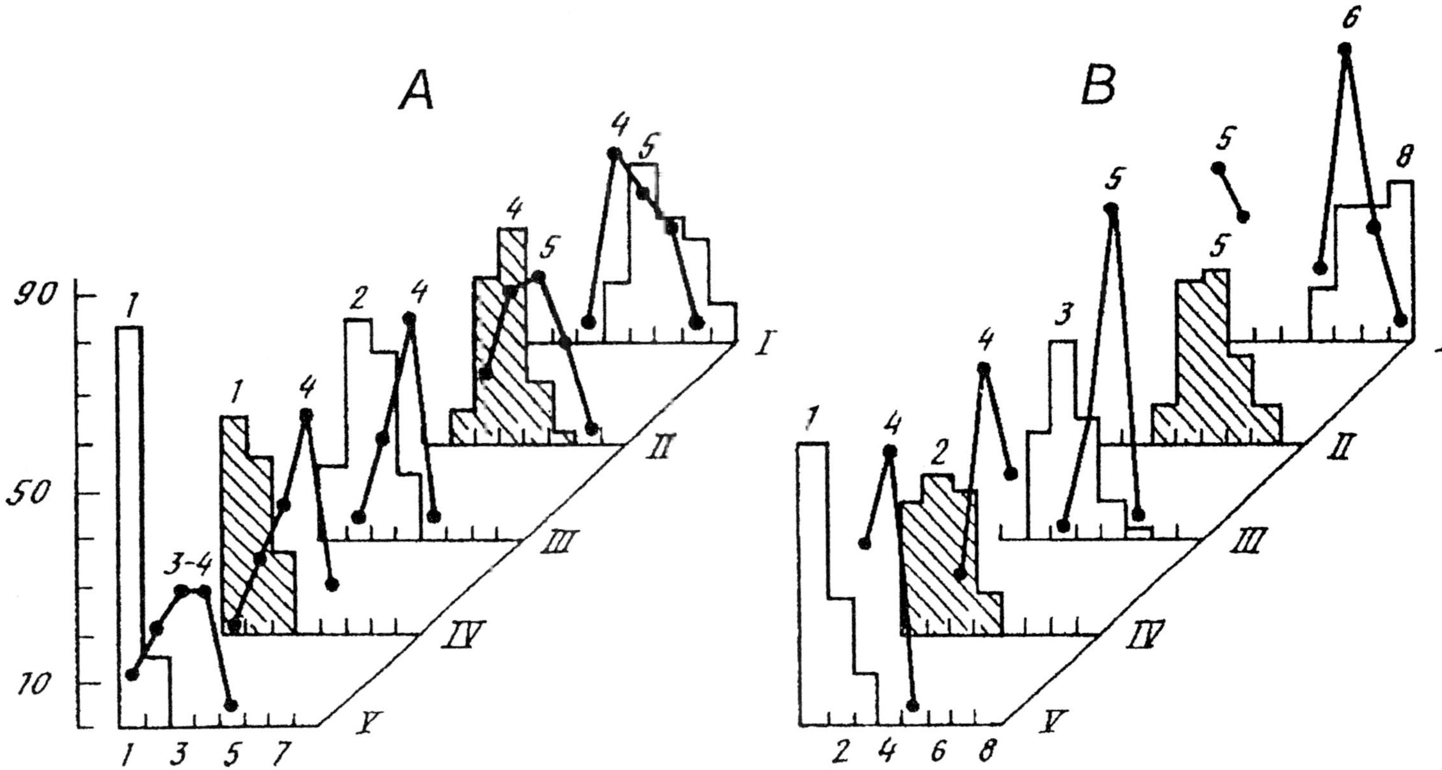

Figure 7.3. Distributions of eight power levels for each of 20-Hz sub-ranges (I–V) of the EA of motor **(A)** and auditory **(B)** areas before (histograms, n = 144) and after (variation curves, n = 280) conditioning in one of the dogs. **Abscissa:** power levels (logarithmic scale, from –100 to –30 dB with step 10 dB); **ordinate:** number of cases (%); **along Z-axis**—sub-ranges of frequencies: **I**—1–20, **II**—21–40, **III**—41–60, **IV**—61–80, **V**—81–100 Hz. Numbers above the distributions—power levels for the modes.

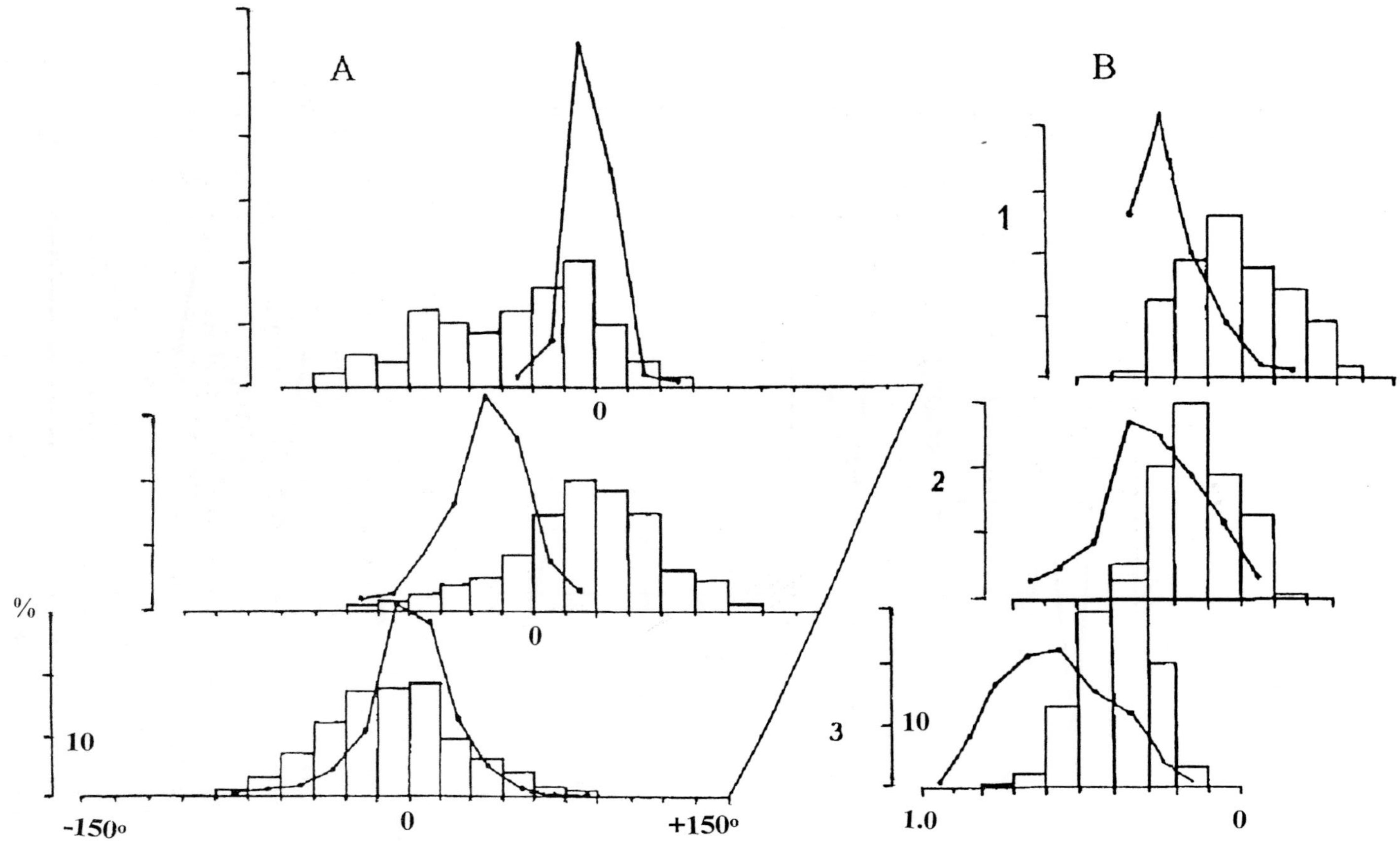

Figure 7.4. Distribution of values of phase shifts **(A)** and coherence **(B)** for three pairs of the areas **(1,2,3)** before the beginning of conditioning (histograms) and after conditioning of a stable motor skill (variation curve) in a dog. **Abscissa:** values of phase shifts (A, step 15°) and coherence (B, step 0.1; increasing from right to left); **ordinate:** number of cases (%). Pairs of the areas: **1**—visual-parietal, **2**—motor-parietal, **3**—motor-visual; n = 18 spectra (or 1800 values for each distribution).

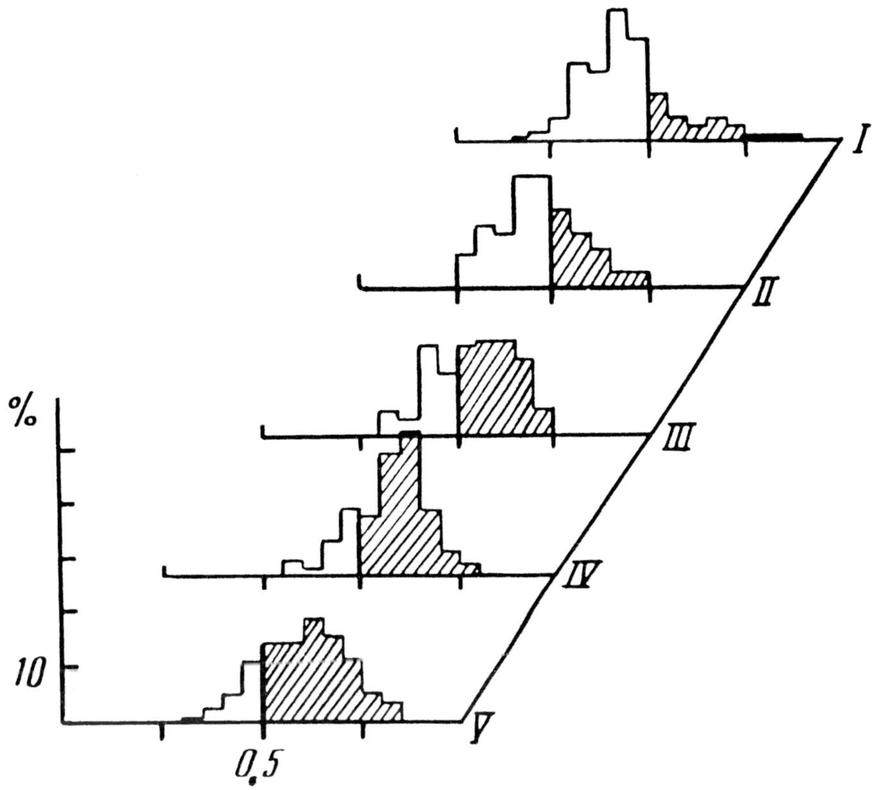

Figure 7.5. Distribution of values of coherence between potentials of the motor and the auditory areas for each of 20-Hz sub-ranges (I–V) in the 1–100 Hz band, in a dog at the stage of attainment of a stable skill. **Abscissa:** values of coherence (0–1.0); **ordinate:** number of cases (%); **along Z-axis**—sub-ranges of frequencies (as in Figure 3). n = 26 coherence spectra (or 520 values for each distribution).

Thus, in the course of learning, the HF components appear to be the most dynamic part of the spectra. Their parameters are changed to a considerably greater degree than the parameters of the more powerful fluctuations in the 1–30 Hz band. This suggests that the HF components have a high information content. Another feature of the HF fluctuations consists of a more local representation of these fluctuations in the neocortex, in comparison with frequencies in the 1–30 Hz band. Large distinctions in power, coherence and phase between EA of neighbouring points of the cortex (3–5 mm apart) testify to this (Dumenko, 1988, 1992). In the process of elaboration of a motor skill, the power level of the HF fluctuations significantly increases for the whole interstimulus interval. In addition, the share of high coherence in a range 40–170 Hz increases as the small phase shifts become dominant (the phenomenon of "*synphase HF-fluctuations*").

The dynamics specified above are not observed globally in the whole HF band: high values of peaks, in relatively narrow bands, are identified in autospectra, as well as in coherence-phase spectra, indicating functional heterogeneity of the HF band. These results have allowed us to draw conclusions concerning the formation of relatively narrow localized sub-bands of frequencies, during the course of learning. On the basis of such evidence subsystems defined by coherence and phase are formed, which in total create an integrated spatio-temporal organization of brain potentials, characteristic of a given learning paradigm (Dumenko, 1992).

Difficulties of evaluating the features which spectra have in common, or which distinguish them (including the above mentioned phenomenon of "narrow sub-bands") arose from their natural variability, and the inevitable smoothing during averaging. These difficulties were aggravated by possible inhomogeneities of in the gamma-frequency band. In order to obtain reliable proof of the functional heterogeneity of the HF band we have employed one of the methods of multivariate statistics—factor analysis (FA).

2.1.2. Studies of EA by means of the FA-technique

The FA models are directed to "compression" of information, creating a certain idea concerning the internal structures of the investigated phenomenon. For FA of spectral density values in the bands 1–100 and 1–200 Hz we used a standard method of principal components analysis, with subsequent rotation of the factors (Dumenko, *et al.*, 1995a, 1996) provided by the programs "Statistics for Windows" (resolution 1 Hz). Based on the total variance values, a number of factors were chosen which explained about 80% of the total variance values of spectral density. Our data show that this part of the total variance depended on 20 factors, but the first 5 factors accounted for most (up to 60%) of the variance. It was characteristic that, after a rotation procedure, this part decreased especially prominently for the HF band. This emphasises both the efficiently of the rotation procedure, and the necessity in these conditions to use a considerably greater number of factors (up to 20).

In this review, the frequencies of EA identified as having high factor loads (0.6 and above), are presented. Each of these frequencies correlated with high values of loads of one factor, and with low ones of others. Results are not given for the not-infrequent cases of factor "splitting", when high loads of the same factor fall on non-adjacent frequencies ("splitting" of factors). Cases, in which allocated frequencies had similar loading values on several factors, were also not taken into account (as being "unstable" factors). In the paper those groups of frequencies are presented which have been evaluated as comparatively narrow peaks (as a rule, 2–3 points) giving high loadings on only one factor ("stable" factors).

By the use of the frequency band 1–100 Hz the largest part of the variance (30–40%) fell on the first factor in all cases. Its high loadings (0.7–0.9) were distributed monotonically in the HF range (40–100 Hz), although without prominent peaks. Therefore in estimation of the factor structure of EA, it was not taken into account.

High loadings of other factors, which explained essentially smaller parts of the total variance (9–1%), defined values of spectral density with relatively narrow peaks (Figure 7.6[A]). High loads of these factors were distributed in the frequency bands 8–24 Hz. Thus, in the EA of a number of areas, it was possible to make a definite division into two or three factors: 8–10 (alpha), 13–15 and 20–24 Hz (the dominant frequency in the EA of some areas in awake dogs). In addition, a low-frequency (1–5 Hz) factor was always defineable. It is interesting that a frequency of 40–42 Hz also was often observed.

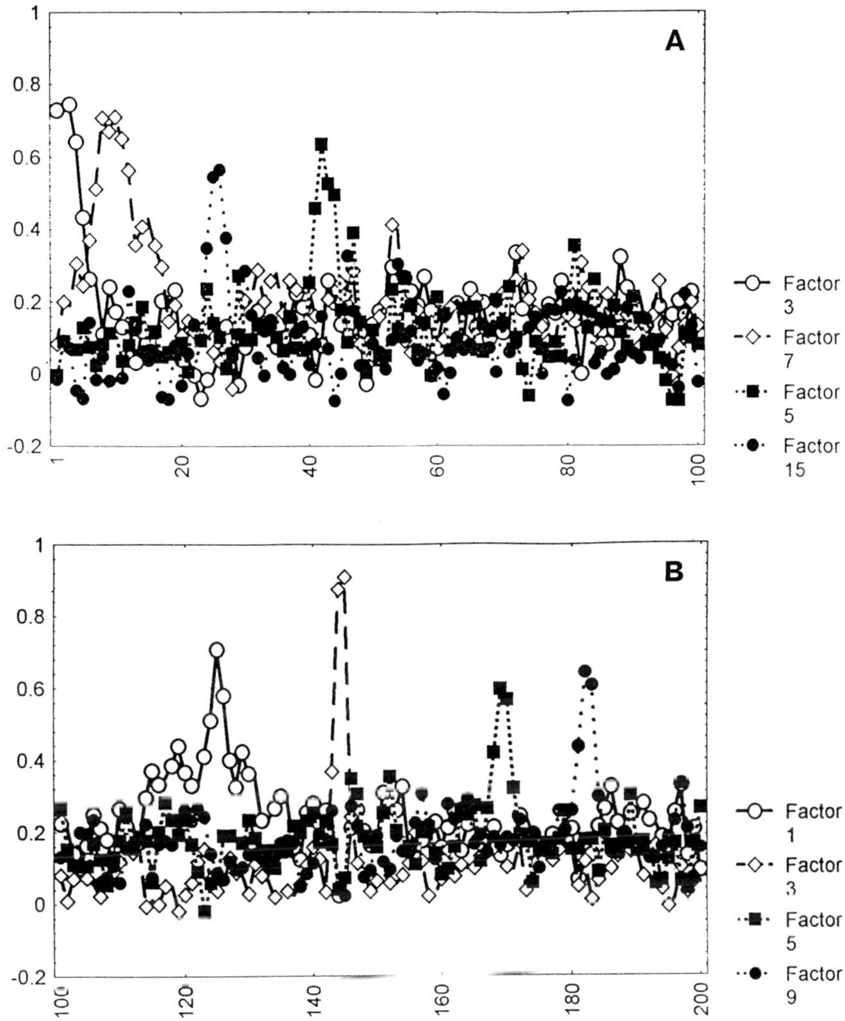

Figure 7.6. Distribution of the high loading on factors of frequencies in electrical activity of the motor area in a dog. **Abscissa:** frequencies (A: 1–100 Hz band, B: 101–200 Hz band); **ordinate:** values of factor loads.

As an exception to the analysis of the most powerful frequencies of the traditional range of 1–30 Hz (i.e. in the bands 41–140 and 101–200 Hz) the factor structure of the HF-part of spectra emerges more distinctly (Figure 7.6[B]). It is described by a larger number of peaks of high factor loads, falling within this range.

At the stage of attainment of a stable motor skill, the factor structure of EA during interstimulus intervals was essentially changed. First of all, the number of factors was increased (Dumenko *et al.*, 1995a). Thus, before the beginning of conditioning, 5 factors were derived from the EA of the motor, orbital and visual areas, for three frequency bands (1–100, 41–140 and 101–200 Hz), and these explained from 62 to 67% of the total

Table 7.2. Proportion of total variance (%) which is explained by the first 5 factors in EA of different areas, for three frequency bands (1–100, 41–140, 101–200 Hz) before (to the left) and after (to the right) conditioning in a dog

Areas	*Parts of total variance, (%)*		
	1–100 Hz	*41–140 Hz*	*101–200 Hz*
The motor	63–42	67–48	66–47
The orbital	62–42	66–52	65–50
The visual	64–45	67–53	65–50

variance. After training these relations changed to 42–53% (Table 7.2). 10 factors were derived, which explained up to 70% of the total variance, and only 20 factors explained up to 80%. At this stage, the number of "stable" factors increased, but the number of "unstable" and "splitting" factors was reduced.

The increase in the number of the "stable" factors was expressed by their being allocated to narrower frequency sub-bands. In the majority of cases, each of them had a high loading (higher than 0.6), confined to a single factor, and a low one on other factors. This permited us to achieve an unequivocal identification of frequencies for each factor, and to speak about some frequency patterns in the factor structure of EA in various areas. By comparison of the factors characteristic of EA with a superposition of appropriate autospectra, it was shown that the analytical opportunities of the FA-technique were essentially greater than those of spectral analysis (Dumenko *et al.,* 1995a).

A clear-cut increase of the regional features of EA was observed on completion of training. In Figure 7.7, autospectra of EA of the five different areas in the 1–200 Hz band are located in a two-dimensional plot of the two most significant factors. It is of great interest to see the pronounced clustering of the vectors, corresponding to the areas analyzed in the state of focused attention while expecting the conditional stimuli. Such a functional mosaic is a result of learning, since before the beginning of training the appropriate vectors appeared markedly intermingled (Dumenko *et al.*, 1996).

Thus, functional inhomogeneities in the HF range are revealed with the help of the FA technique. Multi-factor determination of EA is not restricted to the traditional 1–30 Hz band, but prominently involves HF components. Moreover, the HF band turned out to be the most dynamic part of spectrum, it is functionally inhomogenous, and is determined by combinations of a number of the separate factors. Divisions of the HF range into separate, relatively isolated narrow sub-bands is shown, the basis of whose formation were independent factors, which obviously indicate the possibility that separate sources exist for each of them. These analyses confirm completely the data received earlier with the help of the FFT technique, and testify to the formation of a complex mosaic of EA, formed by variously located HF components, with specific parameters (Dumenko *et al.*, 1996).

2.1.3. Study of EA with the help of the non-harmonic analysis

The large data sets, obtained using the FFT technique created confidence in the functional importance of the HF components (Dumenko, 1992, 1995b). However, by using a harmonic FFT for their evaluation, one could not completely eliminate the possibility of

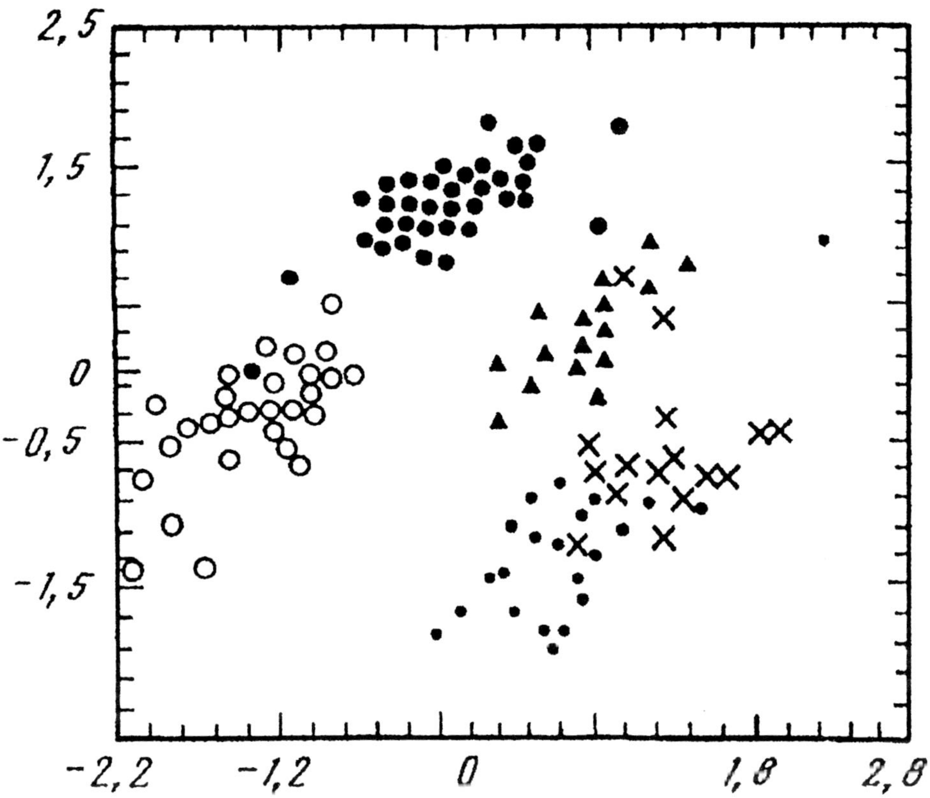

Figure 7.7. Distribution of the data on a 2-dimensional plot of scores of the 1st (abscissa) and 2nd (ordinate) factors for autospectra (1–200 Hz) of electrical activity from five brain areas in a dog during the stage of the stable conditioned reflexes. Areas: motor—open circles; olfactory bulb—closed circles; visual—triangles; orbital—squares; auditory—points; n = 120 autospectra.

mathematical artefacts, in the form of superharmonics, in so far as the shape of EA-fluctuations differs from that of a sinusoidal wave. The non-traditional method of "expansion" of EA-fluctuations, in a system of half-waves, suggested by M.K. Kozlov (Kozlov and Dumenko, 1990), takes into account the shape of fluctuations in a way which, to some degree, eliminates one of the restrictions of FFT. From the parameters of designated half-waves, this method provided an opportunity to construct an exact restoration of the analyzed curve (Dumenko and Kozlov, 1993, 1995, 1997). On the basis of appropriate parameters for these designated half-waves it was possible to construct their distributions in an amplitude-frequency plane, in the form of point and density maps, as well as histograms in three-dimensional space, in the form of surfaces (relief maps) (Figure 7.8[A]). These distributions permitted one to reveal distinctions between the relief maps in different areas, especially between closely neighbouring points (separation of 3–5 mm) within the limits of the motor area (Figure 7.8 [1,2]), that reflected a comparatively local representation of fluctuations. After training (Figure 7.8[B]) the HF-components were

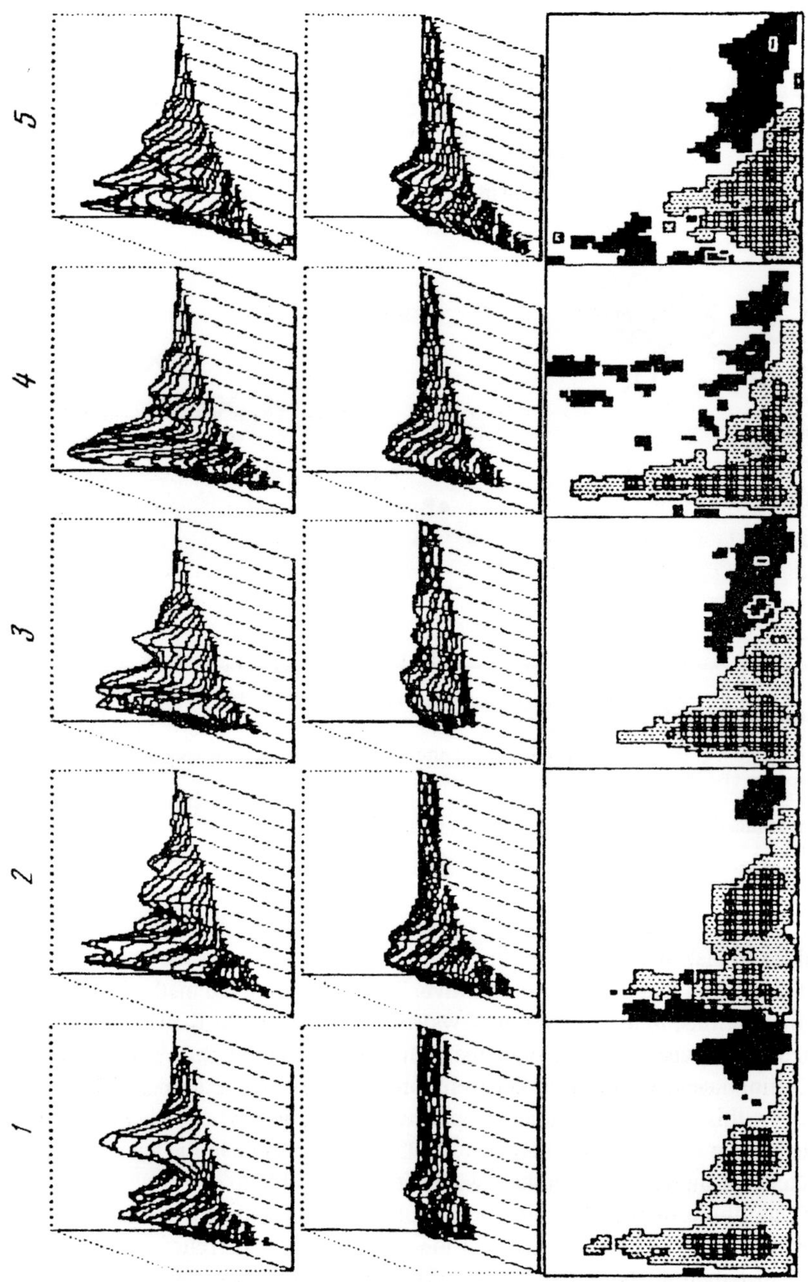

Figure 7.8. Relief maps of the densities of amplitude-frequency distributions for electrograms of five cortical regions in a dog before the beginning (**A**), and after (**B**) conditioning; and maps of the difference in the densities for the corresponding regions (**C**). **Abscissa:** frequencies from 1–200 Hz; **ordinate:** amplitude (arbitrary units); **along the vertical:** summary amplitude of all fluctuations falling in the same cell of the matrix. In maps of the differences in densities (C = A–B) the black colour corresponds to negative values (i.e. elements of B); positive values (two levels) corresponding to the difference (i.e. elements of A) are enclosed by a contour. Areas: **1,2**—two neighboring points of the motor; **3**—auditory; **4**—orbital; **5**—visual.

intensified in all areas, this being seen especially clearly in differences of the relief maps (Figure 7.8[C]) (Dumenko and Kozlov, 1995, 1997).

Therefore, the results based on the non-harmonic form of analysis, testify that the HF components are intensified during the course of conditioning, which completely confirms our data obtained by means of the FFT technique. The confirmation of results, obtained by alternative methods of analysis has allowed us to draw conclusions on *the real existence* in the brain EA of low-amplitude HF-components, which are not dependent on the more powerful fluctuations in the range 1–30 Hz. These results, obtained for the first time, can be considered as important confirmations of the functional significance of HF components in brain activity.

2.2. Short-Term (1s) EA-Reactions to the Conditional Signal

2.2.1. Analysis by means of the FFT-technique

At the stage of attainment of a stable motor skill, a clear-cut pattern of EA reactions is observed, in the form of powerful high-frequency synchronized activity on presentation of the conditional stimulus. This preceded the conditioned pressing of the pedal (Dumenko and Kozlov, 1989). The amplitude of these reactions exceeds by several times the dominant fluctuations of EA in the interstimulus intervals. In the majority of cases these reactions preceded by 40–300 ms the initial changes in EMG for the paw pressing of the pedal. As a result of analysis by means of the FFT technique, distinctions were found between the parameters of EA reactions to stimulus presentations and those of the EA in interstimulus intervals. These distinctions consisted mainly of the fact that, for the EA reactions, the share of larger phase shifts (±30–90°) between potentials of various areas, was increased against a background of high coherence (Figure 7.9). In other words, the phenomenon of *"synphase HF-components"*, which were dominant in the interstimulus intervals, was replaced in response to the conditional signal by the phenomenon of *"out-of-phase HF-components"* (Dumenko, 1992). On the basis of these distinctions the conclusion about the difference of these states and of mechanisms forming them, is justified (Dumenko, 1995).

3. HIGH-FREQUENCY ACTIVITY OF HIGH AMPLITUDE

HF components of the EA whose amplitude substantially exceeds that of the dominant background fluctuations, very often have a burst structure. It has been shown in the laboratory of W. Freeman that, in different species of animals, regular bursts of HF components in the range 35–85 Hz dominated in the EA of the olfactory part of the brain. The amplitude of these bursts increased during motivated behaviour (Bressler and Freeman, 1980; Freeman and Schneider, 1982).

In the EA of the hippocampus in cats, the occurrence of spindles, consisting of waves with frequency of 85–135 Hz was noticed by some authors (Buzsaki *et al.*, 1992; Kanamori, 1985). In awake rats, EA spindles in the hippocampus were recorded which consisted of waves of frequency 125–250 Hz (Suzuki and Smith, 1988; Bragin *et al.*, 1995; Ylinen *et al.*, 1995). The authors consider this form of EA to be a reflection of highly synchronized burst discharges of a large number of pyramidal cells.

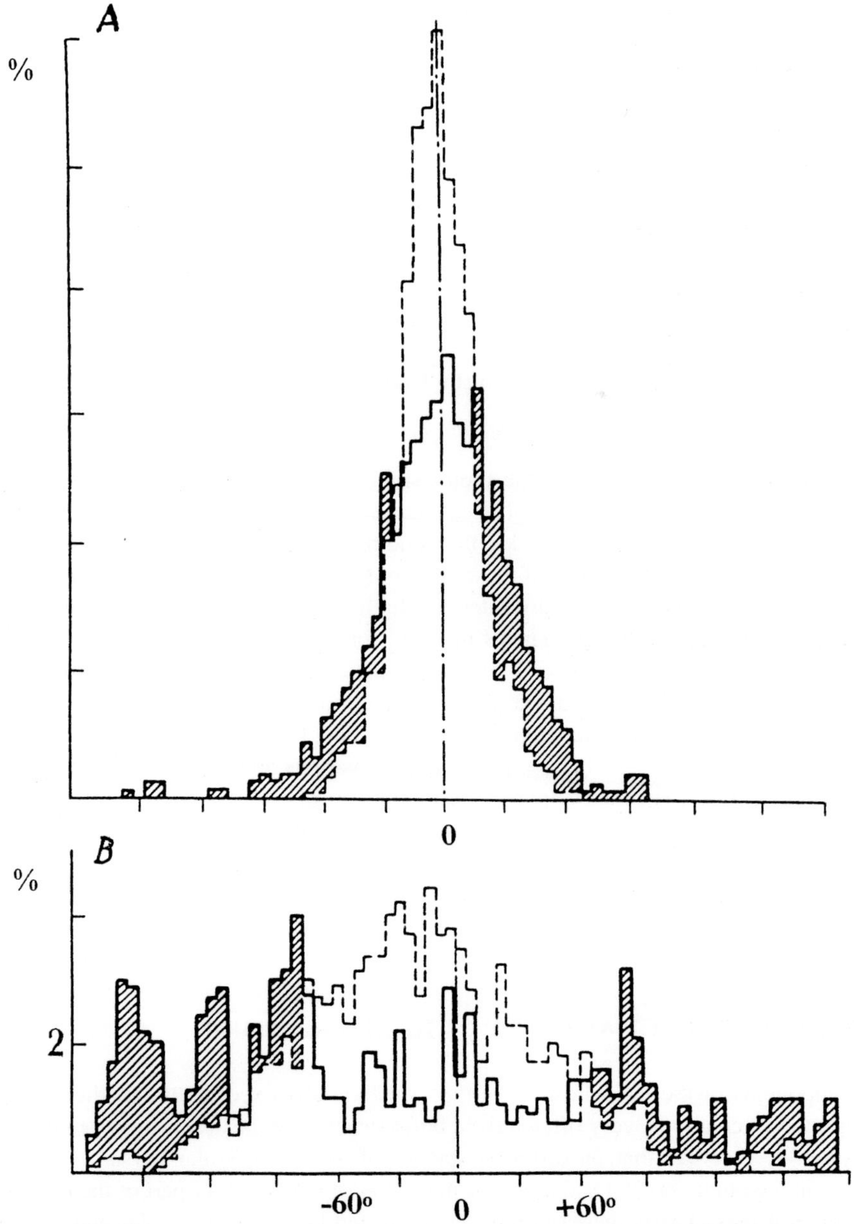

Figure 7.9. Examples of distributions of phase shift values between potentials of the motor and the orbital areas **(A)**, the motor area and the olfactory bulb **(B)** for interstimulus intervals (dotted line) and during reactions to presentation of conditional stimuli (solid line). **Abscissa:** values of phase shifts (steps of 5°); **ordinate:** number of cases (%). n = 11 phase spectra (or 1100 values for each distribution).

Such a type of EA is not characteristic of the neocortex. However, comparatively recently, in the visual cortex of anaesthetized cats, synchronized fluctuations 40–90 Hz in response to light stimuli have been observed (Singer, 1993). In awake monkeys 60–90 Hz fluctuations of high amplitude were observed in the visual cortical area in response to a light stimulus (Eckhorn *et al.,* 1993).

Previously we described patterns of HF bursts, which occurred against a background of HF components of low amplitude in the neocortical EA in dogs in the course of instrumental conditioning (Dumenko, 1977; Dumenko and Kozlov, 1993). As a rule, these patterns were characteristic of interstimulus intervals for unstable conditioned reflexes (the stage of generalization). Very often they preceded arbitrary motor reactions of pressing the feeder pedal, without presentation of the conditional stimuli. The frequency of fluctuations in the burst was from 70–80 up to 150–170 Hz. They were characterized by a rather complex structure. By using the method described above, of "expansion" of EA fluctuations in half-wave systems (see Section 2.1.3.), we showed that the HF range of bursts lacks uniformity (Dumenko and Kozlov, 1997).

4. FUNCTIONAL SIGNIFICANCE OF THE HF-COMPONENTS OF EA

The parameters of EA during interstimulus intervals in the course of training corresponded to characteristic behaviour of the animals; during the interstimulus interval of 2–3 minutes they were in a state of *selective* attention and *expectation* of the conditioned signal. This tonic state, noted repeatedly by many earlier physiologists as the factor determining the character of forthcoming behaviour (Anokhin, 1968; Asratyan, 1963; Bernstein, 1966; Kostandov, 1983), can be considered as a sort of "internal image" of situations (Freeman and Schneider, 1982; Freeman, 1991; Sokolov, 1996), which contains "knowledge", accumulated across all action sequences (Dumenko, 1995). This "knowledge" *ensures preparation* for the realization of goal-directed motor actions—e.g. to press the pedal with a certain force. The "images" are reflected in patterns of EA, which represent specialized spatio-temporal organization of potentials, constructed, according to our data, from narrow frequency sub-bands (Dumenko, 1992, 1995; Dumenko *et al.,* 1995a, 1996).

HF fluctuations appeared to be the part of the spectrum which reveals the most dynamic changes in the course of training (Dumenko, 1992). The HF band is functionally inhomogeneous, and is divided into a number of separate frequency sub-bands (Dumenko *et al.*, 1995a, 1996). An increase in the number of significant factors during training permits one to suggest the incorporation of some new sources of neurophysiological activity during learning. Each of them probably reflects highly synchronized activity of neuronal elements involved in EA generation in narrow frequency sub-bands.

The results presented support the view that the background EA in interstimulus intervals is not passive physiological noise, and plays an important role, "indicating the order in one's own house" (Dumenko, 1977, 1992). Amongst the many conditions determining this order, internal processes of stabilization of neuronal systems are very important. These processes are expressed in the form of a change in the pattern of impulses, observable on training (Dumenko and Sachenko, 1980; Livanov, 1979; Gasanov, 1991; and others). It is important to note that at present the processes of spatial synchronization are considered as mechanisms of co-operative activity of neurones in networks, confirming the known concept formulated considerably earlier by M.N. Livanov (1977). The dynamics of spatio-temporal characteristics of HF fluctuations in the course of learning reflects an intensification of such synchronization of activity of neuronal groups. Investigation of the HF components allowed us to expand the concept of spatial synchronization (Dumenko, 1977; Livanov and Dumenko, 1988), which was originally developed on the basis of studies of a frequency range 1–20 Hz (Livanov, 1977).

Use of various methods of analysis has allowed us to show elements of EA organization at the level of HF fluctuations. This was expressed not only in regional features of brain EA, but also in the differences of power and coherence of HF components in closely neighbouring points (within 3–5 mm of each other) in the neocortex. In the state of quiet wakefulness before conditioning the phase shifts between these potentials regularly increased with frequency, whereas in the 1–30 Hz band, phase shifts are practically identical (Dumenko, 1992). These distinctions of HF parameters in near points indicate the neuronal nature of HF components, especially the low-amplitude ones.

The neuronal origin of HF bursts of high amplitude is supported by convincing evidence obtained by other authors, who described the bursts in EA of the olfactory system and the hippocampus (Bragin *et al.*, 1995; Buzsaki *et al.,* 1992; Eeckman and Freeman, 1990; Ogawa and Motokisawa, 1990; Suzuki and Smith, 1988; Ylinen *et al.*, 1995).

Local representation in the neocortex of the HF components testifies, probably, to their sources having a variety of localizations. In this connection the possibility of studying the participation of HF band in functional mosaics in the brain becomes obvious.

Results concerning the endogenous nature of HF pacemaker potentials (40–200 Hz) support a neuronal origin for the HF fluctuations. These pacemaker potentials were found in cultures of frontal cortex slices (Llinas *et al.,* 1991), in neurones of the olfactory bulb (Eeckman and Freeman, 1990; Ogawa and Motokisawa, 1990), as well as in cortical neurones of the visual area (Singer, 1993) and neurones of the hippocampal structures (Buzsaki *et al.*, 1992; Ylinen *et al.*, 1995). A number of researchers (Bressler, 1990; Sokolov, 1996) consider these data as the basis for assigning to the gamma-frequencies a specific functional role in cognitive activity of humans.

Data from W. Singer's laboratory, concerning synchronization of impulse activity for neurones in the visual area in the course of perception of the whole objects (Singer, 1993), permit one to consider the phenomenon of "synphase HF fluctuations" not only as an electrographical correlate of a "cognitive image", but as an essential mechanism for formation of "integrity". Complex systems of interaction between various brain structures provide the basis for formation of an "image". Appropriate patterns of "image" differ in principle from general arousal reactions, and represent a specialized spatio-temporal organization of potentials, formed in relatively narrow frequency sub-bands (Dumenko, 1992, 1995).

On the basis of distinctions between the patterns of HF-fluctuations in the interstimulus intervals and those of EA reactions to the conditional signal it is possible to draw a conclusion about the distinguishing features of these states. If the phenomenon of "*synphase of HF components*" is a reflection of a *cognitive image*, which ensures the internal visualization of a particular situation, the phenomenon "*out-of-phase HF components*" can be considered as a sort of the "*motor program*" ("efferent synthesis" according to P.K. Anokhin) (Anokhin, 1968).

5. CONCLUSIONS

Using various methods of analysis of neocortical EA, we have described a number of new parameters, reflecting dynamics of low-amplitude (up to 10 μV) and high-amplitude HF fluctuations (40–170 Hz) during instrumental conditioning in dogs. By comparison of results obtained by means of FFT and FA techniques, the functional inhomogeneity of the HF band has been revealed. The subdivision of this band, as a result of training, into a number of comparatively narrow frequency sub-bands which often differ in various areas, probably underlies the formation of system of interregional relations, characteristic for a given paradigm of learning.

Comparison of harmonic FFT and the alternative to it—a non-harmonic form of "expansion" of EA fluctuations acording to their shape—has allowed us to draw the conclusion that the phenomena of HF components do not depend on the more powerful fluctuations in the 1–30 Hz band. Both methods demonstrate increases of the HF components in the course of conditioning, their larger information capacity, and their more local representation in the cortex, in comparison with fluctuations in the 1–30 Hz band. These conclusions apply not only to various areas of the neocortex, but also to closely-neighbouring points within the limits of one area (Dumenko, 1992; Dumenko and Kozlov, 1995, 1997). The data obtained testify to the formation of a complex mosaic pattern of EA, which is characteristic of the model of conditioning employed. This principle obviously underlies the formation of neurophysiological mechanisms responsible for processes of perception and learning. The HF range opens up wide prospects for studies of functional mosaics, with considerably higher resolution than is possible in the traditional 1–30 Hz band.

Features of the HF components have allowed us to expand essentially the opportunities for their functional evaluation (Dumenko, 1995b). The phenomenon of "synphase HF components" in interstimulus intervals is a reflection of the cognitive activity of animals, an electrographic correlate of the "internal image", ensuring *preparation* for forthcoming activity. The phenomenon of "out-of-phase HF components" of EA reactions to the conditional signal reflects "the motor program", ensuring *realization* of the goal-directed motor reaction of pedal pressing.

In so far as the HF components (corresponding to "cognitive images") arise during interstimulus intervals, recent data about the important role of the hippocampus in the process of perception of contexts for a specific learning situation are interesting. The tonic state, serving as a background for the following phase of responding, is created by these contextual stimuli (Myers and Gluck, 1994; Rudy, 1993; Freeman, 1991). These results help our interpretation of data on the intensification of HF components in interstimulus intervals, when conditional stimuli are absent, suggesting that they possess important functional significance.

Our data on the functional role of the HF components in the brain EA during training are in line with researches of the last few years, directed at the study of the functional significance of gamma-frequencies in the human EEG during perception (Bressler, 1990; De France and Sheer, 1988; Freeman, 1991; Loring and Sheer, 1984; Pulvermüller *et al.*, 1995; Ray and Cole, 1985; Sokolov, 1996), and even in the formation of processes of consciousness (Barinaga, 1990; Crick and Koch, 1990).

Two principal conditions form a basis for our researches concerning the dynamics of the HF components in the brain EA: The use of a wide frequency band for recording EA, and the features of the chosen model of instrumental conditioning. The fact that the animal has to overcome a number of restrictions during pedal pressing, requiring a definite effort, results in the formation of a high level of emotional-motivational tension, which is the basic condition for the formation of active purposeful behaviour (Simonov, 1981, 1987). The HF components are the electrographic correlates of these states. It should be noted once more, that revealing the HF components is ensured by intensification of the synchronizing factor in the activity of neuronal systems during the course of learning, a principle that was repeatedly emphasized earlier by M.N. Livanov (Livanov, 1977, 1979, 1988).

REFERENCES

Adrianov, O.S. and Mering, T.A. (1959) *Atlas of the Dog's Brain* (in Russian). Moscow: Medgiz.

Anokhin, P.K. (1968) *Biology and Neurophysiology of Conditioned Reflex* (in Russian). Moscow: Medizina.

Asratyan, E.A. (1963) Tonic conditioned reflexes as form of integrative brain activity. *Zhurnal Vysshey Nervnoy Dejatelnosty*, **13**, 781–788.

Barinaga, M. (1990) The mind revealed? *Science, New York*, **249**, 856–858.

Bernstein, N.A. (1966) *Sketches on Physiology of Movements and Physiology of Activity* (in Russian). Moscow: Medizina.

Boeijinga, P.H. and Lopes da Silva, F.H. (1988) Differential distribution of beta and theta EEG activity in the entorhinal cortex of the cat. *Brain Research*, **448**, 272–286.

Boujer, J.J., Montaron, M.F., Vahnee, J.M., Albert, M. and Rougel, A. (1987) Anatomical localization of cortical beta-rhythm in cat. *Neuroscience*, **22**, 863–869.

Bragin, A., Jando, G., Nadasdy, Z., Hetke, J., Wise, K. and Buzsaki, G. (1995) Gamma (40–100 Hz) oscillations in the hippocampus of the behaving rat. *Journal of Neuroscience*, **15**, 47–60.

Bressler, S. (1990) The gamma wave: a cortical information carrier? *Trends in Neuroscience.*, **13**, 161–163.

Bressler, S. and Freeman, W.J. (1980) Frequency analysis of olphactory system EEG in cat, rabbit and rat. *Electroencephalography and Clinical Neurophysiology*, **50**, 19–24.

Buzsaki, J., Horvath, Z., Urioste, R., Hetke, J. and Wise, K. (1992) High-frequency network oscillations in the hippocampus. *Science, New York*, **256**, 1025–1027.

Crick, F. and Koch, C. (1990) Forwards a neurobiological theory of consciousness. *Seminars in the Neurosciences*, **2**, 263–275.

De France, J. and Sheer, D.F. (1988) Focused arousal 40-Hz and motor programming. In: D. Giannitrapani, F. Murri (eds) *The EEG of Mental Activities*, pp. 153–168 New York, Plenum Press.

Dumenko, V.N. (1977) *Background Neocortical Electrical Activity in Dogs in Some Integrative Types of Behavior* (in Russian). Kiev: Naukova Dumka.

Dumenko, V.N. (1985) Correlation-spectral characteristics of cortical potentials in a wide frequency range in dogs in the process of elaborating a lever-pressing alimentary conditioned response. *Physiologia Bohemoslovaca*, **34**, 29–32.

Dumenko, V.N. (1988) Spatial synchronization of cortical potentials and high-frequency components of neocortical electrical activity during learning. *Neuroscience and Behavioral Physiology*, **18**, 179–187.

Dumenko, V.N. (1992) *Learning and High-frequency Components of Electrical Activity of the Brain* (in Russian). Moscow: Nauka.

Dumenko, V.N. (1995a) Dynamics of parameters of EEG traditional frequency range during conditioning in dogs. *Neuroscience and Behavioral Physiology*, **25**, 353–358.

Dumenko, V.N. (1995b) Dynamics of high-frequency components of electrical activity dog's brain in the process of cognitive activity. *Zhurnal Vysshey Nervnoy Dejatelnosty*, **45**, 835–847.

Dumenko, V.N. and Kozlov, M.K. (1989) Spectral-correlative characteristics of EEG-reactions preceeding conditioned motor reactions in dogs. *Zhurnal Vysshey Nervnoy Dejatelnosty*, **39**, 458–468.

Dumenko, V.N. and Kozlov, M.K. (1993) A computer analysis of EEG-intersignal reactions during the development of motoric alimentary conditioned reflexes in dogs. *Neuroscience and Behavioral Physiology*, **23**, 142–151.

Dumenko, V.N. and Kozlov, M.K. (1995) Study by nontraditional analytic methods of features of cortical potentials, taking high-frequency components into account, in dogs during instrumental learning. *Neuroscience and Behavioral Physiology*, **25**, 150–157.

Dumenko,V.N. and Kozlov, M.K. (1997) Study of the EEG phenomenon of high frequency bursts in the neocortical electrical activity of dogs in the process of alimentary learning by means of newly designed methods. *Experimental Brain Research*, **116**, 539–550.
Dumenko, V.N., Kozlov, M.K. and Kulikov, M.A. (1995a) Study of high-frequency components of dog's electrical activity by means of factor analysis. *Zhurnal Vysshey Nervnoy Dejatelnosty*, **45**, 107–118.
Dumenko, V.N., Kulikov, M.A. and Kozlov, M.K. (1995b) Are gamma frequencies of brain electrical activity homogeneous? *Doklady Biological Sciences*, **339**, 541–544.
Dumenko, V.N., Kozlov, M.K. and Kulikov, M.A. (1996) The dynamics of high-frequency brain activity (up to 200 Hz) in the course of conditioning reflects the functional mosaic organization of the neocortex. *Zhurnal Vysshey Nervnoy Dejatelnosty*, **46**, 719–730.
Dumenko, V.N. and Sachenko, V.V. (1980) Relation background and evoked activity of neurons of the auditory area in cats in the process of elaboration of defensive conditioned reflexes. *Neurophysiogiya*, **12**, 227–237.
Eckhorn, R., Frien, A., Bauer, R., Woelbern, T. and Kehr, H. (1993) High frequency (60–90 Hz) oscillations in primary visual cortex of awake monkey. *NeuroReport*, **4**, 243–246.
Eeckman, F.H. and Freeman, W.J. (1990) Correlations between unit firing and EEG in the rat olphactory system. *Brain Research*, **528**, 238–244.
Efremova, T.M., Morozov, A.T., Sokolov, S.S. and Schlichthaar, R. (1981) Spectral-correlation analysis of high-frequency components of EEG in rabbits during an orienting reaction and defensive conditioned reflex. *Zhurnal Vysshey Nervnoy Dejatelnosty*, **31**, 1207–1218.
Freeman, W.J. (1991) The physiology of perception. *Scientific American*, **264**, 78–85.
Freeman, W.J. and Schneider, W. (1982) Changes in spatial patterns of rabbit olfactory EEG with conditioning to odors. *Psychophysiology*, **19**, 44–57.
Gasanov, U.G. (1991) *Gnostic (Cognitive) Function of Cortical Neuronal Networks*, Moscow: Nauka.
Kanamori, N.A. (1985) A spindle-like wave in the cat hippocampus: a novel vigilance level-dependent electrical activity. *Brain Research*, **334**, 180–182.
Kostandov, E.A. (1983) *Functional Hemispheric Asymmetry and Unconscious Perception* (in Russian). Moscow: Nauka.
Kozlov, M.K. and Dumenko, V.N. (1990) New method expansion of EEG-waves into a system of oscillations providing for the analysis of EEG-phenomena of various duration. *Zhurnal Vysshey Nervnoy Dejatelnosty*, **40**, 1004–1012.
Livanov, M.N. (1977) *Spatial Organization of Cerebral Processes*. New York: Wiley.
Livanov, M.N. (1979) Factor of time in activity of neurons of the brain cortex during learning and trace states (in Russian). *Gagrskie Besedy*, **7**, pp. 8–18.
Livanov, M.N. and Dumenko, V.N. (1988) The phenomenon of spatial synchronization of brain potentials in investigations into the systemic organization of the activity of the brain. *Soviet Science Reviews, Section F, Physiology and General Biology Reviews*, **2**, pp. 403–452.
Llinas, R.R., Grace, A.A. and Yarom, Y. (1991) *In vitro* neurons in mammalian cortical layer 4 exhibit intrinsic oscillatory activity in the 10 to 50 Hz frequency range. *Proceedings of the National Academy of Science, U.S.A.*, **88**, 897–901.
Loring, DW. and Sheer, D.E. (1984) Laterality of 40 Hz EEG and EMG during cognitive performance. *Psychophysiology*, **21**, 34–38.
Myers, C.E. and Gluck, M.A. (1994) Contex conditioning and hippocampal representation in animal learning. *Behavioral Neuroscience*, **108**, 835–847.
Ogawa, J. and Motokisawa, F. (1990) Spontaneous discharge patterns of output neurons in the cat olphactory bulb. *Japanese Journal of Physiology*, **40**, 193–195.
Pulvermüller, F., Lutzenberger, W., Preisel, H. and Birbaumer, N. (1995) Spectral responses in the gamma-band: physiological signs of higher cognitive processes? *Neuro Report*, **6**, 2059–2064.
Ray, W.Y. and Cole, H.W. (1985) EEG alpha activity reflects attentional demands and beta activity reflects emotional and cognitive processes. *Science, New York*, **288**, 750–752.
Rudy, J.W. (1993) Contextual conditioning and auditory cue conditioning dissociate during development. *Behavioral Neuroscience*, **107**, 887–891.
Simonov, P.V. (1981) *The Emotional Brain* (in Russian). Moscow: Nauka.
Simonov, P.V. (1987) *The Motivated Brain* (in Russian). Moscow: Nauka.
Singer, W. (1993) Synchronization of cortical activity and its putative role in information processing and learning. *Annual Reviews of Physiology*, **55**, 349–374.
Sokolov, E.N. (1990) Neurophysiological mechanisms of consciousness. *Zhurnal Vysshey Nervnoy Dejatelnosty*, **40**, 1049–1059.
Sokolov, E.N. (1996) Problem of gestalt in neurobiology. *Zhurnal Vysshey Nervnoy Dejatelnosty*, **46**, 229–240.
Suzuki, S.S. and Smith, G.K. (1988) Spontaneous EEG spikes in the normal hippocampus. Relations to synchronous burst discharges. *Electroencephalography and Clinical. Neurophysiology*, **69**, 532–540.

Vanderwolf, C.H. and Baker, G.B. (1986) Evidence that serotonin mediates non-cholinergic neocortical low voltage fast activity, non-cholinergic hippocampal rhythmical slow activity and contributes to intelligent behavior. *Brain Research*, **374**, 342–356.

Ylinen, A., Bragin, A., Nadasdy, Z., Jando, G., Szabo, I., Sik, A. and Buzsaki, G. (1995) Sharp wave-associated high-frequency oscillations (200 Hz) in the intact hippocampus: network and intracellular mechanisms. *Journal of Neuroscience*, **15**, 30–46.

8 EEG Mapping in Emotional and Cognitive Pathology

V.B. Strelets

Institute of Higher Nervous Activity and Neurophysiology, Russian Academy of Sciences, Moscow, Russia
e-mail: strelets@aha.ru

The spectral power of EEG rhythms in different cortical areas and EEG connections between these areas were studied in depressive and schizophrenic patients in four schematically outlined main cortical quadrants. Studies of depressive patients as well as data from the literature provided evidence that two cortical quadrants (right anterior and left posterior) appeared to be predominantly involved in regulation of negative emotions. The decrease of spectral power of the alpha-rhythm in these areas pointed to their relative hyperactivity in depression, and the decrease of cortical connectivity between these and other areas indicated their functional isolation. Schizophrenic patients were divided into two groups, with predominance of either positive or negative symptoms. In the first group decrease, in comparison with the normal case, of spectral power of most EEG rhythms was found only in parietal areas, while in the second group it was found in all cortical areas. Important results were obtained in comparisons of both EEG indexes between the two hemispheres, revealing patterns of asymmetry typical of each group. Thus there was significant asymmetry of spectral power of the alpha-rhythm, differing in the two groups. In schizophrenics with positive symptoms, the decrease of spectral power of the alpha-rhythm was expressed more in the parietal and occipital areas of the right hemisphere than of the left one, while in patients with negative symptoms, the asymmetry was in the opposite direction. These results point to a relatively higher activation of the right posterior quadrant in the former group and of the left posterior quadrant in the latter one. However, in the anterior cortical quadrants intracortical connections were higher in the left hemisphere in patients with positive symptoms, and in the right one for those with negative symptoms. Thus, in schizophrenia there were different patterns of asymmetry in anterior and posterior cortical quadrants, these asymmetries being in opposite directions in patients with positive and negative symptomatology. The fact that the discrepancy between the activity levels was revealed by different indexes, i.e., spectral power and cortical connections respectively in posterior and anterior areas could point to the "multidimensionality" of the impairment of brain mechanisms in schizophrenia, correlating with disturbances of mental functioning in this disease.

KEYWORDS: schizophrenia, depression, EEG-rhythms mapping, asymmetry

1. INTRODUCTION

At the present time there is significant progress in study of the pathogenesis of mental illnesses, in comparison with previous work, carried out using traditional EEG methods. This progress is connected with new methods for analysis of brain functions, including positron emission tomography (PET), functional magnetic resonance imaging (fMRI) and single photon emission tomography (SPET), these being methods of "living brain imaging". EEG mapping is also regarded as one of these methods. Using this method we have revealed some inter- and intrahemispheric deviations of different EEG rhythms in patients with depression and schizophrenia, characterized (respectively) by predominance either of emotional or cognitive disturbances.

In depression emotional disturbances are obviously primary and dominant. Recently a division of schizophrenia into two types—with the predominance of "positive" or "negative" symptoms was recommended by the World Health Organization as the most adequate for study of the pathogenesis of this disease. "Positive" symptoms of schizophrenia (delusions and hallucinations) were regarded by K. Schneider (1957) as primary ones, connected with perceptual disturbances which are rather specific for schizophrenia, in which cognitive components play the main role. Sims (1991) thinks that the cognitive content, and in particular the "significance" of delusions and hallucinations, turns the "internal" neurophysiological events into the outer psychopathological manifestations. Crow (1980) believes that the special significance of "positive" symptoms determines the primary character of cognitive disturbances in schizophrenia and the specific connection of these disturbances with the schizophrenic process, although pathological emotions can also already appear at this stage of the disease. Thus, schizophrenia can be considered as the model of a predominantly cognitive disturbance, and depression as that of an emotional disturbance, in spite of the fact that in schizophrenic patients emotional disorders, and in depressive patients some cognitive disorders can also take place (Burkhart and Thomas, 1993).

The cerebral cortex can be divided schematically into four main quadrants, playing different functional roles. The posterior quadrants, which include parietal and occipital areas, perceive, and carry out the primary processing of information. The anterior quadrants, which include frontal, fronto-temporal and central areas, are responsible mostly for decision making and motor control. In our previous EEG mapping studies (Strelets, 1989, 1993) analyzing the spectral power of EEG rhythms and EEG intracortical connections (Ivanitsky, 1997), we have shown that in patients with cognitive and emotional pathology there are some inter- and intrahemispheric disturbances. This paper summarizes these results, with emphasis on the typical neurophysiological mechanisms underlying some types of mental pathology: endogenous *vs.* reactive depression as a model of emotional disturbance, and positive *vs.* negative schizophrenic disorders as models of primary cognitive disturbances.

2. METHODS

Six groups of right-handed subjects were studied: two normal groups (36 subjects in total, divided into two subgroups, these being the controls for the groups of depressive and schizophrenic patients) and four groups of patients: those with endogenous depression (ED) (36 patients), reactive depression (RD) (68 patients), and schizophrenics with a predominance of positive symptoms (27 patients) and negative symptoms (31 patients). The age of the subjects was 20–50 years (average 35). All schizophrenic patients and the control group for these patients were males. Among the groups of depressive patients as well as their control groups, females prevailed. All patients having neurological signs of organic brain lesions were excluded from the study. Most of the patients did not take psychoactive medicines; those who did were medication-free for at least 7 days before the study.

The mental state of patients with ED was characterized mainly by the syndrome of severe unipolar depression, ideomotor inhibition, stable depressed mood, melancholy, apathy, and sometimes these symptoms were accompanied by anxiety, disturbances of sleep and appetite and somatic complaints. The mental state of patients with RD included psycho-emotional tension and anxiety, emotional instability, and sometimes a tendency to development of melancholy.

Schizophrenic patients (as already stated) were divided into two groups depending on the predominance in their mental state of either positive or negative disturbances. In the first case the main symptoms were delusions and hallucinations, emotional excitability and impulsivity; in the second case patients were characterized by withdrawal, autism, apathy, emotional retardation and social isolation.

During the experiment the subjects sat comfortably in a light- and sound-proof room. The EEG was recorded from 16 derivations according to the 10–20 system: anterior frontal (FP1, FP2), frontal (F3, F4), central C3, C4), parietal (P3, P4), occipital (O1, O2), frontal-temporal (F7, F8), temporal (T3, T4) and posterior temporal (T5, T6).

Reference electrodes were placed on the ear lobes. High frequency filters were set at 70 Hz, and a time constant of 0.3 s was used. EEG traces of 100 sec were analyzed by the EEG mapper from the company "Medicine-Biology-Neurophysiology" (MBN), Russia. Subsequent analysis of each EEG segment consisted of selection of five 10-seconds EEG fragments, free from artefacts. These fragments underwent Fast Fourier Transform, the results of this procedure being averaged afterwards.

In the first stage, the spatial distribution of the spectral power of the biopotentials was studied. For each derivation in every frequency band (delta, theta, alpha, beta-1 and beta-2) spectral power was calculated with a resolution of 0.2 Hz; these figures were used afterwards for construction of brain maps, by the interpolation method. After quantitative analysis of the spectral power of EEG-rhythms, the ANOVA method was used for comparison between groups. In addition, for RD patients and schizophrenics with positive and negative symptoms, t-test comparisons were done using the index of Interhemispheric Asymmetry (IA):

$$IA = (S - D)/(S + D)*100\%,$$

where S and D are the measures of spectral power of EEG-rhythms for symmetrical points of the left and right hemispheres. The results obtained from different subjects were summarised by computer, which also created averaged maps for each of the investigated groups.

In the second stage a new methodological principal for the "mapping the living brain" was used—Intracortical Interactions Mapping, IIM (Ivanitsky, 1997, see also his paper in this book). In comparison with the usual methods of EEG mapping, showing the potential or its spectral power distribution over the cortical surface, this method reveals the inner connections between cortical areas under different experimental conditions. The method is based on Livanov's (1972) idea that synchronization of potentials from different cortical areas provides evidence for functional connections between these areas. This notion enables one to consider connections between cortical areas as an index of their functional involvement in brain processing.

Experimental neuropsychological tasks using mental arithmetic and imaginary thinking (in which the subject had to imagine his way home from work) were delivered to the group of patients with RD and their normal control subgroup. Patients with ED were studied only in resting conditions, due to the severity of their melancholy. Only one "joint" test, including both mental arithmetic and imaginary thinking, was administered to schizophrenic patients. Subjects were asked to count orally the number of hours on an imaginary clock dial, for example, from 5 o'clock of "to-day evening" until 8 o'clock of "yesterday morning".

3. RESULTS

3.1. The Study of the Spectral Power of EEG Rhythms

In each normal group, averaged brain maps, characterising spectral power distribution of alpha- and beta- rhythms in the resting condition (with closed eyes) were rather symmetrical (Figures 8.1[A1, 2] and Figure 8.2 [A1, 2]). In these figures, the maps were created by different interpolation methods and looked different, although the distribution of spectral power of alpha- and beta-rhythms was symmetrical with both methods. It is known, that during task fulfillment a small asymmetry can occur, the left hemisphere being activated more in mental arithmetic, and the right one in imagery thinking tasks. These data will be presented in the section where comparison with the patients groups is described.

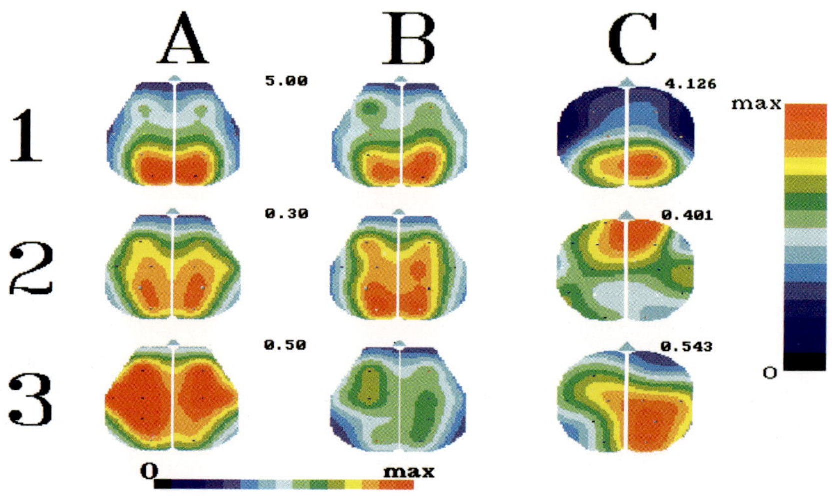

Figure 8.1. Spectral power distribution of the alpha rhythm (1), of the beta rhythm (2) and intracortical interactions in the alpha-rhythm (3). Subject groups: Normal: (A); patients with ED: (B); and patients with RD: (C). The scale for the maps in (1) and (2) are in mV^2/Hz, the scale for the maps in (3) are normalized number of connections. Maximal values on the scale for each picture are indicated by numbers above the corresponding picture.

3.1.1. Depression

In patients with ED, two foci were revealed on EEG maps, showing decrease of spectral power of the alpha-rhythm in comparison with the norm. The first focus was located in anterior areas of the right hemisphere, alpha-rhythm spectral power in these areas being decreased not only in comparison with the norm but also with the symmetrical zones of the left hemisphere; in frontal areas the difference was significant ($p < 0.01$). The second focus of higher activation was located in posterior areas of the left hemisphere, where spectral power in the alpha-rhythm was also decreased in comparison with the same areas in the norm, and the symmetrical zones of the right hemisphere; in occipital areas the difference was significant ($p < 0.05$) (Figure 8.1 [B1]). Zones symmetrical to those activated in ED patients were, on the contrary, relatively inhibited, as spectral power of the alpha-rhythm was increased there. In the beta rhythm, there were no significant differences from the norm, but in the norm this rhythm was evenly distributed across all four cortical quadrants (Figure 8.1 [A2]), while in ED patients it was expressed only in posterior quadrants (Figure 8.1 [B2]).

In patients with RD the first focus was revealed not as a decrease of spectral power of the alpha-rhythm but as a spectral power increase in the beta-rhythm. This focus was located in the right anterior quadrant, its maximum being in the right frontal area (Figure 8.1, [C2]). The spectral power of the beta-rhythm was characterized by a negative IA, that is, by an increase in the right frontal area, compared with the symmetrical area on the left side. (In these patients IA = –22.5, and in the norm IA = 0; $p < 0.01$). The second focus in the left posterior quadrant was revealed only in patients with severe forms of RD, clinically similar to ED (Figure 1 [C1]). The IA in the alpha-rhythm in parietal and occipital areas was negative, indicating the relative activation of the left hemisphere, in comparison with symmetrical zones of the right hemisphere. (In RD patients, IA = –25.6, in the norm = –11.5; $p < 0.05$).

Some differences were also revealed between the norm and RD patients in other rhythms. The total spectral power (from all derivations) of all rhythms except the theta-rhythm in these patients appeared to be significantly decreased, in comparison with the norm. The total theta-rhythm spectral power in these patients was, on the contrary, significantly higher than in the norm.

Additional deviations from the norm were found in RD patients during task performance. The decrease of spectral power of the alpha-rhythm with eyes open and during mental arithmetic in these patients was significantly lower than in the norm, pointing to reduced reactivity in depression. In these situations in RD patients there was also a significant decrease of delta- and an increase of theta-rhythm spectral power. The task requiring right hemisphere activation (imagining the way home from work) in the norm was accompanied by a significant increase of delta- and a decrease of beta-rhythm spectral power, while in depressive patients there were no differences from the resting condition during task performance.

3.1.2. Schizophrenia

In schizophrenic patients in resting conditions the following statistically significant differences from the norm in spectral power in EEG rhythms were found (Table 8.1).

Table 8.1. Spectral power of EEG rhythms in normal subjects, and in schizophrenic patients in resting conditions

Derivat.	*Delta*			*Theta 2*		
	Norm	*Schiz. with pos. sympt.*	*Schiz. with neg. sympt.*	*Norm*	*Schiz. with pos. sympt.*	*Schiz. with neg. sympt.*
F3	2.72	2.81	2.02	2.54	1.40	1.05**
F4	2.65	2.54	2.23	2.86	1.69	1.03**
C3	2.28	1.68	1.39	2.58	1.46	1.01***
C4	2.05	1.63	1.56	2.79	1.48	1.08***
T3	1.70	1.31	0.84	1.10	0.75	0.40**
T4	1.36	1.46	1.16	1.20	0.72	0.47**
P3	2.41	1.32**	1.45	2.66	1.27**	0.84***
P4	2.09	1.38**	1.39**	2.82	1.31**	0.88***
O1	2.38	1.48	1.09***	2.32	1.71	0.67***
O2	1.93	1.27	1.48	2.44	1.27	0.85***
Derivat.	*Alpha*			*Beta 1*		
	Norm	*Schiz. with pos. sympt.*	*Schiz. with neg. sympt.*	*Norm*	*Schiz. with pos. sympt.*	*Schiz. with neg. sympt.*
F3	2.73	1.65	1.38***	0.50	0.32**	0.29***
F4	2.90	2.04	1.40***	0.52	0.41	0.27***
C3	3.88	2.42	1.63***	0.58	0.38**	0.34***
C4	3.94	2.29	1.65***	0.57	0.41	0.32***
T3	1.52	1.23	0.71***	0.36	0.36	0.19***
T4	1.71	1.22	0.77**	0.34	0.34	0.18***
P3	6.97	3.61**	2.33***	0.66	0.39**	0.38***
P4	6.73	3.58**	2.56***	0.66	0.42	0.38**
O1	9.06	6.81	3.41***	0.81	0.47	0.45**
O2	8.55	4.99	4.40**	0.81	0.44*	0.55

Significance of differences (for comparisons with norm):
*—P < 0.05; **—P < 0.01; ***—P < 0.001
Only significant results (P < 0.05)are discussed

Spectral power of all EEG rhythms except beta-2 was decreased. In the group of patients with positive symptoms this decrease was revealed only in both parietal areas (P3, P4) for delta-, theta-, and alpha-rhythms; in the group of patients with negative symptoms it was found in all cortical regions with the exception of the delta-rhythm, whose power was decreased only in right parietal (P3), left central (C3) and occipital (O1) areas.

Significant asymmetry was found in spectral power in the alpha-rhythm in occipital zones, but this asymmetry was different in the two groups: Spectral power in the alpha-rhythm was higher in the left occipital area (O1) in patients with positive symptoms, and in the right occipital area (O2) in patients with negative symptoms. The IA coefficients for occipital areas in patients with positive and negative symptoms were 9.0 and –14.5, respectively, differing significantly and in opposite directions from the norm, where IA = 1.9 (not significant). With eyes open the asymmetry also became significant in parietal areas.

Thus, asymmetry of the alpha-rhythm in the posterior cortical quadrants was typical of schizophrenic patients. In the group with positive symptoms, alpha-rhythm spectral power was higher in the left posterior cortical quadrant, than in the right one (Figure 8.2 [B1]), this asymmetry pointing to the higher activation of the right posterior quadrant. In patients with negative symptoms, on the contrary, alpha-rhythm spectral power was higher in the right posterior quadrant (Figure 8.2 [C1]), pointing to an increased activation of parieto-occipital areas of the left hemisphere. Important differences from the norm in spectral power of the beta-1 rhythm were found in patients with positive symptoms (Figure 8.2 [B2]). This index was significantly decreased in these patients in the left frontal (F3), left central (C3) and left parietal (P3) areas. Thus, spectral power of the beta-1 rhythm in schizophrenics with positive symptoms was higher than in the norm in the right anterior quadrant and right parietal area, evidently pointing to higher activation of the right hemisphere. Spectral power in the beta-1 rhythm in these patients was, however, decreased in the right occipital area (O2), where, judging from the decrease in alpha-rhythm spectral power, we had concluded, that activation was higher than in the left hemisphere. In patients with negative symptoms, on the contrary, spectral power in

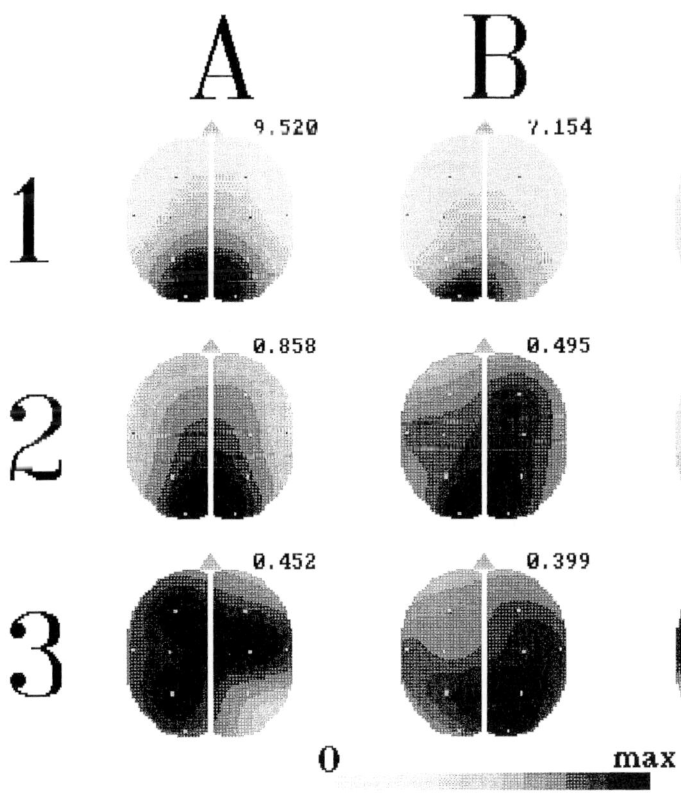

Figure 8.2. Spectral power distribution of the alpha rhythm (1), the beta rhythm (2) and intracortical interactions in the alpha rhythm (3). Subject groups: Normal: (A); schizophrenic patients with positive symptoms: (B); and schizophrenic patients with negative symptoms: (C). The scales on maps in (1) and (2) are in mV^2/Hz, the scale on maps in (3) are normalized number of connections. Maximal values on the scale for each picture are indicated by numbers above the corresponding picture.

the alpha-rhythm was higher in occipital areas of the right hemisphere, in comparison with the left one (Figure 8.2 [C1]). The same asymmetry pattern was typical of these patients in the beta-1 rhythm, which was higher in occipital areas of the right hemisphere, than in those of the left one (Figure 8.2 [C2]). Beta1-rhythm spectral power was also, as already mentioned, decreased in comparison with norm, in all areas. The pattern of asymmetry typical of this group of patients did not change during fulfilment of experimental tasks. In schizophrenics with negative symptoms, activation, revealed from the alpha-rhythm spectral power was lower in the right hemisphere than in the left one, while beta-rhythm spectral power, was higher in the right hemisphere. Normally the classic response to opening of the eyes (activation) is blocking of the alpha- rhythm and beta-rhythm increase. Thus, normally in this situation there is the change of alpha- and beta- activities in opposite directions, while in schizophrenics these two change in the same direction. This points to a discordance of the data, showing on the one hand, a higher, and, on the other hand, a lower level of activation.

In all groups, opening of the eyes was accompanied by statistically significant decreases of power for most EEG-rhythms (Table 8.2); the exceptions were delta- and

Table 8.2. Spectral power of EEG rhythms in normal subjects and in schizophrenic patients with eyes open

Derivat.	*Delta*			*Theta 2*		
	Norm	*Schiz. with pos. sympt.*	*Schiz. with neg. sympt.*	*Norm*	*Schiz. with pos. sympt.*	*Schiz. with neg. sympt.*
F3	2.74	2.67	1.77***	1.85	1.04*	0.71***
F4	2.62	2.93	1.82	1.83	1.15	0.75***
C3	2.33	1.81	1.32***	1.74	1.04	0.75***
C4	2.25	1.74	1.55	1.72	1.06	0.82***
T3	1.51	1.19	0.86*	0.82	0.57	0.33***
T4	1.52	1.17	1.01	0.75	0.56	0.38**
P3	1.92	1.48	1.41	1.40	0.85	0.60***
P4	2.12	1.54	1.33*	1.34	0.91	0.68***
O1	1.77	1.46	1.10**	1.18	0.78	0.51***
O2	1.85	1.38	1.41	1.12	0.72	0.67
Derivat.	*Alpha*			*Beta 1*		
	Norm	*Schiz. with pos. sympt.*	*Schiz. with neg. sympt.*	*Norm*	*Schiz. with pos. sympt.*	*Schiz. with neg. sympt.*
F3	2.48	0.86*	0.65***	0.49	0.29**	0.24***
F4	2.54	1.04	0.68**	0.48	0.33	0.24***
C3	2.95	1.05*	0.73***	0.49	0.30**	0.26***
C4	3.00	1.14	0.79***	0.47	0.31	0.27***
T3	1.32	0.65	0.41*	0.35	0.30	0.19***
T4	1.23	0.63	0.42*	0.28	0.25	0.20
P3	2.86	1.54	0.89***	0.48	0.32	0.31*
P4	3.07	1.61	0.92***	0.49	0.31	0.33
O1	3.96	2.05	1.42*	0.44	0.32	0.39
O2	4.03	1.70*	1.73*	0.47	0.28*	0.48

Significance of differences (for comparisons with norm):
*—$P < 0.05$; **—$P < 0.01$; ***—$P < 0.001$
Only significant results ($P < 0.05$) are discussed

beta-2 rhythms. This decrease was expressed more in normal subjects and in patients with positive symptoms, than in patients with negative symptoms. It is of interest, that due to the decrease, the differences between EEG spectral power in patients with positive symptoms and normals in the resting condition mainly disappeared. Such differences remained only in the alpha and beta rhythms, which were decreased in comparison with the norm in anterior areas of the left hemisphere (F3, C3). Thus, in these two areas in schizophrenics with positive symptoms with eyes open, as in the resting condition, spectral power in both alpha- and beta-rhythms was decreased, again pointing to a discrepancy between the indices of power of alpha- and beta-rhythms. In patients with negative symptoms the decrease of alpha-rhythm spectral power was found in all cortical areas, and, simultaneously, there was the decrease of beta-1 rhythm spectral power in anterior areas of both hemispheres. Thus, in this group of patients, the indices of spectral power of the two rhythms in both anterior cortical quadrants were also incompatible.

The task of counting hours on an imaginary clock dial, in contrast to opening eyes, was accompanied by a significant increase of spectral power of all rhythms in all groups.

Table 8.3. Spectral power of EEG rhythms in normal subjects, and in schizophrenic patients in the test of counting the hours on an imaginary clock

Derivat.	*Delta*			*Theta 2*		
	Norm	*Schiz. with pos. sympt.*	*Schiz. with neg. sympt.*	*Norm*	*Schiz. with pos. sympt.*	*Schiz. with neg. sympt.*
F3	5.85	3.53*	2.80***	4.73	1.57***	1.10***
F4	6.18	2.95**	2.58***	5.12	1.84**	1.20***
C3	5.48	1.95***	1.87***	4.76	1.54***	0.93***
C4	5.49	1.94***	1.84***	4.77	1.55***	1.00***
T3	3.19	1.75	1.03***	1.77	0.83*	0.37***
T4	4.19	1.75*	1.19***	2.72	0.78**	0.45***
P3	4.70	1.61***	1.51***	3.91	1.38**	0.78***
P4	5.20	1.77***	1.56***	4.16	1.38**	0.83***
O1	3.98	1.69**	1.45***	3.18	2.03	0.66***
O2	4.05	1.64***	1.60***	3.42	1.30**	0.88***

Derivat.	*Alpha*			*Beta 1*		
	Norm	*Schiz. with pos. sympt.*	*Schiz. with neg. sympt.*	*Norm*	*Schiz. with pos. sympt.*	*Schiz. with neg. sympt.*
F3	8.18	1.79***	1.58***	1.31	0.40**	0.32***
F4	9.12	2.12***	1.61***	1.42	0.45**	0.33***
C3	11.93	2.03***	1.73***	1.55	0.42**	0.35***
C4	10.91	2.00***	1.78***	1.57	0.42**	0.37***
T3	4.57	1.45*	0.79***	0.91	0.55	0.29*
T4	4.31	1.15***	0.99***	1.05	0.32**	0.26***
P3	19.04	3.37**	2.28***	2.06	0.48**	0.39***
P4	18.27	3.46**	2.50***	2.11	0.48**	0.42***
O1	18.42	6.18**	3.34***	1.95	0.63**	0.49***
O2	22.13	4.89**	4.11***	1.91	0.53**	0.62**

Significance of differences (for comparisons with norm):
*—$P < 0.05$; **—$P < 0.01$; ***—$P < 0.001$
Only significant results ($P < 0.05$) are discussed

However, this increase was expressed more in both patient groups than in the norm (Table 8.3), causing, on the one hand, an increase of the differences between the norm and patients, and, on the other hand, disappearance of differences between the two groups of patients in practically all cortical areas. In the left temporal area (T3), however, the differences from the norm in patients with positive symptoms were absent in all rhythms, and in patients with negative symptoms they were absent in the spectral power index for the beta-2 rhythm, pointing to the relatively greater differences from norm in the right temporal area. Thus, one can suppose that, during task fulfillment, schizophrenic patients differed from the norm more, the asymmetry of the difference being revealed mostly in the right temporal area (T4). These results correspond to data from other authors, obtained by highly technological methods, for example by fMRI (Erkwoh, 1998).

Summarizing, it should be noted that in depressive patients there are two foci of increased activation—in the right anterior and left posterior cortical quadrants. In both groups of schizophrenic patients there is the decrease in comparison with the norm for spectral power of all rhythms with the exception of the beta-2 rhythm. In patients with positive symptoms the decrease of spectral power is revealed in parietal areas, while in patients with negative symptoms it is found in all brain areas. In schizophrenic patients there is also a significant asymmetry of alpha-rhythm spectral power, this asymmetry being different in the groups with positive and negative symptoms. Finally, results obtained in schizophrenic patients, give rather contradictory indications of the level of activation of some cortical zones, using spectral power indexes of alpha and beta-1 rhythms.

3.2. The Study of Intracortical Interactions

In the norm, in the "eyes open" condition the interaction map was rather symmetrical, although the connection level for the alpha-rhythm in the left hemisphere was higher than in the right (Figure 8.1[3A]; Figure 8.2[3A]). In resting conditions the posterior cortical quadrants had higher connectivity, than the anterior ones, and during performance of the task of counting hours on an imaginary clock dial, the connection level was higher in both anterior cortical quadrants.

In depressive patients intracortical connections for the alpha and beta rhythms with eyes open were significantly decreased in both the foci of higher activation. i.e. right anterior and left posterior cortical quadrants (Figure 8.1[B3, C3]).

In schizophrenics with positive symptoms, opposite to the norm, the connectivity level for the alpha-rhythm was higher in the right hemisphere in the resting condition, as well as with eyes open (Figure 8.2 [B3]). However, during performance of the clock task, the connectivity level was higher in the left hemisphere. In patients with negative symptoms, connectivity for the alpha-rhythm in the resting condition dominated in the left anterior quadrant. Thus, the asymmetry in resting conditions was the opposite of that in patients with positive symptoms. With eyes open the pattern of cortical connections in this group remained the same (Figure 8.2 [C3]), and during task performance symmetrical connectivity in both anterior quadrants was revealed. Symmetrical connections in both anterior cortical quadrants in schizophrenics with negative symptoms resembled that in the norm, but in patients the foci of interactions were located more posteriorly. These data correspond to Buchsbaum's (1995) MRI results, revealing in young normals a higher metabolism in anterior brain areas, and in more posterior ones in young schizophrenics.

In the beta-2 rhythm in the norm the connections in resting conditions dominated in both posterior quadrants and were expressed more in the right hemisphere. With eyes open the focus of interaction shifted to the left quadrant, and, during the task, connections became rather symmetrical again, but the right anterior and left posterior quadrants were slightly (not significantly) more connected with other areas than corresponding ones on the other side.

In patients with positive symptoms, in all three experimental situations, connections for the beta-2 rhythm dominated in the left hemisphere (Table 8.4). Statistically significant differences from the norm (an increase) took place in the left frontal (F3), right frontal (F4), left central (C3) and left temporal (T3) areas. Thus, in these patients, for the beta-2 rhythm, the index of cortical connections was increased in anterior areas of the left hemisphere, in comparison with that of connections in anterior areas of the right hemisphere. It is of interest, that in patients with positive symptoms, connections in the left temporal area were higher than in the norm, and spectral power of the beta-2 rhythm was also higher than in the norm.

In patients with negative symptoms, connections for the beta-2 rhythm were the opposite of the group with positive symptoms, increased in right anterior areas, in comparison with the connectivity level in the left posterior areas (Table 8.4). A significant increase of the index of intracortical connections took place in the right frontal (F4) and the right temporal (T4) areas. An increase of cortical connections level in beta-2 rhythm in the right hemisphere took place during "eyes open" in comparison with the resting condition. During the task the focus of interactions in this rhythm moved to the left posterior quadrant, the connectivity in the left hemisphere becoming higher than in the right during task performance, and higher than the connectivity in the left hemisphere in the resting condition.

Thus, in patients having signs of relative inhibition of the left posterior quadrant in the spectral power index for the alpha-rhythm (schizophrenia with positive symptoms) there was a pattern of activation of the left anterior quadrant in the beta-2 rhythm cortical connection index, which was pathological (because it was absent in norm). Patients having

Table 8.4. Intracortical interactions in the beta-2 rhythm in normal subjects, and in schizophrenic patients with eyes open

Derivation	*Norm*	*Schizophrenics with positive symptoms*	*Schizophrenics with negative symptoms*
F3	0.40	0.72*	0.65
F4	0.31	0.80**	0.74***
C3	0.52	0.95*	0.69
C4	0.45	0.86	0.77
T3	0.38	0.78**	0.62
T4	0.32	0.61	0.62***
P3	0.59	0.95	0.82
P4	0.53	0.78	0.72
O1	0.46	0.76	0.62
O2	0.53	0.77	0.62

Significance of differences (for comparisons with norm):
*—$P < 0.05$; **—$P < 0.01$; ***—$P < 0.001$
Only significant results are discussed ($P < 0.05$)

relative inhibition of the right posterior cortical quadrant in the spectral power index for the alpha-rhythm (schizophrenia with negative symptoms) were characterized by a pattern of pathological activation of the right anterior quadrant in the beta-2 rhythm cortical connection index. In other words, in the group of patients with negative symptoms, as well as those with positive ones, there is the "discrepancy" between different indexes of activation.

In schizophrenics there were also significant differences from the norm in cortical connections for the theta-rhythm. In patients with positive symptoms, the connections in theta-rhythm were decreased, in comparison with the norm, in the left temporal area (T3) in the resting condition, and in the right temporal area (T4) during task performance. In patients with negative symptoms the connections in the theta-rhythm were significantly increased in the right parietal area (P4) during "eyes open".

4. DISCUSSION

Disruption of the brain mechanisms responsible for higher nervous functions evidently involves a process of cortical activation, in particular an imbalance between activation of the four main cortical quadrants, which each play definite roles in information processing. The two posterior quadrants, as already mentioned, are predominantly responsible for perception and primary evaluation of incoming information. The left posterior quadrant is more important for perception of verbal information, and the right posterior one takes part predominantly in perception of spatially organized information. The anterior quadrants control planning, decision making, and motor responses. In these functions the anterior cortical zones use information coming from posterior regions, as well as from subcortical structures, including the limbic ones. There are now convincing arguments that the anterior quadrants are also involved in emotional reactivity: The left quadrant controls positive emotions, and the right one negative emotions (Davidson, 1993; Davidson *et al.,* 1985; Heller, 1993; Wheeler *et al.*, 1991).

Our data show, that in the normal case, in the resting condition, both indices (spectral power of brain rhythms, and intracortical connection level) are rather symmetrical for the group as whole. Activation is usually expressed slightly more in the left hemisphere in comparison with the right one. In the counting task alpha-rhythm power in both anterior cortical quadrants decreases, and delta-rhythm power increases. These results correspond to data from other authors (Harmony *et al.*, 1994; Schwarz *et al.*, 1995; Tauscher *et al.,* 1995).

In depressive patients, whom we consider as the model of primary emotional disturbances, there is a relative increase of activation in the right anterior and left posterior quadrants, these data also coinciding with data from the literature (Davidson, 1993; Davidson *et al.,* 1985; Heller, 1993; Schneider *et al.,* 1995; Wheeler *et al.,* 1991).

Thus, in both hemispheres there is a gap between activation of posterior and anterior cortical quadrants in depression, which is not observed in the norm. We called this phenomenon "transverse functional blockade" (Strelets, 1989, 1993; Strelets *et al.*, 1996a). In anterior areas of the left hemisphere, activation is decreased, and in posterior areas it is increased, and *vice versa* in the right hemisphere. A high level of activation of the anterior areas of the right hemisphere, which are connected with negative emotions, evidently plays an important role in the mechanisms of depression. Moreover, one may regard activation of this quadrant as the mechanism initiating depression. The results for the reactive depression study confirm this, as high activation of the anterior cortical pole is found in the initial stage of this disease (Strelets *et al.*, 1996a). Interestingly, in students before taking their examinations, we also observed focal activation of this area.

The increase in activation of the right frontal area during stress and depression is compatible with P.V. Simonov's informational theory of emotions (1994). According to this author's point of view, negative emotions arise in situations when the information necessary for the satisfaction of needs exceeds the information available, the former being integrated in the right frontal area, the latter in the left frontal area.

In depression, in contrast to the norm, the mental arithmetic task causes only slight decrease of alpha-rhythm spectral power, and no increase of delta-rhythm spectral power. There is, in contrast, a significant increase of spectral power of theta and (to a lesser degree) of beta rhythms. Thus, reactivity in the alpha-rhythm in depressive patients is decreased, but the significant decrease of spectral power in theta- and beta-rhythms can be regarded as substituting for the alpha-rhythm reaction.

Cortical connections in the alpha-rhythm in foci of increased activation in depressive patients are, however, decreased. Thus, these foci of activation are rather disconnected from other cortical areas, which can be an additional factor interfering with their normality of function.

In schizophrenic patients, which we consider to be a model of primary cognitive disturbances, these disturbances are much more complicated. The indices characterizing topographical distribution of spectral power of all brain rhythms (except beta-2), differ from the norm, these differences being in opposite directions in the two groups of patients.

From the point of view of the pathogenesis of schizophrenia, it is important that at the acute stage, i.e. in patients with positive symptoms (delusions and hallucinations), a decrease of spectral power of brain rhythms (with the exception of the beta-2 rhythm) takes place only in parietal areas. This conclusion confirms that cognitive disturbances in schizophrenia are of primary origin, since parietal areas play a great role in cognitive estimation of the significance of a stimulus, and in decision making (Strelets, 1989). In the chronic stage, in patients with negative symptoms, who reveal severe emotional and personality retardation, the decrease of spectral power of brain rhythms involves all brain areas, i. e. the deficit of spectral power of EEG rhythms becomes generalized.

The fact that, with eyes open, the differences between spectral power of brain rhythms between the norm and patients with positive symptoms disappear, can point to the relative adequacy of activation in the initial, acute stage of the schizophrenic process. In chronic schizophrenia decrease of activation in all cortical zones takes place even in this situation. However, during performance of a more complicated task (counting hours on the imaginary clock dial), which puts demands on the left as well as the right hemispheres, a deficit of activation is observed in both subgroups of schizophrenic patients.

It is well known that the alpha-rhythm is determined to a significant degree by thalamo-cortical connections. The decrease of spectral power of this rhythm in schizophrenic patients could be compared with the data about the decrease of the volume of thalamus (Danos *et al.*, 1995) and left ventricle (Schwarz *et al.*, 1995) in this disease, especially in chronic patients. These data can thus point to disturbances of genesis of rhythms in schizophrenia.

Asymmetry of the alpha-rhythm indicates that in patients with positive symptoms, there is relative prevalence of excitatory processes in the posterior areas of the right hemisphere, and relative inhibition of the posterior zones of the left (communicative) hemisphere. On the contrary, in patients with negative symptoms, posterior areas of the right hemisphere are relatively inhibited in comparison with the left ones. One may conclude that, in patients with positive symptoms, adequate perception of verbal information is disturbed.

In schizophrenics with positive symptoms it is well known that personality defect is not as severe as in schizophrenics with negative symptoms. Our data on the asymmetry of spectral power of the alpha-rhythm in the posterior cortical quadrants, and the data of Persaud and Cutting (1991) can indicate that a major role in integration of personality is played by the right hemisphere, which, according to our data, is disturbed less in patients with positive symptoms, than in patients with negative symptoms.

In schizophrenics, the fact that differences from the norm in spectral power were absent only in beta-2 rhythm, and the fact that differences in cortical connections were present for the beta-2 rhythm can indicate that disturbances of this rhythm play an important role in schizophrenia, and it is therefore designated as a "rhythm of interest". Our data concerning the increase of intracortical connections in the beta-2 rhythm in schizophrenia patients with positive symptoms in the left anterior quadrant, and for those with negative symptoms in the right anterior quadrant, provides evidence for the presence of activation in the anterior quadrant of the hemisphere, such activation being pathological (since it is absent in the norm). This contrasts with the decrease of activation in posterior quadrants. Thus, schizophrenia is characterized by a combination of the relative inhibition in the posterior areas (according to the alpha-rhythm index of spectral power) with some kind of pathological activation in anterior areas of the same hemisphere (in the beta-2 rhythm index of cortical connections). Interestingly, these pathological manifestations, which are in opposite directions in the two groups of patients, could be revealed only by combination of the two different kinds of measurements, i. e. "in two different dimensions".

Our findings—inhibition of the left postcentral zones, including Wernicke's area, and excitation of the left frontal area (F7) including Broca's area, controlling motor speech function, in "positive" schizophrenics—are possibly important for understanding the mechanism of their pathological verbal production (delusions, hallucinations).

The results from the study of intracortical interactions (in the beta-2 rhythm) can be compared with the data from J. Gruzelier *et al.* (1988). For schizophrenics with positive symptoms, these authors have revealed higher performance than in the norm in verbal short term memory task, while in spatial short term memory tasks their performance was lower than normal. In patients with negative symptoms higher than normal performance was found in the spatial task, while in the verbal task performance was lower than normal. These data are quite compatible with our results, which show the higher connectivity in anterior areas of the "verbal" left hemisphere in patients with positive symptoms, and of the "spatial" right hemisphere in patients with negative symptoms.

Thus, the study of intracortical interactions provides the possibility of revealing areas where the disturbances, typical of schizophrenia, take place. These "regions of interest" include, first of all, frontal and temporal ones. It is important that the data obtained by the method of intracortical interactions corresponds to the results of other authors, obtained by different methods—PET, fMRI, SPECT (Atzor *et al.*, 1995; Erkwoh *et al.*, 1995; Hajek *et al.*, 1995; Sharma *et al.*, 1995; Vieweg *et al.*, 1995). Thus, by this method it is possible to discover neurophysiological deviations, that cannot be revealed by standard methods of EEG analysis, but can be found by highly technological methods.

It is obvious that the picture of EEG findings in schizophrenic patients is very complicated and contradictory, especially when the results obtained by different methods are compared. This gives us the possibility of unravelling the complexity and contradictions of the neurophysiological damage in schizophrenia, that are probably connected with the disintegration of cortical-subcortical connections (Bauer *et al.*, 1995; Dolan 1995), or the imbalance between frontal-central and parietal areas (Strelets *et al.*, 1996b).

Abnormalities of limbic structures, and of associative and temporal cortex, revealed in schizophrenia by many authors, have been associated by some authors with disturbances of memory. Ivanitsky and Strelets (1977) found that informational synthesis of the physical parameters and meaning of stimuli retrieved from memory underlies the formation of mental images of external stimuli, and that this synthesis is disturbed in schizophrenics, due to incorrect estimation of the significance of stimuli. Similar ideas are expressed in the "comparatory model" of schizophrenia (Gray, 1995). Interesting data concerning local and generalized EEG disturbances were found in aggressive schizophrenics (Wong *et al.*, 1994). Localized temporal lobe changes were seen in the most violent group, suggesting an association with aggression and violence. The generalized EEG abnormalities were, however, very similar among the groups with different violence ratings. Non-specific white matter changes in MRI, and generalized cortical hypometabolism in PET were found in the group of repetitive violent offenders as well as of non-repetitive ones, while asymmetric gyral patterns in the temporo-parietal region were particularly common in the first group and absent in the second (Wong *et al.*, 1997). The authors conclude that different structural and metabolic changes in the brain are associated with different patterns of violent offending. They also think that the complex interactions between violent behaviour, clinical features and neuroimaging findings in schizophrenia require further studies.

All the above-mentioned facts confirm the ideas about the significant complexity of the disorganization of brain rhythms in schizophrenia, and these disturbances of rhythms, moreover, are multidimensional. The multidimensional character of neurophysiological disturbances in schizophrenia can also be connected with brain malfunctions and odd behaviour of patients.

Schizophrenia has now been studied for several centuries, without a clear understanding of the pathogenesis of this disease. It is obvious now, that the neurophysiological disturbances of schizophrenia cannot be simple and uniform; there are many facts, both clinical and experimental, that cannot be explained convincingly. We can agree with Lehmann (1996) that the intensive study of this disease with highly technological methods of "living brain imaging", including EEG-mapping, makes the results obtained by different authors, more valid and reliable.

ACKNOWLEDGEMENT

The work of the author is supported by INTAS Foundation (Project no 1421) and partly by the Russian Foundation for Humanitarian Research (Project no 97–06–08184).

REFERENCES

Atzor, K.R., Gansicke, M., Franke, P., Falkai, P., Erb, F., Maier, W. and Stoter, P. (1995) Correlation of regional brain volumes (MRT volumetry) and neuropsychological test patterns in first episode schizophrenics. In: *Abstracts: Fourth International Symposium on Imaging of the Brain in Psychiatry and Related Fields, November 1–4, 1995*, Frankfurt/Main, Germany, p. 4.

Bauer, U., Schueler, G., Gallhofer, B., Puille, M., Rossmejer, B., Bauer, R. and Rappelsberger, P. (1995) Lateralization in schizophrenic patients: correlation between I123-IBZM-SPECT and QEEG. In *Abstracts: Fourth International Symposium on Imaging of the Brain in Psychiatry and Related Fields, November 1–4, 1995*, Frankfurt/Main, Germany, p. 6.

Buchsbaum, M., Hazlett, E., Shihabuddin, L., Wei, T., Haznedar, M., Schnur, D., Machac, J., Vallabhajosula, S., Knesaurek, K., Mohs, R. C. and Davis, K. (1995) Metabolic changes with age. In *Abstracts: Fourth International Symposium on Imaging of the Brain in Psychiatry and Related Fields, November 1–4, 1995*, Frankfurt/Main, Germany, p. 10.

Burkhart, M.A. and Thomas, D.G. (1993) Event-related potential measures of attention in moderately depressed subjects. *Electroencephalography and Clinical Neurophysiology*, **88**, 42–50.

Crow, T.J. (1980) Positive and negative schizophrenic symptoms and the role of dopamine. *British Journal of Psychiatry*, **137**, 386–396.

Danos, P., Bauman, B., Falkai, P., Henning, H., Schneider, T. and Bogerts, B. (1995) Thalamic pathology in schizophrenia—a morphometric and immunocytochemical post mortem study. In *Abstracts: Fourth International Symposium on Imaging of the Brain in Psychiatry and Related Fields, November 1–4, 1995*, Frankfurt/Main, Germany, p. 13.

Davidson, R.J. (1993) Cerebral asymmetry and the nature of emotion: Conceptual and methodological conundrums. *Cognition and emotion*, **6**, 245–268.

Davidson, R.J., Schaffer, C.E. and Saron, C. (1985) Effects of lateralized presentations of faces on self-reports of emotion and EEG asymmetry in depressed and non-depressed subjects. *Psychophysiology*, **22**, 353–364.

Dolan, R.J. (1995) Is schizophrenia a manifestation of abnormal neural integration. In *Abstracts: Fourth International Symposium on Imaging of the Brain in Psychiatry and Related Fields, November 1–4, 1995*, Frankfurt/Main, Germany, p. 18.

Eckert, J., Dierks, T., Northoff, G., Weber, B. and Maurer, K. (1995) Readiness potential in catatonic syndromes. In *Abstracts: Fourth International Symposium on Imaging of the Brain in Psychiatry and Related Fields, November 1–4, 1995*, Frankfurt/Main, Germany, p. 21.

Erkwoh, B., Sabri, O., Steinmeyer, E.M., Bull. U. and Sab, H. (1995) SPECT-findings and psychiatric scores in never treated schizophrenics. In *Abstracts: Fourth International Symposium on Imaging of the Brain in Psychiatry and Related Fields, November 1–4, 1995*, Frankfurt/Main, Germany, p. 21.

Gray, J. (1995) The contents of consciousness: A neurophysiological conjecture. *Behavioral and Brain Sciences*, **18**, 659–676.

Gruzelier, J., Seymour, K., Wilson, L., Jolley, A. and Hirsch, S. (1988) Impairment of Neuropsychologic Tests of Temporohippocampal and Frontohippocampal Functions and Word Fluency in Remitting Schizophrenia and affective Disorders *Archives of General Psychiatry*, **45**, 623–629.

Hajek, M., Boehle, C., Huonker, R., Volz, H.-P., Nowak, H. and Sauer, H. (1995) Functional and structural changes in schizophrenia assessed by MEG and MRI. In *Abstracts: Fourth International Symposium on Imaging of the Brain in Psychiatry and Related Fields, November 1–4, 1995*, Frankfurt/Main, Germany, p. 101.

Harmony, T., Fernandez, T., Reyes, A., Silva, J., Rodriguez, M., Marosi, E.and Bernal, J. (1994) Delta Activity: a Sign of Internal Concentration during the Performance of Mental Tasks. In: *Abstracts: 7th International Congress of Psychophysiology of the Inter-national Organization of Psychophysiology (I.O.P.) 27 Sept to 2 Oct, 1994*. Thessaloniki, Greece, p. 49.

Heller, W. (1993) Neurophysiological Mechanisms of Individual Differences in Emotion, Personality and Arousal. *Neuropsychiatry*, **7**, **4**, 476.

Ivanitsky, A.M. (1997) Informational synthesis in crucial cortical areas as the brain base of the subjective experience. *Zhurnal Vysshey Nervnoy Dejatelnosty*, **47**, 10–21. (English version of the journal for participants of XXXIII International IUPS Congress).

Ivanitsky, A.M. and Strelets, VB. (1977) Brain evoked potentials and some mechanisms of perception. *Electroencephalography and Clinical Neurophysiology*, **43**, 397–403.

Lehmann, D. (1996) EEG in schizophrenia. In: *Third International Hans Berger Congress "Quantitative and Topological EEG and MEG analysis. October 3–6*, p. 31. Jena, Germany.

Livanov, M.N. (1972) *Spatial distribution of the brain processes*. Moscow: Nauka (in Russian).

Persaud, R. and Cutting, J. (1991) Lateralized anomalous perceptual experiences in schizophrenia. *Psychopathology*, **24**, 365–368.

Schneider, K. (1957) Primary and secondary symptoms in schizophrenia *Fortschritte Neurologie Psychiatrie*, **25**, 487–490 (H. Marshall: trans; 1974 In: Hirsch,S.R. and Shepherd, M. (eds) Themes and Variations in European Psychiatry: an anthology. University Press of Virginia, Charlottesville, 1974.

Schneider, F., Grodd, W., Gur, R.E., Klose, U., Alavi, A. and Gur, R.C. (1995) PET and fMRI in the study of emotions. In *Abstracts: Fourth International Symposium on Imaging of the Brain in Psychiatry and Related Fields, November 1–4, 1995*, Frankfurt/Main, Germany, p. 76.

Schwarz, A., Pester, U., Lerche, J., Isensee, T., Kropf, S., Brosz, M., Wurthmann, C., Danos, P. and Bogerts, B. (1995) EEG-power in schizophrenics: correlates with psychopathology, brain morphology and psychometry. In *Abstracts: Fourth International Symposium on Imaging of the Brain in Psychiatry and Related Fields, November 1–4, 1995*, Frankfurt/Main, Germany, p. 79.

Sharma, T., Levis, S.W.L., Sigmundsson, T., Lancaster, E., Barta, G. and Pearlson, H. (1995) Loss of cerebral asymmetry in familial schizophrenia—a volumetric study using unbiased stereology. In *Abstracts: Fourth International Symposium on Imaging of the Brain in Psychiatry and Related Fields, November 1–4, 1995*, Frankfurt/Main, Germany, p. 85.

Simonov, P.V. (1994) Functional asymmetry of frontal neocortex and emotions. *Doklady Academii Nauk*, **338**, 698–699. (In Russian).

Sims, A. (1991) An overview of the psychopathology of perception: first rank symptoms as a localizing sign in schizophrenia. *Psychopathology*, **24**, 369–371.

Strelets, V.B. (1989) Disturbances of physiological mechanisms of perception, emotions and thinking in some types of mental pathology.*Fisiologija cheloveka*, **5**, 135–144. (In Russian).

Strelets, V.B. (1993) Inter- and intrahemispheric disturbances in some types of mental pathology. *Zhurnal Vysshey Nervnoy Dejatelnosty*, **43**, 267–270. (In Russian).

Strelets, V.B., Ivanitsky, A.M., Ivanitsky, G.A., Arzeulova, O.K., Novototsky-Vlasov, V.Y., Golikova, J.V. (1996a) Cortical processes organization disturbances in depression. *Zhurnal Vysshey Nervnoy Dejatelnosty*, **46**, 274–281.

Strelets, V.B., Aljoschina, T.D. and Igoshkin, A.V. (1996b) EEG abnormalities and functional disturbances in schizophrenic patients. In *8th World Congress of IOP. June 25–30*, p. 57. Tampere, Finland.

Tausher, J., Rappelsberger, P., Neumeister, A. Gruppe, H., Bauer, U., Galhofer, B. and Kasper, S. (1995) Changes of EEG—coherence during cognitive activation in schizophrenic patients vs. healthy controls. In *Abstracts: Fourth International Symposium on Imaging of the Brain in Psychiatry and Related Fields, November 1–4, 1995*, Frankfurt/Main, Germany, p. 93.

Vieweg, T., Volz, H.-P., Rzanny, R., Gasser, C., Kaiser, W.A. and Sauer, H. (1995) Neuropsychological and NMR—volumetric findings in positive and negative schizophrenic patients In *Abstracts: Fourth International Symposium on Imaging of the Brain in Psychiatry and Related Fields, November 1–4, 1995*, Frankfurt/Main, Germany, p. 96.

Wheeler, R.E., Davidson, R.J. and Tomarken, J. (1991) Frontal brain asymmetry and Emotional Reactivity: a Biological substrate of Affective Style. *Psychophysiological Research*, **30**, 82–89.

Wong, T.H., Lumsden, J., Fenton, G.W. and Fenwick, P. (1994) Electroencephalography, computed tomography and violence ratings of male patients in a maximum-security mental hospital. *Acta Psychiatrica Scandinavica*, **90**, 97–101.

Wong, T.H., Fenwick, P., Fenton, G., Lumsden, J., Maisey, M. and Stevens, J. (1997) Repetitive and Non-repetitive Violent Offending Behaviour in Male Patients in a Maximum Security Mental Hospital—clinical and neuroimaging findings. *Medicine, Science and the Law*, 37, **1**, 150–160.

9 Brain Organization of Selective Tasks Preceding Attention: Ontogenetic Aspects

N.V. Dubrovinskaya, R.I. Machinskaya and Yu.V. Kulakovsky

Institute of Developmental Physiology, Russian Academy of Education, Moscow, Russia
nvd@idprae.msk.ru.

Intrahemispheric functional organization was studied during a period of task-expectancy, with special reference to attentional mechanisms. Estimates of coherence of functionally identical rhythmic EEG components were made, to characterize intracortical integration. Several factors influencing the possibility of making an adequate prediction, and of confirming it, were varied. Different types of task were used. Subjects of different ages (7, 9–10 years, young adults), and children of the same age differing in their level of brain maturity were under study. It was shown that all factors studied had a definite influence on the brain organization underlying attention preceding the task. Clear age differences, as well as a lag between the possibility of formation and confirming a prediction in children were observed. Alternative "strategies" used at different ages to facilitate task performance were analysed; underlying mechanisms are discussed.

KEYWORDS: EEG, children, intracortical integration, attention, development

1. INTRODUCTION

The ontogenetic development of information processing, which forms the basis of cognitive activity, changes along with the possibilities of modulation of such processing, arising from enhanced CNS plasticity. Modulatory influences originate from the activating system of the brain. As part of the main ontogenetic trend, activation processes acquire control functions, and the ability to selectively modulate the functional state of task-relevant cortical regions. The controlling function is exercised by gradual maturation of the executive centres in the anterior associative areas (frontal lobes) (Beteleva *et al.*, 1977; Farber and Njiokiktjien, 1993). Selectively-controlled activation, appearing as a result of analysis of the prevailing situation ensures informational and mobilizing effects on current brain activity (i.e. selective attention)(Machinskaya and Dubrovinskaya, 1994). Attention controls all stages of non-automated performance (real or mental). However, anticipatory attention involved in expectancy (Machinskii *et al.*, 1988; Tecce, 1972) or priming (Posner *et al.*, 1989; Sheppard and Boyer, 1990) acquires special significance. At least two interacting processes should be involved in producing facilitatory effects for such prestimulus attention. These are: (1) the formation of an adequate prediction of the situation, and (2) triggering of the appropriate mechanisms of activation, on the basis of such a prediction. A complicated organization of brain dynamics is needed to realize these processes. This depends on several important factors: Depending on the type of task and the mode of its presentation (repetition, interval stability, warning signal) prediction may be facilitated. The other group of factors includes parameters of the reactive system, determined (in our study) by brain organization dependent on age and individual characteristics. All these factors may mediate both the formation of a model of expected events, and the actual production of activation influences.

Based on these statements, we carried out an electrophysiological study of the brain organization underlying processes preceding attention directed at task performance, varying the factors mentioned above. Two types of task were used, differing in complexity and level of prediction (either a simple sensory or a complicated verbal task). Subjects of different ages (children 7–8 and 9–10 year old, and adults), and children of the same age (7–8) with different levels of brain maturity took part in the experiment.

2. BRAIN ORGANIZATION OF ATTENTION DURING EXPECTANCY OF A SENSORY TASK

The task (Machinskii *et al.*, 1988) consisted of an intramodal binary classification of tactile and auditory stimuli, according to their duration. Clicks were presented via earphones to the right (left) ear; tactile signals were presented through a vibrator to the right (left) index finger. "Short" (5–10 ms) and "long" (15–25 ms) stimuli were used. Preliminary training made it possible to adjust stimulus duration within the abovementioned limits, and thus to equalize the probability (0.6–0.8) of correct responses across all subjects. However, stimulus duration values remained unchanged during the actual experiment. Four classes of stimuli were presented randomly : auditory short and long, and tactile short and long, grouped in two series of left-sided and right-sided stimulation (LSS and RSS respectively). Two types of warning stimuli (WS) (i.e. letters on the monitor screen) preceded presentation of the go-signal. In response to the first of these—the letter "T" (indicating "task")—the subject prepared for the task and signalled his readiness by pressing the triggering button. The other WS was either "E" (indicating "ear") or "H" (indicating "hand") which specified the modality of the forthcoming signal. The expectancy interval was varied randomly from 2.5 to 3.5 s, thus encouraging subjects to sustain attention to a specific modality. The results obtained confirmed that attention was important for correct performance.

Thus, only the duration of the stimulus remained unknown to the subject; all other task parameters (stimulation side, modality) were constant or, in the case of expectancy interval, were varied slightly, to prevent time-locked attention. In addition, sustained attention and an appropriate level of motivation were encouraged by visual feedback (letters "C" for correct, and "M" for mistake). Thus, the emphasis on "certainty" in the task structure was designed to facilitate prediction.

EEG recordings were made from 7 symmetrical scalp sites (O, P, TPO, C, T, Fi, F) and from the vertex (V), referenced to linked ear lobes. For each subject 20 artefact-free 2s EEG-epochs were sampled both under background conditions (rest, eyes closed) and under task conditions (i.e. the attention period between presentation of the WS and the go-signal). Task conditions were then classified according to response type.

Data processing. Spectral EEG analysis was carried out. EEG coherence (Coh) computed for all possible intrahemispheric pairwise combinations of channels served as an index of hemispheric functional organization. Statistical analysis of individual data was performed for each series, stimulation modality and hemisphere. Coherent rhythmic alpha-components were referred to functionally identical subdivisions of the alpha frequency band. These could vary slightly in frequency from individual to individual, but showed similar situation-dependent dynamics. Maximum Coh-values for all two-channel combinations were compared between task and background conditions (type I comparison) and between correct *versus* erroneous conditions (type II comparison) using nonparametric Wilcoxon and Wilcoxon-Mann-Whitney criteria. Statistically significant individual differences were then summarized to obtain group characteristics. The data were also verified by means of group averaging. Coh-enhancement was recognized as a sign of functional integration of cortical regions where coherent activity was generated.

Subjects. All subjects were right-handed. Children aged 9 (n = 10) and 7–8 (n = 20) took part in the experiment. The younger children were divided in two subgroups: (1)—10 subjects with high academic achievement and high intellectual ability (established by professional psychological testing) and with a brain maturity level appropriate to their age (according to EEG structural analysis—Lukashevich *et al.*, 1995); (2)—10 children with low academic achievement level and EEG—patterns suggesting functional immaturity of brain regulatory structures. The EEG of these children with learning difficulties was characterized by bilaterally synchronous bursts of regular theta-waves (4–7 Hz) in anterior locations. Data obtained in normal children aged 5–8 (Machinskaya *et al.*, 1997) and in patients with thalamic lesions (Lukashevich and Sazonova, 1996) permitted one to recognize deviant anterior activity as an EEG-sign of immaturity of the fronto-thalamic system (FTS) responsible for selectivity of attention (Skinner and Lindsley, 1973; Batuev, 1987).

Ontogenetic data were compared with results obtained in adult subjects (Machinskaya *et al.*, 1993).

In adults, and in children without CNS immaturity, the EEG-characteristics of prestimulus attention showed that prediction was occurring. Task specificity—for the expected stimulus modality—is reflected in the parameters of functional integration. Selective alpha-Coh enhancement was observed in pairs of cortical leads showing obligatory participation of the primary projection areas of the expected modality (Machinskaya and Dubrovinskaya, 1994, 1996; Machinskaya *et al.*, 1993). Expectancy of the tactile task resulted in functional integration centred around the central region, while the temporal cortex became the integration centre for auditory attention. In our opinion such organization can be interpreted as an effect of controlled activation (Dubrovinskaya, 1995), providing the informational aspect of prestimulus attention (Machinskaya and Dubrovinskaya, 1994), that is, the selective modulation of activity in the task-relevant cortical regions. At the same time it is one of the ways of formulating a prediction. The mobilizing effect of prestimulus attention manifested itself as a broad, modality-independent integration of cortical regions, focusing on vertex—the projection area for non-specific activation influences. The significance for task performance of the EEG organization described above was tested by means of type II comparisons (see methods). The results of evaluation of alpha-Coh differences showed that a correct decision is provided by the informational and mobilizing effects of selective attention during expectancy (respectively by selective modality-specific functional systems and diffuse, modality-independent functional integration).

Between-group comparisons revealed the specificity of EEG organization for pre-stimulus attention in adults and children. In adults, the two kinds of facilitation effects (informational and mobilizing) are hemisphere-dependent. The first one is located in the left hemisphere (LH), the second in the right hemisphere (RH). It seems as if, in the LH, "the model" of an expected stimulus is formed. At the same time, in the RH, the characteristics of the whole experimental situation are present (shown by interaction of all association regions, the projection areas and the vertex). These facts, described elsewhere (Machinskaya *et al.*, 1993; Dubrovinskaya *et al.*, 1993) are illustrated partly in Figure 9.1[A,1] and Figure 9.2[A]. It turned out in the experiment that in both cases (RSS and LSS) only the processes taking place in the hemisphere which is addressed by stimulation are responsible for the correct response. The EEG-organization of the "passive" ipsilateral hemisphere does not differ between epochs before correct and erroneous decisions (Figure 9.1[A,2], RH; Figure 9.2[A] LH). The efficiency of such a strategy, based on the key features of LH and RH, was supported by the results of binary classification of visual

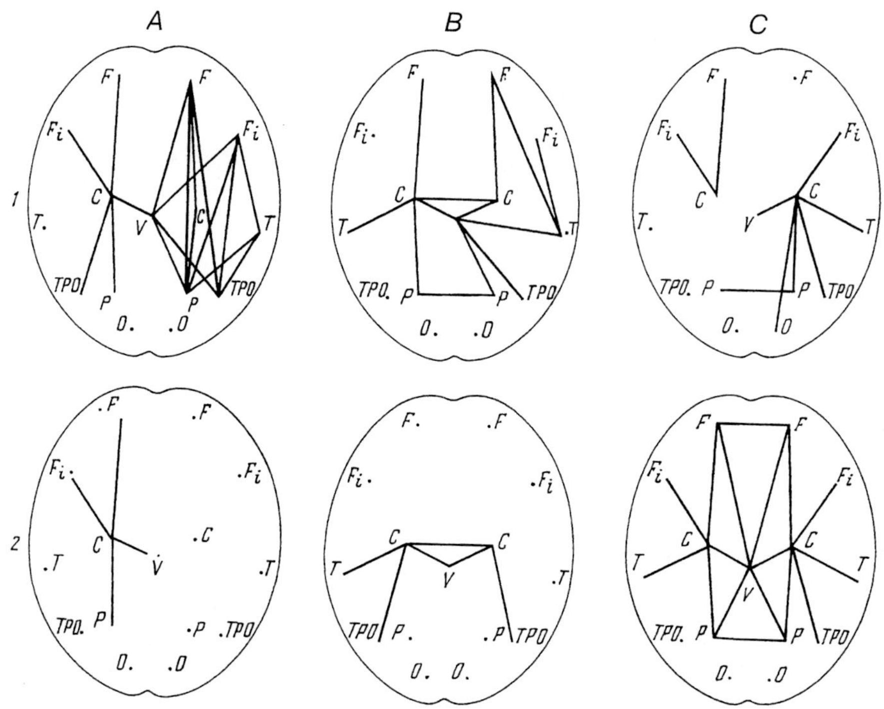

Figure 9.1. Topography of enhancement of alpha coherence during expectancy of the tactile task (LH stimulation, RSS series). Lines connecting pairs of cortical leads show the sites where alpha-Coherence values are higher during expectancy *versus* background (**1**, comparison type 1), or are higher in expectancy followed by correct response *versus* expectancy resulting in error (**2,** comparison type II). **A**—adult subjects; **B**, **C**—children aged 9 and 7–8 respectively. Letters in the schemes: registration points. All data represent group averages.

stimuli presented in the centre of the screen (control) and randomly included in the RSS and LSS series. The hemispheric dichotomy was observed in this case as well: Integration of cortical regions around the occipital cortex was seen in the LH, and diffuse intercentral interaction was noted in the RH (Machinskaya *et al.,* 1993; Dubrovinskaya *et al.,* 1993).

A substantially different picture was observed in children. First of all, the absence of hemispheric dichotomy should be emphasized. In both hemispheres of a child's brain the transformations which took place were similar. Bilateral modality-specific integration appeared in children during expectancy of the tactile task in the RSS series (Figure 9.1[B,C,1]), instead of the essentially different intrahemispheric organization in the adult brain (Figure 9.1[A,1]). A difference between the groups of children was also noted. The

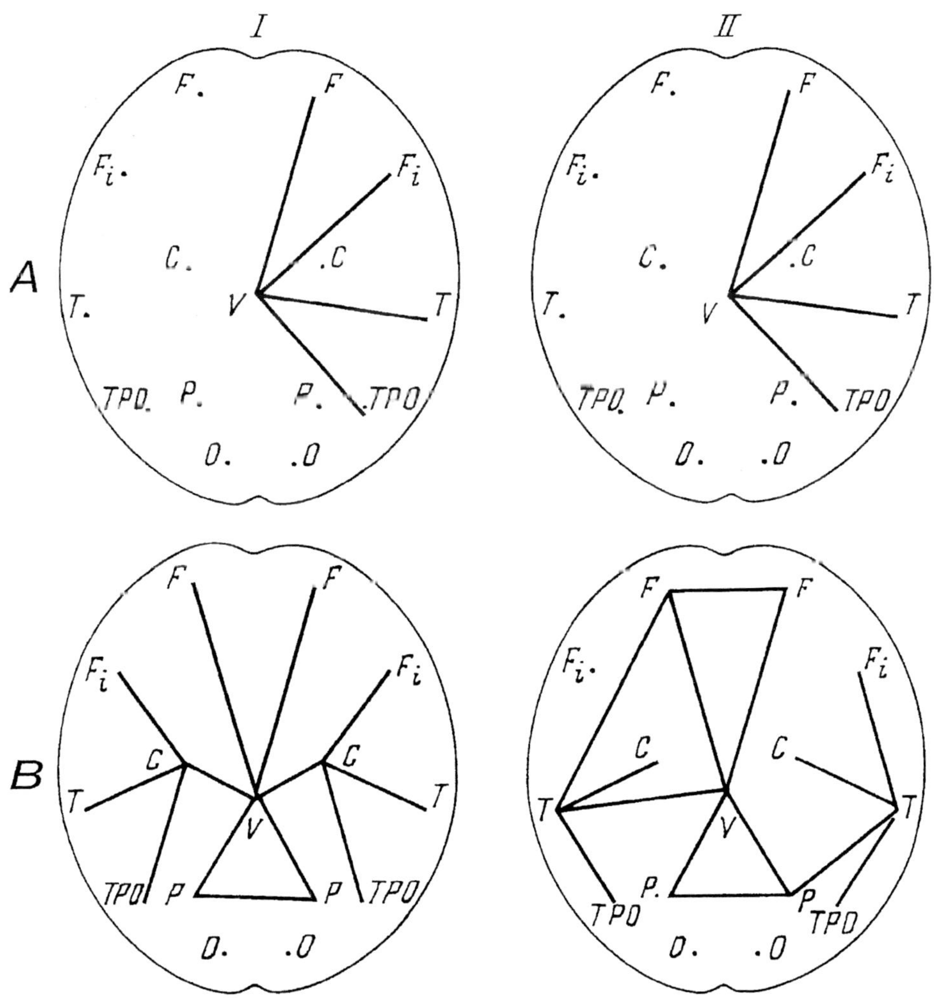

Figure 9.2. Topography of Coherence enhancement during effective prestimulus attention (comparison type II). Tactile (I) and auditory (II) stimuli are addressing RH (LSS series). Other indications as in Figure 1.

selectivity of modality-specific integration is much more pronounced in younger children, especially in the ipsilateral RH (Figure 9.1[C,1]). In 9-year-olds (Figure 9.1[B,1]) the LH characteristics are closer to adults. The RH organization is manifested as a combination of modality-specific integration with non-specific functional systems ("foci of interconnected activity"—FIA centred on the vertex and Td). Bilaterality of hemispheric organization in children is also seen clearly, as a result of type II comparisons (Figure 9.1[2]; Figure 9.2): Correct decisions in the RSS (Figure 9.1[B,C,2]) and the LSS (Figure 9.2 [C]) series are mediated by both hemispheres, in contrast to adults. In 7–8-year old children functional integration is completely independent of the side of stimulation. Almost mirror symmetry of functional organization is formed during expectancy, in both series. Bilateral modality-specific FIA, in central (Figure 9.1[C,2]; Figure 9.2[C,1]) and temporal (Figure 9.2[C,II]) regions are seen. In addition, bilateral modality non-specific integration between the vertex, and frontal and parietal areas was necessary for a correct response. In older children intercentral integration has mixed properties, and is dependent on the side of stimulation. Expectancy of tasks addressing the LH (i.e. the RSS series) produces bilateral modality-specific integration (Figure 9.1[B,2]tact). In the LSS series, generalized modality-independent transformations are revealed, differing however from those typical of adults.

Thus at the ontogenetic stages under study (7 and 9 years of age) the prediction of a forthcoming task is achieved in both hemispheres, as a duplication of the facilitating influences providing the informational and mobilizing effects of selective attention.

The subgroup of 7-year-olds, with brain immaturity focused on the fronto-thalamic system responsible for regulation of selective attention, demonstrated intrahemispheric organization different from that in children with a normal development course (see Figure 9.3). Modality-specific integration is absent in these subjects, pointing to a lack of both an appropriate preparatory set, and of the informational effects of selective attention. Indeed, the hemispheric organization of children with fronto-thalamic immaturity is characterized by bilateral modality-non-specific integration around the parietal regions (shown by the results of both type I and type II comparisons). The topography of integration suggests the probable involvement of the posterior attention system (Posner *et al.*, 1988,1989; Dujardin *et al.*, 1995). Constancy of the visual warning and feedback stimuli necessary for task fulfilment, and the prevention of distraction are problems of utmost importance for these children. Nevertheless, such an alternative strategy may be considered as adaptive in this case. In any case, due to the probability of the correct decision being above the chance level, one can suggest that modality-specific transformations may be timed for the poststimulus period.

The data under consideration, characterizing the brain organization underlying prestimulus attention in children and adults, are the result of the course of brain development, and of the age-specificity of brain functions. Predictive ability relies upon the remarkable "leap" in maturation of the frontal lobes (Luria, 1973; Stuss, 1992) which makes simple planning and prediction possible. On this basis, the triggering of certain components of the activation system can be accomplished via the descending frontal pathways, thus providing the appropriate preparatory set. Its lack in children with immaturity of the fronto-thalamic system supports this statement. Maturing frontal lobes acquire not only a predictive function but also an executive one—modulation of the functional state of task-relevant cortical regions.

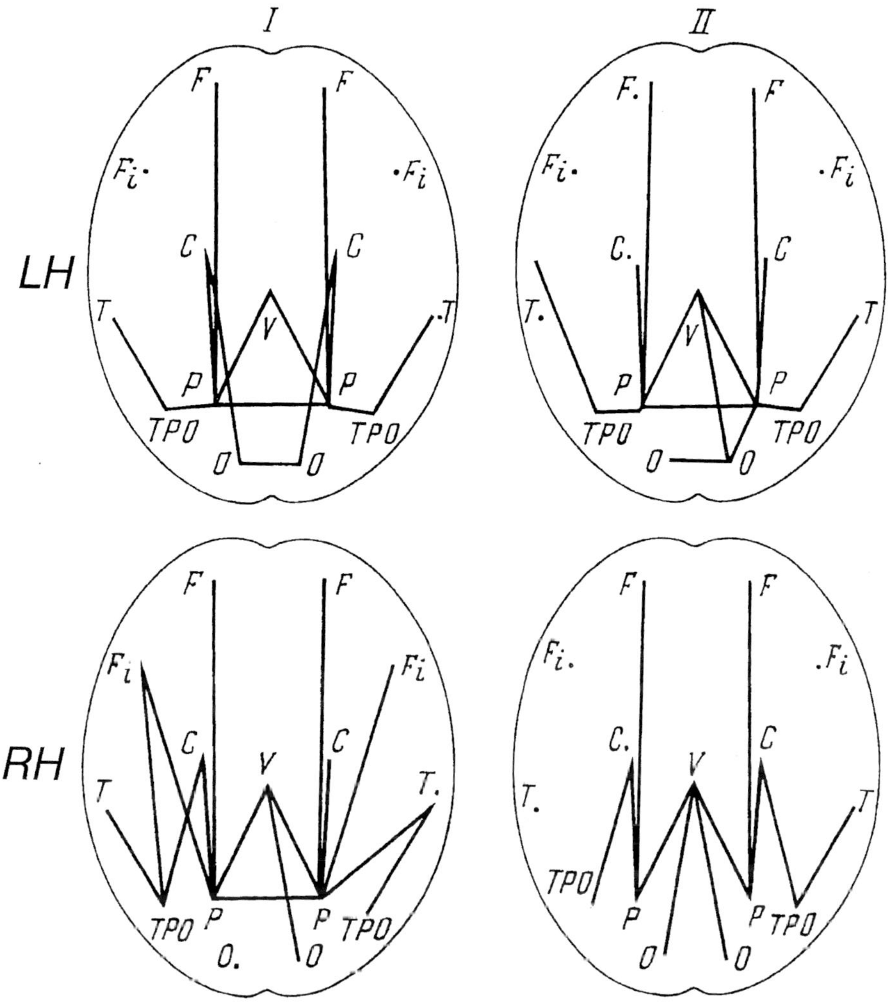

Figure 9.3. EEG organization of the task preceding attention in 7-year old children with CNS immaturity. **I**—tactile, **II**—auditory task. LH, RH—stimulation addressed to left and right hemispheres.

Age-specificity for the confirmation of predictions should be emphasized. At the age of 7, lack of hemispheric dichotomy, and bilateral symmetry of intrahemispheric organization reflect the long-lasting non-linear course of hemispheric specialization, that is the alternation of periods when local and diffuse cortical regional interaction takes place (Thatcher, 1994): According to the results of Coh analysis carried out by Thatcher it is at this age that a period of local interaction in the RH is present. Data obtained in our study are supported by the well known persistence of ipsilateral projections at this ontogenetic stage (Dennis, 1976), and by corpus callosum immaturity leading to a deficiency of all aspects of interaction betweeen hemispheres (Yakovlev and Lecours, 1967; Levy, 1985). By 10 years commissural organization comes closer to the mature level (Yakovlev and

Lecours, 1967; Dennis, 1976; Levy, 1985). Moreover, according to our data, the EEG-characteristics of hemispheric functions are changed. However, despite certain progressive developmental trends timed to the interval from 7 to 9 years, the informational and mobilizing effects of selective attention directed towards sensory parameters of stimuli are located in both hemispheres.

3. BRAIN ORGANIZATION OF ATTENTION BEFORE A VERBAL TASK

The same experimental paradigm (WS, expectancy, directed attention) was used to study the influence of the task type on predictive ability and preparatory set formation. The only difference consisted of the presentation of a more complex and uncertain verbal task, instead of a simple sensory one. In this study not only pre- but also post-stimulus EEG-reorganizations were compared to evaluate the attentive forecasting effects.

Task.: The subjects have to form words from horizontal letter strings, presented on the display, and to refer them to animal/nonanimal categories by pressing the corresponding button. The principle for rearrangement of letter patterns was changed from presentation to presentation to prevent the development of a constant algorithm. Selection of verbal material in preliminary sessions was made according to the time needed for decision making (5.5–6 s), to equalize the task difficulty across subjects. Thus, 45 three-, four- or five-letter words were presented respectively to children of two age groups and to adult subjects. WS (an exclamation mark) preceded task presentation by 3–4.5 s.

EEG recordings were made from 7 symmetrical parasaggital scalp sites (O, P, TPO, C, T, Fi, F) and from two saggital leads (Fz and Cz) referenced to linked ear lobes. The EEG was registered at rest (eyes closed), during attention (2.5 s expectancy), and during task performance (2×2.5 s). We are aware of debates concerning the appropriate control situation used in studies with human subjects (Medvedev *et al.*, 1996a). Nevertheless we believe that a comparison of resting and performance conditions makes it possible to reveal the operations immediately connected with the task, regardless of other forms of verbal activity typical of human beings in any condition.

Data processing. The general approach to data analysis did not differ from that described above. Due to the interindividual variability of components belonging to the commonly-distinguished alpha frequency subdivisions (Inouye *et al.,* 1986; Farber and Vildavsky, 1996), special attention was paid to choosing functionally identical components of the alpha-rhythm. The statistics of individual data were used to take this into account. For group statistics, equal frequency boundaries were chosen for all subjects, to compare Coh-values in the task and attention conditions *versus* the resting condition. This method allowed us to obtain more stable and similar changes, with less impact of the individual strategy.

Subjects. Healthy right-handed children aged 10 (n = 18) and 7 years (n = 12) and young adults (n = 17) were under study.

From group data in adults, intrahemispheric brain organization during verbal task expectancy (Figure 4.4[A,I]) was represented by interaction of the vertex, the LH-association structures, and right frontal cortex (F4). Data from the literature concerning brain organization of verbal function (Luria, 1973; Posner *et al.*, 1988,1989; Ivanitsky and Ilyuchenok, 1992; Medvedev *et al.*, 1996b) and our own results (Kulakovsky and

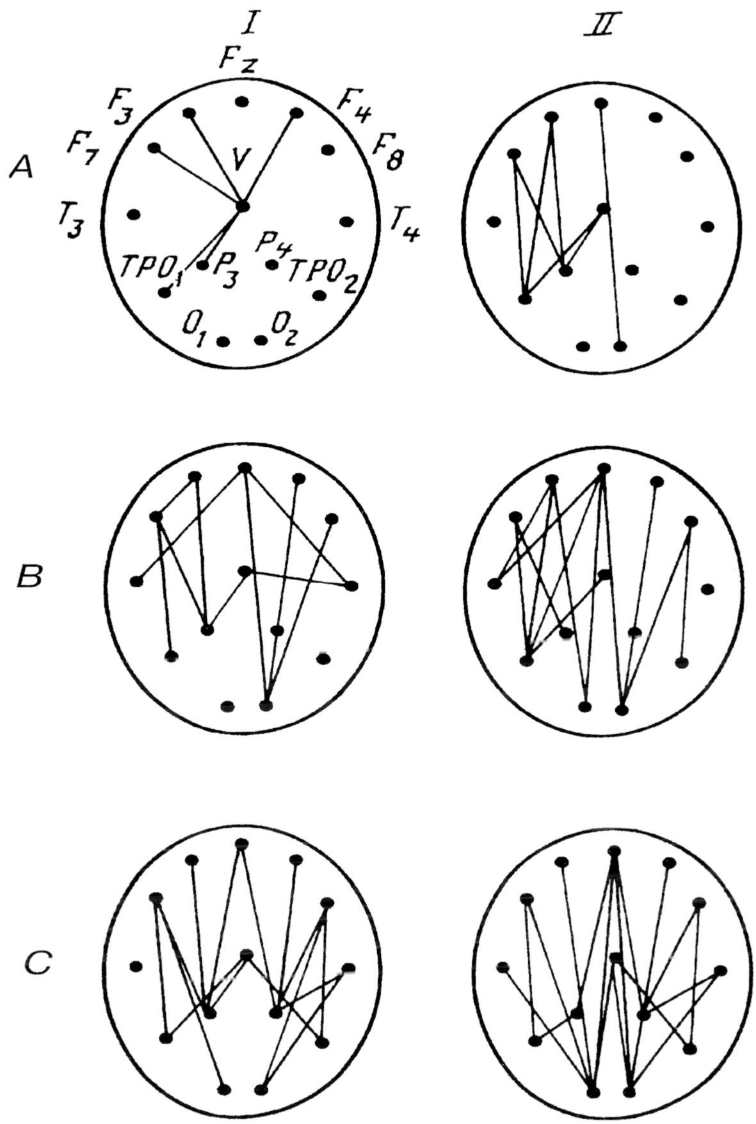

Figure 9.4. EEG organization of expectancy (attention) (**I**) and performance (**II**) of a verbal task in adult subjects (**A**) and in children aged 10 (**B**) and 7 (**C**) years.

Dubrovinskaya, 1997) testify that cortical regions active during attention are involved in verbal task performance (see also Figure 9.4 [A,II]). Only the interaction type differs, thus pointing to the situational specificity of brain organization, and to the ability to predict only the most probable components of the future system. It seems that an "activation field" is formed during expectancy, containing task-relevant regions of LH and RH-frontal cortex involved in mediation of sustained attention (Wilkins *et al.*, 1987).

Immediate verbal processing in the post-stimulus period is provided by another type of functional integration: Frontal, parietal, TPO regions in the LH are coupled with one another and with Cz. Participation of caudal (P, TPO) and rostral (F3, F7) cortical parts in verbal processing ensures the recognition and analysis of task specific material (letters on the screen), semantic decoding and categorization (Posner *et al.*, 1988,1989; Ivanitsky and Ilyuchenok, 1992). The results of group statistics showed that preparation for verbal processing and its realization are mediated exclusively by LH mechanisms. Functional integration in the LH is selectively formed on the basis of coherent high-frequency alpha-components (12–13.5 Hz), thus reflecting the effort of this intellectual operation (Giannitrapani, 1985; Krause *et al.*, 1997).

Hence in adults, predictions formed on the basis of existing information corresponded to the degree of certainty, and were realized through the specific EEG reorganization in the LH. (Participation of RH-mechanisms, not obvious in averaged group data, was clearly demonstrated using individual analysis, thus pointing to variability of the RH-"strategy".)

In children aged 10 (Figure 9.4 [B,I]), the functional system produced during attention (frontal cortex of both hemispheres and right occipital region) is probably mediated by the instruction content. The activity of executive centres of the activation system, and the involvement of the occipital cortex for primary perception of visual information, support this view. Similar patterns of activity persisted into the poststimulus period (Figure 9.4 [B,II]), with additional involvement of the left temporo-parieto-occipital cortex in verbal processing.

In 7-year old children, attention preceding the task is based on a generalized intercentral interaction, with slight RH dominance. Both parietal regions are intensely involved in such functional integration. The posterior attention system might be responsible for these transformations, providing the same function as in children with frontothalamic immaturity. The difficulties of formation of a preparatory set in the younger children are probably due to the uncertainty of specific characteristics of the task material. Functional interaction is enhanced during performance of the verbal task, without any remarkable change in the pattern of integration.

Comparison of brain organization underlying attention before simple sensory tasks and before verbal processing has shown that, along with the enhanced complexity and uncertainty of the task, not only the manner of confirming a prediction, but also its content was changed. The first parameter is determined by the degree of cortical regional maturity, by the dynamics of hemispheric specialization, and by the function of the activation system. The content of prediction depends on the ability to single out constant features in the task structure, according to the hierarchy of their significance. So the "specific components" of expected verbal processing are predicted in adults. In 10-year old children the most important instruction-mediated processes (mobilization and sensory analysis) are present. At the age of 7 task- or instruction-specific parameters are not obtained in the preparatory set. A similarity of pre- and post-stimulus brain organization was noted. Probably in the younger children a tonic set (general mobilization) was formed, lasting at least from the WS to the motor response. Based on this set, task fulfilment is accomplished by means of age-dependent mechanisms.

4. CONCLUSIONS

A study of the brain organization of attention, involving preparation for different kinds of performance, has been carried out. Evaluation of its dependence on several factors, such as the type of task, and the subjects' functional abilities (age, individual features) allowed us to observe adequate alternative brain strategies directed at facilitation of task performance. These data testify to the plasticity and dynamic character of functional brain organization. Ontogenetic transformations of attentional effects consist of their differentiation (informational and mobilizing aspects) and the enhanced selectivity of spatial and frequency organization. This ontogenetic trend is determined by the maturity level of the system providing predictive formation, and by the mechanisms of confirming predictions, involving various activation influences.

REFERENCES

Batuev, A. (1987). *High Integrative Systems of the Brain.* New York, London, Paris, Montreal, Tokyo, Melbourne, Gordon and Breach, 296 pp.

Beteleva, T., Dubrovinskaya, N. and Farber, D. (1977) *Sensory Mechanisms of Developing Brain.* Moscow, Nauka Publishing House, Chapt. 8 (in Russian).

Dennis, M. (1976) Impaired sensory and motor differentiation with corpus callosum. Agenesis: The lack of cortical inhibition during ontogeny? *Neuropsychologia*, **14**, 455–469.

Dubrovinskaya, N. (1995) On the problem of cortical activation: The role of certainty/uncertainty of situation. *Zhurnal Vysshey Nervnoy Dejatelnosty*, **45**, 638–646. (in Russian).

Dubrovinskaya, N., Machinskaya, R. and Savchenko, E. (1993). In: *Developing Brain and Cognition.* D. Farber, Ch. Njiokiktjien (eds). Amsterdam, Suyi Publ., pp. 11–133.

Dujardin, K., Bourriez, J. and Guieu, J. (1995) Event-related desynchronization (ERD) patterns during memory processes: Effects of aging and task difficulty. *Electroencephalography and Clinical Neurophysiology*, **96**, 169–182.

Farber, D. and Njiokiktjien, Ch (1993) *Developing Brain and Cognition.* Amsterdam, Suyi Publ.

Farber, D. and Vildavsky, V. (1996) Heterogeneity and age dynamics of the alpha rhythm of the electroencephalogram. *Human Physiology*, **22**, 517–524. (Translated from Russian).

Giannitrapani, G. (1985) *The Electrophysiology of Intellectual Functions.* Basel, Kargel, 247 p.

Inouye, J., Shinosaki, K., Yagasaki, A. and Shimizu, A. (1986). Spatial distribution of generators of alpha-activity. *Electroencephalography and Clinical Neurophysiology*, **63**, 353–360.

Ivanitsky, A. and Ilyuchenok, I. (1992) Brain biopotentials mapping during verbal task performing. *Zhurnal Vysshey Nervnoy Dejatelnosty*, **42**, 627–635. (in Russian).

Krause, Ch., Porn, B., Lang, A. and Laine, M. (1997) Relative alpha desynchronization and synchronization during speech perception. *Cognitive Brain Research*, **5**, 295–299.

Kulakovskii, Yu. and Dubrovinskaya, N. (1997) Age-dependent peculiarities of Brain Organization of Verbal Activity: An Electrophysiological Analysis. *Human Physiology*, **23**, 378–380 (Translated from Russian).

Levy, J. (1985) Interhemispheric collaboration: Single mindedness in the asymmetric brain. In: C.N. Best (ed.) *Hemispheric Function and Collaboration in the Child.* New York, Academic Press, pp. 11–31.

Lukashevich, I., Machinskaya, R. and Fishman, M. (1995) Determination of brain function in young schoolchildren with learning problems. *Human Physiology*, **20**, 353–358. (Translated from Russian).

Lukashevich, I. and Sazonova, O. (1996) The Effects of Lesions in Different Thalamic Regions on the Character of Brain Bioelectrical Activity in Human. *Zhurnal Vysshey Nervnoy Dejatelnosty*, **46**, 866–874 (in Russian).

Luria, A.R. (1973) *The working Brain.* Harmondsworth, Penguin Books, 192 pp.

Machinskaya, R. and Dubrovinskaya, N. (1994) Age-dependent differences in functional brain organization of selective attention: Perceptual task expectancy. *Zhurnal Vysshey Nervnoy Dejatelnosty*, **44**, 448–456 (in Russian).

Machinskaya, R. and Dubrovinskaya, N. (1996) Brain hemispheres functional organization during selective attention in 7–8 year old children. *Zhurnal Vysshey Nervnoy Dejatelnosty*, **46**, 437–446 (in Russian).

Machinskaya, R., Lukashevich, I. and Fishman, M. (1997) Dynamics of Brain Electrical Activity in 5- to 8-Year-Old Normal Children and Children with Learning Difficulties. *Human Physiology*, **23**, 517–522. (Translated from Russian).

Machinskaya, R., Machinsky, N. and Deryugina, E. (1993) Functional organization of the right and left hemispheres during directed attention. *Human Physiology*, **18**, 77–85. (Translated from Russian).

Machinskii, N., Machinskaya, R. and Trush, V. (1988) Alpha diapason of the EEG at directed attention. *Neuroscience and Behavioral Physiology*, **18**, 216–222.

Medvedev, S., Bekhtereva, N., Vorobjev, V., Pakhomov, S. and Rudas, M. (1996a) Human brain processing of different characteristics of visually presented words studied with positron emission tomography. *Human Physiology*, **22**, 174–181 (Translated from Russian).

Medvedev, S., Bekhtereva, N., Vorobjev, V., Pakhomov, S. and Rudas, M. (1996b). Human brain processing of different characteristics of visually presented words studied with positron emission tomography II. Brain system of word reading. *Human Physiology*, **22**, 263–268. (Translated from Russian).

Posner, M., Petersen, S., Fox, P. and Raichle, M. (1988) Localization of Cognitive Operations in the Human Brain. *Science, New York*, **240**, 1627–1631.

Posner, M., Sandson, J., Dhawan, M. and Shulman, S. (1989) Is word recognition automatic? A Cognitive-Anatomical Approach. *Journal of Cognitive Neuroscience* **1**, 50–60.

Sheppard, W. and Boyer, R. (1990) Pretrial EEG Coh as a predictor of semantic priming effects. *Brain and Language*, **39**, 57–68.

Skinner, J. and Lindsley, D. (1973) The unspecific mediothalamic-frontocortical system: Its influence to cortical activity and Behavior. In: K. Pribram, A. Luria (eds) *Psychophysiology of the Frontal Lobes*. Academic Press, New York, London, pp. 185–212.

Stuss, D. (1992) Biological and Psychological Development of Executive Functions. *Brain and Cognition*, **20**, 8–23.

Tecce, J. (1972) Contingent negative variation (CNV) and psychophysiological processes in man. *Psychological Bulletin*, **68**, 20–27.

Thatcher, R. (1994). Cyclic Cortical Reorganization: Origins of Human Cognitive Development. In: G. Dawson, K. Fisher (eds). *Human Behavior and Developing Brain*. New York, London, The Guilford Press, pp. 232–269.

Wilkins, A., Shallice T. and McCarthy, R. (1987) Frontal lesions and sustained attention. *Neuropsychologia*, **25**, 359–365.

Yakovlev, P. and Lecours, A.R. (1967) The myelogenetic cycles of regional maturation of the brain. In: A. Minkowski (ed.) *Regional Development of the Brain in early Life*. Oxford, Oxford University Press, pp. 3–70.

10 Formation and Realization of Individual Experience in Humans and Animals: A Psychophysiological Approach

Yu. I. Alexandrov, T. N. Grechenko, V. V. Gavrilov, A. G. Gorkin, D. G. Shevchenko, Yu. V. Grinchenko, I. O. Aleksandrov, N. E. Maksimova, B. N. Bezdenezhnych and M. V. Bodunov

Laboratory of Neural Basis of Mind, Institute of Psychology, Russian Academy of Sciences, Moscow, Russia
nyualex@psychol.ras.ru

A systemic methodological approach to psychophysiology is described. In the framework of this approach a wide range of experimental data are analyzed, including the results of neuronal recordings *in vitro*, and in awake normal and pathological animals, performing both complex instrumental and simple behavioural acts. Also included are data from experiments with human subjects in tasks involving categorization of words, skilled performance, participation in game activity in groups, and completion of psychodiagnostic questionnaires. On the basis of these analyses, qualitative and quantitative descriptions of the principles of formation and realization of individual experience are suggested within the framework of a unified methodology.

KEYWORDS: psychophysiology, functional system, individual experience, systemogeny, learning, memory, humans, animals, neuronal activity, event-related potentials, individuality

1. INTRODUCTION

Discovering the principles of organization of behaviour, based on experience accumulated by an individual, and the laws governing the formation of such experience is a multidisciplinary task. This general problem poses the majority of the specific questions of psychology, neurosciences, developmental biology, and genetics. At the same time, the solution of the general problem can be based only on synthesis of the achievements of a wide range of disciplines. Such synthesis is hampered by obstacles resulting from attempts to create a unified description from diverse data relating to humans and animals, of an individual synapse, or a neurone, or a whole organism, complex unlocalized mental processes and local physiological phenomena. The aim of the present article is to suggest a system of views, based on the literature and our own experimental data, within the framework of which such obstacles may be overcome.

In order to describe the cerebral basis of formation and realization of *individual experience* (IE), we define first the *elements* of IE (EIE). Today only a few researchers question the conclusion that the "properties ... of a brain are emergent" and are "systemic", not "just the sum ... of properties" of neurones, but a specific quality that emerges as a result of "dynamic interaction" of neurones within the system (Mountcastle, 1995, p. 294).

Analysis of possible levels of behaviour suggests that the level of "a unified group of neurones" is the most elementary level of analysis where the corresponding behaviour may still be described as an emergent function (Bottjer *et al.*, 1994). A cerebral equivalent of EIE, which is established during the formation of a new behaviour, and realized during its subsequent performance, may be defined as an organization of a group of neurones, constituting the corresponding system. The question of what is meant by a "system" must be answered before we can use an understanding of EIE to describe the formation and realization of IE.

From our point of view, the most well-developed and un-contradictory version of the systemic approach to analysis of neuronal basis of behaviour, is the theory of functional systems elaborated by P.K. Anokhin and his school (Anokhin, 1973). The major distinguishing characteristic and advantage of this theory is the definition of a system-creating factor—the *result* of a system, which is understood as a desired relation between an organism and environment, achieved through the realization of that system. In other words, the principal determinant of a system is an event which is not in the past with respect to behaviour, that is, a stimulus, but which occurs in the future, a result. Thus a system is understood as a dynamic organization of activity of components with different anatomical localization, the interaction of which takes the form of *mutual facilitation*, in the process of ensuring a result that is adaptive for an organism.

It was demonstrated that the mutual facilitation in achieving any behavioural outcome is ensured by uniting synchronously-activated neurones situated in different brain structures (Shvyrkov, 1990). There is increasing evidence for this suggestion (Bullier and Nowak, 1995). The evidence is also increasingly important for the understanding not only of a specific behaviour, but also of learning. The association of synchronously active cells may ensure the achievement of the result even during the first trial, and may serve as a base for further consolidation: "Neurons wire together if they fire together" (Singer, 1995, p. 760).

In addition to the systemic idea described above, another important premise of the theory of functional systems is the idea of development (Shyleikina and Khayutin, 1989). Both ideas are merged in the concept of systemogeny, which states that, during early ontogeny, those differently localized elements undergo selective and accelerated maturation that is essential for achieving the results of the systems, providing for the survival of an organism at the early stages of individual development (Anokhin, 1973). Nowadays it is commonly accepted that many regularities of modification of functional and morphologic characteristics of neurones, as well as of control of gene expression, serve as a basis for the formation of adaptive behaviour in adults, and are comparable to those found at the early ontogenetic stages (Anokhin and Rose, 1991; Bottjer *et al.*, 1994; Singer, 1995). The idea that systemogeny takes place not only during the early ontogenetic period, but also during adult development was formulated within the framework of the theory of functional systems nearly 20 years ago (Shvyrkov, 1978; Sudakov, 1979). This idea arose because the formation of a new behavioural act is always a formation of a new system. Later it was suggested that an understanding of the role of different neurones in the organization of behaviour depends on the history of behavioural development (Alexandrov, 1989; Alexandrov and Aleksandrov, 1982), or in other words, the history of the successive systemogenies, and the system-selective concept of learning was inferred (Shvyrkov, 1986). The latter concept is in line with the modern idea of "functional specialization" which substituted the idea of "functional localization" (Mountcastle, 1995) and with the idea of the selective, rather than the instructive, principle of learning (Edelman, 1987). This concept considers the formation of a new system as the fixation of the stage of individual development—the formation of a new EIE during learning. The base of this process is the specialization of some "reserve" of (silent) neurones, but not the change of specialization of previously specialized units. Thus, the new system becomes an "addition" to the existing EIE (Shvyrkov, 1986, 1995). The selection of particular neurones from the reserve is governed by their individual features, that is, by the characteristics of their "metabolic needs" that are genetically determined. Newly formed systems do not substitute previously existing ones, but are "superimposed" over them; the appearance of neurones with new specializations results in the increase of the total number of units activated, whereas the number of neurones with old specializations does not decrease (Gorkin, 1988; Shvyrkov, 1986). The suggestions that the number of active neurones is increased during learning, and that learning involves new neurones rather than "re-learning" of the old ones has recently been confirmed by data from other laboratories (Bradley *et al.*, 1996; Wilson and McNaughton, 1993).

What does it mean—"to superimpose, but not to substitute"? Many experiments in our laboratory have demonstrated that a complex instrumental behaviour is mastered not only through the realization of new systems (Figure 10.1, new systems), that were formed during the process of learning the acts comprising the behaviour, but also by the simultaneous realization of older systems (Figure 10.1, old systems), that had been formed at previous stages of individual development. The latter may be involved in the organization of many behavioural patterns, that is, they belong to EIE that are common to various acts (Figure 10.1). Therefore, it appears that the realization of behaviour is the realization of the history of behavioural development, that is, of many systems, each fixing a certain stage of development of the given behaviour.

These ideas are fundamental for *systemic psychophysiology*, which suggests the following solution to the psychophysiological (mind-body) problem. The organization of physiological processes into a system is based on specific systemic processes. Their substrate is physiological activity, whereas their informational content is psychical. In other words, psychical and physiological are different aspects of the same systemic processes (Shvyrkov, 1995). From this point of view, mind may be considered as a subjective reflection of the objective relation of an individual to the environment. That is, mind is considered as a structure represented by systems accumulated in the course of evolutionary and individual development. Relations between these systems (intersystem relations) may be described qualitatively, as well as quantitatively. The range of problems of systemic psychophysiology includes studies of formation and actualization of systems (EIE), studies of their taxonomy, and dynamics of intersystem relations in behaviour and activity. Thus, it may be concluded that investigation of the formation and realization of an IE is the task of systemic psychophysiology. It should be carried out at different levels, ranging from cellular and subcellular to complex human activity.

2. INDIVIDUALITY OF A NEURONE

As noted above, the system-selective concept of learning is based on the following suggestion: Neurones are originally diverse in their genetic and, consequently, in their

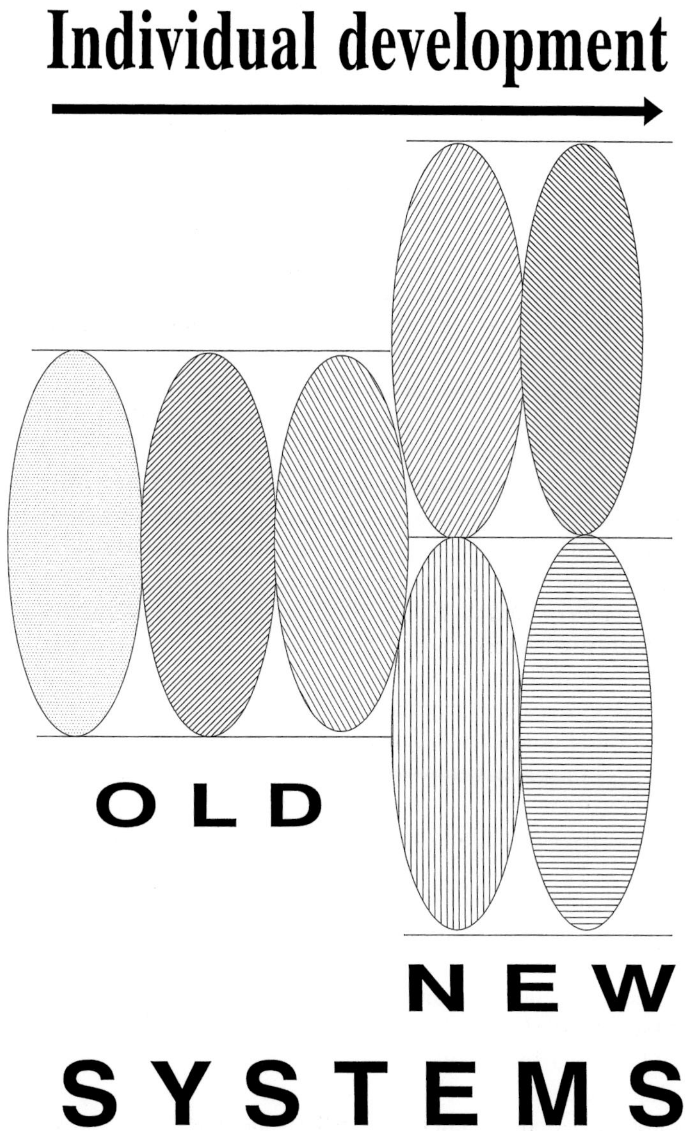

Figure 10.1. The scheme of the systemic structure of behaviour. (See text for explanation).

metabolic properties, and, during learning, only neurones with specific properties are incorporated into a system's organization. The stability of these properties was demonstrated in experiments with completely isolated nerve cells, by using the methods of mechanical and fermentative treatment.

When working with isolated neurones, cells keep the specific properties of background activity that they used to have in the nervous system. Alving (1968) used a mechanical method of isolation to demonstrate that the spontaneous electrical activity of isolated nerve cells which she recorded before isolation, stayed similar. Chen *et al.* (1971), using fermentative treatment, found that completely isolated identified neurones maintained, after isolation, the main electrophysiological characteristics, such as the level of membrane potential, rhythm and patterns of spontaneous and elicited activity. Chemosensitivity was also stable. Isolated neurones were characterized by chemosensitivity to the same neurotransmitters that were effective before isolation. Our experiments have been performed on completely isolated neurones of the snail *Helix pomatia*. The results confirmed the stability of individual electrophysiological characteristics of identified cells. Not only were the background activity and chemosensitivity found to be stable, but also the dynamics of complex forms of neuronal plasticity remained similar prior to and after isolation (Grechenko, 1993). So, from the comparison of these individual characteristics of the same cells *in vitro* and *in vivo* it can be concluded that the analyzed properties of neurones in adult animals are stable.

Culturing identified isolated neurones *in vitro*, and analyzing their properties after involvement in the formation of new neuronal networks, allows us to find out if these properties stay stable in such a new neuronal organization. Syed *et al.* (1990) have done the experiments by culturing neurones of Aplysia, which formed new interneuronal connections. During these experiments the authors tried to describe the modifications of the neuronal electrophysiological characteristics, but no modifications were elicited by the procedure of culturing or axonotomy, the main parameters of electrophysiological activity remaining constant. It is necessary to note that each neurone formed new synaptic connections similar to those functioning *in vivo*.

Similar results were obtained in two artificial neuronal nets: respiratory and motor networks. In the latter, the transformation of action potentials was explored in the course of associative learning. These modifications were similar to the experiments *in vivo* and *in vitro*. Individual properties of neurones were stable in the neurotransplantation experiments. The stability of structural, intrinsic neurotransmitter, and electrophysiological characteristics of graft transplanted neural tissue have also been shown (Vinogradova, 1994). These findings confirm the stability of individual properties of a neurone, and support views on the regularities of learning suggested by the system selection concept.

3. RESULTS OF BEHAVIOUR AS A DETERMINANT OF FORMING OF INDIVIDUAL EXPERIENCE

Within the framework of our approach, the specialization of neurones is considered to be a systemic one instead of "sensory" or "motor". Thus, we assume that even in conditions of "sensory deprivation"—for example, cessation of contact with the visual environment—neuronal activity in "visual" structures is necessary for achievement of results of behaviour. Indeed, it was found that the activity of neurones in visual cortex, in retina and lateral geniculate body (Alexandrov and Aleksandrov; 1982; Alexandrov and Jarvilehto, 1993) is related to the realization of food-procuring behavioural acts in animals, both with "open" eyes and with eyes closed with light-tight covers. According to the same logic, it should be assumed that during the formation and realization of behaviour under "motor" deprivation, and even in combination of "sensory and motor" deprivation, the activity of neurones is related to the realization of systems aimed at achieving the results of behaviour as well. This is shown by the fact that if an animal is restricted from moving voluntarily, but is nevertheless able to achieve some behavioural results during passive movement within an experimental arena, then a specific IE is formed, which corresponds to the analogous behaviour in freely moving animals. That is, neurones specialized according to the elements of this IE can be found.

This last assumption was tested in experiments (Gavrilov *et al.*, 1994, 1996) with single unit recordings from CA1 complex-spike cells in awake rats, slightly restrained in a sling, and placed on a computer-driven robot. A rat was moved within a square arena (3 m × 3 m), from one corner to another, along the walls and diagonally. A drop of water was delivered (as a "reward") every time the rat approached one of the corners, this contingency remaining the same throughout the experiment. We found that about a half of the neurones increased their firing rate significantly while the rat was passively transported in particular parts of the arena, although these neurones had "spatial specificity" of low resolution, that is, their "firing fields" were larger compared to the those found in freely moving animals (O'Keefe and Nadel, 1979; O'Keefe and Recce, 1993). Some of these neurones maintained the same spatial selectivity of discharge when the rat was displaced on the robot in total darkness.

These results could be interpreted in terms of the currently-dominant views on the hippocampus as a pivotal structure for forming high-level representations of space on the basis of convergence of multimodal sensory information (O'Keefe and Nadel, 1979). From our point of view, these data support the idea of the determining role of results of behaviour in the formation of elements of IE. Representation of space is considered to be a reflection of the environment divided into elements according to the results achieved in this environment ("space of outcomes") on the base of some sensory "modalities". Formation of this representation is the formation of EIE. This also means that the existence of various spatially selective neurones which are active when the rat approaches one of the corners, irrespective of the direction and speed of passive displacements (i.e. irrespective of different means of attainment of the animal's contact with a particular place of the arena) is due to the fact that this place is in a constant spatial relation to that corner in which the animal was "rewarded" with water. Disappearance of the specific activation of hippocampal neurones when a restrained rat was placed into the "firing fields" of these neurones, that is, into the areas of the arena where these neurones had increased discharge activity when tested in freely moving rats (Foster *et al.*, 1989), appears to be related to the change of behaviour from food-procuring to defense, and hence, to a change in the set of elements of the IE involved in the realization of the behaviour. Context-dependence of behaviour for spatial selectivity of discharges of the hippocampal neurons was shown earlier by Alexandrov *et al.* (1993) and Wiener *et al.*, (1989).

In sum, the results described above offer good support for our assumptions. Even in restrained animals, passive transportation within the "space of outcomes" results in the activation of neurones in relation to the realization of EIE, which reflect the subjective "division" of the environment according to the results achieved in the environment. The result is similar to the findings for freely moving animals, although the structures of IE (both a set of elements and relationships between the elements) are probably different.

4. LEARNING HISTORY AND SYSTEMIC ORGANIZATION OF BEHAVIOUR

From the assumption that the structure of IE is determined by the history of its formation, one may suppose that the systemic organization of the same behaviour, formed by different learning strategies, differs between individuals because the different history means the formation of a different IE structure. The role of learning history was demonstrated in our experiments. Rabbits were trained to perform a food-procuring instrumental behaviour in a cage with two feeders and two pedals in the corners (Figure 10.2). At any given moment, only one pedal was effective—pressing that pedal switched on a feeder positioned near the same wall. Two different strategies were used during the training of the animals. The animals of one group were trained to execute the whole behavioural cycle along one wall of the cage (pressing the pedal, coming to the feeder and seizing food, pressing the pedal, and so on), then along the other wall. The animals of the second group were trained to obtain food from one feeder, and then from the other; to press one pedal, and then to press the other one (Gorkin and Shevchenko, 1991a,b, 1995).

The reflection of the learning history in patterns of specialized neurones' activity was studied in experiments by recording the activity of limbic cortex neurones (area retrosplenialis) in rabbits. The averaged frequency of activity and the activation probability were calculated for each behavioural act. Each of two behavioural cycles (along a concrete wall of the experimental cage) was divided into five stages (behavioural acts): seizing food in a feeder, turning a head to a pedal, approaching a pedal, pressing a pedal, approaching a feeder. So, all food-procuring behaviour in the cage turned out to be presented in ten stages: 1st to 5th on the left side of the cage and 6th to 10th on the right side. For each stage we have defined the mean frequency of neuronal activity during the time of its recording, and the distribution of frequencies composed a pattern of neuronal activity in behaviour (Figure 10.2).

For further analysis we selected neurones specialized for new systems of acts of approaching and/or pressing pedals ("pedal" neurones), as well as acts of approaching and/or seizing food in one of the feeders ("feeders" neurones). Neurones that showed activation in relation to different movements of the animal were considered to be specialized relative to old systems. Whether their activation appears or not is related specifically to a certain movement but independent of its behavioral context. Activations always appear during the same movement, which is performed for instance in relation to approaching the feeder or the pedal. Some neurones showed activation in relation to novel behavioural acts established late in individual development, such as during animal's learning in the experimental cage (e.g. approaching the feeder, approaching the pedal, pressing the pedal). Whether their activation appears or not is specifically related to a certain behavioural act but independent of its motor characteristics. For example, similar activity of these neurones is recorded when the animal presses the pedal with the left paw, right paw, or both. It appeared during the behavioural act, for which this neurone was specialized. This activation was usually several times greater than the "non-specific" activity of the neurone, which was recorded during other behavioural acts and which, unlike the specific one, was much more variable and appeared in fewer than 100% of cases.

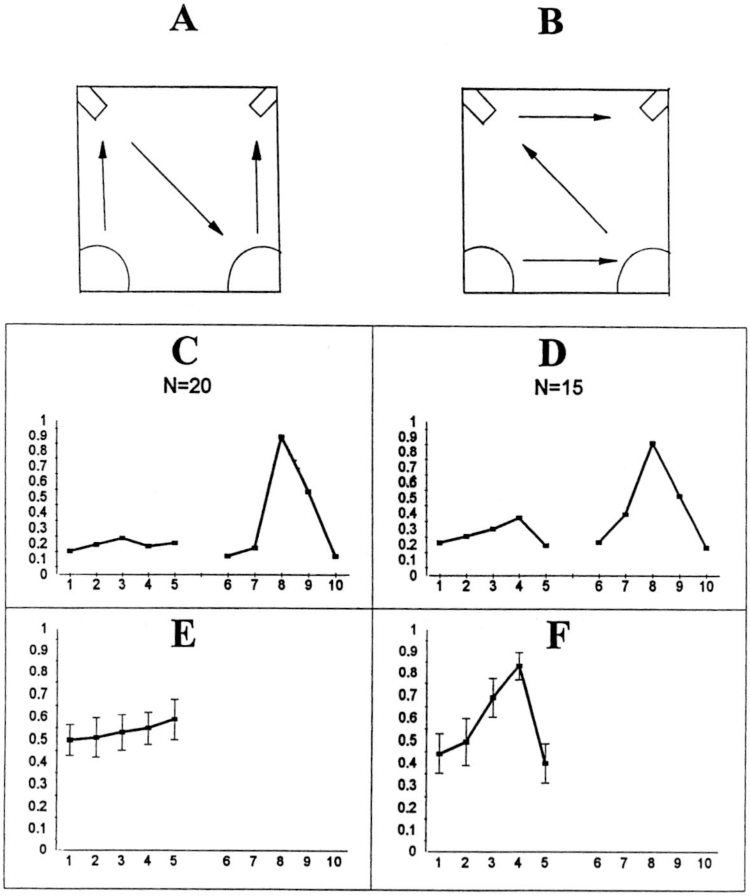

Figure 10.2. Activity patterns of limbic cortex neurons as a function of learning strategy. **A, B**—schemes of learning strategy; the arrows show the sequence of forming of behavioural acts. **C, D**—collective activity patterns of neurones, specialized relative to approaching the first (C) and second (D) pedals in training sequence for rabbits trained by strategies A and B respectively. Abscissa: numbers of the behavioural acts; ordinate: averaged frequency of the activity, normalized with respect to the frequency of impulse activity in the specific act. **E, F**—collective patterns of nonspecific activity, normalized with respect to its maximum, for groups of cells represented on parts C and D respectively.

Comparison of the activity patterns of neurones with similar specialization showed that their "nonspecific" activity differed greatly (Figure 10.2 C,D). However, the distribution of frequencies was not random. There was additional activation of the neurones, specialized relative to the second pedal (with respect to the order of training), when the rabbit pressed the first one. Supplementary analysis involved normalization of the frequency of nonspecific activity with respect to the maximal frequency of activity during nonspecific acts (Figure 10.2 E,F). The analysis allowed this activation to be related to definite strategies of training. It appeared only when the formation of corresponding acts, related to the first pedal, directly preceded the formation of the acts of approaching and pressing the second pedal in the history of training. Thus, among the systems of behavioural acts formed during training one after another, and performed by an animal at different sides of the cage, we found facilitating intersystem relations manifested in a raised degree of actualization of the last-formed system while an animal performed the previous one. Achievement of an act's result is ensured by realization of a specific EIE as well as others. Whereas the specific EIE are realized in the act in all cases, the probability and the degree of actualization of nonspecific ones are considerably lower.

A similar phenomenon was found for "feeder" neurones as well. Additional activation was detected in the nonspecific activity of cells, specialized relative to the second feeder, in respect to the order of training. The particular place where this activation appeared—during approaching and pressing a pedal at the other side of the cage, or during seizing of food from the other feeder—depended on the strategy of training, that is, which act preceded the one specific for this neurone in the animal's training.

Earlier we showed that systemic specialization of a neurone is its permanent characteristic (Gorkin and Shevchenko, 1991a,b). That is why neuronal activity can serve as an index of specific EIE actualization (Shvyrkov, 1995), and "nonspecific" activity of a neurone may indicate the retrieval of a specific system from memory during performance of other behavioural acts. Studies of the activity of system-specific neurones during the performance of cyclic food-procuring behaviour may reveal the relations between the specific system and other functional systems of analyzed behaviour. Thus, identification of intersystem relations can reveal the IE structure acquired by learning. The data obtained confirm the assumption that the IE structure and, consequently, the system organization of behaviour in which this IE is actualized, are determined by the developmental history of the behaviour.

5. "PROJECTION" OF INDIVIDUAL EXPERIENCE UPON BRAIN STRUCTURES AND THE POSSIBILITIES OF ITS MODIFICATION

Spinelli (1978) obtained impressive data, which demonstrated that when kittens were trained to move their front paw in response to stimuli in one orientation, the area of representation of this paw in somatosensory cortex was significantly increased compared to the control animals. Later it was demonstrated that the steady reorganization of receptive fields corresponding to the characteristics of detected objects was induced by learning in adults as well (Tanaka, 1993). An analysis of the literature leads to the conclusion that receptive fields and cortical maps may be modified "at all times between conception and death" (Wall, 1988, p. 549), although the magnitude of these modifications may differ. For example, it was shown that cortical representation of the fingers of the left hand in string players was increased as compared to control subjects, and the increase was greater, the earlier a subject started learning to play (Elbert *et al.*, 1995). Traumatic influences, such as finger-amputation, that force animals to reorganize their behaviour, also induced receptive fields modifications and corresponding changes in cortical maps (reviewed by Wall, 1988).

Analysis of results of our many experiments brought us to the conclusion that testing of receptive fields of neurones reveals their involvement in subserving different behavioural acts (Alexandrov, 1989; Alexandrov and Jarvilehto, 1993; Shvyrkov, 1990). Taking this conclusion into account, it is possible to consider the results of these studies as support for the postulate that the projection of IE upon brain structures in animals and humans changes in the course of individual development, and depends on its characteristics.

The study of the projection of IE upon brain structures in the present framework implies the comparison of patterns of systemic specialization of neurones belonging to these structures at different stages of individual development in normal and pathological subjects. A pattern of specialization of neurones (within the given structure) is defined according to the set of systems with respect to which units of this structure are specialized, and also according to the quantitative relation among neurones belonging to different systems. Comparison of patterns of specialization of neurones in rabbits' limbic and anterolateral motor cortex at successive stages of learning instrumental behaviour revealed that patterns were changed differently in the cortical areas studied (Gorkin, 1988; Shvyrkov, 1986). The change was due to the appearance of a new group of active neurones specifically related to the behavioural act after learning it (for example, pressing a pedal). The number of such new units in the limbic cortex was significantly greater than in the motor one. Thus the resulting specialization pattern of neurones in these structures was entirely different (Figure 10.3, Control): the limbic cortex, as well as hippocampal CA1 and DG, acquired significantly more neurones with new specializations than did the motor cortex. Recording of unit activity in many cerebral structures during instrumental food-procuring behaviour, carried out in our laboratory, demonstrated that, generally, neurones with new specializations were abundant in the cerebral cortex (though different cortical areas may vary with respect to this parameter), whereas phylogenetically archaic and peripheral structures had very few of them, if any (see Alexandrov, 1989; Shvyrkov, 1995). It is reasonable to assume that the specificity of IE projection to cerebral structures is determined by the particular characteristics of neurones composing these structures. These characteristics determine the involvement of neurones of the given structure in the formation of the particular behaviour.

Can anything besides normal learning cause change in IE projection? We found that after recovery of instrumental food-procuring behaviour in rabbits that had been impaired due to bilateral damage of visual cortex, the pattern of neurones' specialization in motor cortex was changed. The percentage of neurones specialized with respect to new systems increased (Alexandrov *et al.*, 1990). Conversely, after acute ethanol administration the portion of active "new" neurones decreased, not however in the motor cortex (Alexandrov *et al.,* 1991) but in limbic structures (Figure 10.3, compare Control and Ethanol) (Alexandrov *et al.*, 1990a, 1993). This effect is due to selective suppression of the activity of neurones belonging to new systems, especially cells located in the upper cortical layers (II-IV). The similar increased sensitivity of relatively new EIE was also found at early ontogenetic stages—in altricial nestlings at the stage of formation of natural behaviour (Alexandrov and Alexandrov, 1993).

In order to test the hypothesis that the age of EIE, actualized during realization of behaviour, is one of the major factors determining the effects of ethanol in humans, we compared the impact of ethanol on event-related potentials (ERP) accompanying actions requiring the use of knowledge that subjects had acquired at earlier and later stages of individual development: at the time of acquisition of native and foreign languages, respectively. After alcohol intake, the amplitude of ERP components decreased, compared to the control condition, in the task of categorization of words of both native and

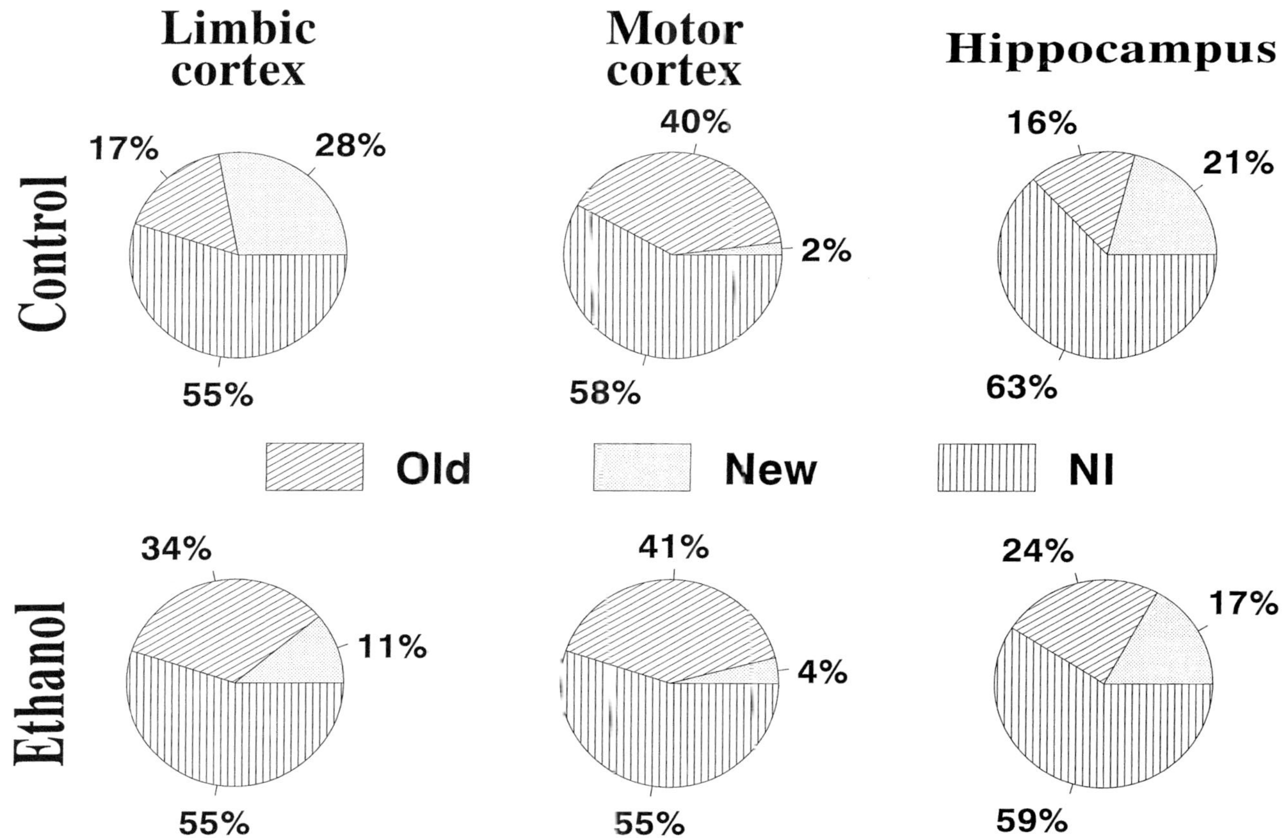

Figure 10.3. Relative numbers of new neurones (cells belonging to new systems formed in rabbits during the learning process in the experimental cage), old neurones (cells belonging to systems formed at previous stages of individual development) and noninvolved neorones (cells displaying no activation in a constant relation with this or that stage of behaviour) in limbic, anterolateral motor cortex and hippocampus in Control (Control) experiments and after acute alcohol (ethanol) administration (1 g/kg, i.p., Ethanol)

foreign languages. However, this decrease was significantly more marked for categorization of words from a foreign language (Alexandrov *et al.*, 1998). Considering the results of our previous studies that have demonstrated the selective influence of ethanol on neurones belonging to newer EIE, we concluded that the basis of the differential influence of ethanol on EIE was its more marked effect on those neurones that subserve actualization of IE, accumulated by subjects at relatively late stages of individual development.

In the case of the acute effect of ethanol, we are dealing with reversible changes of the projections of IE. In the case of chronic alcoholization of rabbits (for 2.5–3 and 9 months), just like the situation of local brain damage, these modifications appear to be irreversible (Alexandrov *et al.*, 1994). We found that the main target of the damaging impact of chronic alcoholization are neurones belonging to new systems; neurones localized in those layers and areas of the brain that are most susceptible to acute ethanol administration. Because of changes in these cells, the numerical density of cortical neurones decreases, and the pattern of specialization changes. In limbic cortex, the quantitative relation between neurones belonging to new and old systems becomes inverted, as compared to the healthy animals: after 9 months of alcoholization, neurones belonging to old systems dominate in the population.

Thus, the projection of IE to cerebral structures depends on the specific characteristics of the neurones in each structure, is determined by the history of learning in a course of individual development, and is modified by pathological conditions.

6. FROM ANIMAL TO HUMAN BEING—A SYSTEMIC PERSPECTIVE

One of the major targets of the studies of the cerebral basis of formation and realization of IE in animals is to determine the principles that could be applied to human studies. However, serious methodological problems arise here (Domjan and Purdi, 1995). One problem is the notion that such principles may significantly differ in humans. That is why Tulving and Markowitsch (1994) suggested that data obtained in animal studies are inadequate for the investigation of specifically human functions such as language use. We do not oppose the view that human experience has special features, and appreciate the necessity of its analysis. However, we think that the above radical point of view, accepted by many scholars, is rooted in structural-functional concepts that correlate the activity of cerebral structures to specific functions, such as sensory processing, generation of motor programs, and construction of cognitive maps. It follows, quite naturally, that in animal experiments it is impossible to study those specific functions that are not linked to underlying special structures and mechanisms. From our point of view, neuronal activity is related to the realization of systems that are subserved by units with different anatomic localization and that, although different in the level, intricacy, and quality of a result achieved, nevertheless conform to common principles of systemic organization. Discovery of these common principles is one of the goals of any systemic study in general, and of systemic psychophysiology, in particular (Alexandrov and Jarvilehto, 1993; Anokhin, 1973; Shvyrkov, 1990). That is why systemic principles revealed by studies of unit activity in animals may be applied to develop views about systemic mechanisms of IE usage in various forms of human activity. For instance, these principles may be used in the aforementioned task of categorization of words of native and foreign languages, as well as in operator tasks, in group game activity and in answering questionnaires for psychodiagnostic research (see below).

Obviously the most adequate method to study human IE, enabling direct description of taxonomy and relations among elements of experience, would be an analysis of the dynamics of activity in neurones specialized with respect to systems of different age (Shvyrkov, 1995). However, for ethical and methodical reasons, the most widely used method of investigation of human cerebral activity is still EEG-analysis, along with other methods of brain mapping. V.B. Shvyrkov substantiated, both theoretically and experimentally, the suggestion that components of ERPs correspond to neuronal discharges, and to dynamics of systemic processes, at successive stages of realization of behaviour, including transitional processes ensuring the change of behavioural acts in continuum. He also showed that brain potentials cannot be classified as sensory, motor, or cognitive (Shvyrkov, 1990). The development of these views helped to show that different ERPs are just fragments or variations of averaged potentials, corresponding to the realization and change of behavioural acts (Maksimova and Aleksandrov, 1987). The relation of neurones of different systemic specialization to EEG waves has also been demonstrated (Gavrilov, 1987). Within our framework, the information on relations of EEG and unit activity to the dynamics of systemic processes, derived from animal experiments, may serve as a foundation for using recordings of gross electric brain activity in studies of the principles of IE formation and realization in humans.

7. MANIFESTATIONS OF THE DYNAMICS OF INDIVIDUAL EXPERIENCE IN THE WAVEFORM OF EVENT-RELATED BRAIN POTENTIALS

A number of studies analyzing the correlation between ERPs and various aspects of behaviour has provided a considerable body of factual evidence. However, a point of particular importance regarding the significance of the ERPs as a tool in psychophysiological research of behaviour is a widely debated topic (Loveless, 1984). It can be argued that the most appropriate way to resolve the matter is to examine the ERPs with respect to (1) the activity of units related to certain EIE, and (2) the dynamics of IE inferred from observable behaviour.

In a signal detection task, human ERPs identified by traditional means were compared with their analogs in rabbits. The similarity of the temporal structure of behaviour in humans and animals gave us the possibility of applying data on the activity of specialized cortical units, and of interpreting ERPs in humans in terms of the IE dynamics. The behaviour of humans and animals was considered as a sequence of two acts ("waiting for a signal" and "report"), maintained by two diverse sets of EIE. During implementation of the "waiting" act the proportion of units specialized with respect to the act increased towards the time of achievement of the result. The transition from the "waiting" to the "report" act coincides in time with concurrent activations of units related to the preceding and following acts. Furthermore, the transformation of sets of active units is in accordance with the ERPs accompanying observable behaviour: Negative-going potential shift corresponds to growth of the proportion of active units specific to the implementing act, whereas high amplitude positivity corresponds to the overlapping of activations of units related to successive acts (Aleksandrov and Maksimova, 1985, 1987).

Our review of the ERPs suggests that the slow negative wave, and following high amplitude positivity, are the basic components of a unified potential. The unified potential accompanied the subject's behaviour in different experimental paradigms, including the behaviour of various species (Maksimova and Aleksandrov, 1987). It is important to note that these components may be conventionally labeled as the CNV-P300 complex, or the readiness potential-motor potential complex, etc. Thus, if the subject performs a task as a succession of two behavioural acts, then a unified biphasic potential waveform will be recorded. Therefore, one could state that the unified ERPs' waveform manifests successive stages of transformation of EIE sets underlying ongoing behaviour. Negative components represent outward growth of the set's consistency, and positive components represent decrease of the set consistency. According to this view, the use of ERPs in the psychophysiological study of behaviour can be extended to human voluntary activity. Clearly, a formal description of overt behaviour as well as of IE dynamics is needed to achieve this.

Within this framework, overt behaviour and the correlation of IE dynamics with ERPs, were analyzed for subjects engaged in a strategic game, with full information and zero sum of two players. Tic-tac-toe, on a 15 × 15 gameboard, was employed as an experimental paradigm (Aleksandrov, 1995; Aleksandrov and Maksimova, 1988; Maksimova, 1995). The subject's play in the game was defined as the relation between two successive moves of his opponent. The formal description of the play included numerical indices of three successive situations on the gameboard: (1) before and (2) after, the player's move, and (3) after the return move of the opponent. Plays with identical numerical indices were assigned to the same type. The protocol of the game described as the serial moves of two players was transformed into two protocols of individual plays. Stable sequences of 2–7 plays were considered as strategies. Protocols of individual plays served as a basis for formal description of the player's IE structure, components of which represented the relation between the plays and strategies.

To select the best description of the structure and dynamics of IE, different multiple regression models of the serial plays were compared. The results of the modelling were as follows. After the player's move, alternative IE sets represented all the likely plays after the opponent's anticipated moves are actualized. The player's move rejects IE sets that contradict an actual situation. At that moment IE sets, other than rejected ones, are actualized and influence decision-making, but are never observed in overt behaviour.

Occurrence of play corresponds closely to ERPs of a unified waveform. Slow negative shifts coincide with sequential plays, and positive waves of high amplitude persists until the opponent's return move. Slow negativity correlates with the transformation of EIE, set initially for the choice of move, that is, the rejection of alternative plays. Fragments of the negative wave reflect consecutive stages of the selection of the EIE specific to an actual move. In contrast, the high amplitude positive potential reflects the actualization of the EIE sets.

To estimate the extent to which amplitudes of ERPs are determined by the dynamics of EIE, a backward stepwise multiple regression procedure was applied. Maximum amplitudes of raw negative and positive potentials recorded in players were used as the dependent variables. Parameters of EIE (the number of components representing the plays, strategies, the entropy of components' sets, and indices of their interrelations) at corresponding intervals, were used as independent variables. Simulation revealed that nearly

40% of the variance of the ERPs' amplitudes can be explained by the EIE dynamics (Aleksandrov and Maksimova, 1996). The analysis can be considered to be valid, taking into account the possibility that a significant part of the variance of the ERPs amplitude is related to postural control, and vascular and metabolic aspects of the neural tissue activity.

The remarkable aspect of the data was that players master the game while they continue to play (Aleksandrov, 1995; Maksimova, 1995). The total number of acquired IE components is fitted well by a power function of the number of moves performed. In the main, the results implied that the rate of acquisition of new IE components can be assessed by the exponent of a power function. For plays and strategies it is less than 1; for interrelations it is significantly greater than 1. It is our view that ERPs may serve as indicators of the course of forming and realization of new IEs (the EIE dynamics).

8. DYNAMICS OF ERP CHARACTERISTICS AND INTERSYSTEM RELATIONS DURING SKILL DEVELOPMENT

With repeated performance of behavioural acts, subjects improve their performance (Adams, 1987), that is, they achieve the ability "to bring about some end result with maximum certainty and minimum outlay of energy, or of time and energy" (Guthrie, 1952, p. 136). The transformation of ERPs' components that accompany the stages of mastering the behaviour (Hansen and Hillyard, 1988) can be considered to be an indicator of the changes of relations among EIE that underpin this behaviour. Hence, the dynamics of the ERPs over the course of mastering a behaviour allows us to investigate modifications of intersystem relations.

In our study subjects were employed in a choice reaction time task which demands prolonged training and, therefore, allowed us to investigate changes in ERPs at successive stages of the task.

In the first series of experiments subjects were presented with equiprobable alternative visual signals in a random sequence. Their task was to release a home button and press a report button corresponding to the presented signal as quickly as possible.

One could recognize at least two steps by which ERPs changed. At first, these changes appeared both in frontal and parietal ERPs, corresponding only to quick and correct reports of a target signal, whereas these changes were only in frontal but not in parietal ERPs corresponding to reports of other signals. At the next stage, the frontal ERPs maintained their changed shape during further training; report time became shorter; the number of erroneous reports decreased; and the shape of parietal ERPs became similar to that of frontal ERPs (Bezdenezhnych, 1993). We proposed that these changes, reflecting improvement in the subject's performance, are related to the changes in well known sequential effects.

The second series was based on the assumption that each current report in this task was influenced by the preceding ones, and these sequential effects were reflected in ERPs and, particularly, in P300. The subject had to release the home button and press the button corresponding to one stimulus after its onset (Quick-go), and the button corresponding to another stimulus after its offset (after 900 ms, Delay-go) as quickly as possible. The ERPs related to these reports were distinguished by their P300: In particular, the mean amplitude and peak latency of P300 corresponding to Delay-go reports were significantly greater than those of P300 corresponding to Quick-go.

Initially both amplitude and latency of P300 changed. In all subjects the P300 amplitude increased when it was related to the report preceded by Delay-go report, and it decreased when it was related to the report preceded by Quick-go report. In four subjects P300 latency decreased, and in two subjects it increased when the preceding report was an alternative to the present one. At a second stage of the experiment the mean report time became significantly shorter, and erroneous reports disappeared. The P300 latency became independent of the preceding report. The P300 amplitude continued to depend on the preceding report. At a third stage of the experiment, report time became more stable, and there were no sequential effects on either the amplitude or latency of P300.

Taking into account the suggestion that P300 amplitude reflects a readout from memory of EIE related to a current act (Aleksandrov and Maksimova, 1985; Maksimova and Aleksandrov, 1987), one can draw the following conclusions: In the course of mastering the tasks, the number of EIE actualized in a current report are reduced at the cost of an "inhibition" of an active EIE corresponding to the preceding response. Mastery is also accompanied by fixing the timing of EIE realizations to the act of reporting. It is possible that the changes in parietal ERPs reflect the processes which belong to the later stages of learning, whereas processes reflected by frontal ERPs belong to the earlier ones. Inasmuch as the representations of new EIE prevail in the structures whose activities are reflected in frontal derivations (Bezdenezhnych, 1993), one can propose that the relations among these EIE determine learning.

9. STRUCTURE OF INDIVIDUAL EXPERIENCE IN DIAGNOSTICS OF PERSONALITY

In the framework of the suggestions presented above, some light can be shed on key aspects of the problem of description and diagnostics of generalized characteristics of individuality. According to generally accepted notions, an individual peculiarity of a person can be considered as a generalized characteristic only at a certain (high) level of constancy, in which it is possible to differentiate two interrelated facets: stability and consistency (Mischel and Shoda, 1995).

The first aspect refers to the temporal constancy of demonstrated characteristics of individuality. The higher the temporal stability of the individual peculiarity, the higher the probability that the given characteristic belongs to the class of essential features of the individual. Under a high level of temporal stability, individual specificity reflects the most stable components of the IE structure, as well as the most strong relations among them.

The second (more important) facet, consistency, is related to the specificity of an individual characteristic in different instances of interaction with environment, that is, in different behavioural acts relevant to this individual characteristic. A high level of consistency of an individual peculiarity reflects the specific organization of generalized experience for coping with the environment. This experience is formed in ontogenesis on the basis of inborn dispositions (Bodunov, 1988). The higher the consistency of an individual, the higher its stability (temporal constancy). The higher level of consistency testifies to the stronger expression of an individual trait. The emergent nature of their genetic determination has been demonstrated for many consistent individual characteristics (Lykken and McGue, 1992).

Widespread methods of diagnosing individual characteristics by means of questionnaires evaluate the consistency of individual characteristics, through items representing typical situations in which the characteristic can be manifest. The primary factor, which mediates the subject's response-selection in personality questionnaires, is the structure of the IE as a whole.

One experimental method for modifying the structure of an actualized IE, is acute alcohol consumption, which, as we showed, suppresses the activity of new EIE (see above). This characteristic of alcohol allowed us to use it as a methodological means of selective inhibitory influences of EIE, which determine the manifestation of consistent individual characteristics.

Computerized versions of the Pavlovian Temperamental Survey, the NEO Five-Factor Inventory and the Structure of Temperament Questionnaire were administered to a sample of subjects. The experimental group of subjects consumed alcohol (1 ml/kg) before performing the test battery. The control group of subjects completed the tests after receiving equivalent amount of nonalcoholic liquid. We found that alcohol does not change the mean values of scales related to generalized characteristics of individuality. The structure of relations among characteristics did not change significantly. However, change in the mean value of one of the characteristics ("Conscientiousness" from the NEO Five-Factors Inventory) can be explained by the supposition that new EIE are reflected in this characteristic to a greater degree. Under the influence of alcohol, changes in the proportion of preferred variants of responses (matrices of responses) for some test items were detected. It was assumed that in this type of item, new IE was actualized to the greatest extent. For a large number of test items of multivariant type, reduction of the latent periods of responses was observed with alcohol. We assume that inhibition of intersystem relations (formed between EIE, which are associated with different variants of responses) plays a leading role in this reduction.

Test items were selected by the authors from personality and temperament questionnaires, on the basis of hypothetical (mainly intuitive) beliefs about the manifestation of individual characteristics in different behavioural situations. In the framework of our approach, the role of EIE of different stages is manifest in the different responses to test items on the personality questionnaires. Thus, not only can we identify features of the structure of IE that mediate individual differences revealed by these questionnaires, but we can also design new methods of generating test items which relate to the EIE formed at different stages of individual development.

10. CONCLUSIONS

The framework of systemic psychophysiology has been described. Data were described from studies of unit activity *in vitro*, and in awake normal and pathological animals learning new behaviours, and realizing previously-learned behaviours (both complex instrumental behavior and simple acts). Data were also described from experiments with human subjects who learned and performed tasks of word categorization and skilled performance, and board games, and who answered psychodiagnostic questionnaires. Based on these descriptions, qualitative and quantitative descriptions of the principles of formation and realization of individual experience were suggested in the framework of a unified methodology. From the authors' point of view, the development of a new system—systemogeny—is considered to be the fixation of a stage of individual development. The performance of a behaviour is the simultaneous realization of many elements of individual experience, formed at different stages of individual development. The dynamics of formation and actualization of the structure of individual experience were thus demonstrated, and the relations among elements of individual experience in different forms of behaviour and at different stages of its formation and realization were described.

ACKNOWLEDGMENTS

The authors wish to express their appreciation to L.I. Alexandrov and O.E. Svarnik for their assistance with preparation of the manuscript. The studies described above were supported by grants provided by the Russian Humanitarian Scientific Fond (NN 99–06–00169, 95–06–17292, 96–03–04627), the Russian Fund for Fundamental Research (NN 93–06–10787, 96–06–80626, 96–15–98641, 97–06–80275), the Finnish Academy of Sciences, The Finnish Foundation for Alcohol Studies, the Foundation pour la Recherche Medicale, Human Frontiers, CNRS Programme Cognisciences.

REFERENCES

Adams, J.A. (1987) Historical review and appraisal of research on the learning, retention, and transfer of human motor skills. *Psychological Bulletin*, **101**, 41–174.

Aleksandrov, I.O. (1995) Assessment of the acquisition rate of procedural and declarative components of individual knowledge. *European Journal of Psychological Assessment*, **11**, Supplement, 66.

Aleksandrov, I.O. and Maksimova, N.E. (1985) P300 and psychophysiological analysis of the structure of behavior. *Electroencephalography Clinical Neurophysiology*, **61**, 548–558.

Aleksandrov, I.O. and Maksimova, N.E. (1987) Slow brain potentials and their relation to the structure of behavior: Data on cortical unit activity. In: R. Johnson, J.W. Rohrbaugh, and R. Parasuraman (eds.), *Current Trends in Event-Related Potential Research* (EEG Suppl. 40) pp. 4–7 Amsterdam: Elsevier.

Aleksandrov, I.O. and Maksimova, N.E. (1988) P300 and validity of psychopysiological description of behavior. *Behavioral and Brain Sciences*, **11**, 374.

Aleksandrov, I.O. and Maksimova, N.E. (1996) Modeling of EEG amplitude in humans using parameters of the individual knowledge structure. In: *8th World Congress of IOP*, June 25–30, Tampere, Finland.

Alexandrov, L.I. and Alexandrov, Yu.I. (1993) Changes of auditory evoked potentials in response to behaviorally meaningful tones induced by acute ethanol intake in altricial nestlings at the stage of formation of natural behavior. *Alcohol*, **10**, 213–217.

Alexandrov, Yu.I. (1989) *Psychophysiological significance of* activity *of central and peripheral neurons in behavior.* (In Russian) Moscow: Nauka.

Alexandrov, Yu.I. and Aleksandrov, I.O. (1982) Specificity of visual and motor cortex neurons activity in behavior. *Acta Neurobiologiae Experimentalis*, **42**, 457–468.

Alexandrov, Yu.I., Grinchenko, Yu.V. and Jarvilehto, T. (1990) Change in the pattern of behavioral specialization of neurons in the motor cortex of the rabbit following lesion of the visual cortex. *Acta Physiologica Scandinavica.*, **139**, 371–385.

Alexandrov, Yu.I., Grinchenko, Yu.V., Laukka S., Jarvilehto T., Maz V. and Svetlaev, I. (1990a) Acute effect of ethanol on the pattern of behavioral specialization of neurons in the limbic cortex of the freely moving rabbit. *Acta Physiologica Scandinavica*, **140**, 257–268.

Alexandrov, Yu.I, Grinchenko, Yu.V., Laukka, S., Jarvilehto, T. and Maz V. (1991) Acute effects of alcohol on unit activity in the motor cortex of freely moving rabbits: Comparison with the limbic cortex. *Acta Physiologica Scandinavica*, **142**, 429–435.

Alexandrov, Yu.I., Grinchenko, Yu.V., Laukka, S., Jarvilehto, T., Maz V. and Korpusova A. (1993) Effect of ethanol on hippocampal neurons depends on their behavioral specialization. *Acta Physiologica Scandinavica*, **149**, 105–115.

Alexandrov, Yu.I. and Jarvilehto, T. (1993) Activity versus reactivity in psychology and neurophysiology. *Ecological Psychology*, **5**, 85–103.

Alexandrov, Yu.I., Korpusova, A.V., Grinchenko, Yu.V, Jarvilehto, T. and Laukka S. (1994) Structural changes and reorganization of cortical neurons' activity in behavior of chronically alcoholized rabbits. *Zhurnal Vysshey Nervnoy Dejatelnosty*, **44**, 1077–1085 (in Russian).

Alexandrov, Yu.I., Sams, M., Lavikainen, Yu., Reinikainen, K. and Naatanen R. (1998) Differential effects of alcohol on the cortical processing of foreign and native language. *International Journal of Psychophysiology*, **28**, 1–10.

Alving, B.O. (1968) Spontaneous activity in isolated somata of Aplysia pacemaker neurones. *Journal of General Physiology*, **51**, 29–45.

Anokhin, K.V. and Rose, S.P.R. (1991) Learning-induced increase of immediate early gene messenger RNA in the chick forebrain. *European Journal of Neuroscience*, **3**, 162–167.

Anokhin, P.K. (1973) *Biology and neurophysiology of conditioned reflex and its role in adaptive behavior.* Oxford: Pergamon Press.

Bezdenezhnych, B.N. (1993) Dynamics of EEG potentials during attention tasks. *Psychologichesky Zhurnal*, **14**, 120–130 (in Russian).

Bodunov, M.V. (1988) Individual rate as a generalized formally-dynamic characteristic of behavior. *Psychologichesky Zhurnal*, **9**, 33–43 (in Russian).

Bottjer, S.W., Braun, K., Cline, H., Dudai, Y., Fernald, R.D., and Hammer, L. *et al.* (1994) Group report: to what extent does experience reorganize the brain and behavior? In: R.J. Greenspan and C.P. Kyriacou (eds.) *Flexibility and constraint in behavioral systems*, Chichester, Wiley, pp. 133–146.

Bradley, P.M., Burns, B.D., King, T.M., and Webb, A.C. (1996) Electrophysiological correlates of prior training: An *in vitro* study of an area of the avian brain which is essential for early learning, *Brain Research*, **708**, 100–107.

Bullier, J. and Nowak, G. (1995) Parallel versus serial processing: new vistas on the distributed organization of the visual system. *Current Opinion in Neurobiology*, **5**, 497–503.

Chen, C.F., von Baumgarten, R.I. and Takeda, R. (1971) Pacemaker properties of completely isolated neurones in Aplysia californica. *Nature, London*, **233**, 27–29.

Domjan, M. and Purdy, J.E. (1995) Animal research in psychology. *American Psychologist*, **50**, 496–505.

Edelman, G.M. (1987) *Neural Darwinism: The theory of neuronal group selection.* New York: Basic.

Eimer, M.(1993) Effects of attention and stimulus probability on ERPs in a Go/Nogo task. *Biological Psychology*, **35**, 123–138.

Elbert, T., Pantev, C., Wienbruch, C., Rockstroh, B. and Taub, E. (1995) Increased cortical representation of the fingers of the left hand in string players. *Science, New York*, **270**, 305–307.

Foster, T.C., Castro, C.A. and McNaughton, B.L. (1989) Spatial selectivity of rat hippocampal neurons: dependence on preparedness for movement. *Science, New York*, **244**, 1580–1582.

Gavrilov, V.V. (1987) Relation of EEG and unit activity in rabbits' behavior. In V.B. Shvyrkov, V.M. Rusalov, and D.G. Shevchenko (eds.), *EEG and Neuronal Activity in Psychophysiological Experiments* (in Russian) pp. 33–44 Moscow, Nauka.

Gavrilov, V.V., Wiener, S.I. and Berthoz, A. (1994) The properties of hippocampal neurons observed during passive displacement of rats on a mobile robot. *Journal of Neuroscience, Supplement 7*, 137.

Gavrilov, V.V., Wiener, S.I. and Berthoz, A. (1996) Discharge correlates of hippocampal neurons in rats passively displaced on a mobile robot. *Society for Neuroscience Abstracts*, 910.

Gorkin, A.G. (1988) Learning and neuronal specialisation. In: V.B. Shvyrkov and R. Naatanen *et al.*, (eds.), *Psychophysiology of cognitive processe* pp. 99–104 Moscow, Institute of Psychology, Academy of Sciences.

Gorkin, A.G. and Shevchenko, D.G. (1991a.) The stability of units behavioral specialization. *Neuroscience and Behavioral Physiology*, **21**, 222–229.

Gorkin, A.G. and Shevchenko, D.G. (1991b) Structure of memory as a determinant of activity of the system-specific neurons. *Soviet Journal of Psychology*, **12**, 60–69.

Gorkin, A.G. and Shevchenko, D.G. (1995) Differences in activity of rabbits' limbic cortex neurons as determined by learning strategy. *Zhurnal Vysshey Nervnoy Dejatelnosty*, **45**, 90–100 (in Russian).

Grechenko, T.N. (1993) Active memory: neurophysiological researching. *Journal of Russian and East European Psychology*, **31**, 55–72.

Guthrie, E.R. (1952) *The psychology of learning.* New York: Harper.

Hansen, J.S. and Hillyard, S.A. (1988) Temporal dynamics of human auditory selective attention. *Psychophysiology*, **25**, 316–329.
Loveless, N. (1984) Book review. *Biological Psychology*, **18**, 77–79.
Lykken, D.T, McGue, M., Tellegen, A. and Bouchard, T.J. (1992) Emergenesis: Genetic traits that may not run in families. *American Psychologist*, **47**, 1565–1577.
Maksimova, N.E. and Aleksandrov, I.O. (1987) Typology of brain slow potentials, neuronal activity and dynamics of system organization of behavior. In: V.B. Shvyrkov, V.M. Rusalov and D.G. Shevchenko (eds.), *EEG and Neuronal Activity in Psychophysiological Experiments* (in Russian), Moscow: Nauka, pp. 44–72.
Maksimova, N.E. (1995) An experimental paradigm for quantitative estimation of the individual knowledge acquisition. *European Journal Psychological Assessment*, **11**, Suppl.
Mischel, W. and Shoda Y. (1995) A cognitive-affective system theory of personality: Reconceptualizing situations, dispositions, dynamics, and invariance in personality structure. *Psychological Review*, **102**, 246–268.
Mountcastle, V.B. (1995) The evolution of ideas concerning the function of the neocortex. *Cerebral Cortex*, **5**, 289–295.
O'Keefe, J. and Nadel, L (1979) The hippocampus as a cognitive map. *Behavioral Brain Research*, **2**, 487–533.
O'Keefe, J. and Recce, M.L. (1993) Phase relationship between hippocampal place units and the EEG theta rhythm. *Hippocampus*, **3**, 317–330.
Shyleinkina, K.V. and Khayutin, S.N. (1989) Modern systemogeny theory. *Zhurnal Vysshey Nervnoy Dejatelnosty*, **49**, 3–19 (in Russian).
Shvyrkov, V.B. (1978) Neuronal mechanisms of learning as a formation of a functional system of behavioral act. In: *Mechanisms of Systemic Cerebral Activity* (In Russian), Gorky, Gorky Medical Institute, pp. 147–149.
Shvyrkov, V.B. (1986) Behavioral specialization of neurons and the system-selection hypothesis of learning. In: F. Klix, and H. Hagendorf (eds.) *Human Memory and Cognitive Capabilities*. Amsterdam: Elsevier, pp. 599–611.
Shvyrkov, V.B. (1990) *Neurophysiological study of systemic mechanisms of behavior*. New Delhi: Oxonian Press.
Shvyrkov, V.B. (1995) Introduction to objective psychology. In: Yu.I. Alexandrov (ed) *Neuronal Basis of Mind* (in Russian) Moscow: Institut Psychology, Russian Academy of Sciences.
Singer, W. (1995) Development and plasticity of cortical processing architectures. *Science, New York* **270**, 758–764.
Spinelli, D.N. (1978) Neural correlates of visual experience in single units of cats visual and somatosensory cortex. In: *Frontiers in Visual Science*, New York: Springer, pp. 674–688.
Sudakov, K.V. (1979) Systemogeny of behavioral act. In: *Mechanisms of Brain Activity* (in Russian), Moscow, Gosnauchtechizdat, pp. 88–89.
Syed, N.J., Bulloch, A.G.M. and Lukowiak, K. (1990) *In vitro* reconstruction of the respiratory central pattern generator of the mollusc Limnaea. *Science, New York*, **250**, 282–285.
Tanaka, K. (1993) Neuronal mechanisms of object recognition. *Science, New York*, **262**, 685–688.
Tulving, E. and Markowitsch, H.J. (1994) What do animal models of memory model? *Behavioral and Brain Sciences*, **17**, 498–499.
Vinogradova, O.S. (1994) Some factors controlling morpho-functional integration on the transplanted embryonic brain tissue. *Zhurnal Vysshey Nervnoy Dejatelnosty*, **44**, 414–430.
Wall, J.T. (1988) Variable organization in cortical maps of the skin as an indication of the lifelong adaptive capacities of circuits in the mammalian brain. *Trends in Neuroscience*, **11**, 549–561.
Wiener, S.I., Paul, C.A. and Eichenbaum H. (1989) Spatial and behavioral correlates of hippocampal neuronal activity. *Journal of Neuroscience*, **9**, 2737–2763.
Wilson, M.A. and McNaughton B.L. (1993) Dynamics of the hippocampal ensemble code for space. *Science, New York*, **261**, 1055–1058.

11 Applicability of the Reinforcement Concept to Studies in Simple Nervous Systems

P.M. Balaban

Institute of Higher Nervous Activity and Neurophysiology Russian Academy of Sciences, Moscow, Russia
e-mail: balaban@ihna.msk.ru

An analysis is undertaken of the applicability of the reinforcement concept to studies of learning in simple nervous systems of invertebrates. Analysis of literature and the author's own results suggest that reinforcement can be regarded as a state of the nervous system which precedes long-term changes of behaviour. Using the example of neurophysiological mechanisms of aversive conditioning to food in the gastropod snail *Helix* it is shown that the state of the network which may correlate with the state of reinforcement can be elicited in this simple nervous system by activation of serotonergic pedal cells, positively modulating the avoidance behaviour of the animal, and considered to be related to motivation. The assumption "reinforcement state = emotional state" suggests the existence of emotions in invertebrates, which is supported by the possible existence of self-stimulation in invertebrates. The conclusion is drawn that, with certain limitations, the reinforcement concept can be used in studies on simple nervous systems.

KEYWORDS: snail, learning, neurones, emotions, self-stimulation

1. INTRODUCTION

The reinforcement concept is widely used in studies of learning and memory mechanisms which have been performed in the last few decades, not only at the behavioural level, but also in model systems, such as brain slices, invertebrate isolated nervous systems and synaptically connected neurones *in vitro* and *in vivo*. In these studies the terminology elaborated for animal behaviour has mainly been used. The present paper is aimed at analysing the limitations and applicability of the behavioural concept of reinforcement to neurophysiological experiments in model systems.

2. REINFORCEMENT IN LEARNING THEORIES

Analysis of the abundant literature concerning theories of learning in psychological terms, (i.e. describing the behavioural events), allows us to find common ideas in each approach, and to relate them to the contemporary concept of reinforcement.

Historically, the term reinforcement has had many differing meanings. Spencer's theory of learning (Spencer, 1870) was the first systematic attempt to offer explanations for the differential strengthening of behavioural patterns. His theory implied that an organism would obviously tend to repeat actions that brought pleasure, and desist from those which brought pain. In cases when pleasure accompanied actions that were beneficial for survival, or pain accompanied injurious actions, the animal had an advantageous position for natural selection. Spencer's theory was the first to imply the existence of a reinforcement process which was necessary for differential change in behaviour.

The term "reinforcement" was first introduced into the psychophysiological arena in the laboratory of I.P.Pavlov, and appeared in the world literature after publication in 1927 of a translation of his "Twenty Years of Experience" first published in Russian in 1923 (Pavlov, 1923). Reinforcement in this theory of learning is, by implication, a property of the unconditioned stimulus which exerts the "reinforcing action" (Figure 11.1). It was noted in the paper devoted to the problems of reinforcement by Pavlov's student Asratyan (1977) that in Pavlov's publications there is no special work analyzing reinforcement as a concept, which suggests that Pavlov never regarded this problem as an independent one.

In the "Law of Effect" formulated by Thorndike (1905) it is clearly stated that the nervous system is so constructed as to lead to the strengthening of those connections which have been active just prior to a satisfying event, and to the weakening of those connections which have been active prior to annoying events. This change of

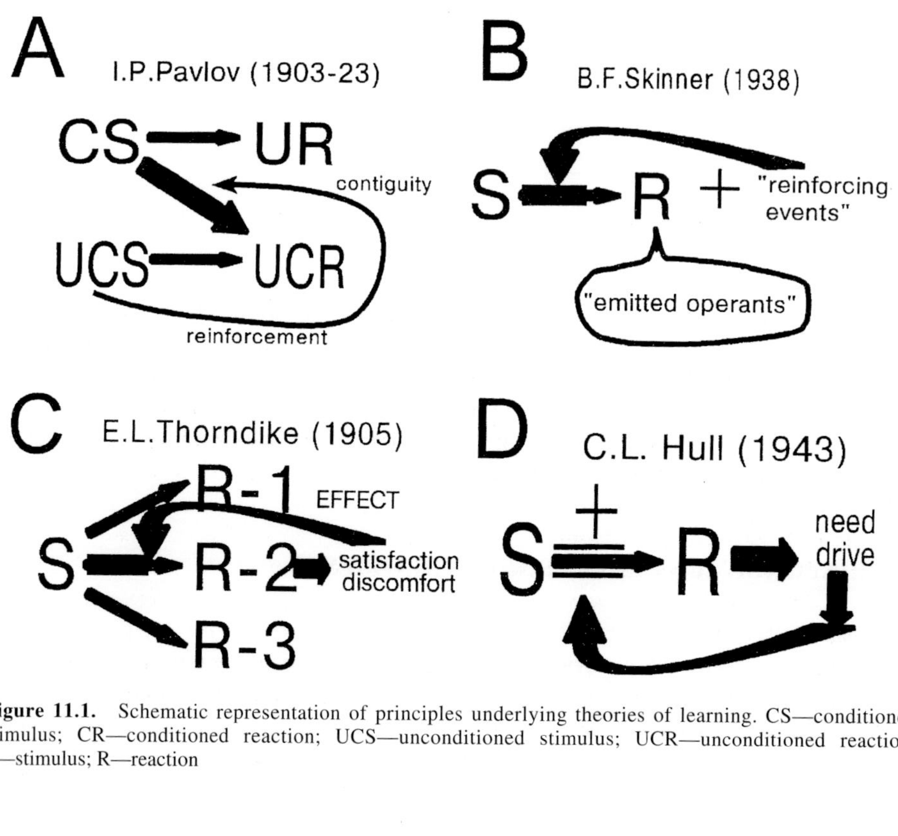

Figure 11.1. Schematic representation of principles underlying theories of learning. CS—conditioned stimulus; CR—conditioned reaction; UCS—unconditioned stimulus; UCR—unconditioned reaction; S—stimulus; R—reaction

effectiveness of stimulus-response (S-R) connections implies the existence of some sort of nervous process influencing the S-R connections, upon which the behavioural effect of contingent stimuli depends.

In the studies by C.L. Hull (1943), which are to some extent a continuation of Thorndike's "Law of Effect", the concept of a biological need and its associated drive were introduced. Drive reduction, increasing the probability of S-R association, has the properties of reinforcement in this theory (Figure 11.1). B.F. Skinner (1938) introduced a concept of events which may serve as "reinforcers", and which are not connected exclusively with stimulus or reaction (Figure 11.1). A concept of "emitted operants", which can be brought under the control of a stimulus by arranging for the emission and reinforcement of the operant in the presence of the stimulus, completed his description of behaviour.

It is essential to note that, in all the theories of learning just mentioned, there are common properties: *Stimulus, Response*, and *Motivation* are mentioned in all of them. The context may differ somewhat, but the involvement in learning of these phenomena is acknowledged. Based on their own experimental experience, different authors assign the reinforcing properties either to Stimulus or Reaction or Motivation, but in all cases the argument is strong and valid. It suggests two possibilities: Only one of the authors is right, which is improbable, or all of them are right, which means that reinforcement cannot be attributed to only one of the phenomena mentioned.

Our next question is: "What is reinforcement? Is it an independent behaviourally-described phenomenon, or is it a state of the organism which can be elicited in different ways?"

3. REINFORCEMENT: AN INDEPENDENT PHENOMENON OR A STATE OF THE ORGANISM?

As has been noted, in publications of I.P. Pavlov, reinforcement was never considered as an independent phenomenon (Asratyan, 1977; Pavlov, 1923) in spite of the fact that its importance was mentioned in each paper. A prominent student of Pavlov's, P.K. Anokhin, also stressed the role of reinforcement (Anokhin, 1968), but never treated it independently in his papers. A brilliant analysis of reinforcement as a concept was made in a chapter of a book "Reinforcement and Behavior" by E. Walker (1969). He compared the influence of reinforcement on studies of learning with "The One Ring" from Tolkien's "The Lord of the Rings". The possessor of the One Ring could exercise mastery over every living creature, but the use of the ring inevitably corrupted the person who used it. In fact, using reinforcement as a concept has some advantages in the interpretation of behaviour, but using it undermines the conceptual framework. E. Walker suggested destruction of this powerful instrument in order not to distort the learning mechanisms with an excessively mechanistic interpretation of its functioning. Indeed, reinforcement is a mechanistic concept used to explain why repeated associations are strengthened. The simplistic way to explain the strengthening is to suggest the existence of some sort of a selective "glue" which is present in the organism. But all searches for this glue as a physical entity in the organism were in vain, as well as the search for the physiological basis of reinforcement which appeared to be too diversified.

From our point of view, it is necessary to define the term reinforcement operationally. Analysis of the literature suggests that either external stimuli, behavioural response, or motivational state of the organism may have "reinforcing" properties. In order to have an optimal description of all known data concerning reinforcement, it is logical to consider a special *state of the organism* preceding changes in behaviour as the "*reinforcing state*", which can be elicited by presentation of stimuli, behavioural responses or changes in motivation, as has been shown experimentally.

4. SIMPLE NETWORKS AND REINFORCEMENT

The aim of the present analysis is to find out whether it is correct and useful to use this behavioural term in investigations of learning mechanisms in model situations when only a small part of nervous system is under the experimenter's control. Even in the relatively simple nervous system of molluscs, tens of thousands of neurones participate in control of behaviour, and, without a conceptual framework of the organization of the behavioural act, it is impossible to approach mechanisms of learning. One of the most complete descriptions of organization of a behavioural act in the traditions of Pavlov's school was presented in the "Conceptual Reflex Arc" by E.N. Sokolov (1985). This concept provides a physiological basis for stimulus perception, integration of information and decision making, realization of motor programs of behaviour, and in addition a neural base for motivational influences is described. In our work in gastropod snails we used this conceptual framework to describe the avoidance reaction and feeding response (Figure 11.2). It is well known that, after pairing of food presentation with noxious stimuli, the snail reacts selectively by avoidance of the type of food associated in time with noxious stimuli (Balaban and Zakharov, 1992). In this short description of the paradigm, we have not used the term "reinforcement". More than that, it is not necessary to use this concept, and we cannot find a place for it in the scheme (Figure 11.2). But when we want to describe the results of the learning procedure we cannot miss the "change in synaptic effectivity" which is in some sense absolutely equal to Thorndike's "strengthening of connections between stimulus and reaction". It is essential to note that we can judge whether there was reinforcement or not only *after* the learning procedure, during testing of the resulting behaviour. In the case of aversive food conditioning, the animal ceases to respond with feeding behaviour to the type of food associated with noxious stimuli, which indicates that the reinforcement state was evoked.

What is the place of reinforcement in this simple example? Actually the concept of reinforcement describes the difference between two possibilities. The first possibility refers to the case when testing shows absence of changes in behaviour. The conclusion is drawn that the reinforcement state was not achieved in the learning procedure. The second possibility is the case in which the learning procedure changed the behaviour, which means that the reinforcing state *was* evoked. Therefore, we conclude that there was a reinforcement only *post factum,* and we cannot "apply the reinforcement", as is often found in literature. It suggests that the reinforcement is a change in a state of the nervous net schematically drawn on Figure 11.2, but we cannot assign it to some particular block on the scheme. Nevertheless, we can investigate this network and try to find out which part of the nervous system creates this "reinforcement state".

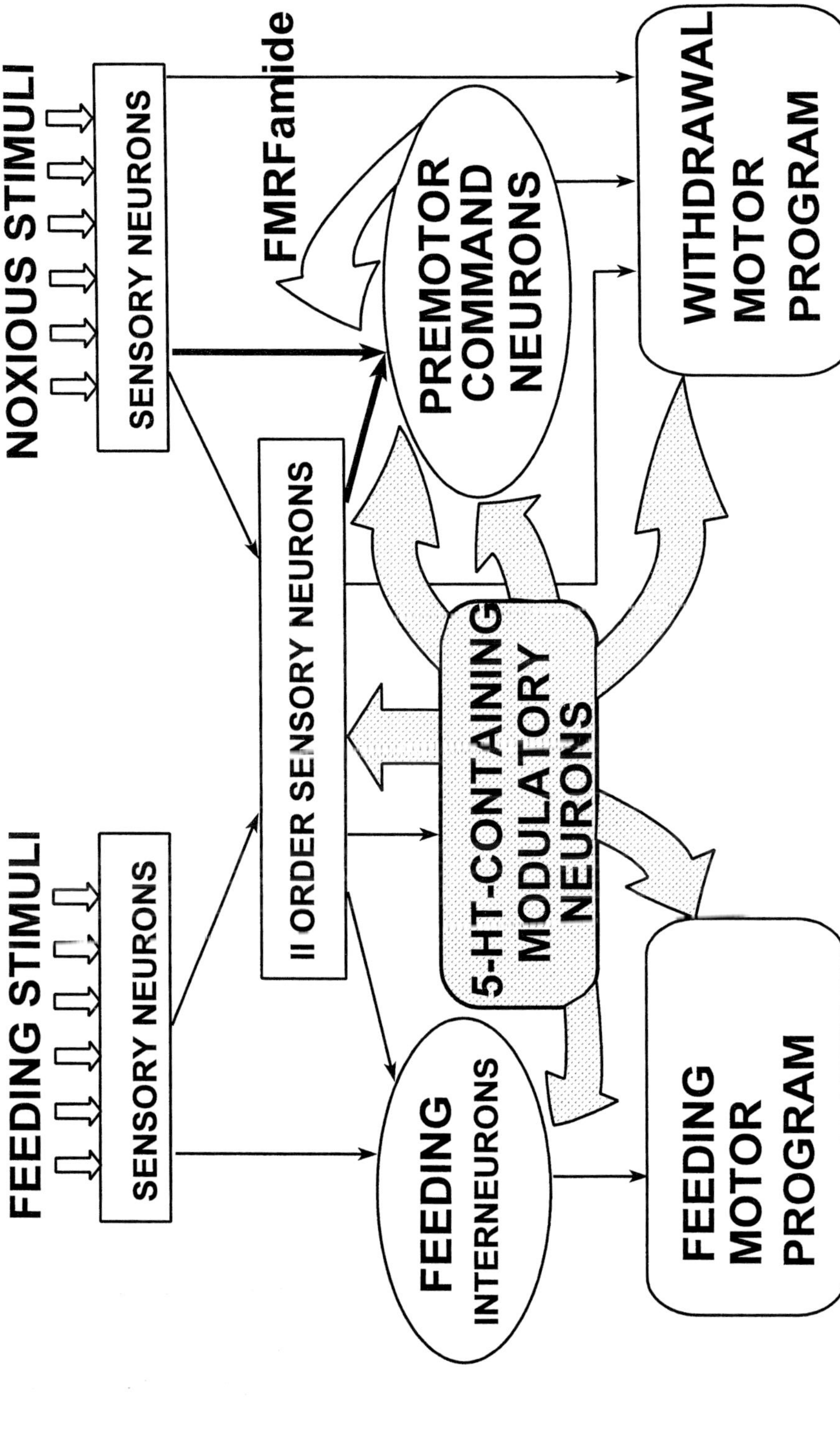

Figure 11.2. Schematic representation of relations between functional groups of neurones involved in avoidance and feeding behaviour, and food conditioning.

5. CELLULAR ORIGIN OF REINFORCEMENT IN A SIMPLE NETWORK DURING LEARNING

There is only one case described in the literature in which intracellular stimulation of an identified neurone substituted for the unconditioned stimulus in associative olfactory learning (Hammer, 1993), thus exerting a reinforcing effect. It is essential for our analysis to note that this identified neurone (in a honeybee) which contained octopamine, was shown to be involved in modulation of behaviour (sensitization) in this animal (Hammer and Menzel, 1995). The neuromodulatory substance serotonin was shown in *Aplysia* to participate in associative conditioning (Hawkins *et al.,* 1993).

In *Helix*, modulatory neurones are also the main candidates for exerting the reinforcing effect. In this section we will try to find out the cellular locus in the network whose activity may correlate with the state of reinforcement, using the example of aversive conditioning in the snail described in detail in the literature (Balaban *et al.,* 1987; Balaban and Zakharov, 1992). Only the main behavioural component of aversive learning will be considered, namely the aversive response to food paired previously with noxious stimuli. As a first step of analysis one should determine the functional group of cells, whose activity changes in correlation with existence of reinforcement.

It is well known that modulatory neurones address their influence to a number of functionally different neurones, therefore the changes of their activity would synergically influence several behaviours, serving as a coordination centre.

As a result of learning according to this scheme, the change should take place only in certain sensory inputs of a command cell. Where can the reinforcement be "applied" in this case? It was established in electrophysiological experiments that, due to the learning procedure, a novel response appeared in command neurons for withdrawal, and a corresponding behavioural withdrawal to food presentation appeared (Balaban *et al.*, 1987). A morphological connection between sensory neurones reacting to food presentation and command neurones existed before the learning procedure, but was weak and evoked only subthreshold responses. The factor which can influence the synaptic input of command neurones may have a relation to reinforcement. The best candidates for strengthening the connection between sensory input and spike discharge in command neurones are serotonergic modulatory neurones, which have been well described in the snail (Zakharov *et al.*, 1995). During elaboration of the conditioned reflex the coincidence of conditioned sensory input and modulatory influence may serve as a condition for selective changes in sensory input of the command neurones. In terrestrial snails it was shown that activity of serotonergic neurones is a necessary condition for the elaboration of the conditioned reflex. However, their activity has no relation to the retention of memory, which suggests that the effects exerted by these cells are very close to those predicted by the reinforcement concept. It is well established that activity of these modulatory cells can be affected by strong external stimuli, and motor performance also depends on their activity (Zakharov *et al.,* 1995). Therefore, in the particular example of aversive learning of the snail, the most likely group of cells—the activation of which may correlate to the existence of reinforcement—is a functional class of serotonergic modulatory cells. These modulatory cells were shown to react to such contextual cues as acidity of the substrate, and the level of glucose corresponding to satiation level (Zakharov, unpublished). Modulatory cells directly participate in elaboration and reproduction of contextual conditioning (Balaban and Bravarenko, 1993). In a more general sense, these cells may be called *motivational*. Therefore, the most likely candidates for neurones *creating the state of reinforcement* are the neuromodulatory cells related to motivation.

6. SPECULATION: REINFORCEMENT = EMOTIONS

Neuropsychologists widely accept that the appearance of an emotional *state of the organism* improves learning and memory. Let us assume that the reinforcement state is the same as the emotional state. In this case the main problem for invertebrates will be the question: Is it possible for the snail to consider a certain situation as emotionally positive or negative? In other words, can the snail feel pleasure? It is evident, that the human experimenter never can place himself in the snail's brain, but in neurobiology an experimental technique—self-stimulation—can provide an answer regarding whether the situation is pleasant for the animal, or causes a distress.

In the phenomenon of self-stimulation, an animal receives direct electrical stimulation of the brain as a consequence of its operating a manipulandum. If the electrode is implanted in certain areas of the brain, the animal repeatedly self-stimulates (Olds and Milner, 1954; Olds, 1958). Since its discovery (Olds and Milner, 1954), numerous experiments have confirmed the phenomenon of self-stimulation in various vertebrate species (Olds, 1977). However, self-stimulation itself, as well as its relation with learning and reward, remains to be explained by mechanisms at the cellular level (Stellar and Stellar, 1985). To develop a new approach to this problem, we investigated whether a snail, with its relatively simple and technically advantageous nervous system, will self-stimulate. Freely moving animals with chronically implanted electrodes were used, and the rewarding properties of a contingent extracellular stimulation of a certain cellular groups in semi-intact preparations were investigated.

The experiments were conducted in 40 min sessions, one per day. The snail was tethered by its shell in a manner allowing it to crawl on a ball that rotated freely in a 0.01% solution of NaCl (Figure 11.3). The ball was laced with bare stainless steel wire, to complete an electrical circuit between the animal's foot and a carbon reference electrode placed in the water. To receive stimulation, the tethered snail was required to displace the end of a rod (pedal), thus closing a switch. Usually, the snail would first sense the rod with its tentacles, then raise its head to explore the rod with its lips and mouth, displacing the rod during exploration.

Each session began with a 20 min period without reinforcement, followed by a 20 min period with reinforcement. An electrical timer automatically switched reinforcement conditions. The general activity of the snail was monitored using a light beam and a photocell.

To determine the rate of touching of the rod in the absence of reinforcement, snails were allowed free access to the rod in a 40 min session prior to any stimulation. The animals were slightly more active in touching the rod early in a 40 min session than at the end, but there was no significant departure from a sustained frequency of contact.

When reinforcement was delivered to snails having electrodes implanted in the parietal ganglion (in which stimulation site are located neurones related to avoidance behaviour), there developed a strikingly consistent pattern of behaviour. After one or two

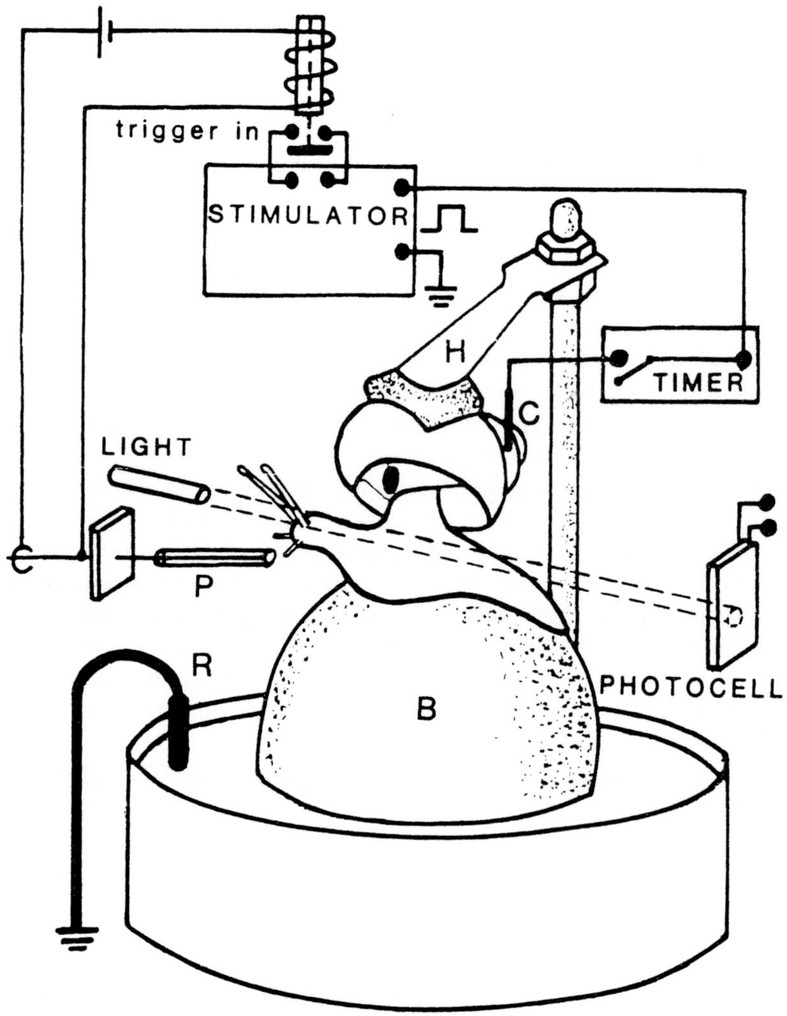

Figure 11.3. Experimental set-up for self-stimulation in the terrestrial snail. Touching the pedal (P) closes the circuit for application of the stimulus . Using the light beam and photocell it was possible to monitor the movements of the snail. B—ball floating in water; R—reference electrode.

reinforcements, the snails remained active but appeared to avoid the rod. Although there was some recovery between sessions, the mean response rate fell nearly to zero by the end of each session with reinforcement.

Quite different results were obtained when reinforcement was delivered via electrodes implanted in the mesocerebrum (in which site of stimulation are located cells related to sexual behaviour). After just a few reinforcements, these snails began to contact the rod with increasing frequency. The magnitude of the effect varied between animals, but in no case was there a decline in response frequency, such as was seen in the snails having parietal electrodes.

Pooled results are shown in Figure 11.4. Comparison of changes during 40 min without reinforcement, and in two different groups of snails (electrodes in mesocerebrum and in parietal ganglia), shows a significant effect of reinforcement on the on-going behaviour.

The opposite direction of changes in the response rate in snails with electrodes in the parietal ganglion vs. those with electrodes in the mesocerebrum provides additional evidence for the difference in behavioural effects of snail brain stimulation in two different zones (Figure 11.4).

Neural substrates mediating reward (or reinforcement), have been identified in vertebrate brains using the self-stimulation procedure. The medial forebrain bundle at the level of the hypothalamus is the most effective site (Gallistel, 1983). However, no complete neural circuits have been delineated, nor has it been possible to identify any individual neurones whose participation is essential. These difficulties have so far prevented any mechanistic explanation of self-stimulation at the cellular level. The identification of the mesocerebrum as a site of reward in snails offers the possibility of cellular studies, because the neurones in this region are large (diameters up to 80 mm) and easily accessible for intracellular investigation. Interrelations between mesocerebral cells and parietal giants were investigated (Balaban and Chase, 1990), and it was shown that stimulation of the mesocerebrum causes suppression of spiking in the parietal command neurones in response to tactile stimulation of the skin. In addition to the effects observed at the level of command neurones, a second, independent control over withdrawal is brought about by the inhibition of neurones which are capable of sensitizing the afferent excitation of the same parietal command neurones (Balaban and Chase, 1990). Unfortunately, these data cannot be linked directly to emotionally positive effects exerted by the same mesocerebral cells, and this impossibility is caused by the absence of a cellular hypothesis for emotional processes, with only a small overlap of behavioural physiology and cellular physiology. We hope that further investigation of emotionally-dependent behaviours in animals with relatively simple behaviours and accessible for cellular analysis nervous system would increase this overlap.

The experiments reported here suggest the possibility that snails learn the rewarding properties of electrical stimulation during the course of a session, or a series of a sessions. It means that the interpretation of reinforcement as equal to emotion is valid for the invertebrate animal as well.

7. CONCLUSIONS

1. Reinforcement is a state of the nervous system which precedes long-term changes of behaviour.
2. During elaboration of aversive conditioning to food in gastropod snails, the state of the underlying network, which may correlate with the reinforcement state, can be elicited by activation of serotonergic pedal cells modulating avoidance behaviour of the animal. The function of these cells may be interpreted as motivational.
3. The appearance of this "reinforcement state" is a condition for memory consolidation, but is not connected with memory retention.

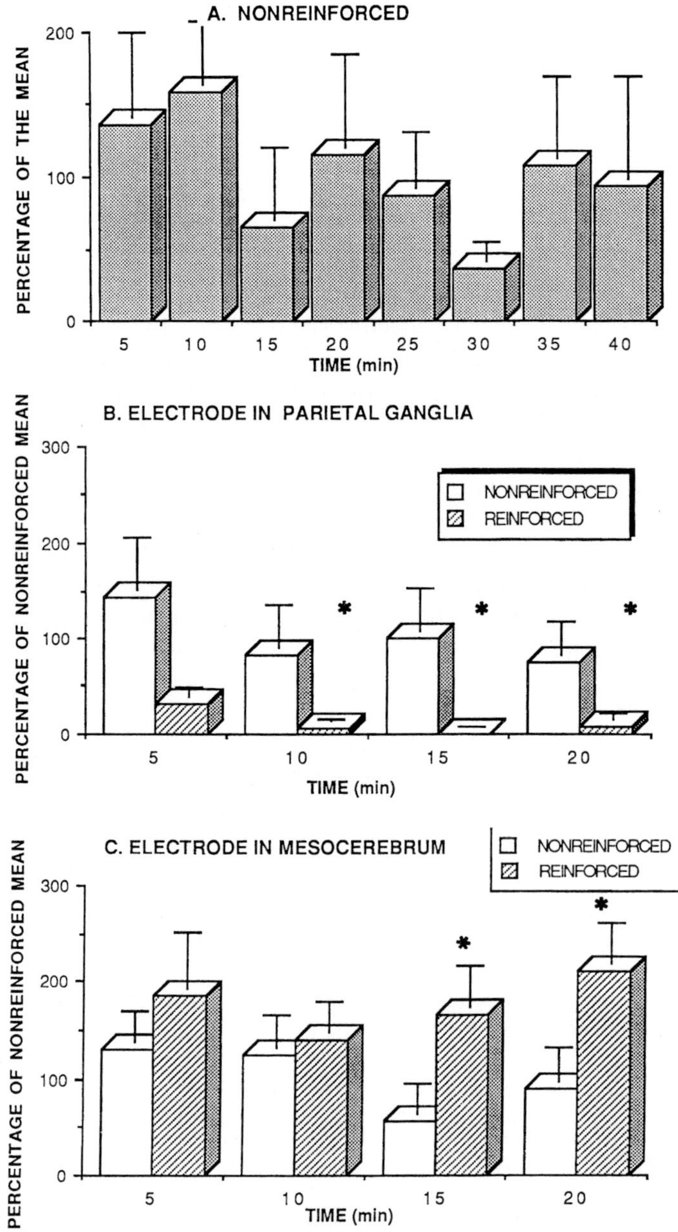

Figure 11.4. Average frequency of contacts with bar (pedal on Figure 3) by freely behaving snails with implanted electrodes. Pooled results. Each histogram shows the number of contacts in 5 min periods, expressed as a percentage of the mean number of bar contacts recorded during all non-reinforced periods. One averaged score was taken for each snail for each 5 min. **A**. Non-reinforced control in 40 min sessions, showing absence of any systematic change in the number of bar contacts during the course of test sessions prior to brain stimulation; 7 snails. **B**. Self-stimulation of parietal ganglion; 3 animals. **C**. Self-stimulation of mesocerebrum; 7 snails. In B and C the data are from two consecutive 20 min periods, the first without reinforcement, the second with reinforcement. The error bars show SEM's. Dotted line separates reinforced from non-reinforced periods.

4. The speculation "reinforcement state = emotional state" suggests the existence of emotions in invertebrates, which may be true, given the ability of invertebrates to show self-stimulation behaviour.

REFERENCES

Anokhin, P.K. (1968) *Biology and Neurophysiology of Conditioned Reflex* (in Russian). Medicine, Moscow.

Asratyan, E.A. (1977) *Essays on Higher Nervous Activity* (in Russian). Armenian Academy of Sciences, Press, Erevan.

Balaban, P.M. (1983) Postsynaptic mechanism of withdrawal reflex sensitization in the snail, *Journal of Neurobiology*, **14**, 365–375.

Balaban, P.M. (1991) Command neuron concept and decision making. In: D.A. Sakharov and W. Winlow (eds), *Simple Nervous Systems*, pp. 375–382, Manchester, New York, Manchester University Press.

Balaban, P.M., Vehovszky, A., Maximova, O.A. and Zakharov, I.S. (1987) Effect of 5,7-DHT on the food-aversive conditioning in the snail *Helix lucorum L. Brain Research*, **404**, 201–209.

Balaban, P.M. and Zakharov, I.S. (1992) *Learning and Development: Common Basis of Two Phenomena* (in Rusian). Moscow, Nauka.

Balaban, P.M., Maksimova, O.A. and Galanina, G.N. (1985) Cellular responses during elaboration of two-way conditioned reaction in snail *Zhurnal Vysshey Nervnoy Dejatelnosty*, **35**, 497–503 (in Russian).

Balaban, P. and Chase, R. (1990) Inhibition of cells involved in avoidance behavior by stimulation of mesocerebrum. *Journal of Comparative Physiology A*, **166**, 421–427.

Balaban, P., Bravarenko, N. (1993) Long-term sensitization and environmental conditioning in terrestrial snails. *Experimental Brain Research*, **96**, 487–93.

Gallistel, C.R. (1983) Self-stimulation. In: J.A. Deutsch (ed.) *The Physiological Basis of Memory*, New York, Academic Press, pp. 269–349

Hammer, M. (1993) An identified neuron mediates the unconditioned stimulus in associative olfactory learning in honeybees. *Nature, London* **366**, 59–63.

Hammer, M. and Menzel R. (1995) Learning and memory in the honeybee. *Journal of Neuroscience*, **15**, 1617–1630.

Hawkins, R.D., Kandel, E.R. and Siegelbaum, S.A. (1993) Learning to modulate transmitter release: themes and variations in synaptic plasticity. *Annual Review of Neuroscience*, **16**, 625–665.

Hull, C.L. (1943) *Principles of Behavior*, Appleton, N.Y.

Maksimova, O.A. and Balaban, P.M. (1983) *Neural Mechanisms of Behavioral Plasticity* (in Russian). Nauka, Moscow.

Olds, J. (1958) Self-stimulation of the brain: Its use to study local effects of hunger, sex, and drugs, *Science, New York*, **127**, 315–321.

Olds, J. (1977) *Drives and Rewards: Behavioral Studies of Hypothalamic Functions*. 140p New York, Raven Press.

Olds, J. and Milner, P. (1954) Positive reinforcement produced by electrical stimulation of septal area and other regions of the rat brain. *Journal of Comparative and Physiological Psychology*, **47**, 419–427.

Pavlov, I.P. (1923) *Twenty Years of Objective Study of Higher Nervous Activity in Animals* (in Russian). State Press, Moscow, Petrograd.

Spencer, H. (1870) *Thee Principles of Psychology*, **1** (2nd ed.), Appleton, N.Y.

Sokolov, E.N. (1985) Conceptual reflex arc. In: E.N. Sokolov, L.A. Schmelev (eds) *Questions in Cybernetics: Neurocybernetical Analysis of Behavior Mechanisms*, p. 5, Cybernetica, Moscow.

Skinner, B.F. (1938) *The Behavior of Organisms; an Experimental Analysis*, Appleton, New York.

Stellar, J.R. and Stellar E. (1985) *The Neurobiology of Motivation and Reward*. 255 p New York, Springer.

Thorndike, E.L. (1905) *The Elements of Psychology*. Seiler, New York.

Walker, E.L. (1969) Reinforcement—"The One Ring". In: J.T. Tapp (ed.) *Reinforcement and Behavior*, p. 47, New York, Academic Press.

Zakharov, I.S., Ierusalimsky, V.N. and Balaban, P.M. (1995) Pedal serotonergic neurons modulate the synaptic input of withdrawal interneurons in Helix. *Invertebrate Neuroscience*, **1**, 41.

12 Sensory Factors in the Ontogenetic Reorganization of Behaviour

V.V. Raevsky, L.I. Alexandrov, T.B. Golubeva, E.V. Korneeva, I.E. Kudriashov and I.V. Kudriashova

Institute of Higher Nervous Activity and Neurophysiology, Russian Academy of Sciences, Moscow, Russia
e-mail: neuroont@neuroont.msk.ru

Data are described that support the proposal that an increase in sensory input during early ontogeny results in a delay in the development of sensory systems. The sensory basis for behavioural patterns becomes ineffective, causing their reorganization and the appearance of new forms of behaviour. Limitation of sensory input during critical periods of development accelerates the manifestation of behavioural patterns. However, this acceleration also has long-lasting negative effects on learning and memory in adult animals.

KEYWORDS: functional development, systemogenesis, birds, visual deprivation, rats

1. INTRODUCTION

The principles underlying the development of an organism are among the most actively studied and discussed issues in biology. Various and sometimes contradictory experimental data have inspired different theories. One such problem has been the focus of research attention during the last two years. In their article entitled "Is neural development Darwinian?" Purves *et al.* (1996) critically assessed the concept of ontogenetic development of the nervous system, first formulated in 1888 by Wilhelm Roux (Roux, 1974) and later developed by Ramon-y-Cajal (1929). According to this concept, during neurogenesis there is a competitive struggle among outgrowths, and perhaps even among nerve cells for space and nutrition. The key point is that the nervous system contains more neural elements (neurones, neuronal branches, synaptic connections, and groups of interconnected neurones) early in development than in maturity. Purves *et al.* supply ample data on the developmental increase in the number of neural elements, synaptic contacts, and neurites' branching. They conclude that an unwarranted enthusiasm for the idea that neural development proceeds by winnowing an initial excess can only obscure the essentially constructionist nature of the development of the mammalian brain, and may impede an effort to understand it. The idea offered by Purves *et al.* was questioned by Sporns and Tononi (1994) who claimed that it ignored other brain theories based on variation and selection (Edelman, 1987). These selectionist theories have three major components: "The first is the generation of variability within populations of neurones, which is manifest structurally through cell replication and cell death, and ongoing neurite extension, retraction and synaptic remodeling, and dynamically through continuous modifications of firing patterns. The second is the interaction of the variable circuitry and firing patterns with the organism's environment. The third is differential amplification or attenuation of the contribution of neuronal or synaptic populations to neuronal function, either by local rules of plasticity based on correlated firing, or by global changes mediated by diffuse ascending (value) systems" (Sporns, 1997, p. 291).

Another actively discussed problem deals with the idea of inborn and acquired behavioural patterns. Consistent with the traditional view, behavioural development involves a process of modification of inborn patterns by environmental influences and through accumulation of individual experience (Lorenz, 1963; Tinbergen, 1951; Thorpe, 1963). The role of genetic factors cannot be disputed. On the other hand, the role of sensory information in the ontogenetic development of functions has been clearly established. Moreover, the analysis of even the earliest ontogenetic stages has not revealed any behavioural patterns independant of learning during development (Gottlieb, 1968, 1981; Fillion and Blass, 1986; Khayutin and Dmitrieva, 1991).

Anokhin (1948) formulated the theory of systemogenesis which describes the main principles of functional development. This concept is based on the idea that, at any stage of normal development, the adaptive capabilities of an organism are completely suited to environmental characteristics. Developmental stages are sensitive to environmental modifications resulting from parturition, hatching or leaving the parents' nest, as well as to changes in interaction with the environment (e.g. eyes opening). Each stage is related to the morphological basis and to the learning of specific behavioural patterns.

All the above approaches to understanding the process of functional development share one common aspect—sensory systems. It was the study of sensory systems that yielded the data basic for neural Darwinism and its criticism. Sensory information plays a crucial role in learning processes. The ability to adequately evaluate the qualitative and quantitative characteristics of the "goal-object environment" (i.e. the environment as perceived by an organism, constructed from elements that can be involved into behavioural organization, and become goals of behavioural acts), as well as the structure of the "goal-object environment" itself, at any stage of ontogeny depends on the degree of maturity of sensory systems.

The sequence of maturation of the analyzers of different modalities is fixed in all birds and mammals (Gottlieb, 1971). However, the rate of maturation of sensory mechanisms is significantly influenced by specific environmental factors. The degree of involvement of analyzers in behaviour is changed during ontogeny. For example, suckling movements in newly born mammals are induced first by chemoreception, and after several days mostly by the information from oral mechanoreceptors (Shuleikina, 1971). Such change of leading afferent control has been described for the early ontogeny of various animals: human infants, newly-born mammals, nestlings (Shuleikina, 1971; Luschekin, 1983; Bogomolova and Kurochkin, 1987; Khayutin and Dmitrieva, 1991).

The aim of our research was to elucidate the mechanisms for change in leading afferent control—and to find the reasons why, at certain ontogenetic stages, a sensory stimulus, in spite of the continuing development of the respective sensory system, becomes ineffective in subserving the given behavioural pattern.

2. CHANGE IN THE SENSORY BASIS OF EARLY BEHAVIOUR

In selecting the model in which to study changes in the sensory basis of behaviour during early ontogeny, we focused our attention on feeding behaviour of the nestlings of a hole-nesting species—the pied flycatcher (*Ficedula hypoleuca*). Systematic studies (Khayutin and Dmitrieva, 1981, 1991) have revealed that the early feeding response (raising of the head and beak opening) in young birds was elicited by the specific acoustic signals of parents arriving with food. At day 4–5, nestlings eyes open, and during 1–1.5 days acoustically-guided feeding behaviour is changed to visually-guided feeding. In other words, nestlings start to beg in response to a short-term luminosity change in the nest-box caused by the parent bird's body closing the entrance hole. The acoustic signal becomes ineffective for feeding behaviour. In a few days, visually-guided behaviour of nestlings becomes more complicated—they start to respond to a moving silhouette of a parent bird, snatching food from its beak (Figure 12.1).

Apparent substitution of acoustically-guided behaviour by the visually-guided one during as little as 1–1.5 days became the predominant model in our studies. The same stages of development of early behaviour were also described for other avian species: owls, herring gulls, jays and ducks (Golubeva, 1994, 1996). Experimental data obtained from studies with various species indicated the existence of common mechanisms for the change in the sensory basis of early forms of behaviour.

Auditory sensitivity of pied flycatcher nestlings was shown to be partially formed (in low- and medium-frequency bands) by the time of hatching. It is characterized by relatively low thresholds in the range of frequencies coinciding with signals emitted by parents arriving with food (Alexandrov and Dmitrieva, 1992). Such increased sensitivity serves as a base for a high reproducibility of the early begging responses of the young.

Morphological study of the higher avian visual structure, Wulst's area, demonstrated that, by the time the eyes open, it is composed of a large number of non-differentiated neurones and of aspiny stellate cells at different stages of maturation (Figure 12.2A). Later, at day 13, the Wulst contains two types of spiny cells—multi-spiny and medium-spiny stellate neurones—and three types of aspiny cells—small stellate-like, reticular-like and arborescent neurones (Korneeva, 1995; Korneeva *et al.*, 1993, 1994; see Figure 12.2B). Thus, Wulst's area in 4–5 day-old nestlings is still a very immature structure at the time of switching from acoustically-guided to visually-guided behaviour, which raises questions about the advantages of such switching.

Some light was shed on this problem by morphological and electrophysiological studies of the auditory system during the period of change in the sensory basis of early nestlings' behaviour. Evoked potentials (EP) recorded in avian auditory structures at the time the eyes were opening, revealed delayed auditory development for the frequency range of the parent's food calls in flycatcher young (Alexandrov, 1995). In herring gull nestlings, auditory thresholds were even increased (Golubeva, 1994). It is important to note that this phenomenon was observed not only in higher structures—field L of caudal neostriatum, but also at the receptor level. Morphological studies demonstrated the delay in the development of major auditory nuclei, the onset of which coincided with opening of the eyes. The duration of such delay for *n. magnocellularis, n. angularis,* and *n. laminaris* ranged between 5 and 7 days (Golubeva and Barsova, 1994; see Figure 12.3). Thus the reorganization of acoustically-guided feeding behaviour is

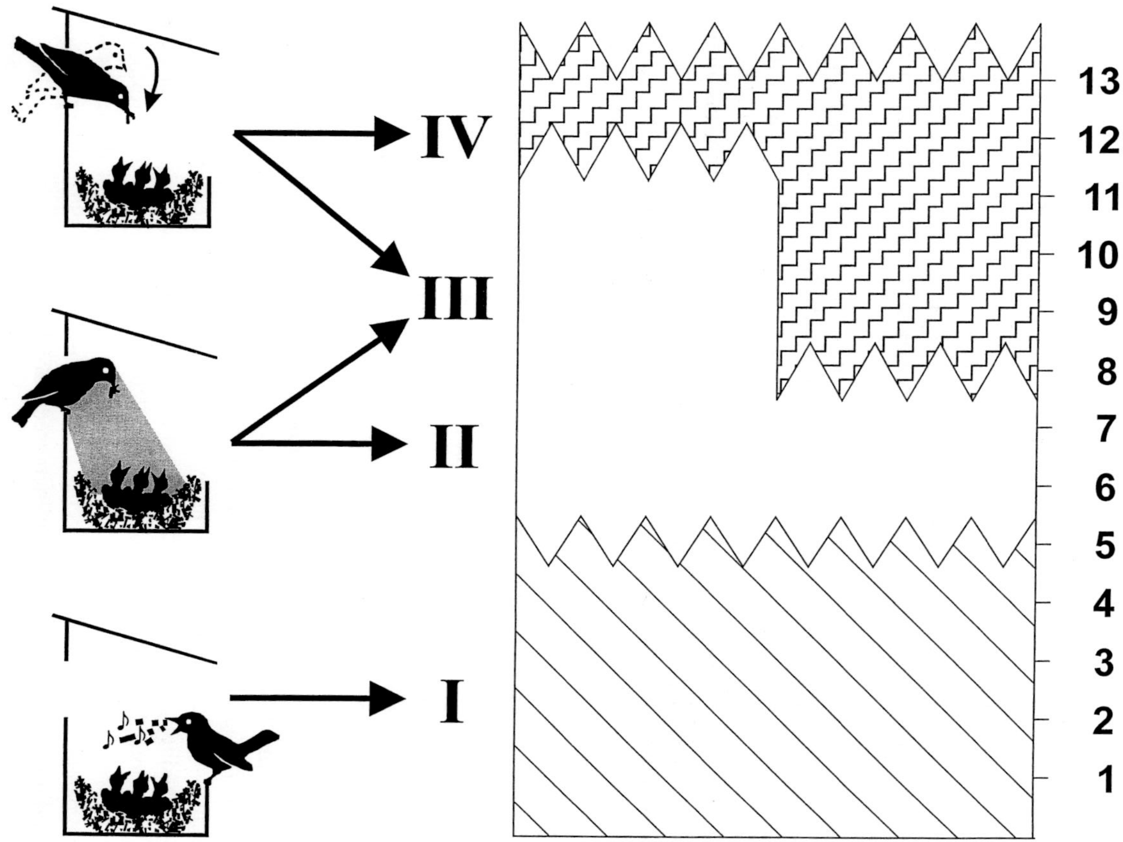

Figure 12.1. The scheme of behavioural development of pied flycatcher nestlings. I–IV—stages of development of feeding behaviour defined on the basis of the leading afferent control. I—begging is elicited exclusively by acoustic signals; II—feeding behaviour is elicited by the diffuse luminosity change caused by a parent blocking the entrance-hole; III—feeding behaviour is triggered by diffuse luminosity change and guided by the moving silhouette of an adult bird; IV—feeding behaviour is both triggered and guided by the moving silhouette of a parent. From stage II onwards, acoustic stimuli loose their efficacy for the natural feeding behaviour.

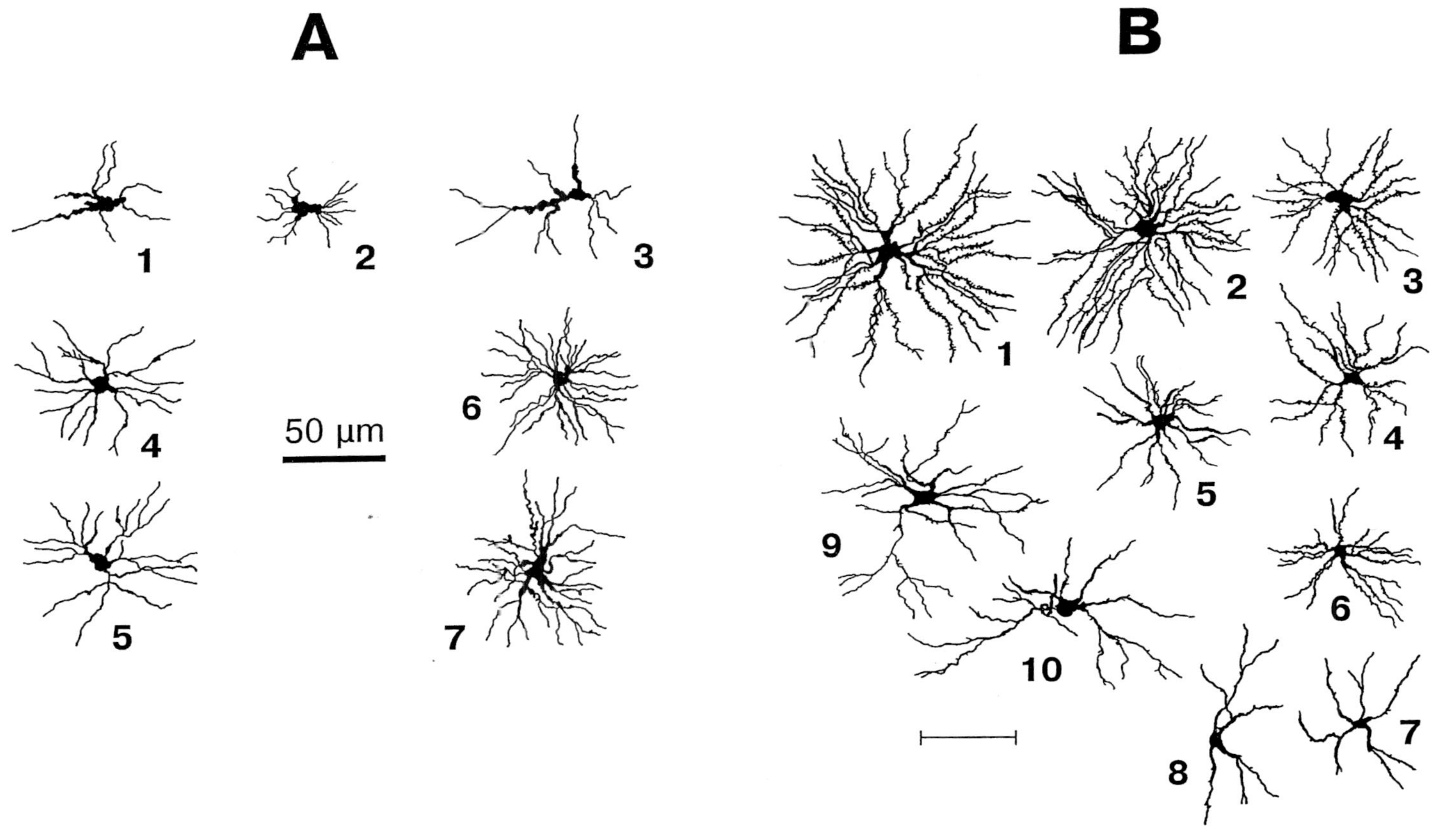

Figure 12.2. Wulst neurones of 6-day-old (A) and 13-day-old (B) pied flycatcher nestlings. **A:** 1–3—non-differentiated neurones, 4–5—less mature spineless stellate neurones; 6–7—more mature spineless stellate neurones. **B:** 1–2—multi-spiny and 3–4—medium-spiny stellate neurones, 5–6—spineless small stellate-like neurones, 7–8—spineless reticular-like neurones, 9–10—spineless arborescent neurones. Golgi staining, camera lucida drawings.

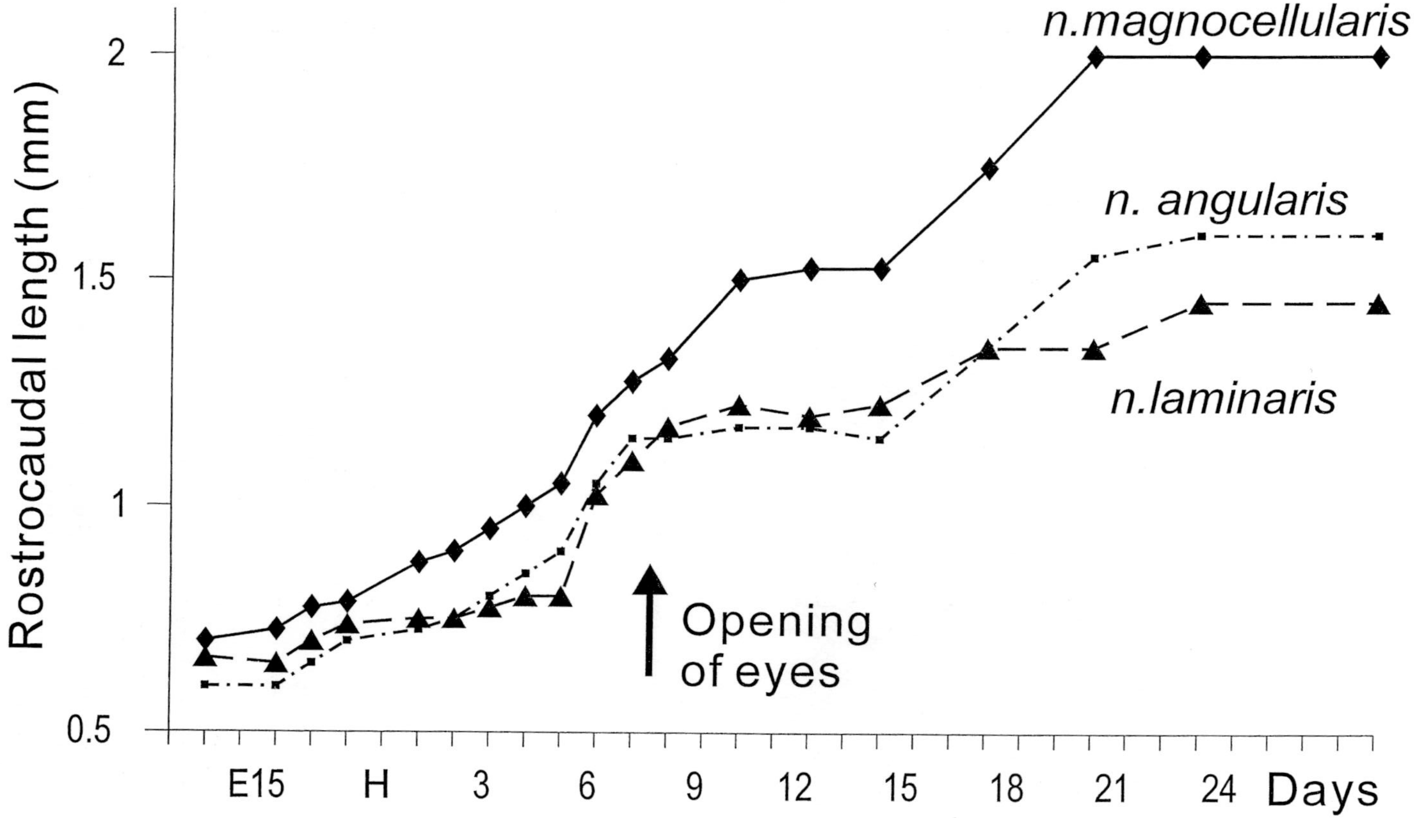

Figure 12.3. The dynamics of the rostro-caudal length of auditory brainstem nuclei in magpie. Abscissa—age, H—hatching.

accompanied by a selective delay in auditory maturation within the range of signals eliciting feeding behaviour in the young.

To understand the causal relations in the process of change in the sensory basis of behaviour, it was essential to find out if the delay in auditory development was the cause of the eyes opening, or whether opening of the eyes was the event inducing a delay in auditory development. If opening of the eyes was a primary factor, the decrease in auditory sensitivity could be explained at least in two ways—firstly, by the onset of functioning of an additional sensory input and, secondly, by the influence of a temperature factor. After opening of the eyes, the thermoemitting surface increased significantly, resulting naturally in the drop of body and brain temperature in 4–5 day-old nestlings with immature thermocontrol mechanisms. The decrease of auditory sensitivity caused by a decrease of brain temperature is a well-known fact (Schermuli and Klinke, 1985).

To solve these problems, an experiment was designed involving visual deprivation. Eyes of 1–1.5 day-old nestlings developing in their native nests were closed by special non-transparent caps. The caps were removed at the age of 13 days, when nestlings' thermocontrol mechanisms are relatively mature. The study of EP from the caudal neostriatum showed that auditory thresholds in 6–9 day-old visually-deprived nestlings were lower than those in the control young (Figure 12.4). When the caps were removed at the age of 13 days, the thresholds of responses to the frequencies eliciting feeding behaviour were temporarily increased. These data suggested that the delay of auditory development is the result of opening of the eyes. The fact that auditory threshold was temporarily increased in 13-day-old nestlings with established thermocontrol indicates that delay of auditory development is the result of the increased inflow of visual information. The temperature factor underlying this delay under natural conditions cannot however be completely ruled out.

The delay in auditory development can explain the shift from auditory guidance of feeding behaviour to visual guidance. But it must be noted that this delay is relatively short, especially as judged by the auditory sensitivity thresholds. It remains unclear why the change of the sensory basis of early feeding behaviour remains irreversible.

Electronmicroscopic study of auditory epithelium hair cells in developing pied flycatcher and ducklings, combined with the analysis of their functional characteristics, revealed a direct relation between the frequency of the auditory signal perceived by a cell and the number of its stereocilia (Golubeva, 1993, 1997). The number of stereocilia on the surface of hair cells was shown to increase in ontogeny. As a result, earlier maturing cells, responding to tones of 1–2 kHz (frequency range of species-typical food call), begin to respond to higher frequencies, whereas the 1 kHz signal excites later developing hair cells, which, correspondingly, having a lesser number of stereocilia (Figure 12.5). The described spatial shift of sensitivity in the receptor region of the auditory system results in the species-typical acoustic signal missing the input of the given system. The entire systemic organization of feeding behaviour, that realized its functions on the basis of a certain set of receptor cells, stops functioning. We think that this should be the cause of complete disintegration of the systemic organization of early feeding behaviour, and should make the shift from auditory-guided feeding behaviour to the visually guided one irreversible.

The reality of disintegration of early functional systems may be confirmed by our previous data in kittens demonstrating the reorganization of brain modulatory systems in early ontogeny of mammals. A conditioning stimulation given to the *locus coeruleus* was found to facilitate neuronal responses to testing stimulation of sensory nerves (lingual and

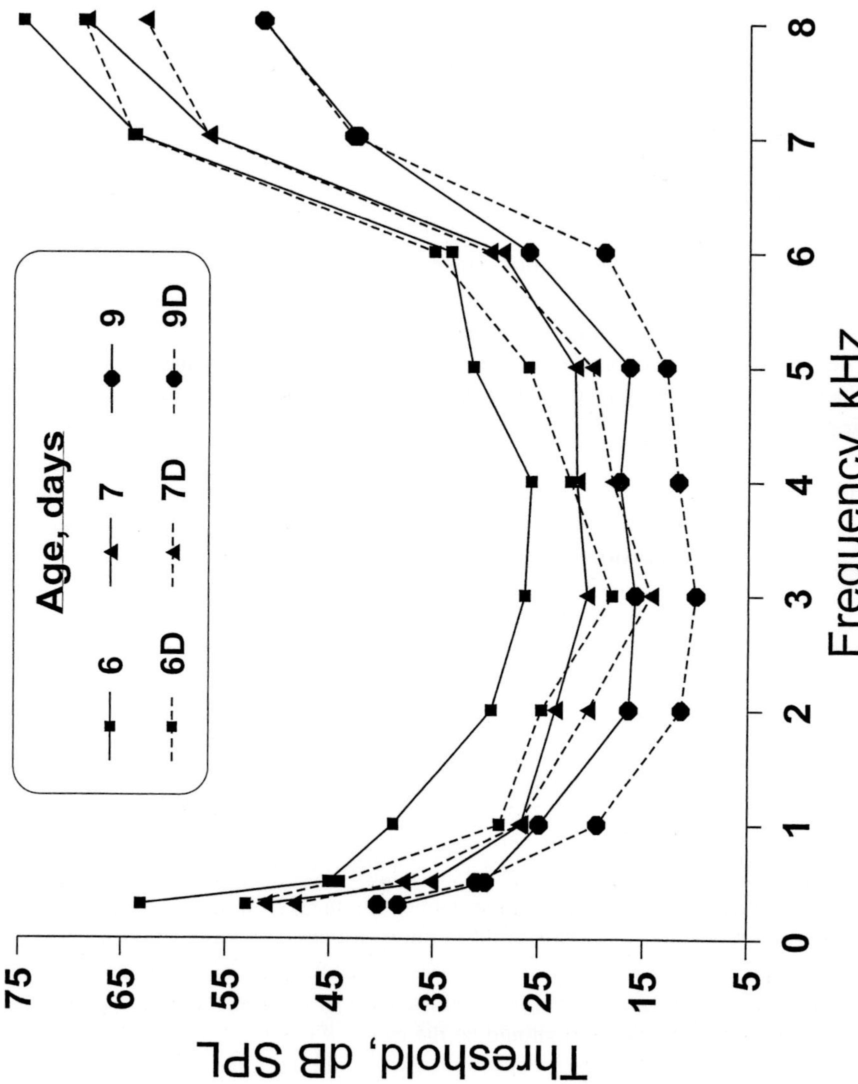

Figure 12.4. Thresholds of field L evoked potentials in response to tone pips in control (solid lines) and visually deprived (dashed lines) pied flycatcher nestlings at the age of 6, 7, and 9 days.

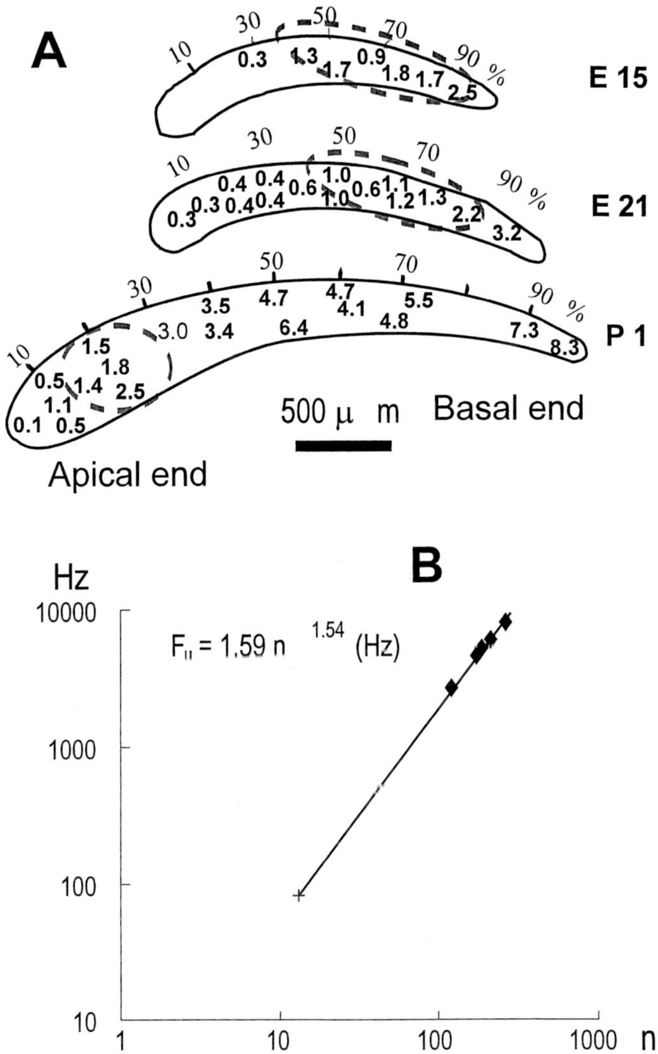

Figure 12.5. Ontogenetic change of the upper limit of the frequency range of auditory hair cells in domestic duck. **A:** schematic drawing of auditory epithelium in domestic duck embryo at the onset of hearing (E15), in 21-day-old embryo (E21) and in newly-hatched nestlings (P1). Digits inside the contour of epithelium—upper limit of frequency range of hair cells in the given location. Digits along the upper outer edge of epithelium—distance from the apical end of the epithelium (as a percentage of its total length). Grey dashed lines encircle the area where hair cells tuned to the frequency range of species-specific signal are located. With age, this area shifts in the direction of the basal end. **B:** The dependence of the higher limit of hearing range upon the maximal number of stereocilia on hair cells of auditory epithelium in the same embryo. Experimental points are approximated by a function $F_u = 1.59\ n^{1.54}$ Hz that was used to calculate frequencies shown in A on the basis of the known number of stereocilia. + indicates the control point with the coordinates calculated in the following way: the known minimal number of stereocilia on a mature hair cell was used to calculate the upper limit of a hearing range using the $F_u = (n)$ function.

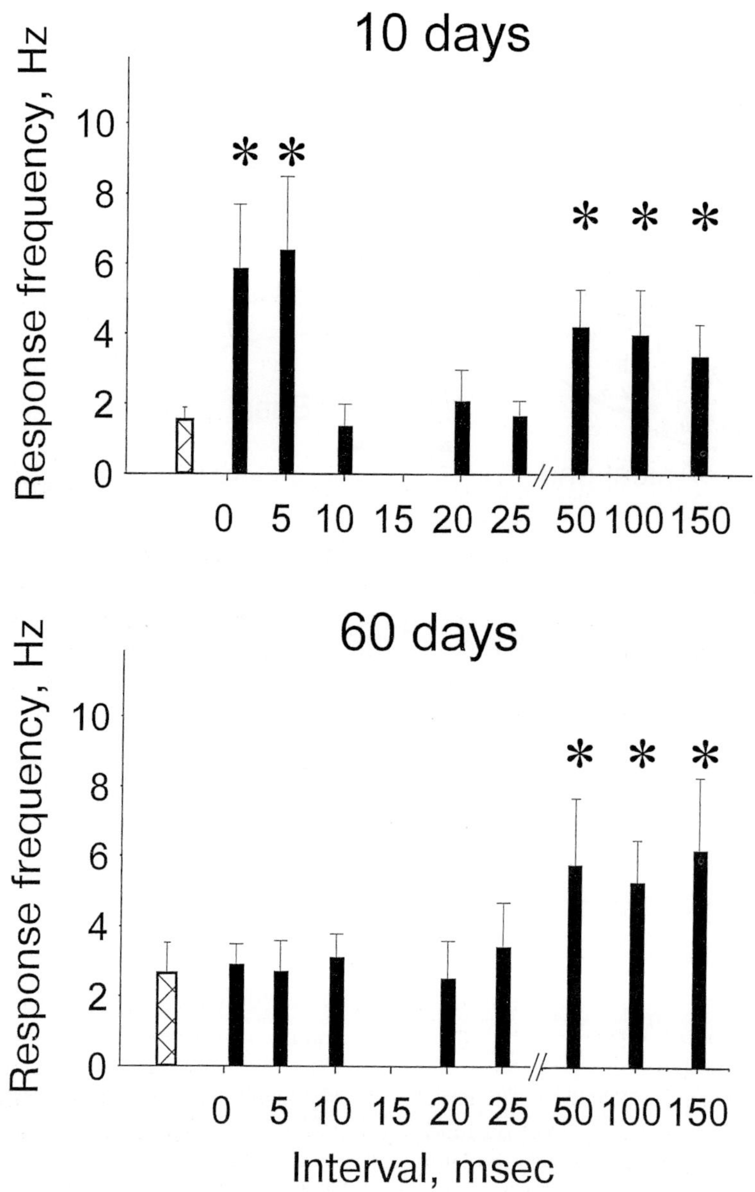

Figure 12.6. Influence of conditioning locus coeruleus stimulation on neuronal responses to ischiadic nerve stimulation in cat's somatosensory cortex. Hatched bars—response to testing stimulus (ischiadic nerve stimulation). Black bars—responses to paired stimulation of locus coeruleus and ischiadic nerve. Abscissa—interval between conditioning and testing stimulus, msec.

ischiadic) in the somatosensory cortex. Changing the interval between the conditioning and testing stimulus revealed a biphasic modulating effect of noradrenergic stimulation in kittens during the first postnatal month: short-latency and long-latency effects. By the age of 2 months, only the long-latency noradrenergic modulation remains (Figure 12.6; see Raevsky, 1991). Thus those forms of neural system organization that had been subserved by the short-latency noradrenergic modulation, during early ontogeny, cannot be reproduced at later developmental stages.

The present data suggest the following hypothesis: The change of behavioural patterns during ontogeny is based on complete or partial disintegration of the systems subserving early adaptive responses. If this hypothesis is correct, then the traditional view of behavioural development as the modification of inborn patterns, induced by the environment and experience, should be reconsidered, since, according to the proposed idea, early patterns disintegrate to be substituted by the ones formed *de novo*. Another consequence of this hypothesis is that the problem of the changing sensory basis of early behaviour loses its initial significance. In this case, during early ontogeny, previously existing systemic organization disintegrates, while the new one is formed, the new one subserving an adaptive behavior using another sensory basis.

Our results indicate that the formation of a new function during the process of individual development may comply to the following scheme: (i) increase of a sensory inflow; (ii) delay in development of a sensory system subserving early behaviour; (iii) disintegration of an old and (iv) development of a new systemic organization (Figure 12.7). According to our hypothesis, an increase of sensory inflow, and its limitation resulting from the delay in sensory development, are consecutive events in the ontogenetic process. It is possible that limitation of sensory inflow at early developmental stages under certain conditions may influence the rate of formation of early behavioural patterns, in the same way that an enriched sensory environment does.

3. INFLUENCE OF LIMITED SENSORY INFLOW ON THE MATURATION OF EARLY BEHAVIOUR

Study of the behaviour of rat pups in the open field during the first postnatal month, carried out in our laboratory by M.L. Pigareva and A.D. Vorobyeva revealed the succession and timing of development of several behavioural patterns (Pigareva and Vorobyova, 1994). By postnatal day 19–21, rats demonstrate a relatively crystallized righting response (standing erect on hind legs), walking on four legs, and washing movements. By day 23, marked manipulatory activity is established.

Partial deafferentation of the forelimbs (transection of *n. medianus)* changed the rate of development of these forms of behaviour. In rat pups, operated on postnatal day 13, 14, or 15, walking, washing movements and righting and manipulatory activity appear simultaneously, on postnatal day 16.

Sensory information and filling of the matured sensory channels with species-typical afferent signals play an important role in the functional maturation of developing brain (Khayutin and Dmitrieva, 1991). Rats' forelimbs are among the earliest-maturing parts of the somaesthetic system. That is why the limitation of sensory input resulting from the partial deafferentation of the forelimbs may cause the significant change in maturational rate.

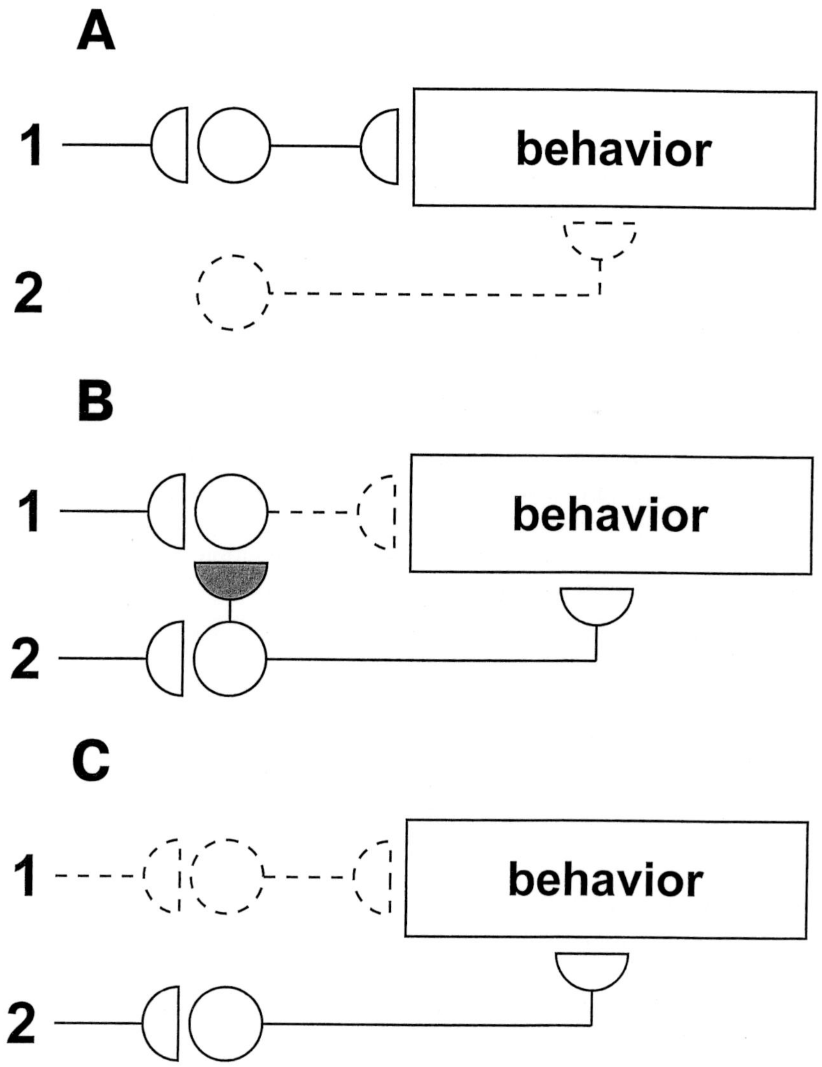

Figure 12.7. Changes of the sensory basis of early behavior. **A**—behaviour is organized on the basis of the sensory input (1) that has matured by that time; **B**—maturation of the new sensory input (2) temporaly inhibits the involvement of the previous sensory input (1) into the behavioural organization; **C**—sensory input 1 is permanently excluded from the behavioural organization.

It was shown that median nerve transection in 15-day-old rat pups induced accelerated appearance of new behavioural patterns (appearing by day 16). Denervation performed in 14- or even 13-day-old rats did not decrease the age when new behaviour appeared any further (Pigareva and Vorobyova, 1994). Thus it is reasonable to conclude that the limitation of sensory input does not affect the rate of morphological maturation, but rather stimulates the earlier functioning of already matured structures.

One of the possible ways to compensate for the deficiency of sensory inflow is to employ those afferent inputs that have not still started to function. It results in an earlier appearance of behavioural patterns that are realized on the basis of information arriving via these other sensory inputs. An actualization of one of such "latent" sensory input was found in the experiments of E.Yu. Sitnikova. It was demonstrated that the number of neurones in rat somatosensory cortex activated in response to stimulation of *n. ischiadicus* (a hind limb nerve) significantly increased in rat pups after median nerve transection on day 13–14 (Sitnikova, 1997).

However, early ontogenetic limitations of sensory inflow not only change the timing and characteristics of formation of sensory systems, but also influence significantly the plasticity of hippocampal neurones, i.e. of the structure closely linked with learning processes. The studies of CA1 population spikes elicited by stimulation of Schaffer collaterals in hippocampal slices prepared from 16-day-old rat pups with *n. medianus* transected at the age of 13 days revealed a significant increase in long-term potentiation (Figure 12.8). The fact that the increase of plasticity in hippocampal neurones after the partial deafferentation of front limbs coincided with the time of appearance of the behavioural patterns studied stresses the role of learning in the development of early behaviour.

The study of the influence of limiting sensory inflow on the development of early behaviour patterns supports the following conclusions:

a) Partial deafferentation of the front limbs in early ontogeny determines the accelerated appearance of several behavioural patterns;
b) The accelerated appearance of behavioural patterns following partial deafferentation of the forelimbs in early ontogeny is the result of actualization of systemic organization of behaviour, but not of the change in the rate of morphological maturation;
c) This acceleration is based on the actualization of previously latent sensory inputs and higher plasticity of hippocampal neurones.

Accelerated formation of sensory systems, increased plasticity of hippocampal neurones, and earlier manifestation of behavioural patterns following partial limitation of sensory inflow could be considered as constructive changes creating additional possibilities for an organism to adapt to its environment. However, to justify this assumption, it was desirable to analyze the long-lasting consequences of early ontogenetic limitation of sensory input.

4. IMPACT OF THE LIMITATION OF SENSORY INPUT DURING THE DEVELOPMENTAL CRITICAL PERIODS ON LEARNING IN ADULT ANIMALS

Study of the influence of limiting sensory input during ontogeny on the development of new behaviour in adult animals was based on the conditioned avoidance model—coupling visual stimuli and electric shock to the feet according to the "hot plate" paradigm. The nest behaviour of experimental rats (after the transection of the median nerve at the age of 13 days) was virtually indistinguishable from that of the control ones. However, the dynamics of learning the conditioned avoidance in experimental rats had some specific characteristics. They learned to avoid electric shock after a smaller number

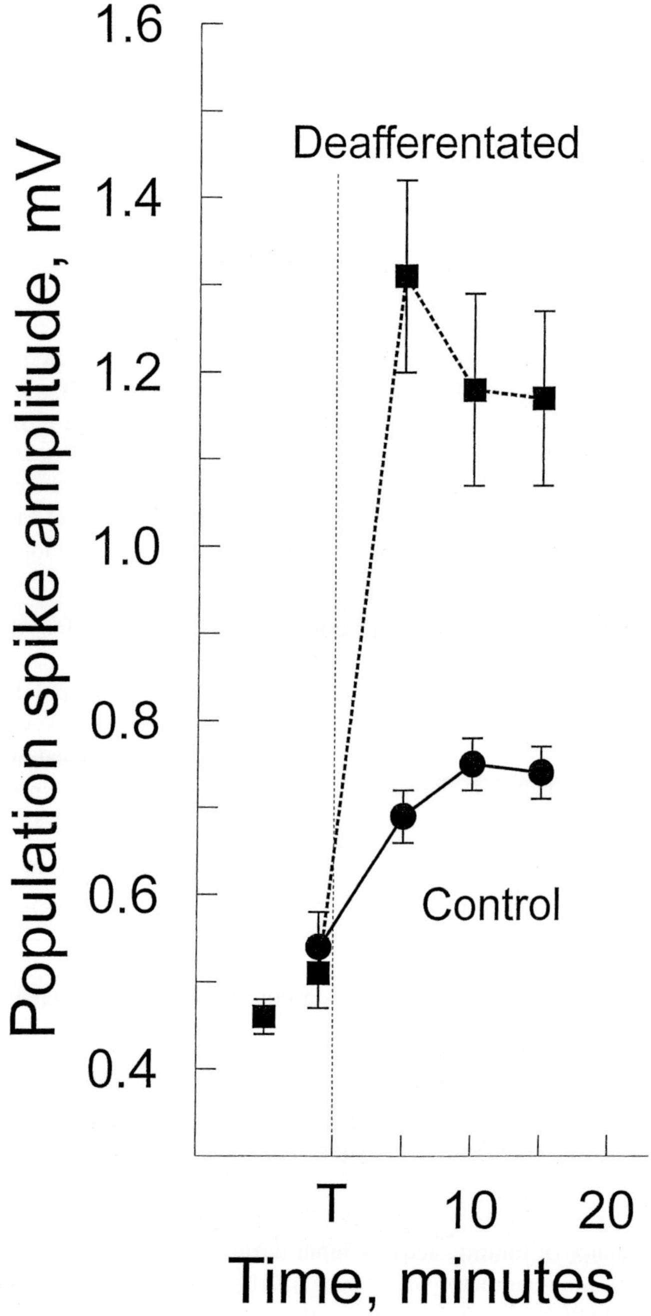

Figure 12.8. LTP profiles in the slices from control and deafferented (on day 13) 16-day-old rats. T—the moment of tetanization.

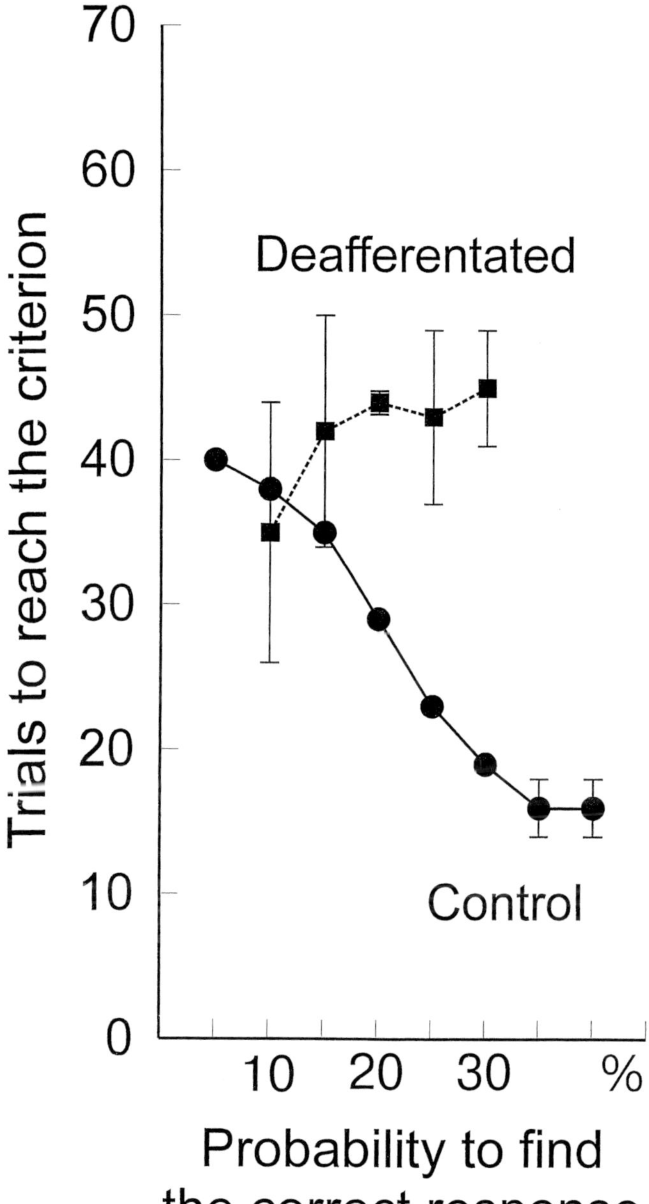

Figure 12.9. The dependence of learning capability (two-way avoidance: light flash + footshock) on the probability of finding the correct response to the conditioned stimulus. The number of trials, required to reach the learning criterion, is plotted against the percentage of correct responses to light flash during learning.

of trials, but consolidation of the response took significantly more time than in intact rats (Figure 12.9).

Corresponding data were obtained by Vorobyeva and Stashkevitch (1995) in studies of instrumental food-acquisition behaviour, namely "reaching" (retrieving food from a horizontal tube). Learning in 1.5–2-month-old rats was characterized by progressive decrease in the time, required to retrieve a food pellet. An experience, accumulated by the rat during the first experimental day, was fixed well, and the timing of the very first trials next day was comparable with the best results that animals had achieved the day before. Rats, subjected to the limitation of sensory inflow in early ontogeny (transection of median nerve at 13–15 postnatal day) also gained skill in retrieving food from the horizontal tube during the first experimental session. However, by the next experimental day they had virtually lost the skill, and had to learn again. The lack of memory consolidation in operated rats was observed during 4–5 experimental days. Thus limitation of sensory inflow during early postnatal ontogeny impairs the processes of consolidation involved in the formation of behaviour in adult animals.

Investigation of the effects of median nerve transection in 13–15-day-old rat pups on the development of some characteristics of synaptic transmission in hippocampus suggested some ideas concerning the nature of this critical period. It was shown in hippocampal slices, that paired-pulse facilitation was significantly higher in denervated animals. This phenomenon was paralleled by marked decrease of population spike amplitudes. It is known that the magnitude of paired-pulse facilitation is inversely proportional to the efficiency of synaptic transmission. This fact may be used to calculate the probability of the mediator release. The result of such calculation in the slices of control and deafferentated animals of different ages is shown in Figure 12.10. Since the amplitude of neuronal responses depends directly on synaptic efficiency and the number of connections, we tried to estimate the difference in the number of connections between control and denervated rats. Figure 12.10 shows that the number of connections was smaller in denervated rats throughout postnatal days 17–20. Thus it is possible to infer that limitation of sensory inflow alters normal synaptogenesis in the hippocampus. Deviations in development of the hippocampus, the structure playing an important role in the realization of cognitive function, should inevitably affect learning process. Considering the fact that the peak of maturation of rat hippocampal neurones corresponds to postnatal days 15–16, it is clear why the limitation of sensory input performed on days 13–15 seriously affects learning in adult animals, while the same operation performed on days 17–19 has no effect on learning.

5. CONCLUSIONS

Our data make it possible to identify some fundamental aspects of the maturation function in the process of individual development.

We propose that the basis of ontogenetic formation of a new function is an increase of sensory inflow. In our experiments this increase was due to the formation of the corresponding sensory systems, but it is natural to assume that an analogous increase of sensory inflow may also be caused by changes in environment such as parturition and leaving the nest. An increase of sensory inflow determines the delay in development of a

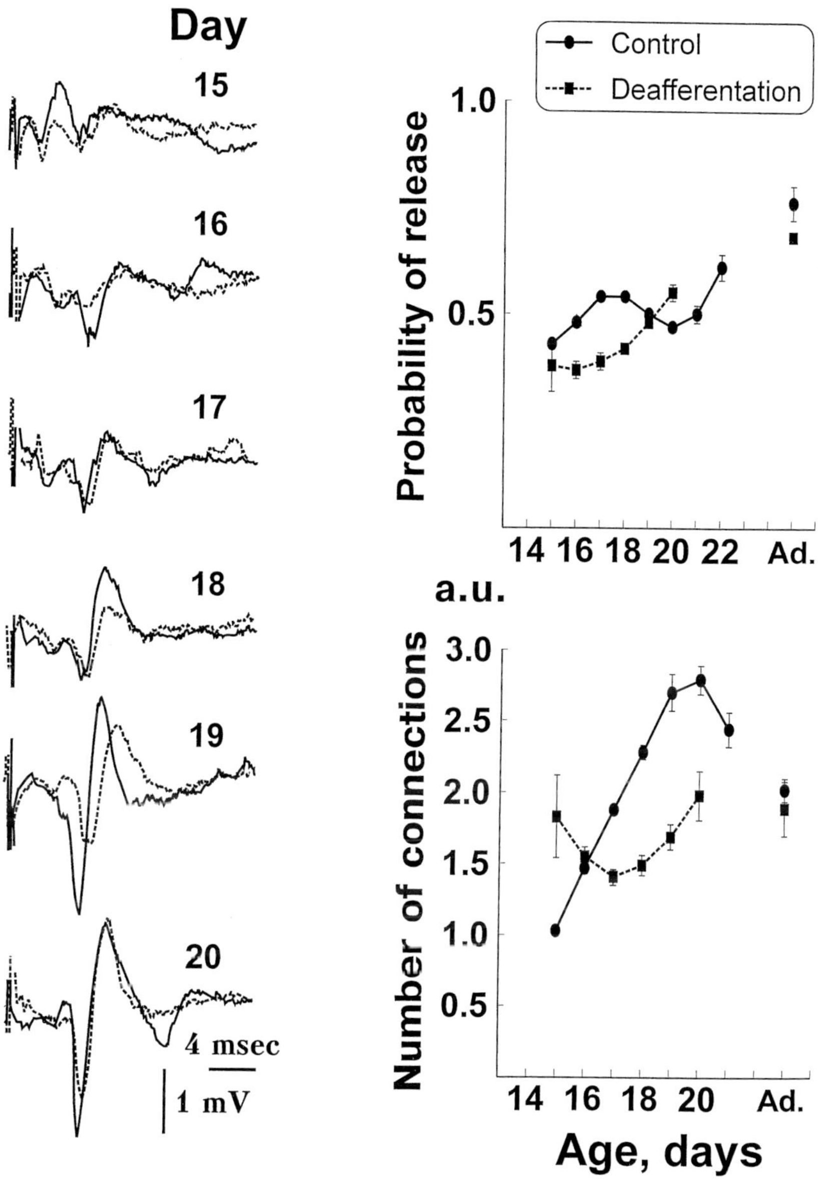

Figure 12.10. Influence of partial front limb deafferentation, performed on day 13, on the development of hippocampal synaptic transmission. Left column—examples of CA1 population spikes induced by the stimulation of Schaffer collaterals at different age (solid traces—control animals, dashed traces—deafferented animals). Right column—influence of deafferentation on plastic characteristics of hippocampal neurones, as revealed by paired-pulse facilitation (PPF) reflecting: top—probability of transmitter release ($p = 1/PPF$), bottom— calculated number of synaptic connections (n is proportional to A/p, where A—amplitude of population spike), arbitrary units. Abscissa—age, days; Ad—adult animals.

sensory system subserving an early behavioural pattern and the appearance of a new function. Later, according to our hypothesis, systems subserving the early adaptive behaviour disintegrate. This hypothesis concerns early behavioural patterns that are in use, involving relatively rigid systemic constructions, the sensory basis of which are conspecific signals that are an inseparable part of the early ontogenetic environment. This may be illustrated by the early feeding responses in altricial nestlings (raising of the head and beak-opening in response to the specific parental vocalization) or mammals (perioral stimulation—suckling). These highly specialized responses are obviously essential for survival during the first days of life, but sooner or later they become an obstacle for the development of more flexible behavioural patterns adequate for new life conditions. The distinguishing characteristic of each successive ontogenetic stage is the increase of the number of external factors that an animal can use, and that it has to use, in the organization of adaptive behaviour, that is, the increased discretization of the perceived environment. We consider such phenomena as the change of qualitative and quantitative characteristics of the "goal-object" environment. This results, we think, in "rejecting" the behavioural patterns adaptive at the early developmental stages and in formation of the new ones that are adequate for new life conditions.

It is possible to suggest that apoptosis—an early ontogenetic phenomenon that inspired the ideas of neural Darwinism—reflects, to some extent, the disintegration of those systemic organizations that subserve adaptive responses specific for a certain stage of ontogeny. If so, apoptosis should be considered as a factor determining the appearance of new behavioural patterns. However, even if the above hypothesis is correct, another aspect of apoptosis should be considered. The study of the hippocampus in mice has demonstrated the dependence of the amount of cell death upon the level of sensory enrichment of early ontogenetic environment (Kempermann *et al.*, 1997). This indicates that sensory information during early ontogeny is a factor determining not only functional capabilities of the nervous system, but the structural basis of the developing brain as well. It may be concluded from our data that early ontogenetic limitation of sensory inflow prevents normal development of synaptic connections in hippocampus, which may become a ground for alterations of cognitive functions at later developmental stages.

ACKNOWLEDGEMENT

This work was supported by the Russian Foundation for Basic Research (grant #96–04–50260).

REFERENCES

Alexandrov, L.I. (1995) Delay in the development of auditory sensitivity and reorganization of feeding behavior, *Zhurnal Vysshey Nervnoy Dejatelnosty*, **45**, 1047–1050 (in Russian).

Alexandrov, L.I. and Dmitrieva, L.P. (1992) Development of auditory sensitivity of altricial birds: absolute thresholds of evoked potentials, *Neuroscience and Behavioral Physiology*, **22**, 132–137.

Anokhin, P.K. (1948) Systemogenesis as a universal tendency in evolutionary process. *Bull. Experimental Biology and Medicine*, **26**, 81–99 (in Russian).

Bogomolova, E.M. and Kurochkin, Yu.A. (1987) Systemogenesis of a behavioral act. *In Functional systems of an organism* (in Russian), pp. 353–372, Moscow: Meditsina.

Edelman, G.M. (1987) *Neural Darwinism: The theory of neuronal group selection*. New York Basic Books.
Fillion, T. and Blass, E.M. (1986) Infantile experience with suckling odors determines adult sexual behavior in the male rats. *Science, New York*, **231**, 729–731.
Golubeva, T.B. (1993) The range of auditory sensitivity and the number of stereocilia on auditory epithelium hair cells in developing bird cochlea. *Doklady Akademii Nauk SSSR*, **332**, 261–263 (in Russian).
Golubeva, T.B. (1994) The delay in auditory development and the change of the leading afferentation in early ontogeny of birds. *Zhurnal Vysshey Nervnoy Dejatelnosty*, **44**, 992–1002 (in Russian).
Golubeva, T.B. (1996) Acoustically-guided behavior in early ontogeny of long-eared owl: characteristics of feeding behavior and parameters of acoustic signals eliciting it. *Zhurnal Vysshey Nervnoy Dejatelnosty*, **46**, 55–62 (in Russian).
Golubeva, T.B. (1997) Development of the basillar papilla and hearing sensitivity in birds. *Physiology and General Biology Reviews*, **12**, 107–201.
Golubeva, T.B., and Barsova, L.I. (1994). The delay in development of auditory brainstem nuclei during the change of the leading afferentation and behavioral patterns in early ontogeny (in Russian). In *II Kolosov Neurohistological Conference*, St-Petersburg, 16.
Gottlieb, G. (1968) Prenatal behavior of birds. *Quarterly Review of Biology*, **43**, 148–174.
Gottlieb, G. (1971) *Development of species identification in birds*. Chicago: Chicago University Press.
Gottlieb, G. (1981) Role of early experience in species-specific perceptual development. In R.N. Aslin (Ed.), *Development of perception 1*. New York: Academic Press, pp. 5–44.
Kempermann, G., Kuhn, H. and Gage, F. (1997) More hippocampal neurons in adult mice living in an enriched environment. *Nature, London*, **386**, 493–495.
Khayutin, S.N. and Dmitrieva, L.P. (1981) *Organization of natural behavior in nestlings* (in Russian). Moscow: Nauka.
Khayutin, S.N. and Dmitrieva, L.P. (1991) *Organization of early species-specific behavior* (in Russian). Moscow: Nauka.
Korneeva, E.V. (1995) Orientation of dendrites in Wulst neurons during the change of forms of visually-guided behavior of pied flycatcher nestlings. *Zhurnal Evoliutsionni Biokhimii i Fisiologii*, **31**, 637–647 (in Russian).
Korneeva, E.V., Gladkovitch, N.G., Vorobyova, A.D., and Shuleikina, K.V. (1993) Neuroontogeny of hyperstriatum during the change of forms of the visually-guided behavior of nestlings. A morphometric study. *Zhurnal Evoliutsionni Biokhimii i Fisiologii*, **29**, 387–397 (in Russian).
Korneeva, E.V., Gladkovitch, N.G., Vorobyova, A.D., and Shuleikina, K.V. (1994) Dynamics of the change of forms of visually-guided behavior of pied flycatcher nestlings and neuroontogeny of Wulst. *Ornitologia*, **26**, 215–217 (in Russian).
Lorenz, K. (1963) *Evolution and modification of behavior*. Chicago, University of Chicago Press.
Luschekin, V.S. (1983). *Sensory factors of spatial orientation in kittens. [Diss. Cand. Biol. Sciences]*, Inst. Vysshey Nervnoy Deyatelnosti i Neyrofiziologii Akad. Nauk SSSR, Moscow (in Russian).
Pigareva, M.L. and Vorobyova, A.D. (1994) Accelerated appearance of behavioral responses in sensory-deprived rats. *Zhurnal Vysshey Nervnoy Dejatelnosty*, **44**, 985–992 (in Russian).
Purves D., White L.E. and Riddle D. (1996) Is neural development Darwinian? *Trends in Neuroscience*, **19**, 460–464.
Raevsky, V.V. (1991) *Ontogenesis of brain transmitter systems*. Moscow, Nauka (in Russian).
Ramon-y-Cajal, S. (1929) *Studies on vertebrate neurogenesis*. Thomas, Springfield, Ill.
Roux, V. (1974) In B.A. Willer and J.M. Oppenheimer (Eds). *Foundations of experimental embryology. Second edition*. Hafner Press, pp. 2–37.
Schermuli, L. and Klinke, R. (1985) Change of characteristic frequency of pigeon primary auditory afferents with temperature. *Journal of Comparative Physiology*, **156**, 209–211.
Shuleikina, K.V. (1971) *Systemic organization of feeding behavior* (in Russian). Moscow, Nauka.
Sitnikova, E.Yu. (1997) Partial deafferentation of the front limbs in rats' early ontogeny increases the reactivity of somatosensory neurons. *Zhurnal Vysshey Nervnoy Dejatelnosty*, **47**, 1051–1054 (in Russian).
Sporns, O. (1997) Variation and selection in neural function. *Trends in Neurosciences*, **20**, 291–292.
Sporns, O. and Tononi, G. (1994). *Selectionism and the brain*. London, Academic Press.
Thorpe, W.H. (1963) *Learning and instinct in animals*. London, Methuen.
Tinbergen, N. (1951) *The study of instinct*. Oxford, Oxford University Press.
Vorobyova, A.D. and Stashkevitch, I.S. (1995) Formation of instrumental food-acquisition behavior in adult rats using the front limb, partially denervated in early postnatal ontogeny. *Zhurnal Vysshey Nervnoy Dejatelnosty*, **45**, 1206–1210 (in Russian).

13 Colour Spaces of Animal-trichromats (Rhesus Monkeys and Carp) Revealed by Instrumental Discrimination Learning

A.V. Latanov, A.Yu. Leonova, D.V. Evtikhin and E. N. Sokolov

M.V. Lomonosov State University, Moscow, Russia
e-mail: latanov@protein.bio.msu.su

Discrimination of colours was studied using an instrumental learning paradigm in monkeys (Macaque rhesus) and fish (*Cyprinus carpio* L.). The confusion matrices composed of probabilities of instrumental responses were treated by factor analysis. The spherical structure of perceptual colour space in both species was similar to that for humans. Four eigenvectors constituting a four-dimensional Euclidean hypersphere corresponded to neuronal channels for 'red-green', 'blue-yellow', 'brightness' and 'darkness'.

KEYWORDS: colour perception, instrumental learning, colour modelling, animals-trichromats

1. INTRODUCTION

Direct estimation of subjective differences between colours is the traditional method of investigation of colour vision in humans. Multidimensional scaling of subjective suprathreshold differences between perceived colours obtained by direct estimations reveals a spherical structure of perceptual colour space in normal subjects. Sokolov and Izmailov (1984) showed that different colour stimuli varying in hue and brightness were located on a spherical surface in a four-dimensional Euclidean space. The distances between points representing colours correlated closely with subjective differences. In the spherical model of colour discrimination, four Cartesian co-ordinates of each colour-point represent excitations of neuronal channels for red-green, yellow-blue, brightness and darkness. Three spherical co-ordinates of the hypersphere correspond to subjective aspects of colours: hue, brightness and saturation (Sokolov and Izmailov, 1984).

Subjective colour differences can also be obtained by using colour naming techniques. In this case each represented colour is associated with a certain colour name with some frequency. Thus all colours in the experiment are represented by specific vectors of their association frequencies with specific colour names. Colour differences are computed as distances between corresponding vectors. Multidimensional scaling of the resulting matrix showed that spherical four-dimensional Euclidean colour space was analogous to the space obtained by direct estimations of colour differences (Sokolov and Izmailov, 1984).

We hypothesise that the structure of colour space in trichromatic animals is based on similar principles as in humans. To test this suggestion instrumental responses analogous to colour categorization in humans were trained with different colours in monkeys (Macaque rhesus) and in fish (*Cyprinus carpio* L.). Using a set of differential stimuli, a set of response probabilities was obtained, which can be considered as a vector corresponding to a reinforced stimulus. Thus different colours were represented by specific vectors of response probabilities. The confusion matrices derived from those vectors were transformed into the correlation matrices which were submitted to factor analysis (Ueberla, 1980) to reveal the basic dimensions of the colour spaces of the animals.

2. METHOD

2.1. Monkeys

Three male monkeys JU, FE and KU (9, 8 and 5 years old, respectively) were trained in a colour discrimination task. Experimental sessions consisted of sequential random presentations of seven stimuli differing in colour generated by an analog colour monitor. The power spectra of colours were measured by a spectrophotometer in order to calculate their CIE-31 chromaticity co-ordinates (Judd and Wyszecki, 1978). The irradiance of spectral compositions was checked by a radiometer. Chromaticity co-ordinates and irradiance of the stimuli are given in Table 13.1.

Colour stimuli (squares of 1.5 degree in size on a dark background) were presented in the middle of the monitor, 57 cm from the monkey's eyes. The experiment was conducted under photopic adaptation with surrounding illumination equal to 50 Lux.

Pressing the lever in response to a particular colour stimulus (the *conditional* stimulus) was reinforced by juice. Responses to six different colour stimuli (*differential* stimuli) in the session were not reinforced. The sequence of stimuli and all behavioural events were controlled and recorded by a computer. The stimuli were presented in a random sequence. In order to maintain the monkey's motivation to drink, the probability of presentation of the conditional stimulus (0.33) was greater than probability of presentation of the differential ones (0.11; stimuli not associated with reinforcement). The interval between trials varied randomly with a range of 4.5 to 10.5 seconds. Between 150 and

Table 13.1. The CIE-31 chromaticity co-ordinates of colours and their irradiation (for monkeys)
The colours are designated by equivalent wavelengths. Asterisks indicate the wavelength of a colour which is complementary to 'purple'.

Colours	*Chromatic co-ordinates* X	Y	*Equivalent wavelength (nm)*	*Irradiation (Watt/sq.m)*
Red	0.644	0.337	605	0.031
Yellow	0.432	0.511	572	0.036
Purple	0.335	0.211	555*	0.076
Green	0.312	0.594	545	0.044
Bluish green	0.241	0.372	490	0.093
Blue	0.145	0.087	470	0.209
White	0.332	0.372	–	0.289

450 trials (total sum of conditional and differential stimulus presentations) were included in each session.

When the percentages of instrumental responses to the conditional and differential stimuli were at a stable asymptotic level, another colour was used as the conditional stimulus and training continued until asymptotic performance was reached again. Each of seven colour stimuli was employed as the conditional stimulus in different conditions.

The set of response probabilities to conditional and differential stimuli calculated over the two or three last sessions of stable performance constituted the specific probability vector. Using different colours as the conditional stimulus, different probability vectors were obtained. The set of the vectors constituted the confusion matrix (see below). Each cell of the confusion matrix was based on 180–300 trials for the conditional stimulus and on 60–90 trials for differential stimuli. The confusion matrices of each monkey obtained under different learning conditions were transformed separately into the correlation matrices which were submitted to factor analysis to reveal the basic axes of colour spaces of animals.

2.2. Fish

Two carp (fish B and fish S) 1.5 and 2 years old, were trained to discriminate colours. Colour stimuli were presented on the analog monitor, placed close to wall of the experimental tank. The power spectra of colour compositions were measured by a spectrophotometer. The spectra at the carp's eyes were corrected to take account of glass and water absorbance spectra. The chromaticity co-ordinates of stimuli were calculated on the basis of the *Cyprinidae* colour triangle, following the method of Judd and Wyszecki (1978) for human standard observers, but using single cone absorbance spectra (Neumeyer, 1986) instead of colour matching functions. The irradiance of spectral compositions was measured by a radiometer. Chromaticity co-ordinates and irradiance of stimuli are represented in Table 13.2.

Two colour stimuli (squares of 40 mm in size on the dark background) were presented 70 mm to the left and right of the middle of the monitor. The conditional stimuli, and one

Table 13.2. The chromaticity co-ordinates of colours on the *Cyprinidae* colour triangle and their irradiation (for carp)

The colour stimuli are designated by equivalent wavelengths. Asterisks indicate the wavelength of a colour which is complementary to 'purple'.

Colours	*Chromatic co-ordinates* *X*	*Y*	*Equivalent wavelength (nm)*	*Irradiation (Watt/sq.m)*
Blue	0.190	0.370	483	0.231
Bluish green	0.230	0.410	490	0.133
Greenish blue	0.260	0.450	502	0.116
Green	0.330	0.520	538	0.146
Yellow green	0.430	0.450	563	0.133
Yellow	0.470	0.430	571	0.222
Orange	0.660	0.280	600	0.178
Red	0.820	0.170	620	0.178
Purple	0.260	0.350	595*	0.127
White	0.330	0.420	–	0.197

of nine differential stimuli were presented on each trial. Left-right position was varied randomly. The experiment was conducted under photopic adaptation, with ambient illumination equal to about 50 Lux on the water surface.

At first the carp were trained to catch one of two beads, suspended in two positions above the stimuli. The carp were rewarded by a little worm if they caught the bead corresponding to the conditional stimulus. Incorrect responses were not rewarded. After any response the stimuli were switched off at once. Onset and offset of stimulus events and behavioural events were controlled and recorded by experimenter and stored on computer disk.

Each session consisted of 200–250 stimulus pairs. When the percentage of instrumental responses to conditional and differential stimuli was stable another colour was used as a conditional stimulus. Eight of ten colour stimuli were employed as conditional stimuli.

As for monkeys the set of response probabilities to differential stimuli taken from last sessions (at the stage of stable performance) constituted the specific probability vector. The set of those vectors corresponding to eight learning conditions constituted the confusion matrix (see below). The confusion matrices for both carp were separately transformed into the correlation matrices, which were submitted to factor analysis to reveal the basic dimensions of the colour space of animals.

3. RESULTS

3.1. Monkeys

Seven series in total were conducted with each monkey. The process of learning was characterized by ascending learning curves for the conditional stimuli, stabilizing at the level of 70–90%. The percentage of responses to the differential stimuli decreased and stabilized at different probability levels, usually close to the stability level for the conditional stimuli, as these stimuli were similar in colour.

The set of stable response probabilities to conditional and differential stimuli constituted the specific probability vector. The set of these vectors obtained in seven series constituted the confusion matrix (Table 13.3).

The confusion matrix of instrumental responses is characterized by greatest probability values in the main diagonal, which corresponded to the conditional stimuli. Table 13.3

Table 13.3. The confusion matrix composed of stable percent response probabilities of Monkey JU
The colours are designated according to Table 13.1 by equivalent wavelengths. Asterisks indicate the wavelength of a colour which is complementary to 'purple'.

Presented colours (nm)	*Reinforced colours (nm)* 605	572	555*	545	490	470	White
605	81.6	26.0	26.6	9.0	22.9	21.1	13.9
572	17.8	90.6	25.0	14.1	43.3	10.6	13.8
555*	23.6	31.6	86.4	17.7	41.8	24.5	19.2
545	12.3	36.8	22.7	79.0	38.5	14.0	20.1
490	20.2	32.7	25.4	22.8	82.2	27.7	29.3
470	21.2	24.4	32.4	31.7	38.5	86.2	32.7
White	12.1	30.7	29.5	15.7	40.8	21.5	72.1

Table 13.4. Cartesian co-ordinates of colour points in the four-dimensional perceptual colour space of Monkey JU

The colours stimuli are designated in accordance to Table 13.1 by equivalent wavelengths. Asterisks indicate the wavelength of a colour which is complementary to 'purple'.

Colours		*Cartesian co-ordinates*			*Length*
(nm)	*X1*	*X2*	*X3*	*X4*	*of radii*
605	0.742	0.318	–0.487	–0.064	0.945
545	–0.925	0.242	–0.177	–0.141	0.983
572	–0.161	0.674	0.179	0.356	0.800
470	–0.046	–0.912	0.058	0.097	0.920
555*	0.143	0.039	–0.076	0.868	0.884
490	–0.224	0.113	0.547	0.457	0.756
White	0.086	0.034	0.944	–0.133	0.958

Mean of radii 0.892 ± 0.032
Variance coefficient 0.095

shows that when opponent colours ('red' *versus* 'green', 'blue' *versus* 'yellow') are the conditional and differential stimuli, the difference in response probabilities is greatest. 'White' as a conditional stimulus differs mainly from the saturated colours.

The correlation matrix obtained from the confusion matrix was submitted to factor analysis, separately for each monkey. Four significant orthogonal factors were revealed by factor analysis for each animal. An example of the confusion matrix for monkey JU is presented in Table 13.3. Four factors with eigenvectors 1.666, 1.473, 1.334 and 0.911, respectively, were revealed from that matrix. Those factors cover 23.8, 21.1, 19.1 and 13.0% of variance in the experimental data, respectively. The factor loads (Cartesian co-ordinates of the stimuli) are presented in Table 13.4. Thus each colour stimulus was described by a four-dimensional vector.

The projection of the four-dimensional vectors representing colours on the X1X2 plane (Figure 13.1) shows that the colours are located in a circular order according to hue, with 'white' close to zero. Thus the axes X1 and X2 could be considered as the activation of 'red-green' (R+G-) and 'yellow-blue' (Y+B-) opponent systems. The points corresponding to 'red' and 'green' have values respectively 0.742 and –0.925 along the X1 axis. At the same time, the points corresponded to 'yellow' and 'blue' have values respectively 0.674 and –0.912 along the X2 axis. However, 'white' is characterized by low values of co-ordinates along both X1 (0.086) and X2 (0.034) chromatic axes.

The projection of points representing stimuli on the X3X4 plane (Figure 13.2) shows colours which are located according to their subjective brightness (for a human observer). The subjectively brightest colour, for example 'white', is near the positive pole of the axis X3 (0.944). But the subjectively darker colour, for example 'red', is moved towards the negative pole of the axis X3. Colours having brightness between 'white' and 'red' are displaced to the positive pole of the axis X4 (the co-ordinate of 'purple' is 0.868). This means that subjective brightness is determined by the ratio of activation of two achromatic subsystems. The angle of vector radius of stimulus on the X3X4 plane can be interpreted as a measure of subjective brightness. 'White' is characterized by a maximal angle value (about 90 degrees), at which the darkest colours ('red' and 'green') have angles

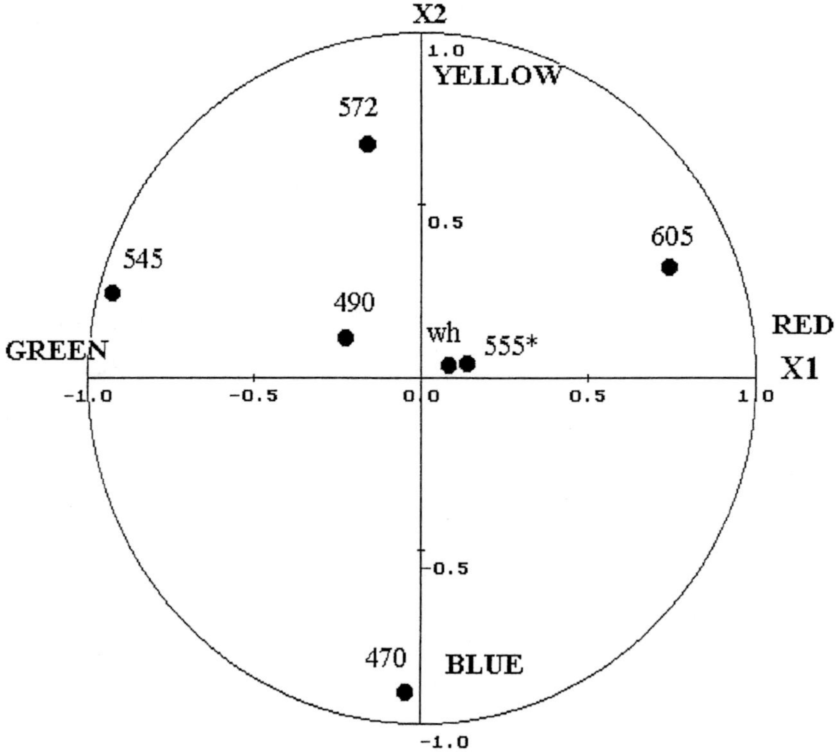

Figure 13.1. The projection of colour points on the X1X2 plane of the perceptual colour space for Monkey JU. The plane is composed of "red-green" (X1) and "yellow-blue" (X2) opponent axes. The colours are designated according to Table 13.1 by equivalent wavelengths. Asterisk indicates the wavelength of a colour which is complementary to "purple".

about –90 degrees. Other colours are characterized by angles distributed in the range between 90 and –90 degrees according to their subjective brightness.

On the chromatic plane X1X2, more saturated colours ('red', 'green', 'blue') are close to the poles of the chromatic axes. On the contrary, less saturated colours are shifted towards the centre of the chromatic plane. This means that lengths of vector radii representing stimuli are proportional to saturation. On the other hand, on the achromatic plane X3X4 the lengths of vector radii representing stimuli are inversely proportional to the saturation. The lengths of vector radii on the chromatic plane X1X2 taken as $(X1^2+X2^2)^{1/2}$ could be considered as a combined chromatic co-ordinate. The lengths of vector radii on the achromatic plane X3X4 taken as $(X3^2+X4^2)^{1/2}$ could be considered as a combined achromatic co-ordinate. The new plane composed of these combined axes represents the compressed four-dimensional space (Figure 13.3). The projection of the stimulus points on these planes shows that the colours lie on an arc according to their saturation. Thus the angle of the vector radius representing stimuli could be interpreted as a measure of saturation.

The mean value of lengths of the four-dimensional vectors representing the colour stimuli is equal to 0.892+/–0.032. The variance coefficient is equal to 0.095. Such a small

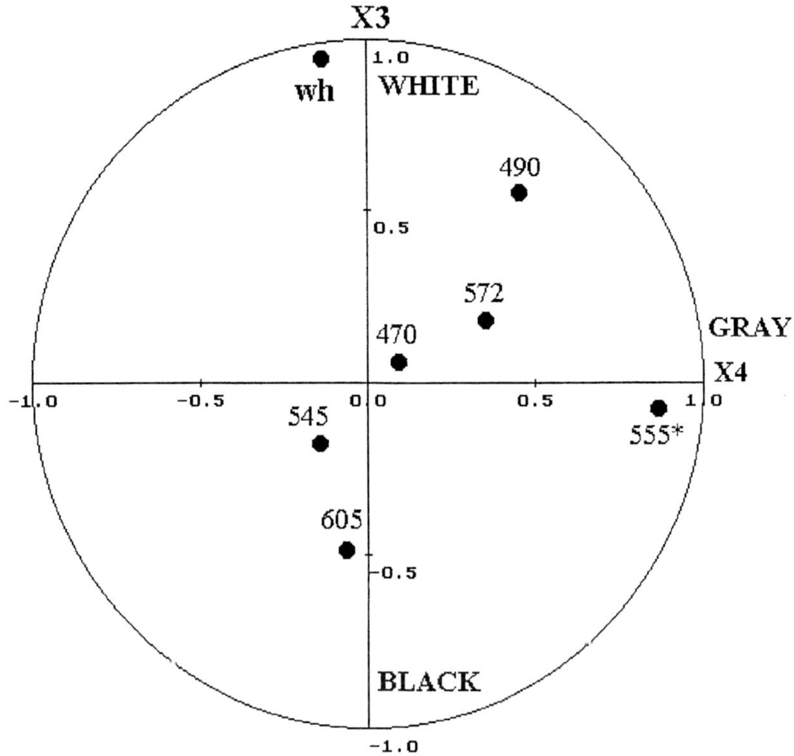

Figure 13.2. The projection of colour points on the X3X4 plane of perceptual colour space for Monkey JU. The plane is composed of "brightness" (X3) and "darkness" (X4) axes. The colour stimuli are designated as for Figure 13.1.

variation in the radii suggests that the points representing colour stimuli are located on the surface of a hypersphere. Spherical co-ordinates (lengths of radii and respective angles) correspond to subjective aspects of colour perception. The angles on the plane X1X2 and on the plane X3X4 correspond to hue and to subjective brightness (lightness), respectively. The angle on the plane constructed from common chromatic and common achromatic axes corresponds to saturation.

3.2 Fish

Two fish were trained in the discrimination task for eight experimental series. In each series a discriminative response to one of the ten stimuli was trained. The response probability to the conditional stimulus presented in combination with differential ones increased over the process of learning. The percentage of responses to the differential stimuli, in turn, decreased and then became stable. The response probability to the differential stimuli decreased to a greater extent for greater differences between the differential and conditional stimuli. After stabilization of response probabilities the next stimulus was used as a conditional one, and the experiment lasted until the next stabilization.

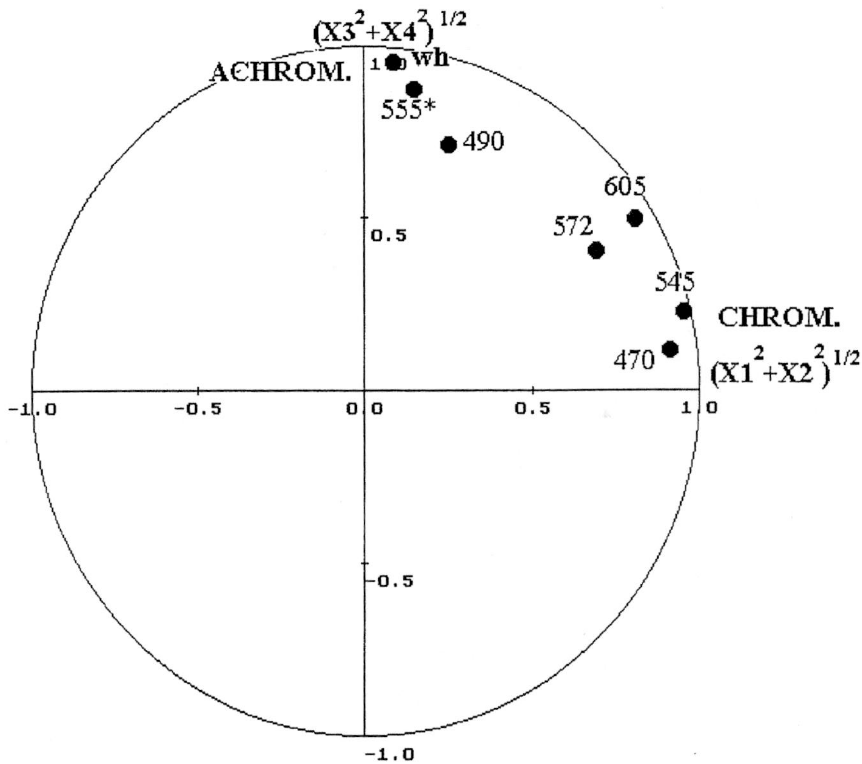

Figure 13.3. The projection of colour points on the plane composed of unified chromatic ($(X1^2 + X2^2)^{1/2}$) and achromatic ($(X3^2 + X4^2)^{1/2}$) axes of perceptual colour space for Monkey JU. The colour stimuli are designated as for Figure 13.1.

Table 13.5. The confusion matrix of response probabilities (percent) for Carp B for the last session of each series

The colours are designated according to Table 13.2 by equivalent wavelengths. Asterisks indicates the wavelength of a colour which is complementary to 'purple'.

Presented colours (nm)	*Reinforced colours (nm)* 483	502	538	563	571	620	595*	White
483	–	5.3	3.7	0.0	0.0	0.0	4.8	12.0
490	15.4	21.1	3.6	0.0	0.0	10.5	0.0	20.0
502	24.0	–	11.1	4.8	8.0	10.0	0.0	38.5
538	3.8	15.8	–	25.0	8.0	0.0	0.0	8.0
563	0.0	11.1	25.0	–	24.0	5.0	0.0	24.0
571	0.0	0.0	0.0	36.4	–	15.0	0.0	20.0
600	4.0	0.0	3.6	0.0	4.3	26.3	5.3	8.0
620	3.8	5.6	3.6	0.0	8.7	–	0.0	4.0
595*	4.0	5.3	3.4	0.0	0.0	15.0	–	23.1
White	3.8	10.5	3.7	4.5	0.0	10.0	15.8	–

Table 13.6. Cartesian co-ordinates of colour points in the four dimensional perceptual colour space for Carp B The colours are designated according to Table 13.2 by equivalent wavelengths. Asterics indicates the wavelength of a colour which is complementary to 'purple'.

Colours (nm)	*X1*	*Cartesian co-ordinates* *X2*	*X3*	*X4*	*Length of radii*
483	–0.119	–0.785	–0.010	–0.392	0.886
502	–0.130	–0.365	0.518	–0.497	0.816
538	–0.991	–0.081	–0.178	–0.024	1.000
563	–0.557	0.621	0.014	–0.191	0.856
571	0.011	0.687	–0.102	–0.274	0.747
620	0.521	–0.059	–0.597	0.093	0.800
595*	–0.036	–0.258	0.221	0.859	0.924
White	0.093	0.004	0.872	0.075	0.880

Mean of radii 0.865 ± 0.029
Variance coefficient 0.094

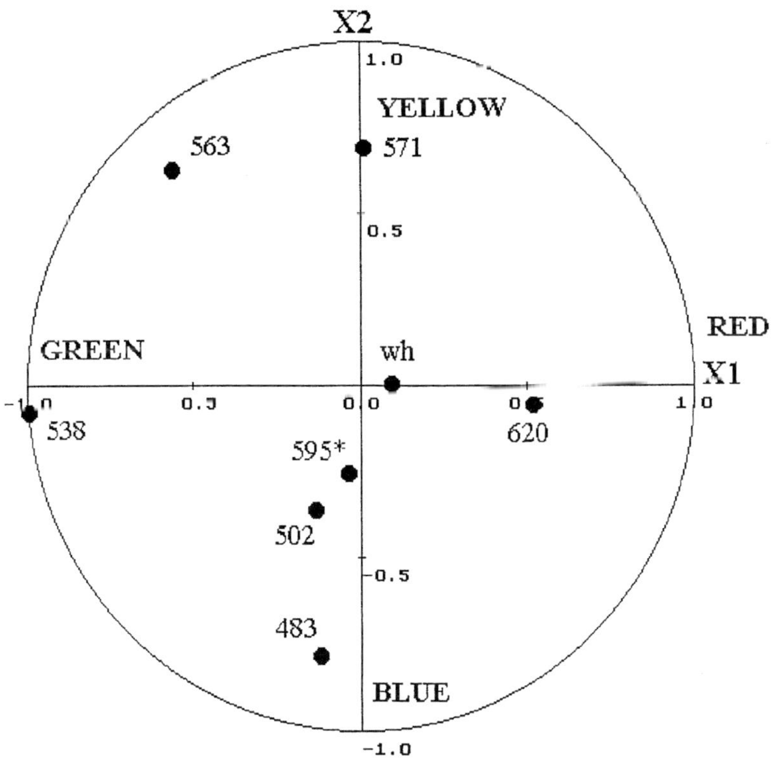

Figure 13.4. The projection of colour points on the X1X2 plane of perceptual colour space for Carp B. The plane is composed of "red-green" (X1) and "yellow-blue" (X2) opponent axes. The colours are designated according to Table 13.2 by equivalent wavelengths. Asterisk indicates the wavelength of a colour which is complementary to "purple".

The results of training an instrumental response to a particular stimulus when the other stimuli were used as differential ones were collated in the confusion matrix composed of the probabilities of the instrumental responses. The data obtained for stable performance were used for further analysis (Table 13.5). In this 8 × 10 matrix each colour stimulus is characterized by a vector-column of the response probabilities. In the matrix the probabilities of the responses to conditional stimuli are omitted because the choice depended on the differential stimulus presented in the pair. The coefficients of correlation between the vectors which represent respective colour stimuli constitute a 8x8 correlation matrix. This matrix was submitted for factor analysis.

Four significant factors were extracted from the correlation matrix, with eigenvalues of 1.907, 1.597, 1.300 and 1.100 contributing 73.9% of the variance of the experimental data. The contributions of the separate factors in variance were 23.8, 20.0, 16.3 and 13.8%, respectively. The factor loads (co-ordinates of stimuli) in the four-dimensional space are given in Table 13.6. In the fish, as in the monkeys, the colour points on the X1X2 plane are positioned according to a Newton circle (Figure 13.4). The 'red' (0.521) and 'green' (–0.991) are located on the opposite sides of X1 axis. The 'yellow' (0.687) and 'blue' (–0.785) are positioned on the opposite ends of X2 axis that is orthogonal to

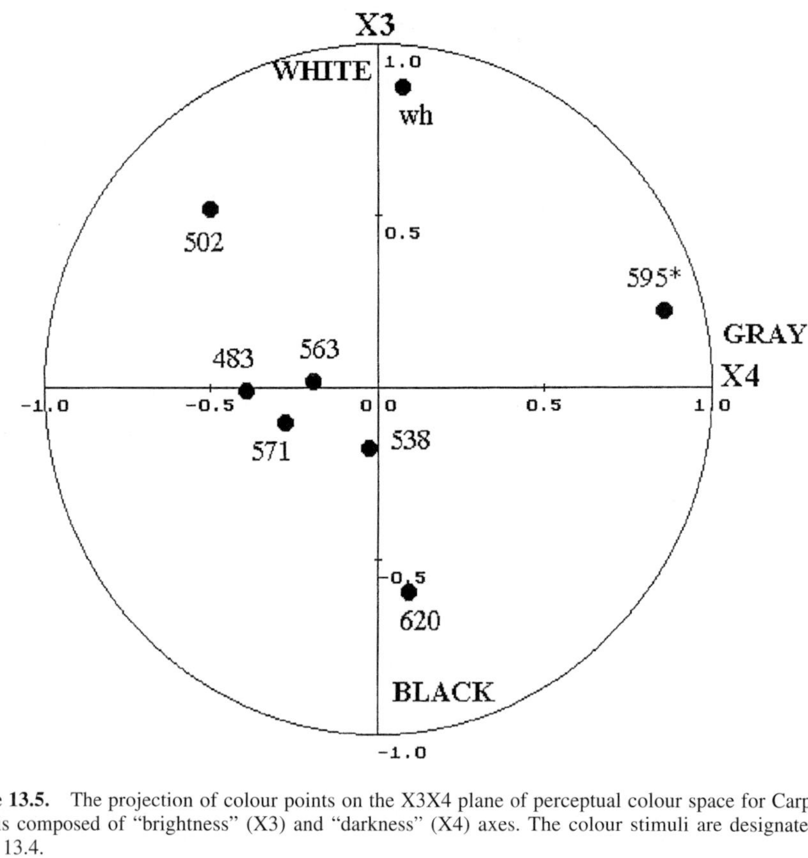

Figure 13.5. The projection of colour points on the X3X4 plane of perceptual colour space for Carp B. The plane is composed of "brightness" (X3) and "darkness" (X4) axes. The colour stimuli are designated as for Figure 13.4.

axis X1. Thus these two factors can be regarded as the opponent 'red-green' (R+G-) and 'yellow-blue' (Y+B-) mechanisms.

On the X3X4 plane, mostly dark 'red' and mostly light 'white' are located on opposite ends of X3 axis (Figure 13.5). 'Purple,' with lightness between 'red' and 'white,' is located near the positive pole of axis X4 that is orthogonal to the axis X3. The highly saturated colours ('blue', 'green', 'yellow-green', and 'yellow') are projected close to the centre of the plane, suggesting that achromatic mechanisms do not contribute significantly to the discrimination of these stimuli. The position of 'bluish green' can be explained by random errors. Thus X3 and X4 axes correspond to contributions of brightness and darkness channels. The angle characterizing positions of the colour points on X3X4 plane relates to subjective brightness.

On the plane constructed from combined axes $(X1^2+X2^2)^{1/2}$ and $(X3^2+X4^2)^{1/2}$, that represents lengths of projections of four-dimensional vectors on respective planes, the colour points are located on an arc with respect to their saturation (Figure 13.6). Basic colours ('green', 'blue', 'red' and 'yellow') are characterized by large chromatic co-ordinates and small achromatic ones. The most desaturated colour ('white') has a maximal value on the achromatic co-ordinate, and a value close to zero on the chromatic co-ordinate. Mixed

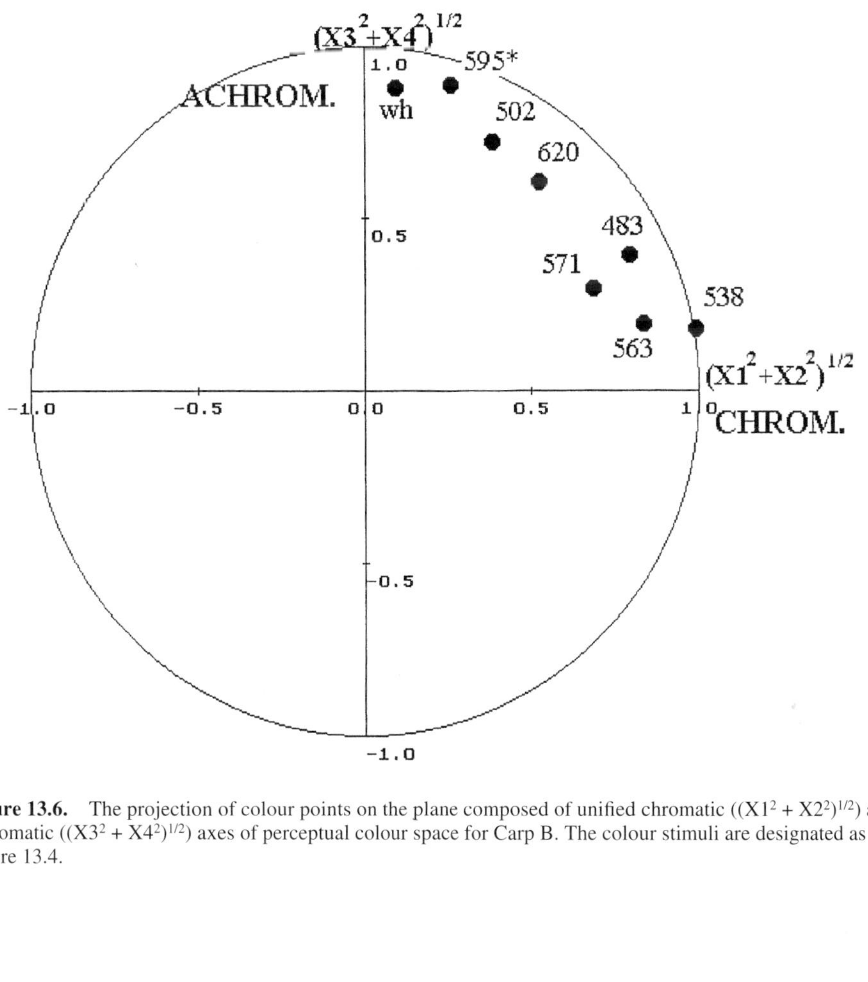

Figure 13.6. The projection of colour points on the plane composed of unified chromatic ($(X1^2 + X2^2)^{1/2}$) and achromatic ($(X3^2 + X4^2)^{1/2}$) axes of perceptual colour space for Carp B. The colour stimuli are designated as for Figure 13.4.

colours ('bluish-green' and 'purple') on this plane are located between extremes. Thus the angle that determines the position of the colour point on that plane changes from 'white' as the least saturated up to 'green' as the most saturated. The characteristics of saturation obtained in this experiment (at least in the range of equivalent wavelength of 530 to 610 nm) corresponds closely to spectral characteristics of saturation demonstrated by Yager (1967).

The eight colours used sequentially as conditional stimuli can therefore be specified by four co-ordinates. The mean value of lengths of vectors in the four-dimensional space is equal to 0.865+/–0.029. The variance coefficient is 0.094. The small variation of radii suggests that the colour points are condensed in a thin spherical layer of the four-dimensional Euclidean space.

4. DISCUSSION

The present results show that the probability of discriminative responses to colour stimuli can provide a measure of the similarity between differential and conditional stimuli. The confusion matrix derived from conditioned response probabilities can be regarded as a colour similarity matrix which contains information concerning neuronal channels that determine the similarity. The structure of perceptual space for the species under investigation was revealed by this matrix using factor analysis. Comparison of the colour spaces of monkeys and carp shows that they are organized according to principles analogous to those for the human colour space. As for humans the colour space for monkeys and fish is defined by four channels working in parallel.

The basis of colour coding channels constitutes colour-opponent neurones located in the retina of carp and monkey (Daw, 1972; Orlov and Maksimova, 1973). The responses of these cells correlate with co-ordinates of colour points on X1X2 plane of the colour space for these animals (Figures 13.1 and 13.4). Neurones having similar colour-opponent characteristics have also been found in the monkey's lateral geniculate body (de Valois *et al.*, 1967).

R. Jung (1973) isolated B- and D-type cortical neurones of cats. B-neurones are characterized by excitatory responses to the onset of light, and remain active during light presentation. Offset of light evokes an inhibitory response with respect to background activity in these cells. D-neurones, in turn, are excited by offset of light and inhibited by onset of light. The responses of these cells in the visual cortex due to superimposed receptive fields are characterized by expressed reciprocal relations with respect to foveal stimulation. Fomin *et al.* (1979) suggested a model linking the functioning of these cells to the discrimination of brightness of local stimuli in the context of simultaneous and successive contrast. The model explains the bipolar structure of the brightness-coding channel. Neumeyer *et al.* (1991) presented data suggesting separate channels for the handling of information related to brightness and hue.

In accordance with the spherical model of colour and brightness coding the four channels considered above converge on selective neurones which can be regarded as elements of a colour-specific projection. Vautin and Dow (1985) have described neurones in the striate cortex that are selectively tuned to narrow ranges of wave lengths around 450, 506, 577 and 656 nm. The maxima of their responses correspond to spectral characteristics of basic colours in humans and monkeys: 'blue' (450 nm), 'green' (506 nm), 'yellow' (577 nm) and 'red' (656 nm). The existence of such neurones in the nervous system of fishes and primates makes it possible to construct a specific network which codes light stimuli according to their hue and lightness.

5. CONCLUSIONS

(1) The matrices composed of probabilities of instrumental conditioned reflexes to colour stimuli contain information about the organization of the colour spaces of fish and monkey.

(2) The set of colours for monkey and fish is represented by points located on the surface of a hypersphere in four-dimensional Euclidean space with axes corresponding to 'red-green', 'yellow-blue', and 'brightness' and 'darkness' neuronal channels.

(3) The three angles of the colour hypersphere in the four-dimensional spaces of carp and of monkey correspond to those aspects of colours that coincide with hue, lightness and saturation in humans.

REFERENCES

Daw, N.W. (1972) Color-coded cells in goldfish, cat and rhesus monkey. *Investigative Ophthalmology*, **11**, 411–417.

Fomin, S.V., Sokolov, E.N., Vaytkyavichus, G.G. (1979) *Artificial sense organs. Modelling of sensory systems* (in Russian).Moscow: Nauka.

Judd, D.B. and Wyszecki, G. (1978) Colour in science and industry (pp. 159–175) (in Russian). Moscow: Mir.

Jung R. (1973) Visual perception and neurophysiology. In R. Jung (ed.), *Handbook of sensory physiology. V. VII/3: Central Visual Information. Part A*. New York: Springer-Verlag, pp. 3–152.

Neumeyer, C., Wiestsma, S., and Spekreijse, H. (1991) Separate processing of "color" and "brightness" in goldfish. *Vision Research*, **31**, 537–549.

Neumeyer, C. (1986) Wavelength discrimination in the goldfish. *Journal of Comparative Physiology*, **158**, 203-213.

Orlov, O.Yu. and Maksimova, E.M. (1973) Evolution of mechanisms of colour vision in vertebrates. In *Functional organization and evolution of visual systems in vertebrates*. Leningrad: Nauka, pp. 12–22.

Sokolov, E.N. and Izmailov, Ch.A. (1984) *Colour vision* (in Russian). Moscow: Moscow State University.

Ueberla, K. (1980) *Factor analysis* (in Russian). Moscow: Statistika.

Valois, de R.L., Abramov, I., and Mead, W.R. (1967). Single cell analysis of wavelength discrimination at the lateral geniculate nucleus in the macaque. *Journal of Neurophysiology*, **30**, 415-433.

Vautin, R.G., and Dow, B.M. (1985) Color cell groups in foveal striate cortex of the behaving macaque. *Journal of Neurophysiology*, **54**, 273–292.

Yager, D. (1967) Behavioral measures and theoretical analysis of spectral sensitivity and spectral saturation of the goldfish (Carasius auratus). *Vision Research*, **7**, 707–727.

REFERENCES

14 Neurobiology of Gestalts

E.N. Sokolov

Department of Psychophysiology, Lomonosov Moscow State University, Moscow, Russia
e-mail:sokolov@cogsci.msu.su

The question of Gestalts is discussed within the framework of their neuronal mechanisms. Two basic hypotheses are considered: 1) that of Gestalts as a result of the hierarchical organization of neurones (gnostic units), and 2) that of Gestalts as a result of the synchronization of neurones of a given level. Analysis of published data led to the conclusion that Gestalts result from vector coding in the hierarchical organization of neurones. High frequency oscillations in the gamma range (40–200 Hz) are of endogenous origin, and their function is to reinforce synaptic inputs to those neurones which are involved in the synthesis of a Gestalt.

KEYWORDS: gnostic unit, gestalt, novelty, long-term memory, vector coding, perception, high frequency oscillations, reinforcement, modelling

1. INTRODUCTION

As analytical studies of neuronal activity progress, the difficulty of understanding integrative processes becomes ever more obvious. In this regard, neurobiologists are paying more attention to the problem of integration of functions, working within the framework of Gestalt psychology.

The term Gestalt (from the German *Die Gestalt*) means "form." The term was introduced because of the need to emphasize the qualitative individuality of the perception of patterns of elements, which cannot be reduced to a simple total. Subsequently, psychological studies focusing on these qualitative individual characteristics of organized groups of elements were included in the term "Gestalt psychology" (Koehler, 1947). A simple example of a Gestalt is provided by a group of points located at the corners of an imaginary square. If these points are located at random positions within the same part of the visual field, the Gestalt disappears. A more obvious example of the Gestalt principle is provided by the constellations of stars: Each constellation is characterized by the qualitative individuality of its perceived shape, as reflected in its name. The basis of Gestalt psychology was laid by the German psychologist M. Wertheimer (1923), whose name is associated with the discovery of apparent movement (the "phi phenomenon"), in which an observer to whom two spatially separated experimental points are presented, sees movement of one of these points along the connecting line. Wertheimer concluded that the perception of movement is a qualitatively individual phenomenon, which cannot be reduced to a simple sequence of spatially distinct stimuli. Qualitative individuality is also an inherent characteristic of the above type of combination of elements (which are called Gestalts). A general theory of Gestalt psychology was developed by K. Koffka (1935).

When the Nazis came to power in Germany, most Gestalt psychologists emigrated to the U.S.A. During their studies in the U.S.A., Gestalt psychologists attempted to define the phenomenology for this area of study by investigating the dynamic properties of the brain's electric field. However, this approach did not lead to a solution to the question of the mechanism underlying Gestalts.

It should be recognized, however, that Gestalt psychologists had no significant influence on the main behaviorist tendency in the development of American psychology, in which the cognitive aspect was ignored for a long time. When cognitive psychology subsequently began to be studied in the U.S.A., the approach relied on the "computer metaphor," in which there was no significant place for Gestalts.

The development of Gestalt psychology took a new turn with the appearance of vector psychology, created by the Swedish investigator G. Johansson (1950, 1994). Professor Johansson was a student of the Gestalt psychologist D. Katz (1911). The vector approach to psychology developed by Johansson is based on detailed study of the perception of movement. Johansson's major experiment concerned an illusion of movement. If two points move towards each other, one along a horizontal and the other along a vertical, the observer perceives this as movement of one point along a sloping line. Johansson explained this effect using the vector concept, and subsequently applied the vector model to the perception of complex forms of movement such as the movement of the human body and limb, as well as the movement of objects in three-dimensional space (Johansson, 1994). How is the Gestalt principle related to the vector theory of perception? At the phenomenological level, the specificity of Gestalts corresponds to a defined vector, which is qualitatively distinct from other vectors in terms of the specific ratios of their components. In other words, the qualitative individuality of a Gestalt is determined by its underlying vectors. However, at the level of phenomenological analysis, it is difficult to answer the question of the real content of the vector interpretation.

2. VISUAL PERCEPTION OF MOTION

A pathway towards further development of vector concepts was provided by studies of neuronal mechanisms. Thus, perception of an object's movement starts with excitation in a pair of neurones which distinguish movement in the horizontal and the vertical directions. These excited neurones are components of the excitation vector, which encodes the particular direction of movement in a local region of the retina. At subsequent stages of information processing, the direction of movement is encoded by a detector which is selectively tuned in relation to a specified direction of movement (Sokolov and Vaitkyavichus, 1989).

A qualitative difference between perception of single dots and perception of visual motion induced by their specific organization in the space domain was discovered by M. Wertheimer (1923), the founder of the Gestalt psychology. Nowadays, after discovery of motion-selective units, this qualitative specificity of motion perception can be explaïned within the framework of separate neuronal channels, differing from the channels extracting stationary patterns.

2.1. The Law of Common Fate

The motion-detecting neurones do, however, also contribute to perception of shapes. This effect was formulated by M. Wertheimer as "The Law of Common Fate". If, in the visual field, there are several dots moving in the same direction and with the same velocity, they are perceived as a unity. The demonstration of this effect is very obvious in the "Dalmatian Dog Percept". The dog is characterized by black patches on a white body. Being motionless, the dog cannot be recognized on the dotted background. It is seen, however, as soon as it starts moving. The dots having a "common fate" are integrated into a moving shape percept (Spillman and Ehrenstein, 1995). The perception of the moving pattern depends on a specific population of cells in the visual cortex, which are stimulated by moving contours (Albright and Stoner, 1995).

Recent studies of Amthor and Oyster (1995) on the rabbit retina have revealed four types of ganglion cells, responding selectively to four directions of movement: upwards, downwards, to the right, and to the left. Introduction of a dye into the ganglion cells of each type showed that there is no overlap between the dendritic fields of cells lying along the same axis, but responding to stimuli which move in opposite directions. At the same time, the dendritic fields of cells lying on orthogonal axes, and responding to stimulus movement within the quadrant formed by these axes, do overlap. Since the dendritic fields of the cells coincide with their receptor fields, stimuli moving within each quadrant elicit excitation in two orthogonally located ganglion cells, which encode the direction of movement as an excitation vector, whose components consist of the excitation of these neurones.

With the development of a wide spectrum of neuroscientific studies, the problem of Gestalt has moved from the purely psychological to the neurobiological, and interest in this question is reflected in the large number of reports in neuroboiology journals. The organization of neuronal ensembles has been found to show vector coding, and the concept of Gestalts has been given a neurophysiological interpretation. The concept of Gestalts is applicable not only to visual patterns, but also to complex sound combinations—speech, music, and sounds of natural origin. The individuality of a sound Gestalt depends both on the elements of which the sound is made, and on its temporal characteristics: the duration over which each element sounds, and the time intervals between them. One feature of sound Gestalts is that their appearance involves short-term memory, which allows the Gestalt to function as an overall unit, despite its extension over time. This has led to the concept of "sound objects" (Fowler, 1994).

It should be emphasized that short-term (iconic) memory is also involved in organizing visual Gestalts. On examining an object, the eyes sequentially fixate the most informative points, and a series of such fixations is integrated into a Gestalt involving iconic memory. Finally, the concept of Gestalt is also applicable to proprioceptive signals arising from limb movements. This is clearest in the case of gestures, when a percepual Gestalt corresponds to each gesture. This approach has been diversified to include speech articulations, which consist of individual gestures with their corresponding kinesthetic Gestalts (Johansson, 1994).

3. GNOSTIC UNITS

Contemporary neuroscience is reaching a solution to the question of the mechanisms of sensory integration, and one can now address the issue of whether a purely psychological concept such as the Gestalt principle could be involved. Two approaches to solving the Gestalt problem are apparent. One posits the need for neuronal structures to be organized hierarchically. This approach was founded by Konorskii (1967), who proposed the concept of *gnostic units*, i.e., neurones which respond to complex stimuli because of the convergence on them of neurones at a more elementary level, the so-called "feature detectors".

Two types of gnostic units can be distinguished: those that are genetically prewired and those acquired in the process of learning. The latter type of neurones, selectively tuned to particular computer-generated visual patterns presented during learning, were demonstrated by Miyashita (1991) in monkey temporal cortex. To understand the process of formation of learning-dependent gnostic units, one has to take into account the formation of orientation-selective feature detectors in kittens during the process of development. The visual stimuli presented to the animal during the sensitive period recruit feature detecting neurones from a pool of potential candidates, by modifying their synapses in accordance with the input stimulus. This process is reinforced by the activating system of the reticular formation (Singer, 1990). It might be assumed that a similar process is present in temporal lobe neurones for formation of gnostic units. The formation of gnostic units occurs from a reserve pool of neurones, which are "standing in line", waiting for subsequent activation by a specific input stimulus. Recruitment of a potential gnostic unit is then performed if the input stimulus is a new one. The novelty signal, generated by hippocampal neurones, stimulates the reticular activating system. The coincidence of input stimulus and reticular activation modifies synaptic contacts of a potential gnostic unit. This transformation of synapses is done during "a short sensitive period", after which the synapses cannot be changed, and gnostic units become a member of the declarative memory system (Sokolov, 1997). The selectivity of gnostic units is due to selective modifications of their synaptic inputs from the relevant feature detectors.

A hierarchical structure for cognitive processes, with representations of external stimuli on a map formed by selective detectors, is based on this concept of gnostic units. The concept of this model is based on neuronal ensembles, which are sets of neurones having a common input and converging on a single neurone at a higher level (Sokolov and Vaitkyavichus, 1989). Stimulus-induced excitation in the elements of an ensemble form the components of an excitation vector. The excitation vector acts in parallel on a population of selective detectors which, according to the hypothesis of Pribram, should be called selectors. Each such selector is characterized by a specific set of synapses which form the connection vector. The excitation vector and the connection vector are proposed to be of constant length.

A selector, produces paired products, obtained by multiplication of the excitations arriving at a synapse and the efficiency (weight) of the synapse. Thus, an operation equivalent to calculating the scalar product of two vectors (i.e., the excitation vector and the connection vector) is carried out. Given that these vectors are of constant length, their scalar product reaches a maximum when the excitation vector and the connection vector are in the same direction. A set of selectors of different orientations, but with connection vectors of equal length, forms a spherical surface, which is a signal-representation map. When the input stimulus changes, the corresponding changes in the excitation vector shift the excitation maximum on the map of selectors, to reflect the change in the stimulus.

Thus, stimuli at the input, acted on by excitation vectors of equal length, are mapped as specific regions on the surface of a sphere. These spherical loci are individual selectors. A shift in the excitation maximum on the sphere corresponds to a shift in the excitation maximum on the map of selectors. A "jump" of excitation from one selector to another corresponds to the threshold for signal discrimination, which is measured on the sphere in the units of "angle". This threshold, which corresponds to the angular distance between neighbouring selectors, is equivalent to a constant relative-difference threshold.

This model was tested using colour vision as an example (Sokolov and Vaitkyavichus, 1989). These studies showed that a whole multiplicity of colours is represented by colour-selective neurones on a hypersphere in four-dimensional space, whose axes corresponded to neuronal excitations: red-green, blue-yellow, brightness, and darkness (see Latanov *et al.*, this volume). A hierarchical model of the Gestalt concept represents a development of such a vector approach.

Unlike the hierarchical model of the Gestalt, the second model for Gestalt formation can be termed the "temporary connection" model. One of its leading proponents is von der Malsburg (Malsburg and Schneider, 1986), who proposed the concept of "temporary connection" as the basis for the formation of a Gestalt, and hypothesized that the brain solves the problem of distinguishing patterns on the basis of a series of correlations of a signal in time. These correlations serve as connections between more elementary symbols (neurones) and more complex structures. These processes also lead to the discrimination of Gestalts. Thus, von der Malsburg's model has no place for "gnostic units". From this point of view, Gestalts are constellations of correlationally-connected elementary symbols (neurones). Von der Malsburg's work is largely in agreement with that of Eckhorn *et al.* (1988). From this point of view, spatial segmentation is achieved on the basis of a spatial "synchronization contrast" appearing between regions representing different objects. The main experimental evidence for this theory consists of "stimulus-specific synchronization" of the higher EEG frequencies (35–90 Hz). Stimulus-specific synchronization in the case of visual stimuli consisting of lines of different orientations is seen in EEG rhythms as a high-level orientational selectivity. One feature of this synchronization is that there is no phase shift. Synchronization in the activity of visual system neurones is postulated as a means for providing spatial connection of characteristics.

In these models, synchronization of neuronal activity is a characteristic of the combination of cells into ensembles. The models explain integrative processes without considering the involvement of any individual integrative neurones. Neurones which are temporarily connected together by synchronization form ensembles. The neurones can "move" from ensemble to ensemble, depending on the phase of oscillation. Oscillation in the gamma range has also been observed in pigeons when presented with visual stimuli (Neuenschwander and Valera, 1994).

These two approaches to the question of how a Gestalt is formed appear at first glance to exclude each other completely. Nonetheless, there are indisputable experimental data supporting both the idea of "gnostic units" and that of high-frequency oscillations showing high levels of correlation between different parts of the cortex. One possible solution to this contradiction is a new idea, regarding the nature of high-frequency rhythms. There is a basis for considering these rhythms to reflect endogenous (pacemaker) oscillations, which are initiated in neurones by an arriving signal. The high level of correlation of these oscillations is then a consequence of their being initiated simultaneously by the arriving signal. Thus, the correlation of high-frequency oscillations is not a means for connecting elements, but is a result of the arrival at these elements of a common signal. The stimulus specificity of these oscillations also becomes understandable in terms of this point of view.

Studies of the brain's responses to sound stimuli, presented at a frequency of 40 Hz, led to the conclusion (Tiitinen, 1994) that this rhythmic activity reflects an increase in the level of vigilance at the time of arrival of the sound stimulus. This rhythm, arising with a latent period of 40 msec, lasts until a point 100 msec after presentation of the stimulus. It has been proposed that there is no particular generator for the 40-Hz rhythm, but that synchronous activity when the stimulus starts is the result of generators which are already active, and become synchronized in response to a stimulus.

High-frequency pacemaker potentials increase the efficiency of the neurones involved in the response, with increases in discharge frequency and duration of the response. In terms of the model of Gestalts based on hierarchical organization of neuronal structure, high-frequency oscillations illuminate (like a projector) and emphasize the architecture of a Gestalt.

The perception of movement can be used as an example of the use of the concept of hierarchical organization of neuronal structures as the basis for Gestalt formation, and the involvement in it of high-frequency neuronal discharges. To be considered first is the hierarchical structure of the visual movement analyzer, which is responsible for producing apparent movement, this being an important part of Gestalt theory.

The visual movement analyzer of primates is known as the magnocellular tract, which begins in the morphologically distinct population of ganglion cells in the retina, whose axons project to the magnocellular layer of the lateral geniculate body. The movement analyzer then includes region V1 of the visual cortex, and the medial temporal visual part of the extrastriate cortex (designated as MT or V5). A characteristic feature of area V5 is that its neurones have directional selectivity in relation to the direction of movement, and do not respond to the shape or colour of the stimulus. Neurones of region V5 are subdivided into two groups: local detectors, which detect movements of points within a limited part of the visual field, and movement detectors for visual pattern (Celebrini and Newsome, 1994). These detectors of moving shapes integrate elements which receive signals from local detectors (Albright and Stoner, 1995). Detectors of moving patterns belong hierarchically to a higher level than local movement detectors, although they are localized in the same V5 region and account for about 25% of all movement detectors. Such integrating detectors, sensitive to moving patterns are also characterized by movement direction selectivity, although the condition required for excitation consists of the presence of a pattern of coherently moving points. These neurones do not respond to local movement of a point. Thus, perception of a moving shape requires the synchronous excitation of local movement detectors at several points of the visual field. Coherent movement of a series of points represents a necessary but not a sufficient condition. Perception of movement of a pattern needs the presence of movement detectors of a hierarchically higher level.

Recent studies have yielded data supporting the hierarchical organization of the visual movement analyzer. Neurones in MT (V5) region have themselves been found to converge on neurones in the medial superior temporal cortex (MST, or V5A). This region of the extrastriate cortex is ventral and anterior to MT. One feature of MST neurones is that they respond to stimuli moving coherently throughout the visual field. Thus, they resemble the integrating MT neurones. The excitation threshold of an individual MST neurone agrees with the behavioural threshold for detecting movement of a visual stimulus in monkeys. This leads to the conclusion that MST neurones transmit the information required for carrying out a behavioural response (Celebrini and Newsome, 1994).

Hierarchical organization of visual functions has been observed in studies of the perception of faces, and of the position of stimuli in space, using positron emission tomography (Haxby *et al.*, 1994). Perception of faces involves not only the striate cortex, but also Brodman fields 19 and 37 of the extrastriate cortex. This agrees with results obtained in microelectrode studies, which have shown these to be the locations of neurones which respond selectively to various aspects of face images.

Perception of different localizations of an object in the visual field involves the striate cortex, along with selective activation of field 19 of the extrastriate cortex and field 7 of the parietal cortex (Eccles, 1994). Colour and movement stimuli activate different regions of the extrastriate cortex, which do not coincide with the regions involved in perception of faces or stimulus positions. This led to the conclusion that the neural mechanisms of perception are organized hierarchically (Eccles, 1994). Different aspects of visual stimulation are represented in different parts of the extrastriate cortex. As information is sequentially transferred from lower levels to higher levels, information pathways diverge. In addition, within each information pathway, the neurones of different levels form hierarchical structures.

An example supporting the hierarchical structure of a Gestalt is provided by prosopagnosia, the selective loss of the ability to recognize faces, associated with local lesions of the brain affecting the associative cortex. Patients with prosopagnosia perceive clearly the individual features of faces (eyes, ears, nose, mouth), but cannot integrate these into a single image, i.e., a Gestalt. Neurones of the associative temporal cortex, involved in face recognition, are in turn subdivided into those neurones detecting individual features of faces, and those neurones selectively responding to a particular face as a whole. The latter neurones can even distinguish two different faces. These integrating neurones thus form a subclass of the broader class of "gnostic units".

Comparison of data obtained from patients with prosopagnosia with the neuronal mechanisms of face perception leads to the conclusion that the simple presence of detectors responding to particular facial features is insufficient: Combining these requires the participation of neurones more highly organized in the hierarchy i.e., gnostic units. Mere synchronization of the discharges of detectors for particular facial features is clearly insufficient for integral recognition of a face.

In addition, since these neurones can discriminate the faces of given individuals, it should be recognized that processes of learning are involved in the formation of the selective responsiveness of these neurons (Creutzfeldt, 1993; Spillman and Ehrenstein, 1995).

A new approach to the study of integral perception has been provided by synergetics, an interdisciplinary field of knowledge involving investigation of cooperative processes, occurring between the individual elements of systems which produce spatially, temporally, or functionally organized structures (Staller and Kruse, 1994). One of the bases of the synergetic approach to perception is the analogy between the *formation* of a Gestalt and its *recognition*. Using stimuli consisting of ambivalent (or multivalent) Rubik figures (e.g. the face/vase figure), factors determining the degree of stability and instability of an image were identified. These factors included contextual and semantic influences.

In terms of the hierarchical approach, the multivalency (multistability) of an image depends on whether the stimulus at the input activates two (or more) Gestalts. Variation in perception from one variant to the other is hypothetically associated with the switching of the "illumination" of the gamma-range high-frequency rhythm from one Gestalt hierarchy to another (see section 5: "The Gestalt pyramid"). If elementary characteristics are sequentially excluded from one of the competing Gestalts, the probability that it will dominate decreases, eventually reaching complete stabilization of perception on one of the variant images.

Support for the hierarchical organization of neuronal structures underlying Gestalts requires a consideration of the function of synchronization of neuronal discharges. If gnostic units are a requirement for the generation of a Gestalt, the function of high-frequency synchronized oscillations is to strengthen signals, by switching the intracellular high-frequency generator at the synaptic level.

4. WHAT IS THE NATURE OF THE ENDOGENOUS OSCILLATIONS? GAMMA PACEMAKER POTENTIALS

Studies on isolated neurones from edible snails have demonstrated the existence of endogenous sub-threshold oscillations in membrane potential at frequencies of 1 to 20 Hz (pacemaker potentials). The mechanism producing these was associated with activation of low-threshold calcium channels resulting in depolarization, followed by hyperpolarization due to activation of calcium-dependent potassium channels. The oscillation frequency increased with membrane depolarization. When the amplitude of pacemaker potentials reached the threshold for spike generation, the neurone generated a series of discharges (Sokolov, 1991).

Pacemaker potentials represent an intrinsic intracellular generator which is controllable: Switching-on of this generator turns a short-lasting synaptic influence into a long-lasting sequence of action potentials. An analogous mechanism occurs in cortical neurones.

The pacemaker potentials are of different frequency. The high-frequency pacemaker oscillations (40–200 Hz) are termed "gamma oscillations". They are induced by interaction of calcium currents (passing via low-threshold calcium channels), and calcium-dependent potassium currents induced by opening of potassium channels. The locus of gamma oscillators is found in dendrites (Llinas and Grace, 1995). The frequency of gamma oscillations is increased by depolarization of a neurone. The function of the gamma pacemaker can be identified by analyzing the hair cells of the inner ear. The selectivity of hair cells with respect to sound frequency is due to a specific gamma oscillation frequency. It can be assumed that gamma oscillations perform a similar function in a neurone which receives synaptic contacts on its dendrites which are selectively tuned to particular input frequency. The adjustment of the phase of gamma oscillations to the phase of the input is achieved by a re-setting effect, when the pacemaker wave is initiated by the input, so that the phase-shift between pacemaker wave and the input excitatory postsynaptic potential becomes equal to zero.

Endogenous high-frequency oscillations (40 Hz) have been described for cortical interneurones (Llinas *et al.*, 1991; Singer, 1990). Their generation requires sequential activation of slow sodium channels, followed by activation of potassium channels. Analogous intracellular gamma oscillations (50–200 Hz) have been observed in neurones of the dentate fascia of the hippocampus. These endogenous high-frequency oscillations are modulated by synaptic inputs from the entorhinal cortex and septum. The entorhinal cortex produces diffuse activation. The synaptic influences from the septum occur at theta frequencies, with the result that the frequency of gamma oscillation spikes correlates with the theta rhythm frequency (Bragin *et al.*, 1995).

Another mechanism for producing endogenous oscillations consists of the "intracellular reverberation of neuronal spikes" also occurring in neurones of the edible snail, and described by Palikhova (1993). The essence of this phenomenon is that the somatic action potential activates the spike trigger zone of the neurone. The action potential generated by this mechanism travels in a retrograde direction, and arrives at the cell body in the form of an A-spike. If the A-spike reaches the triggering threshold, the next spike arises in the cell body, and the process is repeated, leading to a high-frequency discharge, limited only by the refractory period. This rhythm continues when chemical synapses are blocked by cobalt ions, and can even be strengthened by exclusion of calcium-dependent potassium channels.

This mechanism for generation of high-frequency rhythms also occurs in interactions between the dendritic and somatic parts of cortical neuronal membranes in vertebrates. The mechanism by which intracellular oscillations of frequency 40–80 Hz are generated involves potential-dependent tetrodotoxin-sensitive sodium ion channels, which can be detected by immunochemical methods on the cell bodies and dendrites of pyramidal neurones. Fast sodium spikes have been observed in both the cell body and the dendrites of pyramidal neurones. The dendritic spike, which is actively propagated towards the cell body, induces a positive potential, which, on reaching threshold, results in the generation of a somatic spike. The somatic spike in turn is propagated antidromally to the dendrite, and induces a dendritic spike, which can again activate a somatic spike. This process results in the endogenous generation of somatic spikes, with a frequency limited only by the refractory properties of ion channels (Turner *et al.*, 1994). The interaction described between the dendrite and cell body converts the pyramidal neurone into a generator of high frequency spike rhythms at its output. Thus, the potential-dependent sodium channels of dendrites are intrinsic amplifiers of synaptic influences. In addition, the frequency range for spike generation is widened, which increases the efficiency and effect on "executive" neurones.

A possible explanation for the synchronization of cell discharges in distant regions of the cortex, without a phase shift, may be found in the effect of interneurones, which themselves show high-frequency endogenous activity. In this case, the absence of a phase shift between neurones results from the effect of a rhythmic interneurone common to the cells involved. These rhythmic synaptic potentials can be sub-threshold for the entire population of dependent cells: Threshold values in these cells may be achieved when an additional specific sensory stimulus arrives. In this case, the rhythmic interneurone has the function of a "sub-threshold activator."

The limit of high-frequency endogenous oscillations (200 Hz) provides an explanation for the observation of a "time quantum", which determines the fine temporal structure of perception. Studies of apparent motion have established that the elementary time interval (time quantum) is 4.5 msec. Other intervals are multiples of this time period (Geissler, 1987). Analysis of rhythmic activity in the gamma range show that the "time quantum" has the same duration as the relative refractory period. Thus, high-frequency oscillations, functioning as amplifiers, also determine the temporal quantization of perception.

The studies of Miyashita *et al.* (1991) made an important contribution to reinforcing the concept of the "gnostic unit". This group showed that the anteroventral temporal cortex of the monkey contains nerve cells which, during associative learning, become selectively responsive to defined optical stimuli with complex organization (which were generated by a computer). During the experiment the monkey received a total of 97 stimuli, differing in terms of colour, shape, size, and orientation. Training consisted of the initial presentation to the monkey of an image, followed 16 sec later by a stimulus for comparison. The animal received reinforcement when identity between the sample and the test stimulus was identified. It is important to emphasize that this procedure did not involve responses to a specific stimulus: The signal consisted of identity between the sample retained in memory and the test stimulus. This comparison between sample and standard was made using neurones of the anteroventral temporal cortex which, during training, became "selectors" for defined complex stimuli. In order to demonstrate that this selectivity was in fact the result of learning, a further 97 new computer-generated stimuli were used. No neurone was observed with selective tuning for this set of stimuli. Thus, repeated use of the 97 stimuli had the effect that a gnostic unit was formed for each one, with selective responses to each specified stimulus.

It can be suggested that defined "gnostic units" correspond to each given visual Gestalt (Figure 14.1). The anteroventral temporal cortex contains a pool of reserve neurones which have the potential to make weak responses to different stimuli. Stimulus repetition produces some level of increased potentiation of synaptic inputs, after which the neurone becomes a selectively-responding "selector" and stops changing. This tuning process is reminiscent of the formation of detectors during the sensitive period of development. One of the features of gnostic unit formation is that the sensitive period is terminated by the learning process itself.

The individual visual characteristics of a stimulus are analyzed by neurones in the prestriate cortex. The elementary properties are synthesized by neurones involved in long-term memory in the anteroventral temporal cortex, which establish representations of stimuli on a "one gestalt—one neurone" basis. This process involves a high degree of abstraction: Neurones retain the specificity of their responses despite changes of stimulus size, orientation, and colour (Miyashita *et al.*, 1991). When each neurone has a high level of specificity for a given gestalt, stimuli presented in a strict order became associated. The neurone responds not only to its specific stimulus, but also produces some response to the associated stimuli, regardless of their optical content. This suggests that, over a period of time, the neurone partially "fixes" information arriving from subsequent stimuli.

The following *hypothesis* can be suggested: Presentation of a new stimulus activates the next available neurone, for the amount of time needed for fixation of the stimulus. The sensitivity of the neurone then decreases, and subsequent stimuli have little effect on it. This time is the "sensitive period" for this neurone. When the next signal arrives, the next reserve neurone becomes sensitive, for a short period of time.

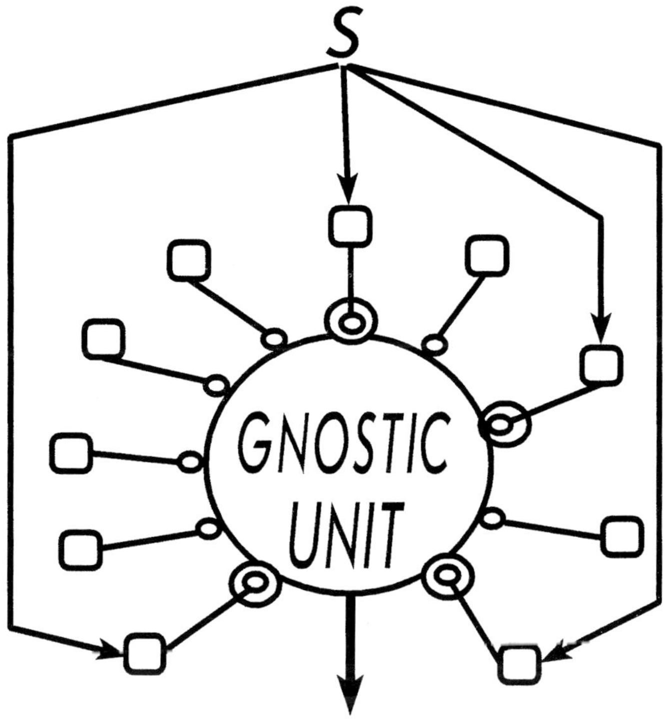

Figure 14.1. Diagram of the organization of the selective inputs in a gnostic unit. S indicates the input stimulus to the neuronal network (squares); double circles indicate altered synapses.

The model for the sequential formation of selectively-responding gnostic units is as follows: There is a pool of reserve neurones, which are activated sequentially by novelty signals from the hippocampus. On activation, the mechanism of plastic rearrangement of synapses is briefly switched on, i.e. for the "sensitive period" of the neurone. The signal acting during this period, processed by the detectors, arrives via channels which increase synaptic efficiency. As a result, the neurone becomes selectively tuned to the stimulus concerned. At the end of the "sensitive period" synaptic efficiency stops changing. If the next stimulus arrives during the "sensitive period," it has a partial effect, and neighbouring neurones will produce partial responses to stimuli which are close together in time. The novelty signal appears when the arriving stimulus "fails to find" a memory neurone corresponding to it. Thus, the novelty signal is a signal produced by agreement between the stimulus and its corresponding gnostic unit (Figure 14.2).

The main principles of organization of a Gestalt are spatial and temporal proximity of the elements forming the Gestalt, as well as their similarity. However, these criteria do not limit the grouping-together of elements to make a whole Gestalt. There is also a more complex principle—the appearance of an illusory contour. One example of an illusory contour is provided by the Kanizsa square (Hirsh *et al.*, 1995), in which the boundaries between objects in the visual field are determined not by different levels of illumination or different colours, but by the effects of boundary-inducing image elements. The

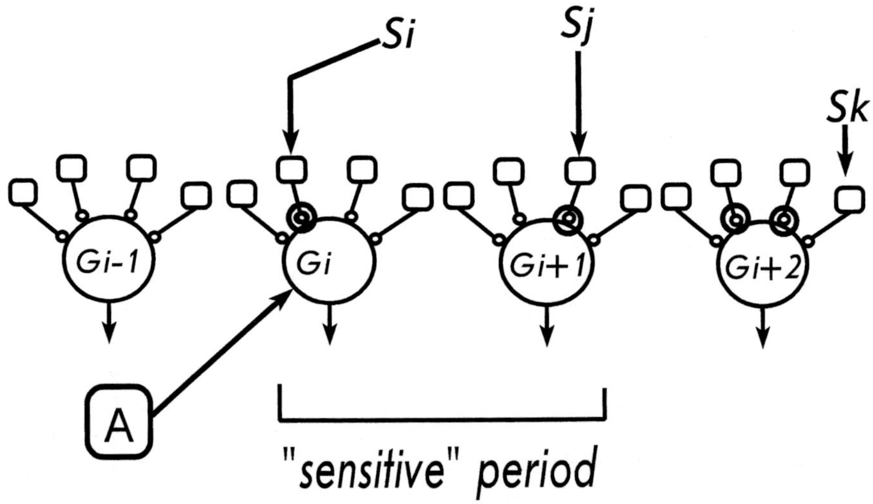

Figure 14.2. Diagram showing the role of the "sensitive period" in forming long-term plastic rearrangements. A is activation, G indicates gnostic units. For other details see the caption to Figure 1.

illusoriness of this type of contour is such that this effect determines the perception of one object covering another (Hirsh *et al.,* 1995). The illusory contour should thus be considered as a form of Gestalt organization, resulting in segmentation and active perception of an object. The illusory contour was found to correspond to a special neuronal mechanism: Cells responding to the illusory contour were observed in the second visual zone (V2) of monkeys. Functional nuclear magnetic resonance studies in humans showed that specific regions of the extrastriate cortex of the right hemisphere (Brodman field 18) responded to the illusory contour, these regions being distinct from areas which responded to a real contour. Thus, when the subject saw the illusory contour, corresponding activity of a specific neuronal population occurred, representing a grouping of local detectors used to make a single percept.

It can be concluded that the formation of a Gestalt involves roles for specific neurones, which are distinct from the neurones which separate particular characteristics such as the orientation of individual lines in the visual field. This evidence also supports the view that Gestalts involve hierarchical organization, which cannot be reduced to a combination of detectors of elementary components by the establishment of single-level connections between them.

The Gestalt concept can also be applied to the problem of the mechanism of subjective states. The concept of gnostic units connects the appearance of a subjective state with a pyramid of sequential neuronal transformations, with activation of a gnostic unit forming the apex. Another approach was developed by Eccles (1994): The essence of this approach is that the dendrites of pyramidal neurones at the level of cortical layer IV are grouped together in a dendritic bundle which ascends to layer I. This bundle of dendrites (from a group of 70–100 neighbouring large and small pyramidal cells) has been termed a "dendron". Each dendron, according Eccles' theory, is unambiguously connected with an elementary mental experience, which is called a "psychon". The multiplicity of psychons covers all the various subjective states, and each psychon is associated with a specific dendron.

The complexity of dendron structure should be emphasized. Each dendron consists of a multiplicity of synapses bearing end plates for terminal axons. The complexity increases at the subsynaptic level because of the discrete quantal release of transmitter molecules into the synaptic cleft, and the impulse nature of ionic currents flowing through the ion channels opened by receptors for the transmitter. Further complexity is introduced into this picture by the fact that cell processes have their own characteristic mechanism of translation, with synthesis of critically important proteins by directed transfer of mRNA from the nucleus to a specific post-synaptic region of the dendrite (Steward and Banker, 1992).

Considering the question of consciousness, it is worth noting that one of the conditions for the appearance of conscious perception is additional activation from the reticular activating system. In the model of gnostic units this activation consists of high-frequency oscillations of the membrane potential. In the dendron model, this activation is known as the reticular effect on the apical dendrites of pyramidal neurones. Thus, both the gnostic unit and the dendron are necessary but not sufficient conditions for the generation of consciousness. Only the combination of specific and non-specific activation of cortical neurones is able to create conscious experience.

5. THE GESTALT PYRAMID

The hierarchical structure of the neuronal network underlying Gestalts can be represented by a pyramid of converging neurones (Figure 14.3). The principles of network organization are common for feature-detectors and gnostic units. Feature-detectors are supplied by excitations of "pre-detectors". The base of the pyramid is represented by feature detectors that stimulate a limited number of "pre-gnostic" neurones. The pre-gnostic neurones contribute an input to a set of gnostic units. The stimulus-relevant gnostic unit is selectively activated in correspondence with its synaptic weights. The selectivity of activation is enhanced by coincidence of the frequencies of inputs with those of postsynaptic oscillations. If the activated gnostic units are linked to a semantic unit, an additional process occurs, the categorization of the input stimulus.

The elementary components of a stimulus are reflected in parallel on maps (screens) of simple detectors, which form the base of the Gestalt pyramid (Figure 14.3). The individual elementary stimulus components are encoded in terms of the position of maximal excitation on the appropriate screen. A stimulus with a series of elementary components is encoded by a whole set of these excitation maxima on a series of such maps.

Maps of complex components are formed from combinations of elementary components. Complex components are encoded by specific sites of maximal excitation on maps of complex detectors. Simple and complex components form synapses on gnostic cells, which are excited by specific sets of elementary and complex stimulus components.

Thus, we propose a neuronal model of the Gestalt as a multi-level structure—a pyramid, in which the apex consists of a gnostic unit, upon which converge the detectors of elementary and complex components. Neurones at different levels of the Gestalt pyramid are "illuminated" by activating influences, when an adequately complex

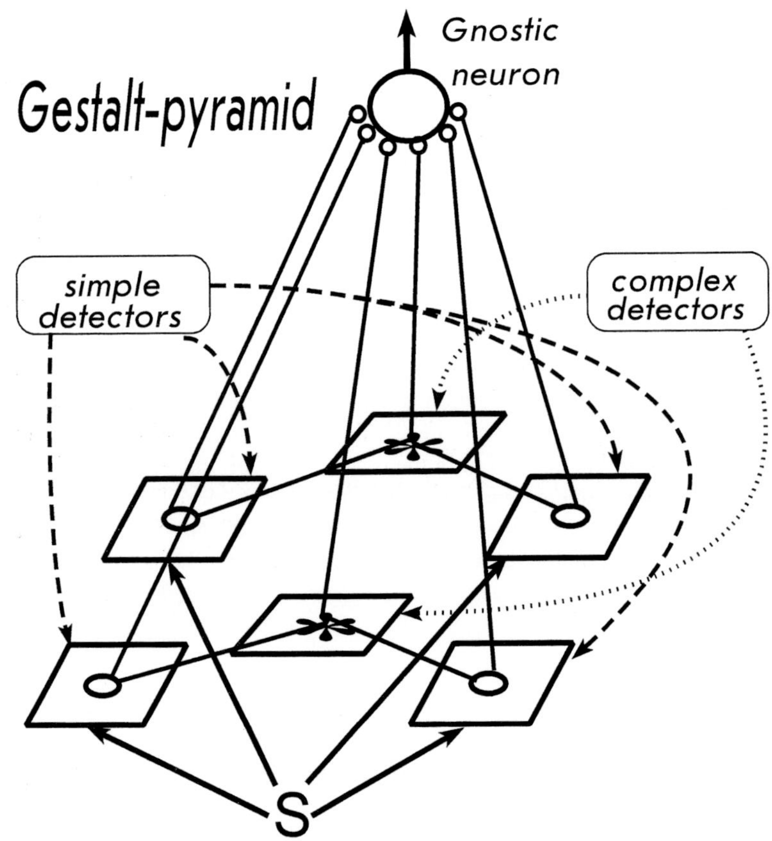

Figure 14.3. Neuronal structure of a Gestalt: the Gestalt pyramid. The lower level consists of simple detectors (circles), and higher levels consist of complex detectors (asterisks). For other details see the caption to Figure 1.

stimulus is presented, these activating influences consisting of high-frequency oscillations of cell membrane potentials. This activation of the Gestalt pyramid of hierarchically organized neurones forms the basic mechanism of consciousness.

5.1. Bottom-up and Top-down Operations

The process of activation of gnostic units, starting from the feature-detector level and ending with symbolic representation of the stimulus at the semantic level, constitute a "bottom-up" operation of stimulus categorization. The search of a particular Gestalt in the visual field requires "top-down" operation. The verbal instructions that characterize the target result in a descending activation from the gnostic unit at the top of the Gestalt pyramid down to the feature detectors. The base-layer of the Gestalt pyramid is continually scanning the visual field until the pre-excited feature detectors are actually activated by the target stimulus, and when bottom-up excitation of the Gestalt pyramid occurs, it would result in target detection.

5.2 Gestalt Representation within Perceptual Space

The excitations of pre-gnostic neurones by an input pattern constitutes an input excitation vector. It activates a particular set of gnostic units having synaptic weights directly proportional to components of the excitation vector. Some other gnostic units would be characterized by different excitation vectors. Thus, sets of gnostic units can be represented by points within perceptual vector space. The multidimensional scaling of subjective differences between Gestalts reveals perceptual space of low dimensionality, determined by pre-gnostic neurones, or modules according to Edelman (1997).

6. CONCLUSIONS

The hierarchical organization of neurones in a Gestalt pyramid is suggested as the neuronal basis for Gestalts. Vector coding realized within such a Gestalt pyramid suggests that Gestalts are represented by points within a perceptual space of low dimensionality. Endogenous pacemaker gamma oscillation enhance the selectivity of neurones within the Gestalt pyramid.

REFERENCES

Albright, T.D. and Stoner, G.R. (1995). Visual motion perception. *Proceedings of the National Academy of Sciences, U.S.A.*, **92**, 2433–2440.

Amthor, F.R. and Oyster, C.W. (1995). Spatial organization of retinal information about the direct of image motion, *Proceedings of the National Academy of Sciences, U.S.A.*, **92**, 4002–4005.

Bragin, A., Jando, G., Nadasdy, Z., Hetke, J., Wise, K. and Buzsaki, G. (1995). Gamma (40–100 Hz) oscillations in the hippocampus of the behaving rat. *Journal of Neuroscience*, **15**, 47–60.

Celebrini, S. and Newsome, W.T. (1994). Neuronal and psychophysical sensitivity to motion signals in extrastriate area MST of the macaque monkey. *Journal of Neuroscience*, **14**, 4109–4124.

Creutzfeldt, O.D. (1993). *Cortex Cerebri: Performance, Structural and Functional Organization of the Cortex*, Springer Verlag, Berlin, Heidelberg.

Eccles, J.C. (1994). Evolution of complexity of the brain with emergence of consciousness. In: K.H. Pribram (ed.) *Rethinking Neuronal Networks: Quantum Fields and Biological Data*, Lawrence Erlbaum Associates, Hillsdale pp. 1–28.

Eckhorn, R., Bauer, R. and Jordan, W. (1988). Coherent oscillation: a mechanism of feature linking in the visual cortex? *Biological Cybernetics*, **60**, 121–135.

Edelman, S. (1997). Representation is representation of similarities. *Behavioral and Brain Sciences* (in press).

Fowler, C. (1994). Auditory 'objects': The role of motor activity in auditory perception and speech perception. In: K. Pribram (ed.) *Origins: Brain and Self-Organization*, Lawrence Erlbaum Associates, Hillsdale, New Jersey pp. 594–603.

Geissler, H.-G. (1987). The temporal architecture of central information processing: evidence for a tentative time quantum model *Psychological Research*, **49**, 99–106.

Haxby, J.V. Horvi, Z.B. and Ungerleider, L.G., Maisaig, J.M., Pietrini, P. and Grady, C.L. (1994). The functional organization of human extrastriate cortex: a PET rCBF study of selective attention to faces and locations, "*Journal of Neuroscience*, **14**, 6336–6353.

Hirsh, J., Delapaz, R.L., Relkin, N.R., Victor, J., Kim, K., Li, T., Borden, P., Rubin, N. and Shapley, R. (1995). Illusory contours activate specific regions in human visual cortex: evidence from functional magnetic resonance imaging," *Proceedings of the National Academy of Sciences, U.S.A.* 92, 6469–6473.

Johansson, G. (1950) *Configurations in Event Perception*, Almqvist and Wiksell, Uppsala.

Johansson, G. (1994) Configurations in event perception In: G. Jansson, S.S. Bergstroem and W. Epstein (eds.) *Perceiving Events and Objects*, Lawrence Erlbaum Associates, New Jersey pp. 29–122.

Katz, D. (1911). Die Erscheinungsweisen der Farben und ihre Beenflussung durch die individuelle Erfahrung, *Zeitschrift fur Psychologie*, Erganzungsband 7, 1–47.

Koffka, K. (1935). *Principles of Gestalt Psychology*, Harcourt Brace, New York.

Konorskii, J. (1967). *Integrative activity at the brain: an interdisciplinary approach.* Chicago: Chicago University Press.

Koehler, W. (1947) *Gestalt Psychology: an Introduction to New Concepts in Modern Psychology*, Leveright, New York.

Llinas, R.R., Grace, H.A., and Yarom, Y. (1991). *In vivo* neurons in mammalian cortical layer 4 exhibit intrinsic oscillation activity in the 10 to 50 Hz frequency range. *Proceedings of the National Academy of Sciences, U.S.A.* **88**, 897–907.

Malsburg C. von der and Schneider, W. (1986). A neural cocktail-party processor *Biological Cybernetics*, **54**, 29–40.

Miyashita, Y, Sakai, K., Miguchi, S.-I. and Masui, N. (1991). Localization of primal long-term memory in the primate temporal cortex. In: L.R. Squire, N.M. Weinberger, G. Lynch, J.L. McGaugh (eds). *Memory: Organization and locus of change*, Oxford University Press, New York, Oxford, pp. 239–249.

Neuenschwander, S. and Valera, F.J. (1994) Visually-triggered neuronal oscillations in the pigeon: an auto-correlation study of tectal activity. In: K. Pribram (ed.) *Origins: Brain and Self-Organization*, Lawrence Erlbaum Associates, Hillsdale, pp. 496–535.

Singer, W. (1990). Search for coherence: a basic principle of cortical self-organization. *Concepts in Neuroscience*, **1**, 1–14.

Sokolov, E.N. and Vaitkyavichus, G.G. (1989). Neurointellect: From Neuron to Neurocomputer [in Russian], Nauka, Moscow.

Sokolov, E.N. (1991) Local plasticity in neuronal learning. In: L.R. Squire, *et al.* (eds.) *Memory: Organization and Locus of Change*, Oxford University Press, New York, p. 364–384.

Sokolov, E.N. (1996). Model of cognitive processes. In: M. Sabourin, F.I.M. Craik, and M. Robert (eds.), *Advances in Psychological Science. Vol. 2, Biological and Cognitive Aspects.* East Sussex, U.K., Psychology Press, pp. 355-379.

Staller, M. and Kruse, P. (1994). Neurodynamics and synergetics, In: K.H. Pribram (ed.) *Rethinking Neuronal Networks: Quantum Fields and Biological Data*, Lawrence Erlbaum Associates, Hillsdale, New Jersey p. 31–??.

Steward, O. and Banker, G.A. (1992). Getting the message from the gene to the synapse: sorting and intra-cellular transport of RNA in neurons *Trends in Neuroscience*, **15**, 180–186.

Tiitinen, H., Sinkkonen, J., May, P. and Naatanen, R. (1994). The auditory transient 40-Hz response is insensitive to changes in stimulus features *NeuroReport*, **6**, 190–192.

Turner, R.W., Maler, L., Deerinck, S.R., Levinson, S.R. and Ellisman, M.H. (1994). TTX-sensitive sodium channels underlie oscillatory discharge in a vertebrate sensory neuron. *Journal of Neuroscience*, **14**, 6453–6471.

Wang, X.J. (1993). Ionic bases for intrinsic 40-Hz neuronal oscillations *NeuroReport*, **5**, 221–224.

Wertheimer, M. (1923). Untersuchungen zur Lehre von der Gestalt (Studies on the theory of gestalt), *Psychologische Fortsetzung*, **4**, 301.

15 The Striatal Cholinergic System and Instrumental Behaviour

K.B. Shapovalova

I.P. Pavlov Institute of Physiology, Russian Academy Science, St. Petersburg, Russia
e-mail: kbsh@infran.ru

New data from the author, and existing data from the literature, are reviewed and interpreted in relation to the participation of the striatal cholinergic system in orienting behaviour and the regulation of attention to meaningful stimuli, which are realized via a number of subcortical structures. Among these structures, the pedunculo-pontine nucleus (PPN) and thalamic intralaminar nuclei (CM-Pf) are of particular significance. An important role for the striatal cholinergic system in control of the indirect efferent pathway is substantiated. Using this mechanism, inhibition of the non-specific sensory afferent activity can occur at different subcortical levels and thereby leads to an enhancement of the signal important for a given situation. This was evident, for instance, as a marked improvement of differentiation of sensory signals after activation of the striatal cholinergic system. On the other hand, restriction of excess motor activity by the caudate nuclei may have an important role in decision-making in a difficult situation, for instance, in a sharp enhancement of the environmental afferents; that was modelled by a stimulation of the CM-Pf complex of the thalamus. The degree of involvement of caudal mechanisms responsible for inhibition of the locomotor activity, inhibition of interstimulus leg raisings, enhancement of the tonic movement component and the postural streamlining and stabilization seems to be determined by the level of activation of the striatal cholinergic system, and to be realized via the subthalamic and pedunculo-pontine nuclei.

KEYWORDS: orientation behaviour, attention, efferent outputs of the striatum, subcortical nuclei

1. INTRODUCTION

In recent years a great deal of attention has been paid to the investigation of the forebrain cholinoreactive systems. These systems are involved in many important processes, such as learning, memory, attention, and the sleep-wake cycle (McCormick, 1990).The sources of cholinergic projections in forebrain structures are the cholinergic depots. The main cholinergic depot for forebrain structures is *Nucleus basalis magnocellularis* (Nucleus of Meynert) (Mesulam *et al.*, 1983).

The striatum—a central integrating system of the basal ganglia—is a large paired formation of the forebrain. According to current neuromorphological, neurochemical, and neurophysiological data, the striatum has some peculiarities that distinguish it from other forebrain structures. The striatum receives no direct sensory inputs: The sensory inputs are mediated by the cerebral cortex and thalamic intralaminar nuclei (CM-Pf) (Figure 15.1). In addition the striatum has no direct outputs to the spinal motoneurones; it affects

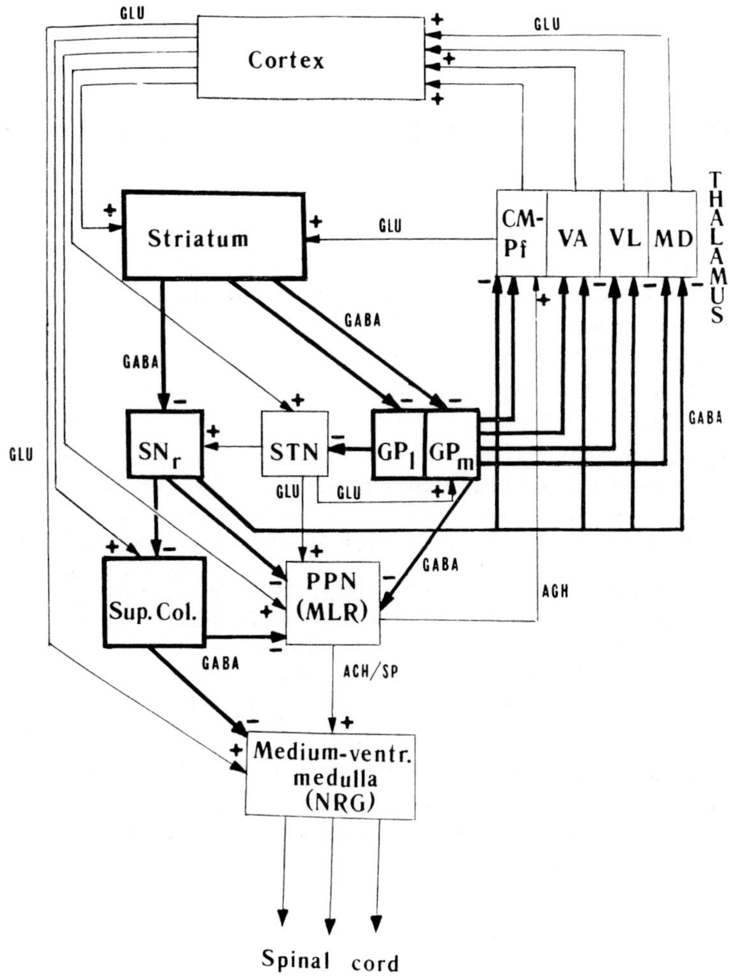

Figure 15.1. Scheme of participation of the striatum in some cortico-subcortico-cortical and subcortico-subcortical loops of sensory and motor information processing. Thick lines—inhibitory connections, thin—excitatory connections. Glu—glutamate, Ach—acetylcholine, GABA—gamma-aminobutyric acid. Thalamic nuclei: CM-Pf—centrum medianum-parafascicular nucelus, VA—nucleus ventralis anterior, VL—nucleus ventralis lateralis, MD—nucleus medialis dorsalis; SNr—substantia nigra, pars reticulata; STN—subthalamic nucleus; GPl—globus pallidus, lateral part; GPm—globus pallidus, medial part; PPN—pedunculopontine nucleus; MLR—mesencephalic locomotor region; Sup.Col.—superior colliculus.

them via a number of subcortical, mesencephalic, and brainstem structures which transmit striatal efferent signals (Figure 15.1). Most inputs to the striatum contain an excitatory transmitter, glutamate, while the outputs mainly use an inhibitory mediator, γ-aminobutyric acid (GABA).

An important mechanism by which the striatum changes activity of its targets—predominantly the globus pallidus medial zone (GPm) and the substantia nigra *pars reticulata* (SNr)—is disinhibition, i.e. an inhibition of the structures which, in turn, inhibit a number of subcortical and brainstem systems (Figure 15.1) (Chevalier and Deniau, 1990).

The striatum, and, first of all, its dorsal part, the neostriatum (nucleus caudatus and putamen) is among the forebrain structures with the highest levels of acetylcholine and the highest activities of acetylcholinesterase and cholineacetyltransferase (Mesulam *et al.*, 1992; Contant *et al.*, 1996). The striatum gets cholinergic innervation from outside, from the laterodorsalis (LDT) and pedunculo-pontine (PPN) tegmental nuclei (Figure 15.1) located in the dorsal part of the reticular formation, at the border of the midbrain and pons Varolii (Leonard and Llinas, 1990). The cholinergic pathways are transmitted via the CM-Pf which send direct, topologically organized projections to the striatum (Berendse and Groenewegen, 1990; Ragsdale and Graybiel, 1991; Sadikot *et al.*, 1992) . However, apart from this innervation, the striatum has an intrinsic source of acetylcholine, peculiar to itself. A small number of striatal neurones have large cell bodies (up to 40 mm) and dendrites without spines, and are known as large aspiny cells (Pasik *et al.*, 1979). These neurones are the striatal cholinergic interneurones (Groves, 1983; Bolam *et al.*, 1984; Gerfen, 1992). This population of interneurones in the striatum is quite distinct neuromorphologically (Pasik *et al.*, 1979; Bolam *et al.*, 1984), neurochemically (Gerfen, 1992), and neurophysiologically (Wilson *et al.*, 1990). Although the cell bodies of the striatal cholinergic interneurones are distributed in a fairly homogeneous manner in the rat and cat striatum (Kimura *et al.*, 1981), the extensive axonal arborizations of these cells are relatively concentrated within the matrix compartment (Hirsch *et al.*, 1989; Kawaguchi, 1992), resulting in a higher density of staining for cholinergic markers in the matrix than in the striosomes.

In spite of the relative rarity of cholinergic neurones, which account (for instance, in rats) only for 5% of the striatal neuronal population, they affect its function very much. First, dendrites of these neurones run for a relatively long distance (up to 1000 μm); the axonal collaterals are spread similarly extensively inside the nucleus (Wilson *et al.*, 1990; Contant *et al.*, 1996). For the last few years, most evidence has pointed towards volume transmission of striatal acetylcholine (Contant *et al.*, 1996).

Second, the striatal cholinergic system is crucial for initiating and modulating one of the striatal efferent outputs, the so-called indirect pathway (see below). It is the intrinsic striatal cholinergic system which may be supposed to play a leading role in performance of the striatal cholinergic function, whereas cholinergic influences from the cholinergic stores (PPN and LDT), which are mediated by the thalamic CM-Pf, only act as modulators of the function of the striatal cholinergic neurones and other striatal transmitter systems (Glowinski *et al.*, 1984; Kilpatrik *et al.*, 1986; Shapovalova, 1993).

Efferent cells of one type (medium spiny cells, S1) whose transmitter is GABA account for 96% of all cells of the striatum (Gerfen *et al.*, 1991). New histochemical and immunohistochemical methods have revealed two main efferent outputs of the striatum to its subcortical targets (GPm and SNr): direct (inhibitory) and indirect (excitatory), the latter transmitted via the globus pallidus lateral part (GPl) and subthalamic nucleus (Gerfen *et al.*, 1991; Gerfen, 1992; Wang and McGinty, 1996). These pathways differ topographically, in the opposite sign of their effect on the targets, and in their different coexisting peptides (Figures 15.2A and 15.3), although their transmitter in both cases is GABA (Kita and Kitai, 1988). Nigrostriatal dopamine and intrastriatal acetylcholine have opposite influences on these efferent striatal pathways, including expression of the pep-

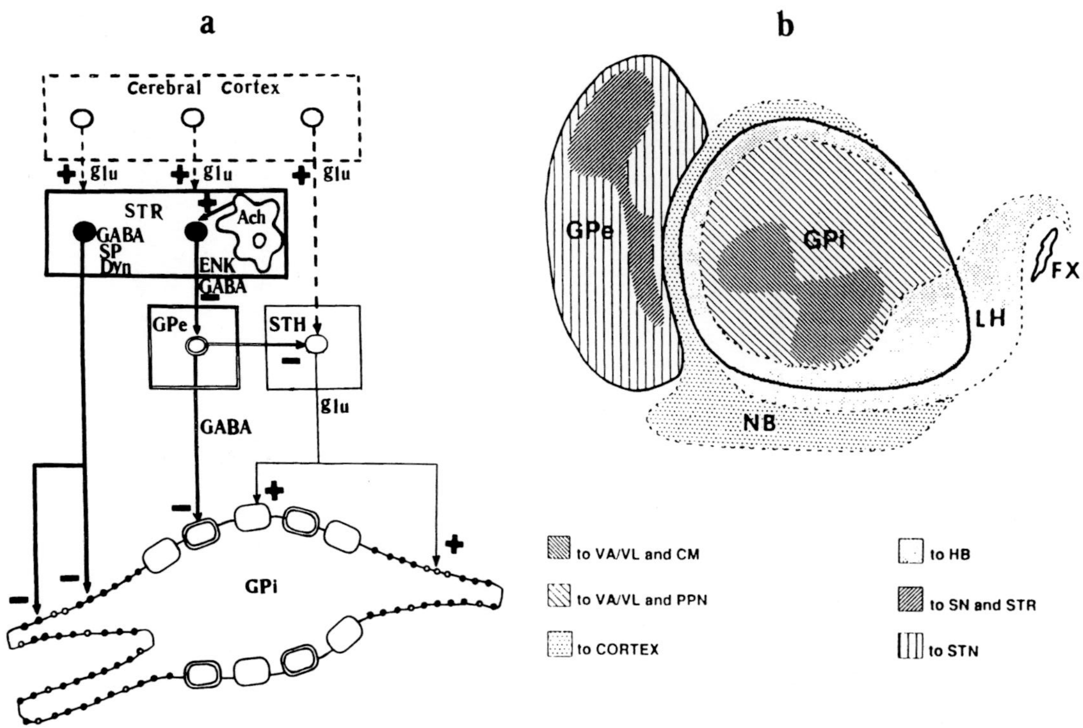

Figure 15.2. **a.** Organization of direct and indirect (through the subthalamic nucleus—STH) outputs of the striatum to neurones of medial part of globus pallidus (GPi). Explanations in the text. STR—striatum; GPe—lateral part of globus pallidus; Ach—large cholinergic interneurone of striatum; Glu—glutamate; SP—subst. P; Dyn—dynorphin; ENK—enkephalin; GABA—gamma-aminobutyric acid (according to Parent and Hazrati [1993]), with some modifications). **b.** A model of the organization of the primate globus pallidus based on retrograde double-labeling studies of its afferent projections. Abbreviations: CM, centromedian nucleus; FX, fornix; GPe, external pallidum; GPi, internal pallidum; HB, habenula; NB, nucleus basalis; SN, substantia nigra; LH, lateral hypothalamus; PPN, pedunculopontine nucleus; STN, subthalamic nucleus; STR, striatum; VA, ventral anterior thalamic nucleus; VL, ventral lateral thalamic nucleus (according to Parent [1990]).

tides responsible for their involvement (Albin *et al.*, 1989; De Boer, 1992; Gerfen, 1992; Parent and Hazrati, 1993; Wang and McGinty, 1996). Thus, an increase in the striatal dopamine activity has been shown to stimulate expression of peptides, dynorphin (Dyn) and the substance P (SP), the expression occurring in the direct efferent striatal pathway. Acetylcholine, on the contrary, increases activity of neurones projecting to GPl (an important link of the indirect pathway), in which an expression of enkephalin occurs (Hong *et al.*, 1985). The same takes place when the striatal dopaminergic activity decreases (Gerfen *et al.*, 1991). Dopamine inhibits release of acetylcholine by affecting D2 receptors and activates the S1 cells via D1 receptors. Striatal cholinergic neurones inhibit gene expression in the striatonigral neurones, presumably via M4 receptors, and stimulates the expression in the striatopallidar neurones via M1 receptors (Wang and McGinty, 1996) (Figure 15.3). These, as well as some other data, have made it possible to put forward a concept emphasising the importance of the dopaminergic/cholinergic interaction in the striatum as a basis for a balanced influence of the striatal efferents on subcortical targets (Scheel-Kruger, 1985; Shapovalova, 1985, 1989, 1993; De Boer, 1992), the influence being necessary for adequate performance, and particularly for acquisition of voluntary motor acts. The direction and degree of this interaction are determined by a number of factors, such as the form of the voluntary movement, the level of afferent activity from the environment, the type of reinforcement, level of motivation, type of higher nervous activity, etc.

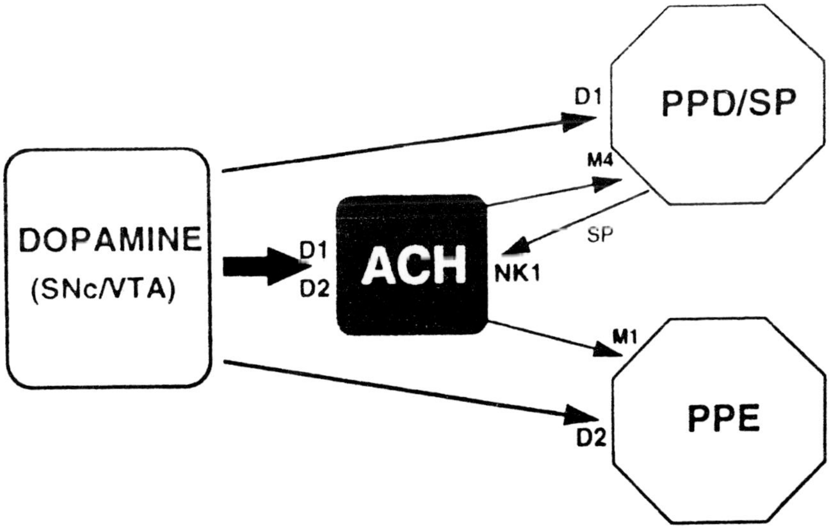

Figure 15.3. Schematic illustration of the acetylcholine/dopamine interactions which are proposed to regulate striatal gene expression. Dopaminergic inputs from the substantia nigra *pars compacta* (SNc) and ventral tegmental area (VTA) stimulate gene expression in striatonigral neurones containing preprodynorphin/substance P via D1 receptors and inhibit that striatopallidal neurones (to GPl) containing preproenkephalin via D2 receptors (according to Wang and McGinty, 1996). Cholinergic interneurones (Ach) inhibit gene expression in striatonigral neurones possibly via M4 receptors and stimulate that of striatopallidal neurones via M1 receptors. Dopaminergic transmission also exerts a D1—dependent facilitatory influence and D2-dependent inhibitory influence on acetylcholine release (after Wang and McGinty [1996]).

All the above allows the striatal cholinergic system to be considered as an important factor controlling one of the striatal efferent pathways, specifically the pathway whose targets are both motor and sensory brain structures. Therefore, by using adequate pharmacological effects on the striatal cholinergic system (for instance, by microinjections of its agonists or antagonists directly into the structure), it is possible to change the degree of its involvement in regulation of motor and sensory components of the voluntary responses, and thereby to modulate motor behaviour.

The striatum is known to be included in a number of parallel loops, starting in several cortical zones, transmitted consecutively in the striatum, pallidum, SNr, a number of thalamic nuclei and ending in one of the zones of the cerebral cortex (Alexander and Crutcher, 1990). Processing of information in these loops is considered to play an important role in the formation and realization of voluntary motor responses. However, the striatum is also included in other, predominantly subcortical, loops involved in sensory and motor information processing, which also include the superior colliculi, the PPN, and the CM-Pf (Chevalier *et al.*, 1984). All these structures are of great importance in orienting behaviour. The present review analyses a possible role of interactions of the striatal cholinergic system, PPN, and CM-Pf (the centrum medianum-parafascicular nuclei) of the thalamus in the realization of sensory and motor components of the acquired (learned) motor response.

2. METHODOLOGICAL APPROACHES

We studied the influence of activation of the dorsal striatum cholinergic system on motor and sensory components of an instrumental defensive reaction, connected with maintenance of a certain flexor posture (Petropavlovsky, 1934). The experiments were carried out on dogs, fixed on a tensoplatform. The task of the animal was, after presentation of the conditional signal (metronome, 130 beats/min: M130) to avoid the electrical current applied to the left hind limb by lifting it to a certain height (8 cm) and to maintain this position during the whole period of the conditioned signal action (10 s). The current was turned on at the fifth second of the conditional signal and acted together with it for 5 s. The study of the striatal cholinergic system participation in analysis of the informative sensory stimuli was carried out using the same models of motor behaviour, with a number of differentiating signals (i.e. ones not reinforced by the current, viz.: M30, M60, M90).

Activation and blockade of the striatal cholinergic system was performed using microcannulae which were implanted into different striatal areas. Application of different doses of cholinolytics and cholinomimetics produced modulations of the participation of the striatal cholinergic system in the regulation of sensory and motor components of voluntary movement in normal and pathological situations.

In experiments before and after these effects on the striatal cholinergic system, the following recordings were made and analysed: tensograms from four legs, myograms of *m. rectus femoris* and *m. semitendinosus* of the working (left) and supporting (right) hind limbs, amplitude of the movement and its duration, latent periods of the movement initiation and solution of the instrumental task, and the number of interstimulus leg-raisings, and of phasic leg-raisings normally interposed with the tonic type of instrumental response. Exposure of the signals and collection of data in a real time scale were performed by commands of a custom made (directing) PC program. Analysis of the data and presentation of the results of experiments were done using original programs for the statistical treatment and graphical presentation of the information. The method of the experiments has been described in detail earlier (Shapovalova, 1978, 1993, 1995; Shapovalova *et al.,* 1997a,b).

3. ROLE OF THE STRIATAL CHOLINERGIC SYSTEM IN REGULATION OF MOTOR COMPONENTS OF LEARNED MOVEMENT

Successful performance of the instrumental task required attention to the conditional stimulus. This was particularly important during the second half of the signal, when it acted together with electrical current. Figure 15.4 demonstrates that the dispersion (variance: var) of the intermediate limb position falls markedly during the second half of the action of the conditioned signal. Microinjections of carbacholine, an acetylcholine agonist, into the dorsal part of the striatum decreased the variance of limb position during the second half of the conditional signal action, i.e. they improved attention to the defensive stimulus. This occurred in all dogs.

The elaborated instrumental response in dogs was composed of two components: The tonic component (maintenance of the hind limb flexion of a definite amplitude and duration) plays a crucial role in solving the instrumental task. The phasic one (a fast jerk of the leg) reproduces a defensive response to the pain stimulation. This component in most animals was of a kind interposed with the instrumental response. The phasic raisings were also present as background levels.

The foregoing sections (see section 1) emphasised the importance of the dopaminergic/cholinergic interaction in the striatum as a basis for a balanced influence of its efferents on subcortical targets. The defensive situation is a powerful activating factor, particularly for excited dogs. This was manifest in our experiments as a form of marked enhancement of the phasic movement component, an increase of intersignal raisings, postural disturbance, and general anxiety of the animals during the experiment (Figure 15.5[1]). This could be due to an imbalance of the output in striatal efferents, resulting from an enhancement of the activating effect of the direct dopamine-modulated pathway on the main targets. In this case, microinjections of carbachol, an agonist of cholinoreceptors, into striatum could compensate this imbalance, whereas their antagonists, scopolamine, or dopamine, could, on the contrary, enhance it.

Indeed, in dogs with a completely established instrumental reflex (maintenance of flexion for 10 s), microinjections of dopamine into the caudate nucleus head has been shown to impair performance of the learned reaction, due to an increase in interstimulus leg raisings and phasic movements interposed with the tonic-type instrumental reaction. As a result, the animal was more often in a "dangerous" zone and received pain reinforcement. Accordingly, there was a statistically significant decrease in correct solutions of the instrumental task. This dopamine-induced imbalance completely recovered after the end of the microinjection experiments (Shapovalova, 1985, 1993; Shapovalova and Yakimowski, 1988).

Similar data were obtained in experiments on cats (Albertin, 1985). The animals were trained to perform two presses on a lever, either brief or long, depending on the duration

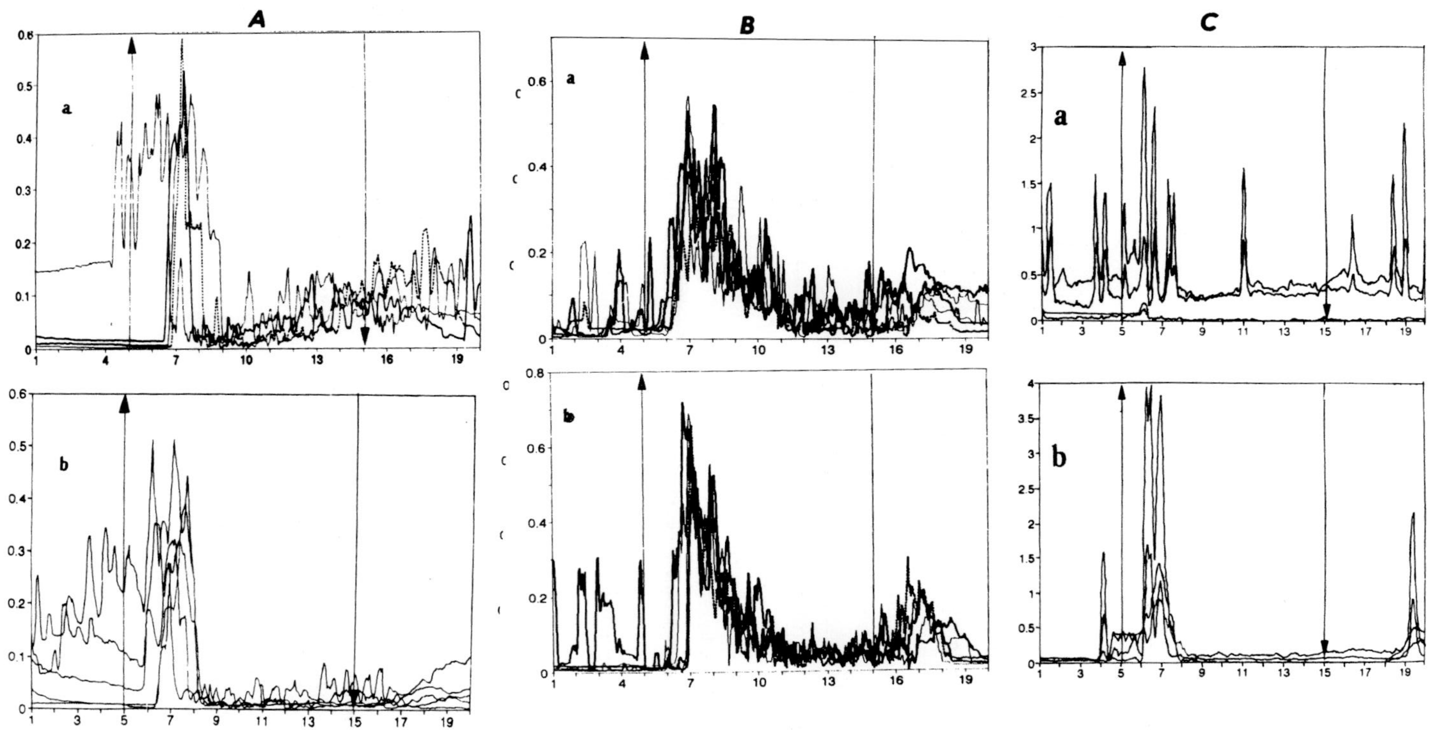

Figure 15.4. Dispersion (var) of the intermediate limb position in three dogs (A,B,C) before (a) and after microinjections of carbacholine into dorso-lateral part of striatum (b). Superpositions of six realizations on each frame. Ordinate: the amplitude of dispersion, cm; abscissa: time, s. Arrows—action of the conditional signal (M130).

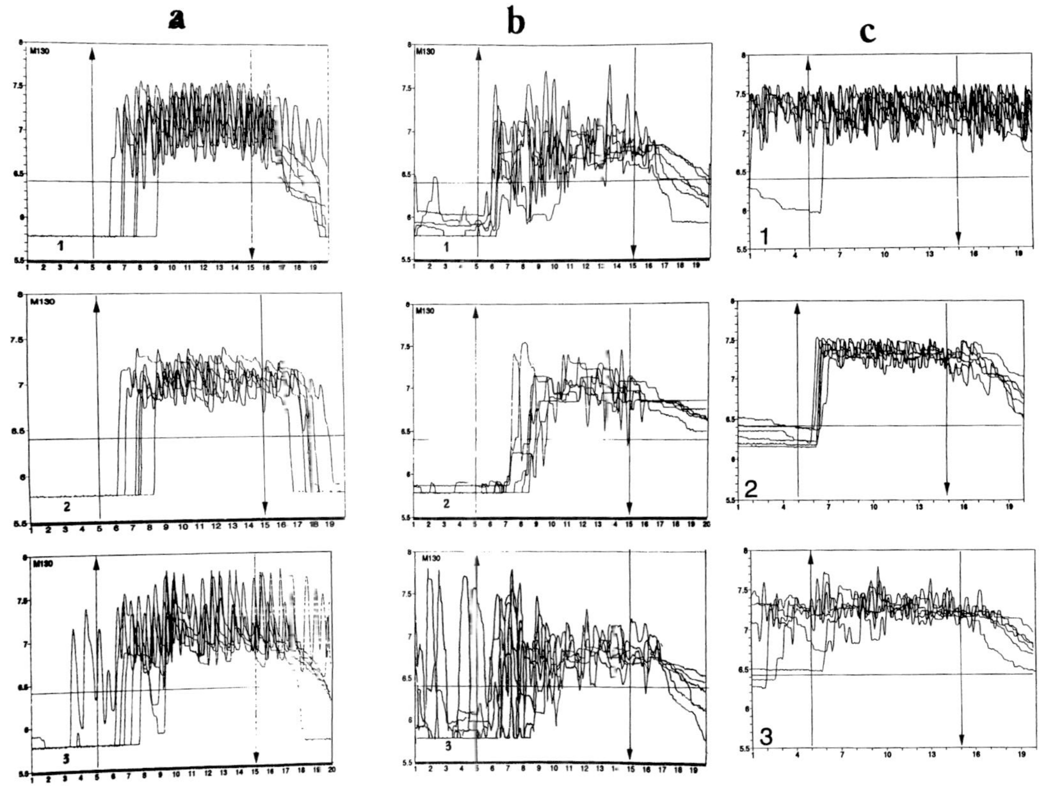

Figure 15.5. Influence of bilateral microinjections of the cholinergic agonist (carbachol) and blocker (scopolamine) on realization of instrumental defensive reaction, connected with the maintenance of flexor posture, in three dogs (a,b,c,). Superpositions of six instrumental reflex realizations on each frame. Ordinate: amplitude of instrumental reaction, cm; Abscissa: time, s. Horizontal line—level of the safety zone. Arrows—the time of action of the conditional signal (M130). • 1—before microinjections, 2–30 min after microinjection of carbachol (0.1 μg for a and b; 0.2 μg for c); 3–30 min after microinjection of scopolamin (0.5 μg).

of the conditional visual signal. Dopamine microinjections into the head of the caudate nucleus were accompanied by a marked increase in the number of phasic interstimulus pressings and in the number of short pressings in response to the conditional signal, as well as by difficulties in performance of the tonic-type instrumental responses.

On the contrary, microinjections of carbachol into the caudate nucleus contralateral to the "working" leg of the dogs resulted in a soothing effect. Interstimulus leg raisings and phasic movement components were inhibited. The instrumental response acquired a kind of "smoothed", ramp-shaped form (Figure 15.5[2]) (Shapovalova, 1993). Microinjections of scopolamine produced an opposite effect (Figure 15.5[3]).

Unilateral (predominantly contralateral) carbachol microinjections into the striatum produced these changes only on the day of the injection. Meanwhile, bilateral microinjections produced long-lasting changes in the formed pattern of the instrumental behaviour. Thus, bilateral carbacholine microinjections (0.1–0.2 μg) led to the following prolonged changes: a decrease in the phasic component and an enhancement of the tonic component of the motor response, inhibition of interstimulus leg raisings, an increase in the movement amplitude (relative to the safety zone), streamlining of the posture and a general soothing of the animal. This prolonged effect was particularly pronounced in one of the animals after an increase in the dose of the administered carbachol up to 0.4 μg. Prior to the experiment with the bilateral carbachol microinjections, the dog was standing during the whole experimental procedure with a raised working leg and performed persistent phasic leg jerks, maintaining, however, a response amplitude that exceeded the disconnection level (Figure 15.6). After the bilateral carbachol microinjections into the striatum, the animals' behaviour changed sharply: The interstimulus raisings stopped, the tonic component of the instrumental response was enhanced, and, what is the most essential, the dog began "standing" well in intervals. For the first time for several years of performing the experiments, a distinctive postural adjustment was recorded prior to the performance of the instrumental movement (Figure 15.6). It should be emphasized that these changes arose after each bilateral microinjection, as if a summation with previous injections took place (Figure 15.6). The after-action was observed for two months.

The effect of the striatal cholinergic system on the postural adjustment was also revealed after unilateral (mainly contralateral) carbachol microinjections into the caudate nucleus. It consisted of a selective prolongation of the main postural adjustment component, the time of unloading of the working leg (Shapovalova, 1993). In dogs, performance of the defensive instrumental flexion reaction is known to be preceded by the so-called postural adjustment cross-pattern (Ioffe, 1991) characterized by changes in pressure on tensoplatforms of all four legs. This might be one of the causes of the significant postural changes observed in our experiments after the bilateral carbachol microinjections as compared with unilateral injections. It should be noted that under all conditions with effects on cholinergic activity, the time of initiation of the postural adjustment did not change. Shortening of the initiation time for postural adjustment was observed after dopamine microinjections into the caudate nucleus, whereas after carbachol injections there was an increase in the unloading time and, hence, in the latent period for the beginning of movement (Shapovalova, 1993).

Thus, the data obtained indicate first, that different components of the postural adjustment as well as the phasic and tonic components of the instrumental movement are controlled by different striatal transmitter systems, connected with the function of different

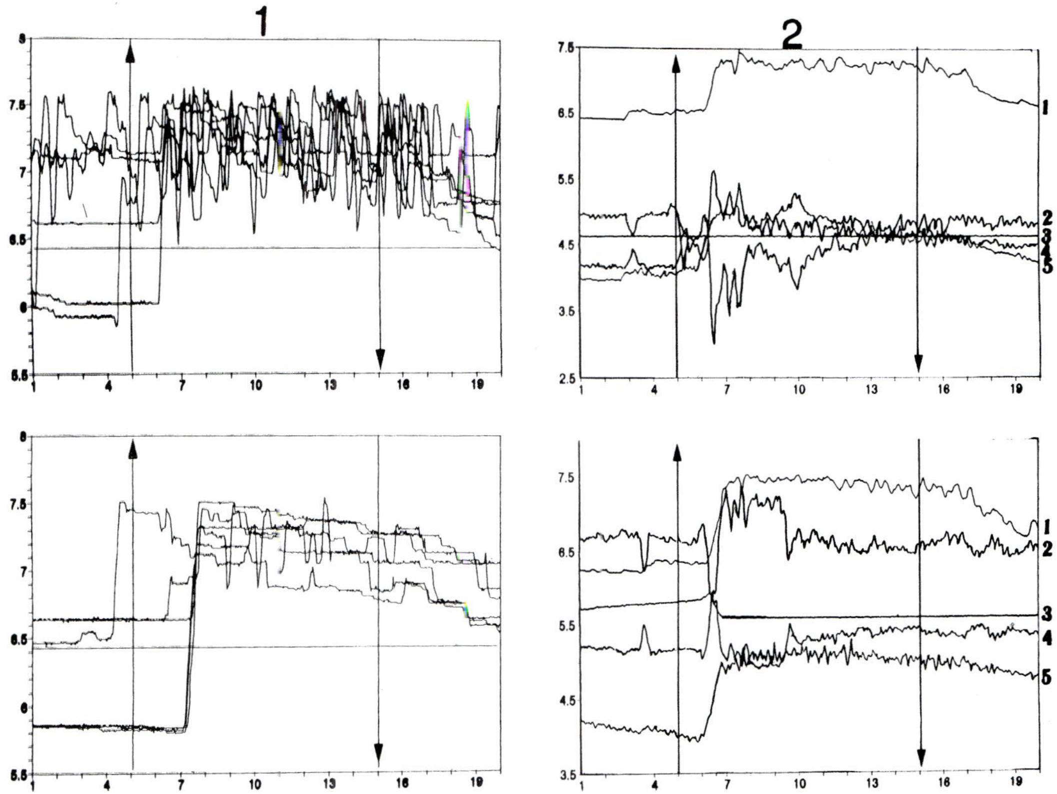

Figure 15.6. Influence of bilateral carbachol microinjections on the tonic component of movement and postural adjustment in an excitable dog. **1, above:** the typical realization of an instrumental reaction in background, before microinjections of carbachol; **below:** 40 min after bilateral microinjection of carbachol (0.4 μg). Superpositions of six instrumental reflex realizations. Other marks as in Figure 5. **2, above:** the typical postural adjustment in a dog before the bilateral carbachol microinjections; **below:** the postural adjustment 40 min after bilateral carbachol microinjection (0.4 μg). On each frame: 1—mechanogram; 2—tensogram of the left forelimb; 3—tensogram of the left hind limb; 4—tensogram of the right forelimb; 5—tensogram of the right hind limb. Summarized data of one experimental session. Other marks as in Figure 15.5.

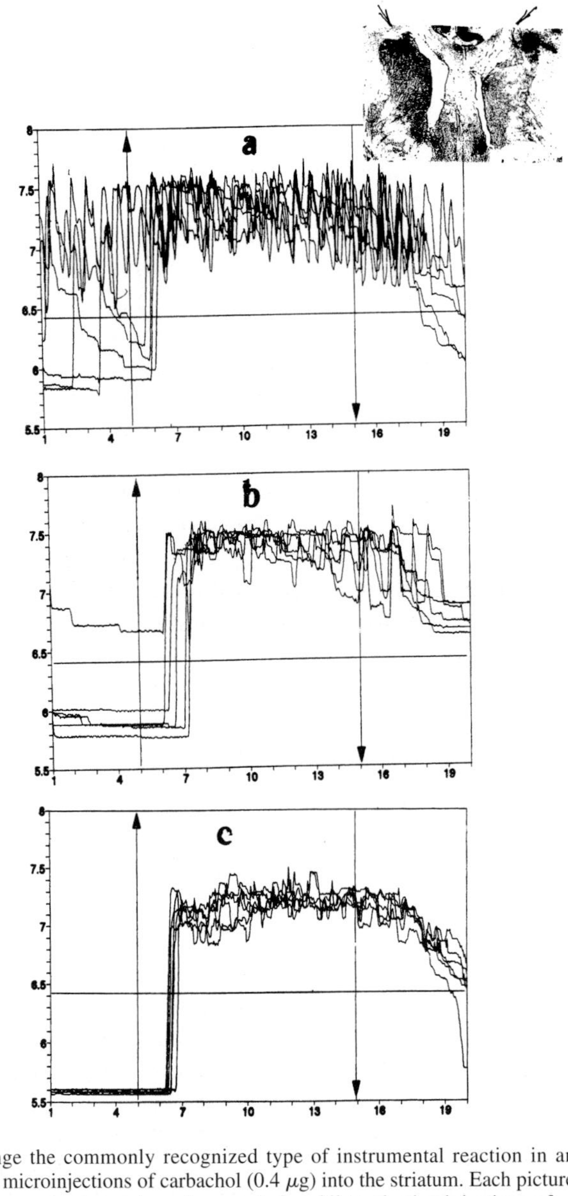

Figure 15.7. Change the commonly recognized type of instrumental reaction in an excitable dog under influence of bilateral microinjections of carbachol (0.4 μg) into the striatum. Each picture shows superpositions of 6 realizations in one experimental session. **a**—before bilateral microinjections ; **b**—3 days after the first bilateral microinjection; **c**—10 days after the third bilateral microinjection of carbachol. Other marks as in Figure 15.5. **Above:** the bilateral localization (arrows) of microcannulae into striatum (nucleus caudatus head).

striatal efferent outputs. Second, it may be supposed that the cholinergic system of the striatum, working through the motor structures involved in the "indirect" efferent pathway, participates in preparing the motor setting necessary and important for successful switching of attention to significant stimuli. The degree of involvement of the striatal mechanisms responsible for inhibition of unwanted movements (Marsden and Obeso,

1992), for inhibition of locomotor activity, general soothing effect, postural streamlining, inhibition of interstimulus leg raisings and enhancement of the tonic movement component seem to be determined by the level of activation of the striatal cholinergic system, these being realized via the PPN.

At the present time, comprehensive experimental material has been accumulated which demonstrates that different transmitter regulation of the tonic and phasic components of movement takes place, both in the direct striatal targets and in those mediated by other structures. Thus, it has been shown that inactivation of the nigrostriatal pathways by muscimol injected into the substantia nigra *pars compacta* (Turski *et al.*, 1982) or destruction of the substantia nigra by ibotenate or 6-OHDA (Ellenbroek, 1988) results in an increase in muscular tone. It has been shown that the disturbances of nigrostriatal dopaminergic or striatal GABAergic projection pathways can act as a trigger in the pathological increase of muscle tonus (Turski *et al.*, 1987). Among the striatal targets, the main regulator of muscular tonus is the substantia nigra *pars reticulata* (Ellenbroek, 1988); its neurones produce a tonic inhibitory (GABAergic) influence on superior colliculus (SC), PPN, and thalamic ventromedial nucleus (Garcia-Rill, 1986). Injections of muscimol into the thalamic ventromedial nucleus, or into the SC, resulted in a dose-dependent increase in the *m. gastrocnemius soleus* EMG activity (Klockgether *et al.*, 1985; Ellenbroek, 1988).

The PPN, a part of the mesencephalic locomotor region (MLR) (Shick *et al.*, 1966), seems to be particularly important for regulation of locomotor activity, muscular tonus, and postural adjustment which are activated by various efferent striatal outputs (Figure 15.1). Parkinson's disease has been shown to be accompanied by pathological destruction of the PPN neurones (Steiniger *et al.*, 1992). The PPN receives direct inputs from the globus pallidus, entopeduncular nucleus, substantia nigra *pars reticulata*, and subthalamic nucleus (Grofova and Spann, 1989). As seen from Figure 15.1, the PPN receives both inhibitory (GABAergic) inputs from the subcortical structures and excitatory (glutamatergic) inputs from the cerebral cortex and the subthalamic nucleus (Garcia-Rill, 1986). The organization and functional significance of the two glutamatergic inputs to the PPN, most likely, should be different. This is confirmed by recent data (Lai and Siegel, 1991), that controls of the muscular tonus and of locomotion are provided by different types of PPN glutamate receptors. Microinjections of agonists of NMDA glutamate receptors in the medial medulla which is a target of the PPN (Figure 15.1) produces locomotion (Kinjo *et al.*, 1990). These areas are identical to those in which the electrical stimulation and activation of non-NMDA glutamate receptors decrease the muscular tonus.

It can also be proposed that the locomotor activity initiated by activation of the cortical glutamatergic input to the PPN can be inhibited, up to the point of complete cessation, by activation of the striatal cholinergic system (Shapovalova, 1993). This activation, as mentioned above, can be connected with an increase of the striatal influences that are realized via an indirect efferent channel, i.e. the subthalamic nucleus (Figure 15.1). This leads, apart from restriction of locomotion (due to an increase of the SNr and GPm GABAergic influences on the PPN neurones) to an increase in muscular tonus owing to direct glutamatergic influences on the PPN from the subthalamic nucleus.

It can be argued that it is these channels that are responsible for the so-called "caudate arrest reaction". We have shown that after an increase in intensity of high-frequency stimulation of the caudate nucleus the so-called "forced" positioning of the leg (i.e. the "arrest reaction") occurs against the background of activation of *m. rectus femoris* in the working leg; such activation of *m. rectus femoris* has never been observed in the background after switching off the conditional signal. This is reminiscent of a pathological pattern of a combination of hypokinesia (inhibition of phasic movements, decrease in the response amplitude, "fixation" with a risen leg) with an extensor muscle hypertonus (Figure 15.8[1]) (Shapovalova, 1995). It is important to note that the preliminary electrical stimulation of the thalamic intralaminar nuclei (centrum medianum-parafascicular nucleus, CM-Pf) decreased the current threshold necessary to get the "caudate arrest reaction" (Figure 15.8[2]). On the contrary, bilateral destruction of the CM-Pf complex resulted in a pronounced, 2–3-fold increase in the current amplitude necessary for the "caudate arrest reaction" (Shapovalova, 1993, 1995).

Normally, the PPN seems to participate in the fine regulation of the locomotor activity and muscular tonus. Thus, it was shown that locomotion could be induced by stimulation of the ventral pontine tegmentum only at a definite muscular activity level (Mori *et al.*, 1986). Injection of NMDA receptor agonists into the PPN produces locomotion only at the average tonus level in the period preceding the injection. These facts may be fundamental to coordination of the postural and locomotor mechanisms during walking. It

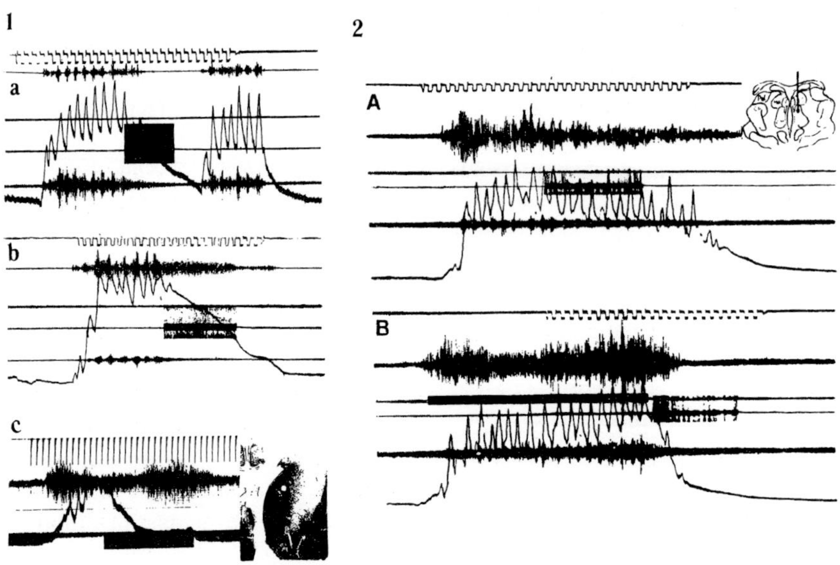

Figure 15.8. Examples of "caudate arrest reaction" in three dogs (a,b,c) and the influence of preliminary stimulation of the CM-Pf complex of thalamus on inhibition, induced by stimulation of nucleus caudatus head **1**: high-frequency stimulation of the caudate nucleus—400 mA (a), 500 μA (b) and 1 μA (c). From above to below: mark of the action of the conditional signal, the EMG of the m.rectus femoris of the working limb (a–c), EMG m.semitendinosus of the working limb (a,b) the mechanogram of the instrumental reflex, the localization of the stimulating electrodes in the caudate nucleus (c). **2:** A—influence of stimulation of the caudate nucleus, which is subthreshold for the elicitation of the caudate arrest reaction on the realization of the instrumental defensive reaction. B—the same after preliminary stimulation of the CM-Pf complex. From above to below: mark of the action of the conditional signal, the EMG of the m.rectus femoris, mark of the nucleus caudatus head stimulation (A,B,), mark of the CM-Pf complex stimulation (B), EMG m.semitendinosus, mechanogram of the instrumental movement (dog a).

should be emphasized that regulation of muscular tonus in higher vertebrates is connected with the necessity not only of maintaining the given posture in the earth's gravitational field but also of fixing the posture urgently before start of locomotion. This fixation is necessary to create the support for the body part moving in space (Gurfinkel *et al.*, 1989). As indicated by the facts presented, the central structures controlling locomotion and those controlling posture providing for maintenance of equilibrium are not the same, although they are closely integrated in the program of the entire motor act (Ioffe, 1991).

Thus, at the PPN level, a separate regulation (including separate transmitters) of the tonic and phasic movement components takes place: Trigger mechanisms of this regulation are provided by activity of different striatal outputs. In this respect, the data obtained under conditions of weightlessness are of great significance, in which a selective impairment of the tonic movement component and of posture was revealed, while the phasic movement components were completely preserved (Kozlovskaya *et al.,* 1986).

The data presented demonstrate with certainty the importance of activation of the striatal cholinergic system for enhancement of the tonic movement component. Since this component was the principal requirement for solution of the instrumental task in experiments on dogs, carbacholine microinjections into the caudate nucleus resulted in a marked rise in the correct response percentage (Shapovalova, 1993), while bilateral carbacholine microinjections into the striatum of excited animals completely changed the form of the instrumental response type for two months or longer (Figure 15.7). In this connection, it is of great importance to note the selected motor behaviour model for the manifestation of effects of microinjections of carbacholine (and other transmitters). In experiments on rats, using a model connected with locomotion (active avoidance in a T-maze), bilateral carbacholine microinjections into the dorsal striatum did not increase the percentage of correct realisations . Meanwhile, dopamine microinjections into the same zones improved performance of behavioural tasks of this type (Shapovalova *et al.*, 1997a,b).

On the other hand, the dorsal striatal cholinergic system may have an important role in restriction of excess (i.e. unwanted) motor activity, by producing (for example) a sharp enhancement of the afferent influence from the environment. An experimental model of this process was achieved by a stimulation of CM-Pf complex of thalamus. The degree of involvement of striatal mechanisms responsible for inhibition of locomotor activity, inhibition of interstimulus leg raisings, enhancement of the tonic movement component and postural stabilization and streamlining seems to be determined by the level of the striatal cholinergic system activation and to be realized via the subthalamic and pedunculopontine nuclei. This mechanism seems to have an important role in motor setting connected with attention.

One of the ways of activating the striatal cholinergic system, as mentioned above, may be to increase activity of the intralaminar thalamic nuclei, that send direct, topically organized projections in striatum. It was shown that Pf nucleus of thalamus sends direct projections to the cholinergic striatal interneurones in the rats (Lapper and Bolam, 1992). In this respect it can be supposed that intralaminar influences on the striatal efferent neurones, which are involved in the indirect efferent pathway, may be mediated by striatal cholinergic cells, because afferents from CM thalamus have practically no direct connections with "indirect" striatal neurones in monkeys, but preferentially innervate striatopallidal neurones projecting to the GPi (Sidibe and Smith, 1996).

4. ROLE OF THE STRIATAL CHOLINERGIC SYSTEM IN REGULATION OF SENSORY COMPONENTS OF LEARNED MOVEMENT

Striatal cholinergic systems have an important role in regulation not only of the motor but also of the sensory mechanisms responsible for realization of the learned movement (Shapovalova, 1995).

We showed that activation of the striatal cholinergic system after microinjections of a choline agonist (carbachol) directly into the structure led to an improvement of the process of differentiation of the signals important in this situation (Figure 15.9). The striatal cholinergic system seems to be involved in this process bilaterally, since the effect of bilateral microinjections was much greater than that of ipsi- or contralateral ones

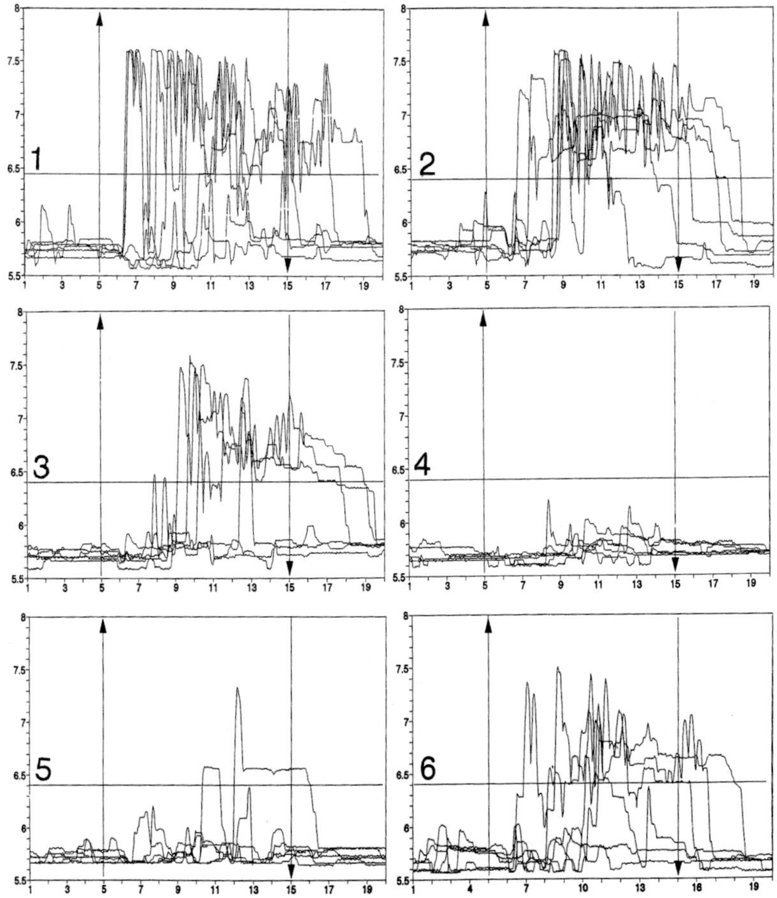

Figure 15.9. Influence of bilateral carbachol microinjections (0.1 μg) in the striatum on the differentiation of M30 in one of the dogs. **1**—background before carbachol microinjection; **2**—10 min after microinjection; **3**—20 min after microinjection; **4**—30 min after microinjection; **5**—on the next day after microinjection; **6**—5 day later after microinjection. Superpositions of six realizations. On each frame, ordinate: amplitude of the movement, cm; abscissa: time, s. Arrows—time of conditional signal action. Horizontal line—safety zone.

(Figure 15.10). Administration of the cholinoreceptor blocker scopolamine produced, on the contrary, a complete disinhibition of differentiations even in a well trained animals (Figure 15.11), which confirms the above conclusion. The control prolonged experiment (during 1 hr) did not change the level of realization of instrumental reaction on presentation of differentiating stimuli (Figure 15.13[C1]) .

In four animals we were able to show that after carbacholine microinjections into the striatum, it is possible to differentiate M90 from M130. A special test for attention—an urgent administration of M130 between the M90 signals—also revealed a clear

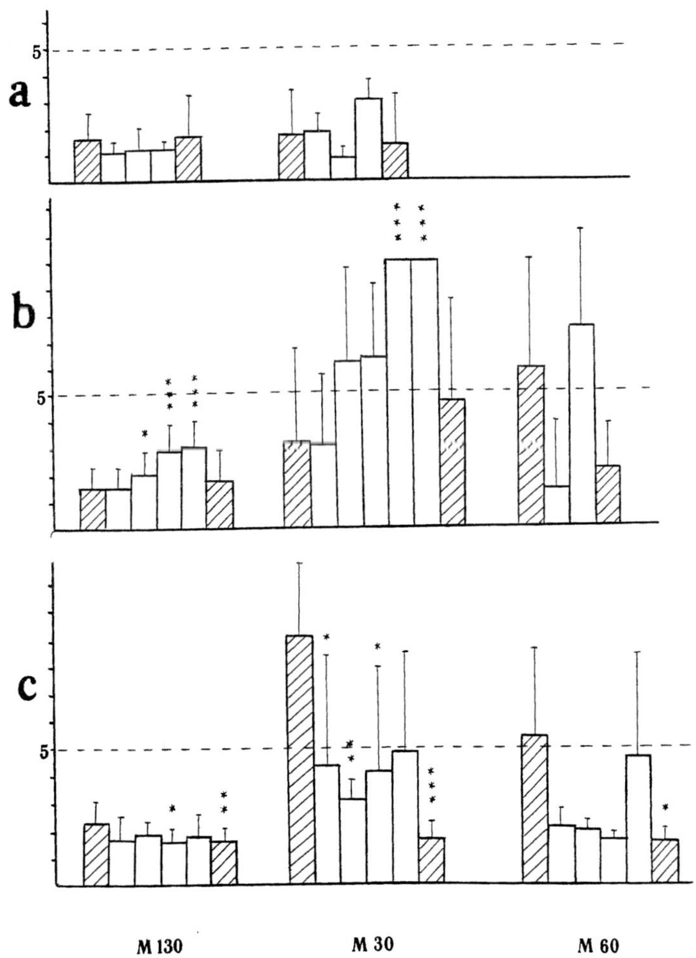

Figure 15.10. Influence of unilateral (**a**) and bilateral (**b**) carbacholine microinjections and bilateral scopolamine microinjections (**c**) on the latent period of instrumental reactions in one of the dogs. Ordinate: the size of latent period of instrumental reaction, s. The hatched columns—the day before and day after microinjection; the white columns—the day of carbacholine microinjection (0.1 μg): successive data every 10 min (during 50 min) after microinjection. Stars—significant changes from background. Each group of columns—summarized data of the two carbocholine microinjections.

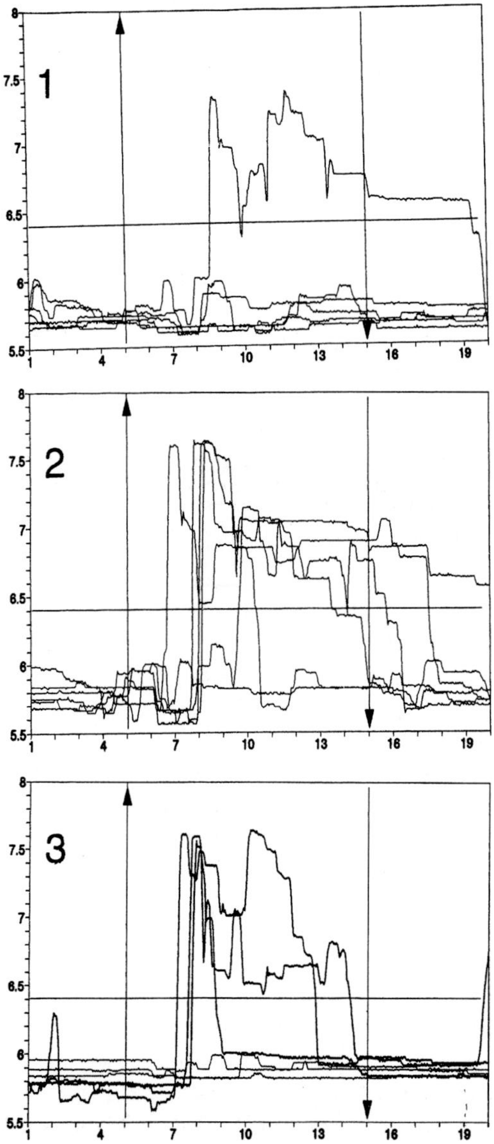

Figure 15.11. Influence of bilateral scopolamine microinjection (0.5 μg) into the striatum on the differentation of M30 in one of the dogs. 1—background before scopolamine microinjection; 2—30 min after microinjection; 3—on the next day after microinjection. Superpositions of six realizations. Other marks as in Figure 9.

differentiation of these signals (Figure 15.12). This was impossible to show both before and after experiments with carbacholine.

There are two possible mechanisms of sensory regulation of the process of attention by the striatal cholinergic system:

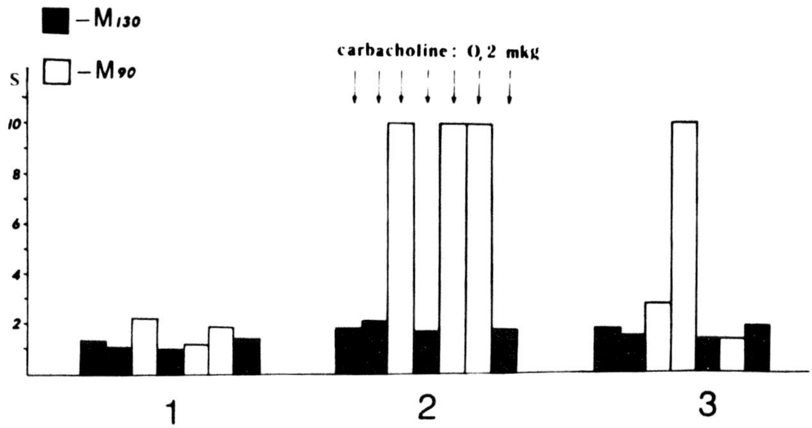

Figure 15.12. Influence of bilateral carbacholine microinjections (0.2 μg) into the dorsolateral striatum on the differentation of M90 in one of the dogs. 1—before; 2—experiment with carbocholine microinjection; 3—in the next day after microinjection. Each column—data summarized from three experimental sessions. Ordinate—latent period of instrumental reaction, s. Other marks on the figure. Explanations in the text.

Activation of the indirect efferent striatal output leads to switching of inhibitory influences from GPm and SNr on its targets (Figure 15.1) and first of all on intralaminar thalamic nuclei (Figure 15.2[B]). This mechanism can provide for inhibition of the environmental afferents, and enhancement of the signal that is informative for this situation. Recordings of electrical and neuronal activity of the cerebral cortex, intralaminar thalamic nuclei and midbrain reticular formation indicate that there is striatal control of the sensory information coming into the cortex, consisting of modulation of afferent input value by inhibition of non-specific afferents (as reviewed by: Arushanian and Otellin, 1976; Shapovalova, 1978). By this mechanism, operating via the indirect striatal efferent output, non-specific "noise" can be reduced, thereby enhancing the signals informative for the current situation.

The second mechanism may be as follows.

Although the neuronal mechanism of activation of the striatal cholinergic system was not studied, it may be comparable with the effects of acetylcholine on other structures. Thus, application of acetylcholine onto cerebral cortical or hippocampal neurones, or stimulation of endogenous acetylcholine release was shown to produce a prolonged potentiation of neuronal activity (Richardson and DeLong, 1988; Jahed *et al.,* 1995). It was also found that the effect of acetylcholine on output currents was mediated intracellularly by cyclic GMP (Woody *et al.*, 1986b) or cGMP-dependent protein kinase (Woody *et al.*, 1986a). These effects reduce postsynaptic conductance and increase excitability of the membrane. Long-term changes in synaptic transmission could result in a longer activation of the excitatory inputs. Cellular conditioning also significantly decreased the total output current. When this was accompanied by application of acetylcholine or cyclic GMP, this decrease of the output current was transformed from a transitory decrease into a more prolonged decrease (Jahed *et al.,* 1995). It was shown, for instance, that brief stimulation of cholinergic pathways from the nucleus of Meynert increased the duration of responses of hippocampal neurones by tens of seconds (Richardson and DeLong, 1988).

An important peculiarity of the responses of cortical and hippocampal neurones to cholinergic stimulation is the dependence of the effect on the state of the neuronal activity of these structures. It turns out that after application of acetylcholine simultaneously with stimulation of the cerebral cortex (application of glutamate, stimulation of excitatory afferent inputs, etc.), the effects observed were recorded for more than one hour, which might be a result of their interaction (Jahed *et al.*, 1995). Thus, long-term changes in the activity of the cortical and hippocampal neurones under the influence of acetylcholine might be an important requirement for their involvement in the learning process (Richardson and DeLong, 1988), and activation of excitatory inputs in forebrain structures might prolong the acetylcholine-induced potentiation of the conditional neuronal activity. This may be confirmed by the results we have obtained using the same model: Improvement of differentiation of informative signals and prolongation of effects were expressed much more after simultaneous activation of the cholinoreactive striatal structures and the CM-Pf than after unilateral carbacholine microinjection into the caudate nucleus head (Shapovalova, 1995). The simultaneous activation of the cholinoreactive structures of the thalamus (CM-Pf) and striatum seems to prolong the effect of acetylcholine in the striatum. Such a suggestion is based on recent neuromorphological and neurochemical data on direct, topologically organized inputs from the CM-Pf to the striatum (Berendse and Groenevegen, 1990; Sadikot *et al.,* 1992), and its compartments (Ragsdale and Graybiel, 1991).

It should be stressed too, that such application of acetylcholine or stimulation of endogenous release of acetylcholine led to long lasting potentiation of cortical or hippocampal neuronal activity only if the latter was involved in a conditioning response (Richardson and DeLong, 1988; Jahed *et al.,* 1995). The level of learning that is manifested in the striatal neuronal activity can also be thought to be a prerequisite for the cholinergic effects on the signal differentiation process. As seen from Figure 15.13, microinjections of carbacholine into the striatum of a dog that did not differentiate M-130 (defensive signal) and M-30 (differentiation signal) in the background produced no improvement of differentiation. In another dog, with a clear differentiation of M-130 and M-30 in the background, such microinjection produced an evident effect. This was particularly evident when two differentiation signals, M-30 and M-60, were used. These animals differentiated well between M-130 and M-30 in the background and did not in fact differentiate M-130 and M-60. Carbacholine microinjections improved differentiation of M-30 but did not change responses to M-60 (Figure 15.14).

Similar data were obtained on rats, using the model of the discriminated conditional reflex of active avoidance (CRAA) in the T-maze. We have found a statistically significant ($p<0.01$) increase in the percentage of correct realizations of the discriminated CRAA after bilateral microinjections of 0.03 μg carbacholine into the rat striatum on the 4th, 5th, and 6th day of training, as compared with results of similar microinjections into the striatum of rats with no previous training experience (Shapovalova *et al.*, 1997a,b). Bilateral lesion of the Pf nucleus resulted in irreversible disturbance of the discriminated CRAA elaborated previously. In rats with the previous bilateral lesion of the Pf nucleus, the discriminated CRAA was not elaborated at all for 10 experimental sessions (160 tasks). Carbacholine microinjections to such animals pro-

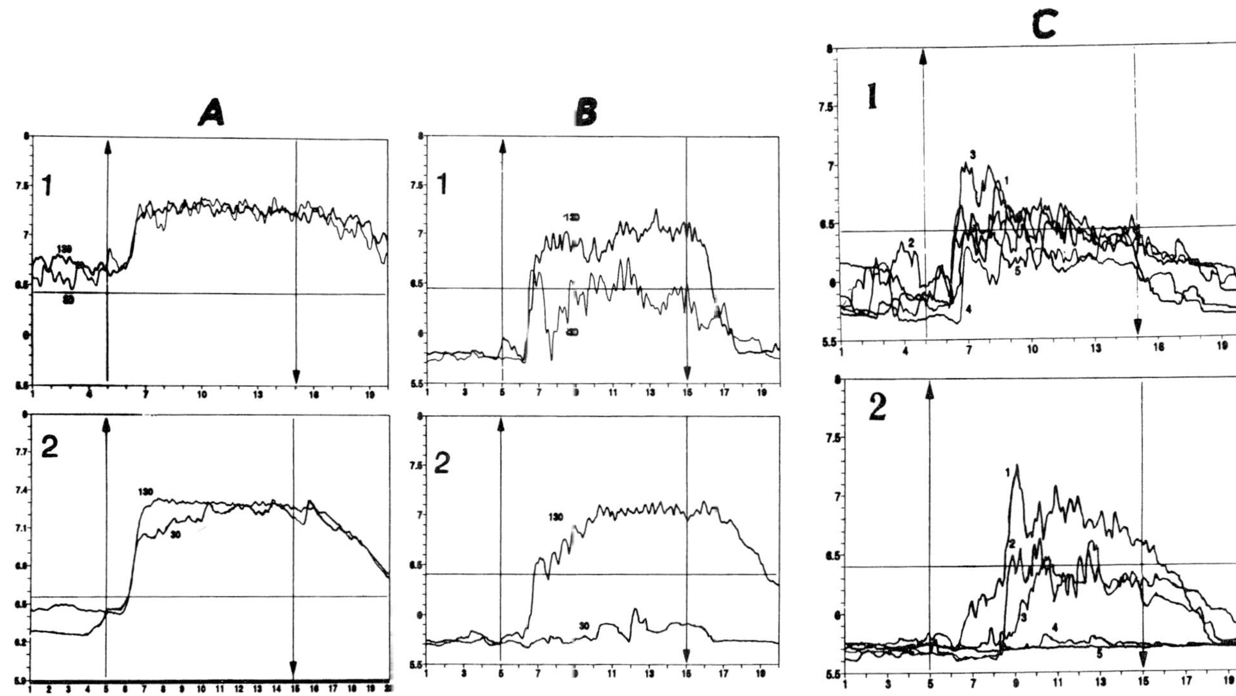

Figure 15.13. Comparison of the effects of bilateral carbachol (0.1 μg) microinjections into striatum on the differentiation of signals in two dogs (A,B) with different degrees of learning in the background. **A** and **B**: comparison of the effects on M130 (defensive signal) and on M30 (differentiated signal) in the background (1) and after bilateral microinjection of carbachol into striatum (2). **C: 1**. Comparison of the effects of the prolonged experimental session (1 hour) and **2.** carbachol microinjection into the striatum on the differentation of M30 in one of the dogs. On each frame, abscissa: time, s; ordinate: the amplitude of movement, cm; arrows against traces: 1–5: successive data every 10 min (starting with 1—at 10th min.). Arrows—time of conditional signal action. Horizontal line—safety zone. Summarized data.

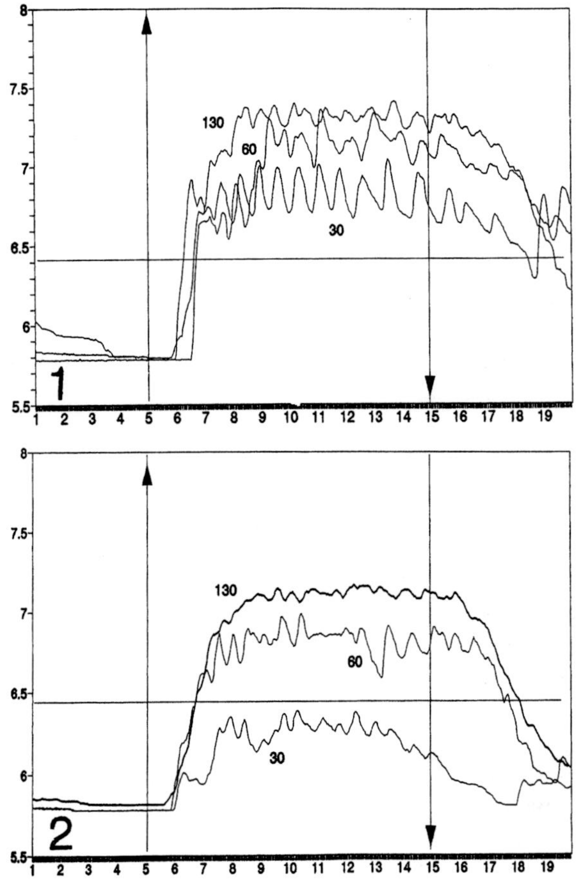

Figure 15.14. Comparison of the effects of bilateral carbacholine microinjections (0.2 μg) into striatum on the differentation of signals M30 and M60. **1**—background (before microinjections); **2**—in the experimental session with carbacholine microinjection. Summarized data. On each frame: abscissa—time, s; ordinate—amplitude of instrumental movement. Arrows—the time of conditional signal action. Horizontal line—safety zone.

duced no effect (Shapovalova *et al.*, 1997a,b). The data obtained also indicate that the integrity of the afferent input from the CM-Pf into the striatum is an important factor in the activation of striatal neuronal background activity necessary for obtaining the effect of striatal cholinergic activation.

The extensive inputs from the CM-Pf into the striatum and the connections of the striatum mediated by the SNr and GPm (Figure 15.1) make it possible to consider the striatal cholinergic system not only as a parallel level of the sensory information procession but also as a system controlling transmission of this information, the control being performed at different levels and directed both to the sensory and motor structures. One of the most important results of this control may be thought to be improvement of attention to informative stimuli.

5. CONCLUSION

Recent neuromorphological and neurochemical data are reviewed concerning the organization of the striatal cholinergic system. A conclusion is made that the nigrostriatal dopamine and intrastriatal acetylcholine have the opposite effects on the main efferent outputs of the striatum, including the expression of peptides. A crucial role is proposed for the striatal cholinergic system in regulation of the so called indirect efferent pathway which controls the main striatal targets: the medial part of globus pallidus and the reticular part of substantia nigra.

As discussed above (see Introduction), the striatum plays an important role in modulation of the cortical motor programs realized via the cortico-strio-thalamo-cortical parallel loops (Alexander and Crutcher, 1990). However, the cognitive processing of information requires a capability for spatial orientation and selective attention to the informative or new environmental stimuli.

The review presents new data and interpretations concerning participation of the striatal cholinergic system in regulation of attention to informative sensory stimuli realized via a number of motor and sensory subcortical structures, among which a particular role is played by the PPN and CM-Pf. It is suggested that the striatum has a crucial role in regulation of the activity of the PPN, a source of cholinergic projections, and, hence, in regulation of the ascending inflow of cholinergic activity afferent to the thalamus. On the other hand, the interaction of the thalamic (CM-Pf) and striatal cholinoreactive structures may be an important factor in enhancing attention to informative stimuli, restricting locomotor activity, increasing the tonic movement component and streamlining posture. In this connection, it is important to note the recently revealed, extensive connections (other than projections to the CM-Pf) of the PPN cholinergic neurones to the SC (Krauthamer *et al.*, 1995). This may indicate, on the one hand, an important role of the PPN cholinergic neurones in the spatial orientation provided by the SC (eye and head movements, changes in posture, etc.), and, on the other hand, an interaction of these information processing systems. Thus, our own data, together with data from the literature reviewed, indicate that the striatum and its cholinergic system are included in regulation of at least two subcortical loops for processing of sensory and motor information dealing with orienting behaviour and attention to significant stimuli.

The extensive inputs from the CM-Pf into the striatum and the connections of the striatum mediated by SNr and GPm make it possible to consider the striatal cholinergic system not only as a parallel level of sensory information processing but also as a system controlling transmission of this information, the control being performed at different levels and directed both to the sensory and motor structures.

ACKNOWLEDGEMENT

This work was supported by the Russian Fund of Fundamental Investigations (grant No. 96–04–48683).

REFERENCES

Albertin, S.V. (1985) Participation of the dopamine reactive system of the Caudate Nucleus in the regulation of the instrumental conditioned reflexes varying in the degree of complexity, *Fiziolicheskii Zhurnal i I.M.Sechenova*, **84**, 87–94 (in Russian).

Albin, R.L., Young, A.B. and Penney J.B. (1989) The functional anatomy of basal ganglia disorders. *Trends in Neuroscience*, **12**, 366–375.

Alexander, G.E. and Crutcher, M.D. (1990) Functional architecture of basal ganglia circuits: Neuronal substrates of parallel processing, *Trends in Neuroscience*, **13**, 366–375.

Arushanjan, E.B. and Otellin, V.A. (1976) *The Nucleus Caudatus*, Leningrad: Nauka, 223 pp. (in Russian).

Berendse, H. and Groenewegen, H. (1990) Organization of thalamo-striatal projections in rat with the special emphasis on the ventral striatum, *Journal of Comparative Neurology*, **299**, 187–228.

Bolam, J.P., Wainer, B.N. and Smith, A. (1984) Characterization of cholinergic neurons in the rat neostriatum: a combination of choline acetyltransferase immunocytochemistry, Golgi impregnationtion and electron microscopy, *Neuroscience*, **12**, 711–712.

Chevalier, G. and Deniau, J. (1990) Disinhibition as a basic process in the expression of striatal functions. *Trends in Neuroscience*, **13**, 277–280.

Chevalier, G., Vacher, S., Deniau, J. and Albe-Fessard, D. (1984) Tonic nigral control of tecto-spinal/tecto-diencephalic branched neurons: a possible implication of basal ganglia in orienting behavior. In: J.S. McKenzie, R.E. Klemm and L.N. Wilcock (eds), *Basal Ganglia. Structure and functions*, New York: Plenum Press, pp. 247–259.

Contant, C., Umbriaco, D., Garcia, S., Watkins, K. and Descarries, L. (1996) Ultrastructural characterization of the acetylcholine innervation in adult rat neostriatum, *Neuroscience*, **71**, 937–947.

De Boer, P. (1992) *Dopamine-acetylcholine interactions in the rat striatum.* Doctoral dissertation thesis. Groningen, The Netherlands. 189 pp.

Ellenbroek, B. (1988) *Animal model of Schizophrenia and neuroleptic drug action.* Doctoral Thesis. Nijmegen. 387 pp.

Garcia-Rill, E. (1986) The Basal Ganglia and the locomotor regions, *Brain Research Reviews*, **11**, 47–63.

Gerfen, Ch. (1992) The neostriatal mosaic: organization of the Basal Ganglia, *Annual Review of Neuroscience*, **15**, 285–320.

Gerfen, Ch., McGinty, P. and Young, W.S. (1991) Dopamine differentially regulated dynorphin, Subst. P and enkephalin expression in striatal neurons: in situ hybridisation histochemical analysis. *Journal of Neuroscience*, **11**, 1016–1031.

Glowinski, J., Besson, M.J. and Cheramy, A. (1984) Role of the thalamus in the bilateral regulation of dopaminergic and GABAergic neurons of the Basal Ganglia. In: D. Evered and M. O'Connor (eds) *Functions of the Basal Ganglia (CIBA Foundation Symposium, no. 107)*, London: Pitman, pp. 150–161.

Grofova, I. and Spann, B. (1989) Involvement of the nucleus tegmenti pedunculo pontinus in the descending pathways of the basal ganglia in the rat. In: *Basal Ganglia. Proceedings of Third Triennial Meeting of International Basal Ganglia Society*, p. 79.

Groves, Ph.M. (1983) A theory of the functional organization of the Neostriatum and neostriatal control of voluntary movement, *Brain Research Reviews*, **5**, 109–132.

Gurfinkel, V.S., Levik, Yu.S. and Lebedev, M.A. (1989) Tonic background as a basis of postural activity. In: K.V. Baev, Yu.P. Limanski, V. Beresovski (eds) *Regulation of sensorymotor function. (Book of abstracts).* Vinnitsa: Bogomolets Physiol. Inst. p. 48 (in Russian).

Hirsch, E.C., Graybiel, A.M., Hirsch, L.B., Duyckaerts, C. and Agid Y. (1989) Striosomes and extrastriosomal matrix contain different amounts of immunoreactive choline acetyltransferase in the human striatum, *Neuroscience Letters*, **96**, 145–150.

Hong, J.C., Yoshikawa, K., Kanamatsu, T. and Sabol, S.L. (1985) Modulation of striatal enkephalinergic neurons by antipsychotic drugs. *Federation Proceedings*, **44**, 2535–2539.

Ioffe, M.E. (1991) *The mechanisms of movement learning.* Moscow: Nauka. 135 p. (in Russian).

Jahed, N., Gruen, E. and Woody, C.D. (1995) Cholinergic dependence of a cortical neuronal mechanism that supports Pavlovian eyeblink conditioning, *Fiziolicheskii Zhurnal i I.M. Sechenova* **81**, 10–17 (in Russian).

Kawaguchi, Y. (1992) Large aspiny cells in the matrix of the rat neostriatum *in vivo*: physiological identification, relation to the compartments and excitatory postsynaptic currents, *Journal of Neurophysiology*, **67**, 1669–1682.

Kilpatrick, I.C., Jones, M.W., Pycock, P., Riches, I. and Philipson, O.T. (1986) Thalamic control of dopaminergic functions in the Caudate-Putamen of the rat. III. The effects of lesion in the parafascicular-intralaminar nuclei on D2 dopamine receptors and high affinity dopamine uptake. *Neuroscience*, **19**, 991–1005.

Kimura, H., McGeer, P., Peng, H., McGeer, E. (1981) The central cholinergic system studied by choline acetyltransferase immunohistochemistry in cat. *Journal of Comparative Neurology*, 200, 151–201.

Kinjo, N., Atsuta, Y., Webber M., Kyle, R., Skinner, R.D. and Garcia-Rill, E. (1990) Medioventral medulla-induced locomotion. *Brain Research Bulletin*, **24**, 509–513.

Kita, H. and Kitai, S.T. (1988) Glutamate decarboxylase immunoreactive neurons in rat neostriatum: their morphological types and populations, *Brain Research*, **447**, 346–352.

Klockgether, T., Schwarz, M. and Turski, L., Wolfarth, S. and Sontag, K.H. (1985) Rigidity and catalepsy after injections of muscimol into the ventromedial thalamic nucleus: an electromyographic study in rats. *Brain Research*, **85**, 559–569.

Koslovskaya, I.B., Kirenskaya, A.V. and Aslanova, I.F. (1986) The influence of imersionnal hypokinesia on characteristics of motor units aktivity. *Fizilogia Cheloveka*, **12**, 627–634 (in Russian).

Krauthamer, G.M., Grunwerg, B.S. and Krein, H. (1995) Putative cholinergic neurons of the pedunculopontine tegmental nucleus projeccting to the Superior Colliculus consist of sensory responsive and unresposive populations wich are functionally distinct from other mesopontine neurons, *Neuroscience*, **69**, 507–520.

Lai, Y. and Siegel J.M. (1991) Pontomedullary glutamate receptors mediating locomotion and muscle tone suppression, *Journal of Neuroscience*, **11**, 29–31.

Lapper, I.R. and Bolam, J.P. (1992) Input from the frontal cortex and parafascicular nucleus to cholinergic interneurons in the dorsal striatum of the rat. *Neuroscience*, **51**, 533–545.

Leonard, C.S. and Llinas, R. (1990) Electrophysiology of mammalian pedunculopontine and laterodorsal tegmental neurons *in vitro*: implications for the control of REM sleep. In: M. Steriade, D. Biesold (eds) *Brain cholinergic systems*. Oxford. New York: Oxford Univ. Press, Pergamon Press. pp. 205–223.

Marsden, C.D. and Obeso, J.A. (1994) The functions of basal ganglia and the paradox of stereotaxic surgery in Parkinson's disease, *Brain*, **117**, 877–897.

McCormick, D.A. (1990) Cellular mechanisms of cholinergic control of neocortical and thalamic neuronal excitability. In: M. Steriade, D. Biesold (eds) *Brain cholinergic systems*. Oxford. New York: Oxford Univ. Press, Pergamon Press, pp 236–264.

Mesulam, M., Mufson, E., Levey, A.I. and Wainer, B.H. (1983) Cholinergic innervation of cortex by basal forebrain: cytochemistry and cortical connections to the septal area diagonal band nuclei, nucleus basalis (substantia innominata) and hypothalamus in the rhesus monkey, *Journal of Comparative Neurology*, **214**, 170–178.

Mesulam, M., Mash, D., Hersh, L., Bothwell, M. and Geula, C. (1992) Cholinergic innervation of the human striatum, globus pallidus, subthalamic nucleus, substantia nigra and red nucleus, *Journal of Comparative Neurology*, **323**, 252–268.

Mori, S., Otha, Y., Sakamoto, T. and Nonaka, S. (1986) Excitability level setting mechanisms in the pons: Their behavioral support in decerebrate reflex standing and freely moving, intact cats. *Brain and Devevlopment*, **8**, 415–420.

Parent, A. (1990) Extrinsic connections of the basal ganglia. *Trends in Neuroscience*, **13**, 254–258.

Parent, A. and Hazrati, L. (1993) Anatomical aspects of information processing in primate basal ganglia, *Trends in Neuroscience*, **16**, 111–116.

Pasik P., Pasik T. and Di Figlia, M. (1979) The internal organization of neostriatum of mammals. In: I. Divac and G. Oberg (eds) *The Neostriatum*. Oxford: Pergamon Press, pp 5–36.

Petropavlovsky, V.P. (1934) To the methods of conditioned movement reflexes, Sechenov Physiol. Journal., **17**, 217–225 (in Russian).

Ragsdale, C.W. and Graybiel, A.M. (1990) Compartmental organization of the thalamostriatal connections in the cat, *Journal of Comparative Neurology*, 311, 134–167.

Richardson, R. and DeLong, M. (1988) A reappriasal of the functions of the nucleus basalis of Meynert. *Trends in Neuroscience*, **11**, 264–267.

Sadikot, A., Parent, A. and Francois, C. (1992) Efferent connections of the centromedian and parafascicular nuclei in Squirell monkey. A PHA-L study of subcortical projections. *Journal of Comparative Neurology*, **315**, 137–159.

Scheel-Kruger, I. (1985) New aspects on functional role of acetylcholine in the basal ganglia. Interaction with other neurotransmitters. In: Singh, M., Warburton, D.M. and Lal, H. (eds) *Central cholinergic mechanisms and adaptive dysfunctions*. New York: Plenum Press, pp 105–140.

Shapovalova, K.B. (1978) *The role of cortical and subcortical structures in sensorymotor integration*. Leningrad: Nauka (in Russian) 180 pp.

Shapovalova, K.B. (1985) The possible neurophysiological and neurochemical mechanisms of striatum participation in regulation and initiation of voluntary movement, *Fiziol. Zhurnal SSSR*, **71**, 537–553 (in Russian).

Shapovalova, K.B. (1989) Neostriatum and the regulation of voluntary behavior in the norm and pathology: facts and hypotheses, *Fiziologia Cheloveka*, **15**, 78–92 (in Russian).

Shapovalova, K.B. (1993) Possible mechanisms of participation of the neostriatum in regulation of voluntary movement. In: T.M. Turpaev (ed.) *Soviet scientific reviews V. 6. part 3*. Harwood Acad. Publ. 90 p.

Shapovalova, K.B. (1995) Afferent and efferent mechanisms of the intensification of neostriatal cholinergic activity. *Neuroscience and Behavioral Physiology*, 25, 71–80.

Shapovalova, K.B., Pominova, E.V., Dubkacheva, T.A. (1997a) The peculiarities of neostriatal cholinergic system influence of rats on active avoidance learning in norm and after lesion of intralaminar thalamic nuclei *Neuroscience and Behavioral Physiology*, 27, 1–12.

Shapovalova, K.B., Zhuravin, I.A., Pominova, E.V. and Dyubkacheva, T.A. (1997b) Role of the cholinergic system of the neostriatum in regulation of several forms of defensive behavior. *Neuroscience and Behavioral Physiology*, 27, 75–81.

Shapovalova, K.B. and Yakimovsky, A.F. (1988) The role of nigrostriatal and mesolimbic dopaminergic brain systems in the control of voluntary movements and postural adjustment in dogs. In V.S. Gurfinkel, Ioffe, M.E., Massion, J. and Rolls, J.P. (eds), *Stance and Motion: Facts and Concepts*. New York, London: Plenum Press, pp. 203–213.

Shik, M., Orlovski, G. and Severin, F. (1966) Locomotion of the diencephalic cat elicited by stimulation of the pyramids. *Biophyzika*, **13**, 127–135 (in Russian).

Sidibe, M. and Smith, Y. (1996) Differential synaptic innervation of striatofugal neurones projecting to the internal or external segments of Globus pallidus by Thalamus afferents in the squirrel monkeys, *Journal of Comparative Neurology*, **365**, 445–465.

Steiniger, T.L., Rye, D.B. and Wainer, B.H. (1992) Afferent projections to the cholinergic pedunculopontine tegmental nucleus and adjacent midbrain extrapyramidal area in the albino rats: Retrograde tracing studies, *Journal of Comparative Neurology*, **321**, 515–537.

Turski, L., Havemann, U., Schwarz, M. and Kuschinsky, K. (1982) Disinhibition of nigral GABA output neurons mediates muscular rigidity elicited by striatal opioid receptor stimulation, *Life Science*, **3**, 2327–2330.

Turski, L., Klockgether, Th., Turski, W., Schwarz, M. and Sontag, K.H. (1987) Substantia Nigra and motor control in the rat: effect of intranigral L-kainate and L-D-glutamylamino-methylsulphonate on motility, *Brain Research*, **424**, 37–48.

Wang, J.Q. and McGinty, J.F. (1996) Muscarinic receptors regulate striatal neuropeptide gene expression in normal and amphetamine-treated rats, *Neuroscience*, **75**, 43–50.

Wilson, C.J., Chang, H.T. and Kitai, S.T. (1990) Firing patterns and synaptic potential of identified giant aspiny interneuron in the rat neostriatum, *Journal of Neuroscience*, **10**, 508–519.

Woody, C.D., Barfai, T., Gruen, E. and Nairn, A.C. (1986a) Intracellular injection of cGMP-dependent protein kinase results in increased input resistance in neurons of mammalian motor cortex, *Brain Research*, **386**, 379–385.

Woody, C.D., Gruen, E., Sakai, H., Sakai, M. and Swartz, B. (1986b) Responses of morphologically identified cortical neurons to intracellularly injected cyclic GMP, *Experimental Neurology*, **91**, 580–595.

16 The Motor Cortex Inhibits Synergies Interfering with a Learned Movement: Reorganization of Postural Coordination in Dogs

M.E. Ioffe

Institute of Higher Nervous Activity and Neurophysiology, Russian Academy of Sciences, Moscow
e-mail: ldo@dataforce.net

A well known function of the motor cortex (MCx) is control of the distal limb musculature. MCx has another function related to motor learning, namely, the suppression of synergies and coordination which interfere with acquisition of performance. This was shown for various limb movements. However, the majority of movement is accompanied by postural adjustment. An experimental model of the rearrangement of an innate pattern of postural adjustment was developed. MCx was shown to suppress the activity of structures providing the innate postural pattern during such learning. In contrast to the control of limb movement, MCx influences were bilateral for postural pattern reorganization. A recent finding that recovery from MCx lesions depends on the sequence of the lesions, suggests, however, a hemispheric difference in compensation.

KEYWORDS: motor learning; motor cortex; reorganization of coordinations; postural adjustments; dog

1. INTRODUCTION

The motor area of the cerebral cortex (MCx) is known to be involved in the control of limb movements, particularly, of the distal musculature, providing hand and finger movements. This function is realized via the pyramidal system, which has not only polysynaptic but also (in primates) monosynaptic connections with distal muscle motoneurones (Baker *et al.*, 1995; Kuypers, 1964; Phillips and Porter, 1977). Accordingly, the motor cortex is known to control precision and fineness of motor actions, particularly isolated finger movements and their coordination, such as during the precision grip. Following motor cortex lesions in primates, separate finger movements disappear and are replaced by a synergistic flexion of all fingers (Bucy *et al.*, 1966; Tower, 1940; Wiesendanger *et al.*, 1994). Compensation for the behavioural deficit after pyramidal tract impairment was shown to be provided by the red nucleus, controlled by the motor cortex and cerebellum, and by the rubrospinal system (Lawrence and Kuypers, 1968; Ioffe, 1973, 1975; Ioffe and Samoylov, 1967) which, along with the pyramidal system, belongs to the lateral descending system (Kuypers, 1964).

However, during the last decade a number of studies showed that the motor cortex is also involved in the process of motor learning. Three groups of findings may be distinguished. First, plasticity of the motor cortical representation map was shown by MCx microstimulation after its partial lesion followed by intensive training of the appropriate limb (Nudo *et al.*, 1996), after training without lesion (Milliken *et al.,* 1992), after microstimulation itself (Nudo *et al.*, 1990), or combined with changes of a limb position (Sanes *et al.,* 1992) or with passive movements (Humphrey *et al.*, 1990) and after section of a peripheral nerve (Sanes and Donoghue, 1992). Some human studies also showed plasticity of the representation map in MCx, as assessed by transcranial magnetic stimulation, after increase of muscle activity (Braille reading) (Pascual-Leone *et al.*, 1993) and during implicit or explicit learning to perform a definite sequence of key pressing by different fingers (Pascual-Leone *et al.*, 1994). It was shown that the plasticity of the map of motor representation in MCx is NMDA-receptor dependent, since the NMDA-antagonist MK-801 blocks the plasticity induced by passive movements in rats (Qiu *et al.,* 1990).

The second group of findings concerns long-term potentiation (LTP) in MCx (Asanuma,1989; Donoghue *et al.*, 1996). Changes in synaptic efficacy may provide the basis of reorganization in MCx during learning. A third group of findings involves brain imaging techniques (PET and fMRI) which reveal activation of MCx during different forms of motor learning (Grafton *et al.*, 1994, 1995; Sadato *et al.,* 1996; Seitz and Roland, 1992). However, in some studies the changes were shown not to be specific for the process of learning and could also be observed during practice of previously learned movements (Jenkins *et al.*, 1994; Karni *et al.*, 1994).

A related issue concerns the definition of motor learning. Adaptation learning, skill learning, and conditional learning are usually separated (Donoghue *et al.*, 1996; Hallett *et al.*, 1996). Assuming that motor learning involves the development or elaboration of new movements, only skill learning could be accepted as real motor learning. However, even for skill learning it is not easy to determine whether it involves learning of a new movement or modification or practice of a previously learned one. To overcome this difficulty, Fitts' law (Fitts, 1954) is used in some studies as a reference point (Donoghue *et al.*, 1996; Hallett *et al.*, 1996). According to Fitts' law, the greater the velocity of a movement, the less its accuracy. Thus, when Fitts' law is applied to repeated trials, learning a new skill is not involved. Only if both parameters are non-reciprocally changed, is it real skill learning (Donoghue *et al.*, 1996). This approach may illustrate the problems concerning the definition of motor learning.

An alternative approach is based on reorganization of coordination in the process of motor learning. During motor learning, some innate or well learned synergies may be incorporated into the pattern of a new movement. However, in some cases innate synergies may interfere with a movement being learned, and thus they have to be inhibited during its performance. The inhibition of interfering synergies or coordinations has to be elaborated in the process of motor learning as well. An example of such a situation is a rotation of both hands or forearms which may be easily performed in the same direction, but requires training for performance in different directions. Natural coordination providing rotation in the same direction is opposite to the coordination required for rotation in different directions, and should be suppressed in the process of learning. Some other examples of such suppression of natural coordinations in bimanual movements may be found in the current literature (e.g., Buchanan *et al.*, 1996; Wiesendanger *et al.*, 1994). In animals, the reorganization of innate motor reactions during instrumental learning was studied by Zelyony *et al.* (1937) who named the elaborated movements "heteroeffector reflexes". Also, neuronal mechanisms of triggering innate synergies (for instance, the placing reaction in cats) from unusual afferent inputs were studied (Kotlyar *et al.,* 1983; Mayorov, 1994).

The motor cortex has been shown to play a crucial role in the inhibition of innate coordination during its reorganization in the process of learning. Figure 16.1 shows two samples of the elaborated motor reactions including reorganized natural synergies or reflexes before and after motor cortex lesions. Panel I of Figure 16.1 shows stages of elaboration of an alimentary instrumental reaction when the dog has to stay in contact with a food cup by lifting the forelimb during eating (Popova, 1970). An innate synergy (lowering the limb during lowering the head into the feeder) interferes with the reaction being learned; thus it is inhibited in the process of learning (Figure 16.1, I; A,B,C), but is disinhibited after a lesion of the motor cortex of the contralateral hemisphere (Figure 16.1, I; D). Another example, a precise avoidance reaction (Ovsyannikov, 1967) is represented in Figure 16.1, Panel II. In the experimental procedure, if the dog lifts the limb above a "safety zone" in response to a conditioned stimulus, it receives an electrical stimulation of the limb, evoking a flexor reflex. To escape the shock, the dog has to inhibit the flexor reflex and to lower the limb into the "safety zone" (Figure 16.1,II; A). This learned inhibition of the flexor reflex is roughly disturbed after lesions of the contralateral motor cortex (Figure 16.1,II; B). In both experimental paradigms intensive retraining for some years does not result in recovery of inhibition of the interfering natural reactions.

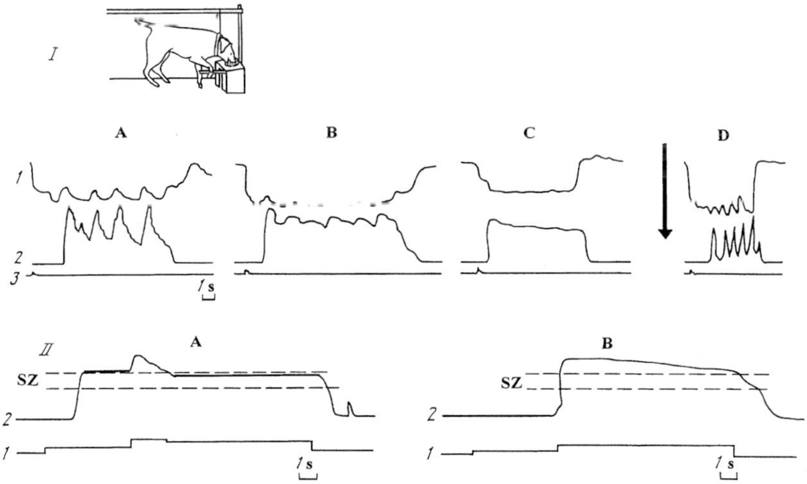

Figure 16.1. Two experimental models of rearrangement of innate coordination by learning in dogs. **I.** Alimentary instrumental reaction of lifting a cup of food and maintaining it during eating by a lifted forelimb; **A, B, C** are sequential stages of training. **D** is after a lesion of the contralateral motor cortex; **(1)** head movements, **(2)** movements of the "performing" limb, **(3)** mark of the moment when the cup of food may be available. **Above:** sketch of the experiment (see Pavlova and Alexandrov, 1992). **II.** Precise avoidance reaction; **A** is after training, **B** is after lesion of the contralateral motor cortex; **(1)** marks of conditioned (sound) and unconditioned (electrical stimulation of the limb) stimuli, **(2)** the limb movement. SZ—safety zone (Balezina, Varga, and Vasilyeva, 1990).

However, performance of the learned movement is in principle possible even after the motor cortex lesions, if the interfering synergies are excluded (for example, the dog is able to eat with the lifted limb from the elevated feeder when it does not have to lower its head [Ioffe, 1991a]).

Thus, besides the control of fine and precise movements of the distal parts of limbs, the motor cortex has another function, namely, inhibition of reflexes and synergies interfering with a movement being learned. This function is specific for the motor cortex; similar disturbances may be observed after combined lesions of parietal and premotor associative cortex (Pavlova *et al.*, 1986) or after cerebellar nuclei lesions (Balezina and Mats, 1995), but they disappear after 3–4 weeks of retraining.

The above data concern the functions of the motor cortex in controlling limb movements. However, limb movements are usually accompanied by postural adjustments, realized by proximal and axial musculature. According to classical neurophysiology, the proximal and axial muscles are controlled by the so-called extrapyramidal system, through the medial descending system including reticulospinal and vestibulospinal pathways (Fulton, 1949; Kuypers, 1964; Lawrence and Kuypers, 1968). Until relatively recently, the motor cortex was assumed not to be involved in the control of postural adjustments preceding and accompanying limb movements. That opinion was based, in particular, on the data obtained by Shumilina (1949) and Koryakin (1958) who observed dramatic disturbances of a learned limb movement but not postural adjustments after bilateral motor cortex lesions. In other experiments stimulation of the motor cortex caused limb lifting without postural adjustment, which resulted in the animal falling (Konorski, 1967; Tarnecki, 1962; Thomas, 1971; Wagner *et al.*, 1967). Various structures were believed to be responsible for the control of postural adjustments, particularly the basal ganglia (Martin, 1967; Shapovalova *et al.,* 1984), cerebellum (Massion, 1979), and brainstem reticular formation (Gorska *et al.*, 1996; Ioffe, 1991a; Koryakin, 1958; Shumilina, 1949). The red nucleus was shown to influence postural adjustment latency (Burlachkova and Ioffe, 1979). However, as early as the 1930's the postural placing reaction was found to be under cortical control (Bard, 1933), but a limb movement is involved in this reaction. Subsequent studies revealed that the motor cortex takes part also in the control of postural reactions which provide a shift of the centre of mass (Birjukova *et al.*, 1989; Ioffe, 1991b; Ioffe *et al.,* 1988; Massion, 1979). Motor cortex influences were found to be bilateral. Unilateral lesion of MCx resulted in changes of the support pressure force latency in all the limbs, though the changes were less pronounced than the latency increase of contralateral limb movements (Burlachkova and Ioffe, 1979). After motor cortex lesions the centre of mass displacement, which was previously ballistic, is realized step-by-step, under permanent afferent control (Birjukova *et al.,* 1989; Ioffe *et al.*, 1988). Figure 16.2 shows changes of the centre of mass displacement, velocity and acceleration in an intact animal (Panel A) and after motor cortex lesions (Panel B). One can see that the shift of the centre of mass was ballistic before the surgery, but after the lesion it had many peaks of acceleration and velocity change. Braking of the centre of mass, that is, stabilization of the body and equilibrium maintenance in the final position, is difficult as well. However, these disturbances of postural adjustment may be compensated after intensive retraining. As the bilateral latency increases even after unilateral motor cortex lesions, this suggests the modulation of cortical influences on some (perhaps, brainstem) structures controlling the postural adjustments. Possibly, the motor

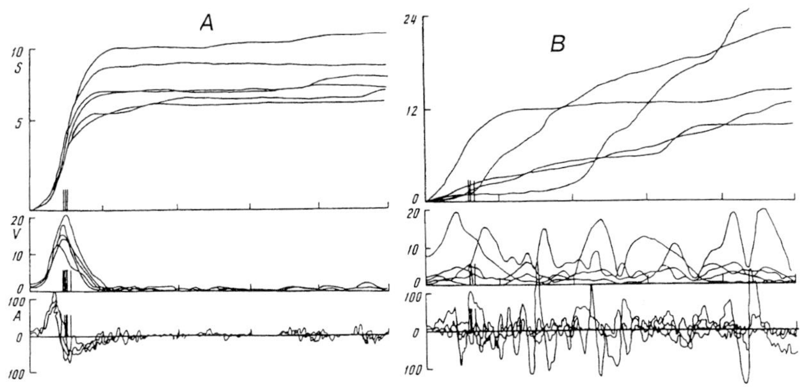

Figure 16.2. Changes of the horizontal displacement (S, cm), velocity (V, cm/s), and acceleration (A, cm/s/s) of the centre of gravity during some trials. **A** shows the well trained avoidance reaction (n = 6), **B** is after lesion of the contralateral motor cortex, before retraining (n = 5). The abscissa is time (s). The vertical bars correspond to the moment of the limb lifting off (Alexandrov *et al.*, 1991).

cortex coordinates and integrates the motor programs of limb movement and postural adjustment, producing a correlation between parameters of limb movement and postural adjustment.

The role of the motor cortex in the reorganization of natural postural synergies in the process of learning is a question of particular interest. As was mentioned above, inhibition of synergies interfering with a movement being learned is a specific function of the motor cortex. This has been shown for limb movements. However, it was not clear whether the motor cortex deals with postural synergies using similar mechanisms, taking into account that it controls mainly limb movement.

This question has been studied by using two experimental models of suppression of natural postural synergies during learning. One of the models is the reorganization of the so-called diagonal pattern of postural adjustment accompanying a limb movement. Usually, the limb diagonally opposite to the lifted one is unloaded whereas the other pair of diagonal limbs is loaded (Figure 16.3, Panel A). Through special training it is possible to train an alternative, the so-called unilateral (ipsilateral) pattern of postural adjustment: the limb ipsilateral to the lifted one is unloaded, whereas the other pair of ipsilateral limbs is loaded (Figure 16.3, Panel B). This pattern may be obtained during elaboration of a kind of avoidance reflex in which the electrical stimulation is applied on two ipsilateral limbs (in Figure 16.3, left forelimb and left hindlimb). The dog can avoid or escape the stimulation of one limb (in Figure 16.3, left forelimb) by its lifting and maintaining above a certain level (usually 5–7 cm) for a definite time (usually 4–5 s) whereas the stimulation of the other (ipsilateral) limb may be avoided or escaped by decreasing the support force of the limb by 10% of the initial level. As a result of such training, the diagonal pattern of the postural adjustment is suppressed and replaced by a unilateral one. The following formula proposed by A.A. Frolov is used for the estimation of degree of "diagonality" (D):

$$D = 1 - \frac{[\Delta F_1 - \Delta F_4] + [\Delta F_2 - \Delta F_3]}{[\Delta F_1] + [\Delta F_2] + [\Delta F_3] + [\Delta F_4]}$$

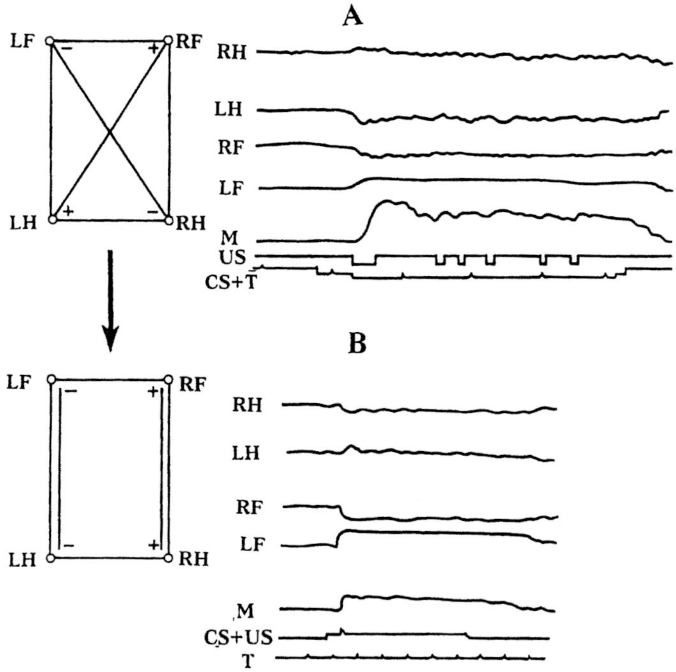

Figure 16.3. Rearrangement of the "diagonal" pattern of postural adjustment accompanying the avoidance reaction into a "unilateral" one. **A** shows the diagonal pattern of postural adjustment (simultaneous unloading vs. loading in pairs of the diagonal limbs); **B** shows the unilateral pattern of postural adjustment (simultaneous unloading vs. loading in pairs of ipsilateral limbs). On the left, schemes of the support forces changes, on the right, fragments of the recordings: LF, RF, LH, RH, force traces of the left and right forelimbs and left and right hindlimbs, respectively, M—trace of the limb movement, CS, US—marks of conditioned and unconditioned stimuli, T—time marks, s (Ioffe, 1991b).

where D is value of the coefficient of diagonality, and F_1, F_2, F_3, and F_4 are values of suppport forces of left and right forelimbs and left and right hindlimbs, respectively. The maximum D value equal to 1 corresponds to the lifting of two diagonal limbs (Gahery *et al.*, 1980).

Figure 16.4 represents the changes in D before (A) and after (B) rearrangement of the diagonal pattern of postural adjustment into the unilateral one, and D the changes during the process of the rearrangement (C). One can see that, before the rearrangement, D is maximal in the dynamic phase of postural adjustment, just before the limb lifts off. Then the value of D decreases, and in the static phase of the movement, during maintenance of the limb-lift and body stabilization the D value is equal to 0.3–0.4. After the end of rearrangement of the pattern of postural adjustment the D value is equal to zero, except for a short (100–200 ms) peak of diagonality in the dynamic phase of the postural adjustment. It was shown that this peak cannot be eliminated completely, for biomechanical reasons (centre of mass projection moving into the so-called "support triangle", arising after a lifting a limb, has to cross the "zone of diagonal pattern" [Frolov *et al.*, 1988]).

Another experimental model of suppression of an innate postural coordination in dogs is the avoidance of electrical stimulation of a limb by increasing the support force of the

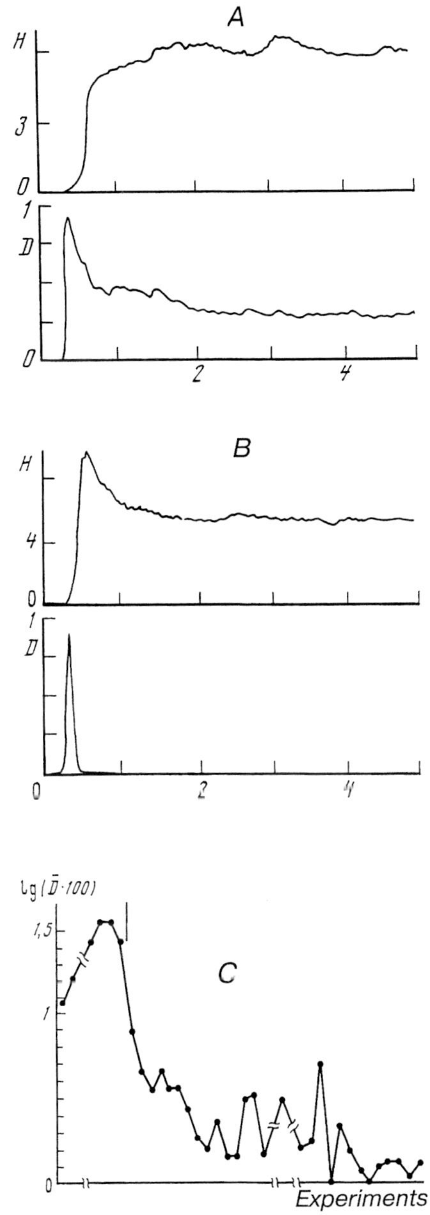

Figure 16.4. Changes of the diagonality coefficient (D) in the course of rearrangement of the "diagonal" pattern of postural adjustment into the unilateral one. **A**—changes of the limb movement amplitude (H, cm) and diagonality coefficient (D) in the well trained avoidance reflex. Average of 10 trials; **B**—the same, after the rearrangement of diagonal pattern of postural adjustment into unilateral. Average of 10 trials. Abscissa—time marks, s (Alexandrov *et al.*, 1991); **C,** changes of D in the course of rearrangement of the diagonal postural pattern into the unilateral one; abscissa—sequential sessions, ordinate—logarithm of mean values of D for a session, per cent (Balezina *et al.*, 1995).

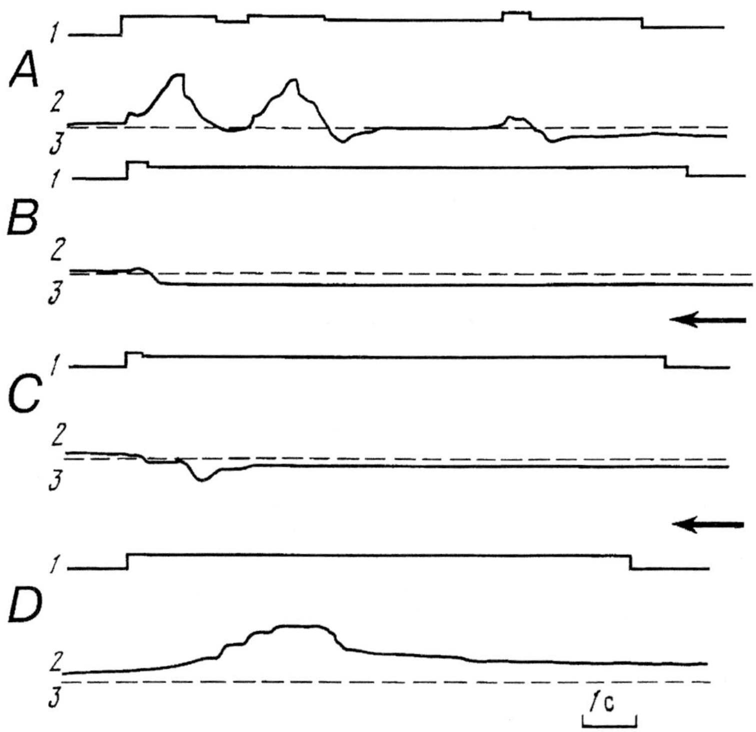

Figure 16.5. Postural escape reaction in the course of training (**A, B**), and after lesion of the motor cortex in the contralateral hemisphere (**C**) followed by the lesion of the motor cortex in the other hemisphere (**D**); 1—marks of the electrical stimulation, 2—support force and limb movement (shift of the line downwards corresponds to increase of the force, shift upwards corresponds to decrease of the force and limb lifting), 3—force level corresponding to switching off the stimulation (Balezina *et al.*, 1990).

limb above a certain level (usually 5 kg), or by a certain percent of the initial support force (Ivanova, 1984; see Figure 16.5 A,B). In this case, the flexor reflex has to be suppressed (as in the above precise avoidance reaction), and the initial pattern of postural adjustment (providing a shift of the centre of gravity from the stimulated limb) is rearranged to the opposite. The centre of gravity has to shift to the stimulated limb. This may be obtained during training, by a passive shift of the animal's body from the stimulated limb during the stimulation. This passive displacement provokes an active resistance reaction, and thus the body shifts to the stimulated limb and the stimulation stops. By the law of effect, the reaction is fixed.

Let us now consider the role of the motor cortex in reorganization of postural coordinations in the described experimental models. Lesion of the motor cortex in the hemisphere contralateral to the stimulated ("active") limb results in temporary disturbances of the reorganized pattern of postural adjustment, which may be compensated by retraining for 3–4 weeks (Figure 16.5 C; Figure 16.6 A). However, the following lesion of the motor cortex in the other hemisphere causes a stable disappearance of the learned postural pattern, and only the diagonal pattern of postural adjustment is observed. No recovery of the unilateral pattern may be obtained by retraining (Figure 16.5, D; Figure 16.6 A). Thus, the

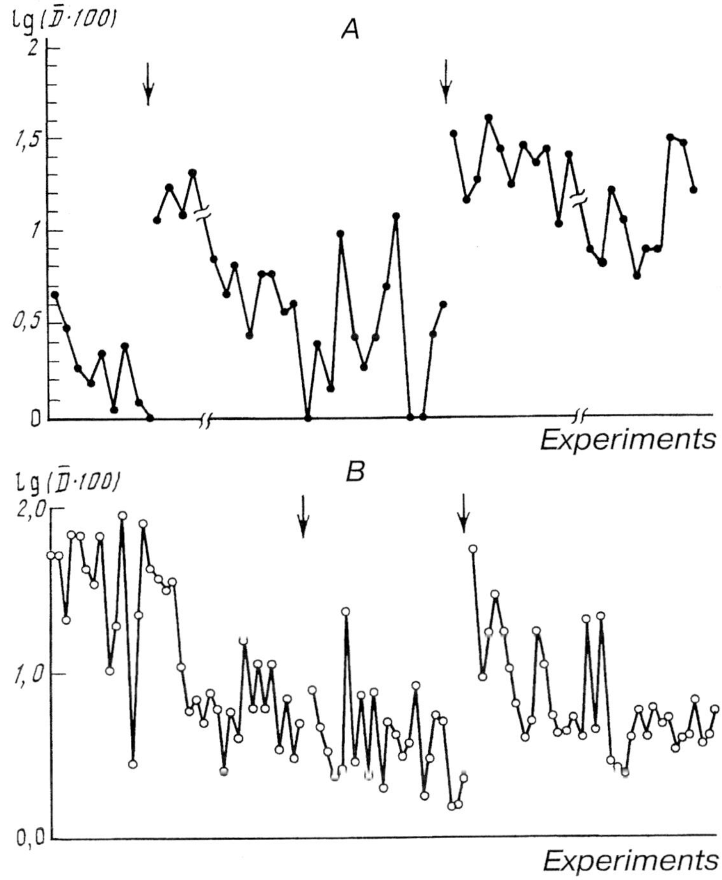

Figure 16.6. Dynamics of the diagonality coefficient before and after bilateral lesions of the motor cortex in different sequences. **A**—after lesion of the motor cortex in the contralateral hemisphere (first arrow) followed by the lesion in the ipsilateral hemisphere (second arrow) (Ioffe *et al.*, 1988); **B**—after opposite sequence of lesions (Ioffe *et al.*, 1996). Designations are as in Figure 16.4C.

motor cortex apparently inhibits the interfering postural coordinations during their rearrangement. However, in contrast to the control of limb movements, influences of the motor cortex during reorganization of postural coordinations are bilateral. This is understandable, taking into account that the pattern of postural adjustment is bilateral as well. Perhaps, in the process of learning, the influences of the motor cortex modulate the activity of brainstem generators, which are responsible for the organization of the pattern of postural adjustment. As a result, the innate postural pattern has been inhibited and substituted by the learned one.

The situation is, however, more complicated if the sequence of the motor cortex lesions after forming the unilateral postural pattern is the opposite (Ioffe, Vasilyeva and Mats, 1996). In this case, unilateral lesion of the motor cortex in the hemisphere ipsilateral to the lifted limb is followed by fast recovery of the learned postural pattern.

However, after the subsequent lesion of the motor cortex in the other hemisphere, recovery is possible as well (Figure 6 B). In these experiments, the movement was always performed by the left forelimb. Possibly, right and left hemispheres play different roles in the compensation. Specific experiments are necessary in order to check this suggestion. Other ways of explaining the phenomenon are possible as well. In any case, the problem of the function of the motor cortex in the reorganization of postural coordinations seems not to be finally solved and needs further investigations.

ACKNOWLEDGEMENT

The work was supported by Russian Foundation of Basic Research, project # 96–04–49412 and by the grant of INTAS-RFBR No. 95–1327.

REFERENCES

Alexandrov, A.V., Vasilyeva, O.N., Ioffe, M.E., and Frolov, A.A. (1991) Some modes of description of different patterns of postural adjustment during motor learning in dogs. *Zhurnal Vysshey Nervnoy Dejatelnosty*, **41**, 937–947.

Asanuma, H. (1989) *The motor cortex.* New York: Plenum Press.

Baker, S.N., Olivier, E., and Lemon, R.N. (1995) Task-related variation in corticospinal output evoked by transcranial magnetic stimulation in the macaque monkey. *Journal of Physiology, London*, **488**, 795–808.

Balezina, N.P., Varga, M.E., Vasilyeva, O.N., Ivanova, N.G., Ioffe, M.E., Pavlova, O.G., and Frolov, A.G. (1990) A study of mechanisms of reorganization of motor coordinations during motor learning. In: M.G. Airapetyants (Ed.), *Brain and behaviour* (in Russian). Moskow: Nauka, p. 105–119.

Balezina, N.P. and Mats, V.N. (1995) Participation of the cerebellar nuclei in elaboration and realization of a learned motor coordination in dogs. In V.V. Fanardjian (Ed.), *Cerebellum and Brainstem Structures* (in Russian). Erevan: Gitutyun, p. 184–191.

Bard, P. (1933) Studies on the cerebral cortex. 1. Localized control of placing and hoppping reactions in the cat and their normal management by small cortical lesions. *Archives of Neurology and Psychiatry*, **30**, 40–74.

Birjukova, E.V., Dufosse, M., Frolov, A.A., Ioffe M., and Massion, J. (1989) Role of the sensorimotor cortex in postural adjustment accompanying a conditioned paw lift in the standing cat. *Experimental Brain Research*, **78**, 588–596.

Buchanan, J.J., Kelso, J.A. and Fuchs, A. (1996) Coordination dynamics of trajectory formation. *Biological Cybernetics*, **74**, 41–56.

Bucy, P.C., Ladpli, R. and Ehrlich, A. (1966) Destruction of the pyramidal tract in monkeys. The effects of bilateral section of the cerebral peduncles. *Journal of Neurosurgery*, **25**, 1–23.

Burlachkova, N.I., and Ioffe, M.E. (1979) The analysis of the postural adjustment accompanying a local movement. *Agressologie*, **20**, 141–142.

Donoghue, J.P., Hess, G., and Sanes, J.N. (1996) Substrates and mechanisms for learning in motor cortex. In J.M. Bloedel, T.J. Ebner, and S.P. Wise (eds.), *The acquisition of motor behavior in vertebrates.* Cambridge, Mass: MIT Press, p. 363–386.

Fitts, P.M. (1954) The information capacity of the human motor system in controlling the amplitude of movement. *Journal of Experimental Psychology*, **67**, 381–391.

Frolov, A.A., Birjukova, E.V. and Ioffe, M.E. (1988) On the influence of movement kinematics on the support pressure pattern during postural adjustment of quadrupeds. In: V.S. Gurfinkel, M.E. Ioffe, J. Massion, and J.-P. Roll (eds.), *Stance and motion: Facts and concepts.* New York: Plenum Press, pp. 227–238.

Fulton, J.F. (1949) *Physiology of the nervous system.* 3rd Ed. New York: Oxford University Press.

Gahery, Y., Ioffe, M.E., Massion, J. and Polit, A. (1980) The postural support of movement in cat and dog. *Acta Neurobiologiae Experimentalis*, **40**, 741–756.

Gorska, T., Ioffe, M., Zmyslowski, W., Bem, T., Majczynski, H., and Mats, V.N. (1996) Unrestrained walking in cats with medial pontine lesions. *Brain Research Bulletin*, **38**, 297–304.

Grafton, S.T., Woods, R.P. and Tuszka, M. (1994) Functional imaging of procedural motor learning: Relating cerebral blood flow with individual subject performance. *Human Brain Mapping*, **1**, 221–234.

Grafton, S.T., Hazeline, E. and Ivry, R. (1995) Functional mapping of sequence learning in normal humans. *Journal of Cognitive Neuroscience*, **7**, 497–510.

Hallett, M., Pascual-Leone, A. and Topka, H. (1996) Adaptation and skill learning: Evidence for different neural substrates. In: J.M. Bloedel, T.J. Ebner and S.P. Wise (eds.), *The acquisition of motor behavior in vertebrates*, Cambridge Mass.: MIT Press, pp. 289–302.

Humphrey, D.R., Qiu, X.Q., Clavel, P. and O'Donoghue, D.L. (1990) Changes in forelimb motor representation in rodent cortex induced by passive movements. *Society for Neuroscience Abstracts*, **16**, 422.

Ioffe, M.E. (1973) Pyramidal influences in establishment of new motor coordinations in dogs. *Physiology and Behavior*, **11**, 145–153.

Ioffe, M.E. (1975) *Corticospinal mechanisms of instrumental motor reaction* (in Russian). Moscow: Nauka.

Ioffe, M.E. (1991a) *Mechanisms of motor learning* (in Russian). Moscow: Nauka.

Ioffe, M.E. (199b) On the mechanisms of motor learning. *Zhurnal Vysshey Nervnoy Dejatelnosty*, **41**, 907–921.

Ioffe, M.E., Ivanova, N.G., Frolov, A.A., Birjukova, E.V., and Kiseljova, N.V. (1988) On the role of motor cortex in the learned rearrangement of postural coordinations. In V.S. Gurfinkel, M.E. Ioffe, J. Massion, and J.-P. Roll (eds.), *Stance and motion: Facts and concepts* (in Russian). New York: Plenum Press, pp. 213–226.

Ioffe, M.E., and Samoylov, M.I. (1967) On the role of pyramidal pathways and red nucleus in movement control. In: *Recovery of functions after lesions of central and peripheral nervous system* (in Russian). Leningrad: Nauka, p. 46.

Ioffe, M.E., Vasilyeva, O.N. and Mats, V.N. (1996) Recovery of a reorganized pattern of bilateral motor cortex lesions depends on the sequence of the lesions. In: G.N. Gantchev, V.S. Gurfinkel and D. Stuart (eds.), *Motor control-YIII*. Sofia:Academic Publ. House, 160–163.

Ivanova, N.G. (1984) Conditional reflex rearrangement of an unconditioned limb flexion into extension in the dog. *Zhurnal Vysshey Nervnoy Dejatelnosty*, **34**, 657–661.

Jenkins, I.N., Brooks, D.J., Nixon, P.D., Frackowiak, R.S.J. and Passingham, R.E. (1994) Motor sequence learning: A study with positron emission tomography. *Journal of Neuroscience*, **14**, 3775–3790.

Karni, A., Meyer, G., Jezzard, P., Adams, M., Turner, R. and Ungerleider, L.G. (1994) Where practice makes perfect: A fMRI study of long-term motor cortex plasticity associated with the acquisition of a motor skill. *Society for Neuroscience Abstracts*, **20**, 1291.

Konorski, J. (1967) *Integrative activity of the brain*. Chicago: University of Chicago Press.

Koryakin, M.F. (1958) On structure of positional excitations in conditional defensive reflex in the dog. *Fiziol. Zhurnal SSSR*, **44**, 393–403.

Kotlyar, B.I., Mayorov, V.I., Timofeeva, N.O., and Shul'govsky, V.V. (1983) *Neuronal organization of the conditional reflex behaviour* (in Russian). Moscow: Nauka.

Kuypers, H.G.J.M. (1964) The descending pathways to the spinal cord, their anatomy and function. *Progress in Brain Research*, **11**, 178–202.

Lawrence, D.G., Kuypers, H.G.J.M. (1968) The functional organization of the motor system in the monkey. *Brain*, **91**, 1–36.

Mayorov, V.I. (1994) Mechanisms of development of neuronal reactions in cat's motor cortex associated with triggering the conditioned placing reaction: a hypothesis. *Zhurnal Vysshey Nervnoy Dejatelnosty*, **44**, 963.

Massion, J. (1979) Role of the motor cortex in the postural adjustment associated with movements. In H. Asanuma and V.J. Wilson (Eds.), *Integration in the nervous system* . Tokyo: Igaku-Shoin, pp. 239–260.

Massion J. (1985) Central coordination of posture and movement. In: A.S. Batuev (ed.) *Associative Systems of the Brain* (in Russian) Nauka, Leningrad, p. 18–24.

Martin, J.P. (1967) *The basal ganglia and posture*. London: Pitman.

Milliken, G., Nudo, R., Grenda, R., Jenkins, W.M., and Merzenich, M.M. (1992) Expansion of distal forelimb representations in primary motor cortex of adult squirrel monkeys following motor training. *Society for Neuroscience Abstracts*, **18**, 214.

Nudo, R.J., Jenkins, W.M., and Merzenich, M.M. (1990) Repetitive microstimulation alters the cortical representation of movements in adult rats. *Somatosensory and Motor Research*, **7**, 463–483.

Nudo, R.J., Wise, B.M., Fuentes, F.S. and Milliken, G.W. (1996) Neural substrates for the effects of rehabilitative training of motor recovery after ischemic infarct. *Science, New York*, **272**, 1791–1794.

Ovsyannikov, A.V. (1967) *The influence of joint and dorsal root deafferentation on motor conditional defensive reflexes of different degree of precision*. Ph.D. Thesis (in Russian) Moscow: Institute of Higher Nervous Activity and Neurophysiology.

Pascual-Leone, A., Cammarota, A., Wasserman, E.M., Brasil-Neto, J.P., Cohen, L.G. and Hallett, M. (1993) Modulation of motor cortical outputs to the reading hand of Braille readers. *Annals of Neurology*, **34**, 33–37.

Pascual-Leone, A., Grafman, J. and Hallett, M. (1994) Modulation of cortical motor output maps during development of implicit and explicit learning. *Science, New York*, **263**, 1287–1289.

Pavlova, O.G. and Alexandrov, A.V. (1992) Head-forelimb movement coordination and its rearrangement by learning in the dog. The role of motor cortex. In: A. Berthoz, W. Graf and P.P. Vidal (eds.), *The head-neck sensory-motor system.* New York: Oxford University Press, New York, pp. 591–596.

Pavlova, O.G., Balezina, N.P. and Ioffe, M.E. (1986). Disorganization of a learned coordination after combined lesion of parietal and premotor associative areas in dogs. *Zhurnal Vysshey Nervnoy Dejatelnosty*, **36**, 450–459.

Phillips, C.G. and Porter, R. (1977) *Corticospinal neurons: Their role in movement.* London: Academic Press.

Popova, E.I. (1970). *Instrumental motor reflexes in aspect of the theory of conditional reflexes.* D. Sci. Thesis (in Russian). Moscow: Institute of Higher Nervous Activity and Neurophysiology.

Qiu, X.Q., O'Donoghue, D.L., and Humphrey, D.R. (1990) NMDA-antagonist (MK-801) blocks plasticity of motor cortex maps induced by passive limb movement. *Society for Neuroscience Abstracts*, **16**, 422.

Sadato, N., Ibanez, V., Deiber, M-P., Campbell, J., Leonardo, and M., Hallett, M. (1996) Frequency-dependent changes of regional blood flow during finger movements. *Journal of Cerebral Blood Flow and Metabolism*, **16**, 23–33.

Sanes, J. and Donoghue, J. (1992) Organization and adaptability of muscle representations in primary motor cortex. *Experimental Brain Research, Supplement*, **22**, 103–127.

Sanes, J.N., Wang, J. and Donoghue, J.P. (1992) Immediate and delayed changes of rat cortical output representation with new forelimb configurations. *Cerebral Cortex*, **2**, 141–152.

Seitz, R.J., and Roland, P.E. (1992) Learning of sequential finger movements in man: A combined kinematic and positron emission tomography (PET) study. *European Journal of Neuroscience*, **4**, 154–165.

Shapovalova, K.B., Yakunin, I.V. and Boiko, M.I. (1984) Participation of head of caudate nucleus in mechanisms of conditional postural displacement. *Zhurnal Vysshey Nervnoy Dejatelnosty*, **34**, 669–677 (in Russian).

Shumilina, A.I. (1949) On participation of pyramidal and extrapyramidal systems in motor activity of a deafferented limb. In: P.K. Anokhin (ed.), *Problems of higher nervous activity* (in Russian). Moscow: AMN SSSR, pp. 176–185.

Tarnecki, R. (1962) The formation of instrumental conditioned reflexes by direct stimulation of sensori-motor cortex in cats. *Acta Biologiae Experimentalis*, **22**, 35–45.

Thomas, E. (1971) Role of postural adjustments in conditioning of dogs with electrical stimulation of the motor cortex as the unconditioned stimulus. *Journal of Comparative and Physiological Psychology*, **76**, 187–198.

Tower, S.S. (1940) Pyramidal lesion in the monkey. *Brain*, **63**, 36–90.

Wagner, A.R., Thomas, E. and Norton, T. (1967) Conditioning with electrical stimulation of motor cortex: evidence of a possible source of motivation. *Journal of Comparative and Physiological Psychology*, **64**, 191–199.

Wiesendanger, M., Wicki, U. and Roullier, E. (1994) Are there unifying structures in the brain responsible for interlimb coordination? In: S.P. Swinnen, H. Heuer, J. Massion and P. Casaer (Eds.), *Interlimb coordination: Neural, dynamical, and cognitive constraints.* San Diego, CA.: Academic Press. Diego, 180–207.

Zelyony, G., Vyssotsky, N., Dobrotina, G., Irzhanskaya, K., Medyakov, F., Naumov, S., Poltyrev, S., and Tuntsova, E. (1937) Kinds and ways of elaboration of associative reflexes. *Proc. YI All-Union Congress of Physiol., Biochem., Pharmacol.* (in Russian). Tbilisi, pp. 165–171.

17 Biochemical Correlates of Individual Behaviour

Gulyaeva N.V. and Stepanichev M.Yu.

Institute of Higher Nervous Activity and Neurophysiology, Russian Academy of Sciences, Moscow, Russia
e-mail: gul@fbns.msk.ru

This article is a review of data (including results of the authors' investigations and data from the literature) concerning the neurochemical correlates of individual behavior in rats. The "emotional resonance" test was used for behavioural selection of rats. Individual behaviour in this test is related to differences in free radical-mediated processes, membrane lipid content, nitric oxide synthase activity and the cAMP pattern in cerebral macrostructures. Behaviour-related differences are sometimes revealed as stable traits. However, most of them are reactive (induced by significant external factors). These differences depend on age. They may be global, specific for selected brain regions, and/or related to interhemispheric lateralization of biochemical parameters.

KEYWORDS: behaviour, behavioural types, stress, brain ischemia, brain biochemistry, free radical-mediated processes, nitric oxide synthase

Individual characteristics of the nervous system become apparent in individual behaviour (IB). Investigation of IB of animals reveals not only individual phenomenology but also the mechanisms underlying IB. Elucidation of molecular mechanisms participating in the formation of IB is important both for basic neurobiology and for applied science. Since rodents are used in most studies, one of the problems related to IB is the search of approaches and methods for evaluating the main parameters which characterize the rodent nervous system.

The "open field" test is widely used as a method of evaluating spontaneous rodent behaviour, based on the exploratory component of behaviour (Bures *et al.*, 1983; Markel, 1981). In this test, levels of locomotor activity (both horizontal and vertical) and of emotionality appear to reflect the functional state of different parts of the nervous system, and to be closely related to other forms of behavior, learning, and memory. Emotionality in this test is evaluated as elements of vegetative activity (defecation, urination), and forms of displacement activity. However, the evaluation of emotionality in the "open field" test is quite difficult because neither the number of defecations and urinations nor motor activity are adequate indices of the emotional status of rodents. The level of motor activity is sometimes used as an index of emotionality, but sometimes results in opposite interpretations of similar data obtained by different authors. D. Kulagin (1975) revealed inverse correlations of locomotor activity with the force of excitatory processes, as well as of emotionality with the average index of conditioning. Whereas this author demonstrated a positive correlation between the locomotor activity and the conditioning rate, G. Chaichenko (1982) revealed a negative correlation between these parameters.

Maximal differences in behaviour may be revealed when animals are exposed to definite biologically significant factors including spatial ones (which are the most important ones in the "open field" test), and the outward appearance and sound signals from another animal. Such types of influence induce qualitatively different responses in different animals, including running away, hiding and aggression (Ajrapetyanz *et al.*, 1980; Khonicheva and Iljna Wiljar, 1981). The "emotional resonance" (ER) test elaborated by P.V. Simonov (1976) is one of the approaches using these influences. This test includes factors of significance for rats: the choice between large and light or small and dark chambers, as well as signals of pain (and maybe also the smell) from another rat. The rat is put into the large light part of the test apparatus, this part, as a rule, being about 37 × 18 × 11 cm. The apparatus also contains an adjacent small dark part (about 18 × 18 × 11 cm). The rat has the option of staying in the light space or leaving it. Each time the rat passes from the light to the dark part, the cry of the "victim" begins. This is emitted by a rat subjected to inescapable footshocks. It is arranged that the cry ceases as soon as the tested rat leaves the dark space. The session usually lasts for 300 sec. The number of passages between parts of the apparatus, and the total time in the light and dark parts are recorded. The IB of rats in the ER test is believed to be a result of differences in the inborn strength of passive avoidance responses. These differences become apparent in rats when they are in conflict with the preference for closed spaces (Khonicheva and Iljna Wiljar, 1981). The willingness of rats to avoid closed spaces may be more typical for some rats than for others, the degree of its expression being dependent on the strain of rat (Meshcherjakova, 1988).

The parameters of behaviour in the ER test make it possible to select quickly and reliably groups of animals with different IB. Correlations between IB in the ER test and food or avoidance conditioning were revealed by Khonicheva (1984). Rats which expressed ER (avoiding the cries of pain of another rat) have a strong, steady and mobile nervous system (according to the results of conditioning with drinking and defense reinforcement), whereas rats without ER have features of strong, steady and inert (conditioning with defense reinforcement) and weak (drinking reinforcement) nervous system (Semagin *et al.,* 1988). Rats with intermittent ER have features of weak (conditioning with defense reinforcement) and unstable excitable (drinking reinforcement) nervous system, their behavior reflecting maximal strain of the emotional conflict (Simonov, 1976). P.V. Simonov (1987) suggests that individual behavior in the ER test is comparable with indices of psychoticism-neuroticism and extra-intraversion of Eysenk's scale. It is this comparability which gives importance to the ER model from the point of view of individual resistance in extreme conditions.

The phenomenon of behavioral "parasitism" (Mowrer, 1940; Masur and da Silva, 1977) is another example of a situation where interaction between two animals depends on specific individual features of the nervous system. Behavior in the situation of competition for food and its procurement correlates with individual behavior in the "open field" and ER tests (Khromova, 1995).

Functioning of the central nervous system is mediated by biochemical processes at the levels of the brain, neurones, and non-neural cells, and at the subcellular level. Animals with different IB appear to differ significantly in energy metabolism, protein metabolism and other metabolic processes in the brain and in other organs (Gershtein *et al.,* 1991; Krakovski, 1987; Krakovski *et al.*, 1986). M. Krakovski (1987) showed that, in rabbits with a strong mobile nervous system, activities of the main enzymes of energy metabolism (pyruvate dehydrogenase, isocitrate dehydrogenase, lactate dehydrogenase, NADH-dehydrogenase, glucose-6-phosphate dehydrogenase and 6-phosphogluconate dehydrogenase) in the neocortex and kidney are higher than in animals with a nervous system of the weak type. In spite of these quantitative differences, functioning of enzymatic ensembles determining the efficacy of energy-providing reactions is well balanced in animals representing different behavioural types. L.M. Gershtein *et al.* (1991) showed that a low level of locomotion of rats in the "open field" test correlates with high activities of aminopeptidase, glucoso-6-phosphate dehydrogenase and glutamate dehydrogenase in the sensorimotor cortex, as well as with low activities of acetyl cholinesterase and monoamine oxidase. Animals displaying high levels of locomotor activity show the opposite biochemical features. In animals with low activity in the "open field" the blood levels of triglycerides, free fatty acids, phospholipids and lipoproteins as well as the level of mitochondrial phosphorylation in liver were low. In these animals the concentrations of protein, cytochromes P_{450} and b_5, as well as liver cytochrome oxidase activity, were lower than those in highly mobile animals (Krakovski *et al.*, 1986).

An animal's behavioural type is also closely related to specific features of monoamine metabolism, and of regional monoamine distribution. The activity of tyrosine hydroxylase, the key enzyme of catecholamine biosynthesis, in the striatum of rats, was twice as high in animals active in the "open field" as in passive rats (Bondarenko *et al.*, 1981). Many results suggest that there exist positive correlations of noradrenergic and dopaminergic brain system activities with locomotor activity in the "open field", as well as negative correlations of cholinergic and serotoninergic system activities with such locomotor activity (Jeste and Smith, 1980; Kulagin and Bolondinski, 1986; Strek, 1989). Turnover of serotonin in the hypothalamus and hippocampus is increased in rats with a pronounced avoidance reaction of their partner's cry in the ER test, in comparison with rats that form this reaction slowly (Getsova and Orlova, 1993). Variation of monoamine content in the brain, induced by administration of the serotonin precursor 5 hydroxytryptophan, or the norepinephrine synthesis inhibitor disulfiram, affects the ER reaction, and dramatically changes an animal's behavioural type (Kruglikov *et al.*, 1995). Conditioning of rats in the ER test is mediated by the cortical cholinergic system, since M-cholinolytics depress such conditioning (Burov and Speranskaya, 1974).

Many groups have studied free radical mediated oxidation (FRMO) processes in brain, blood and other tissues of rats with different IB (Bondarenko *et al.*, 1985; Gulyaeva, 1989; Levshina and Gulyaeva, 1991). "Emotional" rats, selected in the "open field" test, displayed higher FRMO products content in the brain, liver and heart, both before and after the acute stress (as compared with "non-emotional" rats) (Bondarenko *et al.*, 1985). Superoxide dismutase activity was low in "emotional" rats, suggesting a deficiency of antioxidant defence in these rats. Similar results of FRMO investigation in Tryon rats, selected for high (maze-bright) or low (maze-dull) learning ability in a T-maze, gave similar results (Gulyaeva, 1989). "Emotional" maze-bright rats demonstrated low locomotor activity and many defecations in the "open field", and had a higher level of FRMO products in the brain than "non-emotional" maze-dull rats.

Since 1989 we began to study neurochemical correlates of IB systematically. The first investigation to become the basis of this study was carried out mainly in the Laboratory of Functional Biochemistry of the Nervous System (Institute of Higher Nervous Activity, Russian Academy of Science) and was the experiment described by I. Levshina and N. Gulyaeva (1991). They studied the lipid component of cerebral membranes and FRMO, in mongrel white rats with different behaviour in the ER and the "open field" test, as well as changes of FRMO in the brain and blood of these rats during the initial phase of emotionally painful stress. The level of FRMO products was higher in rats demonstrating the ER phenomenon than in rats with weakly expressed ER. Differences in superoxide scavenging activity were not revealed (Levshina and Gulyaeva, 1991). Cholesterol content and cholesterol-to-phospholipid ratio were significantly higher, and the total phospholipid content was lower in the brains of rats displaying the ER phenomenon. Ratios of the above parameters in left and right hemispheres (asymmetry coefficients) were similar in both groups, although the content of FRMO products was higher in the left hemisphere. Footshock induced a decrease of FRMO products, both in the blood and in the brain, the degree of this decrease being similar in both behavioural groups. However, cerebral lateralization of FRMO products changed in opposite directions: In rats with a weakly expressed ER, the left asymmetry of FRMO products and cholesterol increased, and in rats demonstrating ER the content of both FRMO products and cholesterol became higher in the right hemisphere (that is, a stress-induced "right" asymmetry appeared).

Based on these results the following principles were used to reveal some of the neurochemical mechanisms underlying IB:

1) *A single session of ER was used as a tool for quick and effective selection of rats to form groups with different IB* (Simonov, 1976).

The following groups of rats were chosen: rats not displaying the ER phenomenon (type 1), rats demonstrating ER (type 2), rats with an intermittent reaction (type 3), and passive rats that did not choose any behavioral strategy in the ER test (type 4) (Stepanichev, 1996).

2) Significant functional biochemical parameters were used to study the most important processes in the brain: These included FRMO and antioxidant defence systems of the brain, lipid content of cerebral membranes, nitric oxide (NO) synthase activity, and levels of cyclic nucleotides.

FRMO is the natural mechanism of modification of membrane lipids underlying changes of functional membrane characteristics (Halliwell and Gutteridge, 1989). Free radical-mediated regulation of cellular metabolism is closely related to other regulatory systems (Burlakova, 1976).

The study of brain second messenger systems, including the cAMP system, appears to be important in the search for neurochemical correlates of IB. The functional activity of a nerve cell is believed to be determined by the functioning of its receptors. Since most hormones and neurotransmitters have receptors coupled with adenylate cyclase, changes of cAMP content reflect variations of cellular activity as a response to stimulation of its receptors (Salomon, 1991).

Active oxygen species influence mechanisms of signal transduction (Halliwell and Gutteridge, 1989), and special attention is being paid to the role of NO in the brain. NO is a free radical, and therefore should have an influence on the steady state of FRMO processes; however, NO also acts as a second messenger in neurones (Vincent, 1994). Thus, NO represents a link between FRMO and mechanisms of signal transduction.

3) *Since the brain is a heterogeneous organ, the most important macrostructures thought to be related to IB were analyzed.*

P.V. Simonov elaborated a concept concerning the genesis of individual behavioural differences, based on numerous experiments with extirpation and damage of cerebral structures (Simonov, 1987). Simonov suggests that the interaction of the elements of the so-called "informational system" (frontal cortex and hippocampus) and the "motivational system" (amygdala and hippocampus) in the brain determines the specific features of IB. In our studies, investigation of biochemical parameters in these brain macrostructures, and, in the case of cerebral ischemia, also of the cerebellum (which is a main target of ischemic damage) made it possible to reveal the dependence of these parameters on localization in the brain.

4) *Assuming that differences in interhemispheric asymmetry may underlie differences in IB, samples of left and right hemispheres were analyzed separately.*

We suggested that the poor reproducibility of results concerning interhemispheric asymmetry of different biochemical parameters might be caused by the fact that stable (but different) asymmetry is characteristic of different types of IB. The above-mentioned results supporting this suggestion (Levshina and Gulyaeva, 1991) agree with data on the dominance of the right hemisphere in rats which readily formed the ER reaction (Murik, 1990; Bianki *et al.*, 1985). Dominance of the left hemisphere was demonstrated in rats which formed the ER reaction poorly, and, in rats with no dominance of any hemisphere, the ER reaction cannot be formed. Thus, the ER phenomenon is closely related to the problem of brain interhemispheric asymmetry, suggesting the importance of interhemispheric interaction in IB.

5) *It appeared reasonable to reveal reactive neurochemical changes as a response to moderate stress (acute immobilization or footshock) or more extreme factors (global cerebral ischemia resulting from cardiac arrest—a model of postresuscitation pathology).*

IB is a significant determinant of the adaptation of the organism to changing external factors. Specific inborn features of higher nervous activity are believed to determine the activity of the organism under extreme conditions (e.g. under stress) when quick changes of the functional state of an organ or organ system are necessary (Ajrapetyanz and Vein, 1982). Numerous studies have demonstrated that IB in the "open field" or ER tests are closely related to specific features of the stress reaction (Ajrapetyanz *et al.*, 1980; Khonicheva and Iljna Wiljar, 1981; Semagin *et al.*, 1988). IB of rats in these tests can be used to predict individual resistance to stress (Yumatov and Meshcheryakova, 1990). The latter was shown to correlate with quantitative indices of exploratory behavior, and the ratio of vertical to horizontal locomotor activitiy could be used to predict stress resistance (Yumatov, 1986). Differences in stress resistance were determined by specific features of the central neurochemical organization of emotional states, which became apparent in the specific distribution of biogenic amines in different brain structures in stress-resistant, stress-sensitive and adaptable animals. However, animals displayed different resistance and chose different adaptation strategies depending on stress type.

6) *Age was taken into account as a factor directly influencing IB.*

Changes of IB in ontogenesis have not been studied extensively, although there have been several investigations of this topic (Frolkis *et al.*, 1991; Troshikhin *et al.*, 1971).

Changes of IB during aging have been studied least. Aging is accompanied by biochemical alterations at all levels of the organism and by changes in functioning of all physiological systems, including the central nervous system (Cutler, 1985; Frolkis *et al.*, 1991). Age-related changes of morphological and biochemical processes in the central nervous system induce alterations in steadiness and mobility of nervous processes, causing specific age-related features of conditioned reactions. Modification of brain plasticity during aging is the result of structural changes, such as neurodegeneration mediated by active oxygen species (Frolkis *et al.,* 1991; Gershon, 1988). Cholinergic and/or dopaminergic deficit accompanying neurodegeneration results in impairment of learning and memory processes (Abdulla *et al.*, 1995; Frolkis *et al.*, 1991) as well as motor function (Janicke and Wrobel, 1984). Age-related shifts of stable emotional states result in the dominance of negative emotions (Rushkevich, 1989). Thus, age-related changes can cause specific features of IB demonstrated in standard tests.

The use of a complex methodological approach (1–6) made it possible to obtain results forming the experimental basis for the following principal conclusions on neurochemical correlates of IB.

1) IB of rats in the ER test is related to differences in FRMO processes (Levshina and Gulyaeva, 1991; Stepanichev *et al.*, 1995a), *membrane lipid component* (Levshina and Gulyaeva, 1991; Stepanichev, 1995), *NO-synthase activity* (Gulyaeva *et al.*, 1994; Onufriev *et al.,* 1995b; Stepanichev *et al.,* 1997), *and cAMP ratio in brain macrostructures* (Egorova *et al.*, 1995; Stepanichev *et al.*, 1996).

2) Neurochemical differences characteristic of different types of IB may be expressed as stable traits, or be reactive (induced by external factors).

As a rule, many more IB-related neurochemical differences could be demonstrated as reactive changes than as differences in unstressed animals. In unstressed adult and old Wistar rats differences in cortical FRMO and membrane lipids could hardly be demonstrated (Stepanichev *et al.,* 1995a; Stepanichev, 1995), although the cerebral and blood content of FRMO products was higher in adult mongrel rats demonstrating ER, as compared with rats expressing ER poorly (Levshina and Gulyaeva, 1991). Cortical cholesterol content and cholesterol-to-phospholipid ratio were higher, and the total phospholipid content lower in rats with ER.

The type of IB determined the details of brain FRMO changes under stress. Acute immobilization induced an increased generation of active oxygen species (expressed most in the right neocortex), and modifications of membrane lipids dependent on IB in the ER test. Animals that did not choose a behaviour strategy in the ER test (type 4), expressed a stress-response more strongly than rats preferring the dark chamber (type 1) (Stepanichev, 1996).

The cAMP content in hypothalamus, frontal cortex, and amygdala of adult and old Wistar rats with different type of behavior in the ER test did not differ. However, in the hippocampus of old rats preferring the dark chamber cAMP content was higher (Egorova *et al.,* 1995). Stress-induced accumulation of cAMP was related to IB (Egorova *et al.*, 1995; Stepanichev, 1996).

NO-synthase activity was lower, though generation of active oxygen species was higher in brain structures of old Wistar rats expressing ER (type2) as compared to rats without ER (type 1) (Gulyaeva *et al.,* 1994; Onufriev *et al.,* 1995b) (see Figure 17.1). Similar differences of NO-synthase activity were revealed in adult rats (Stepanichev, 1996).

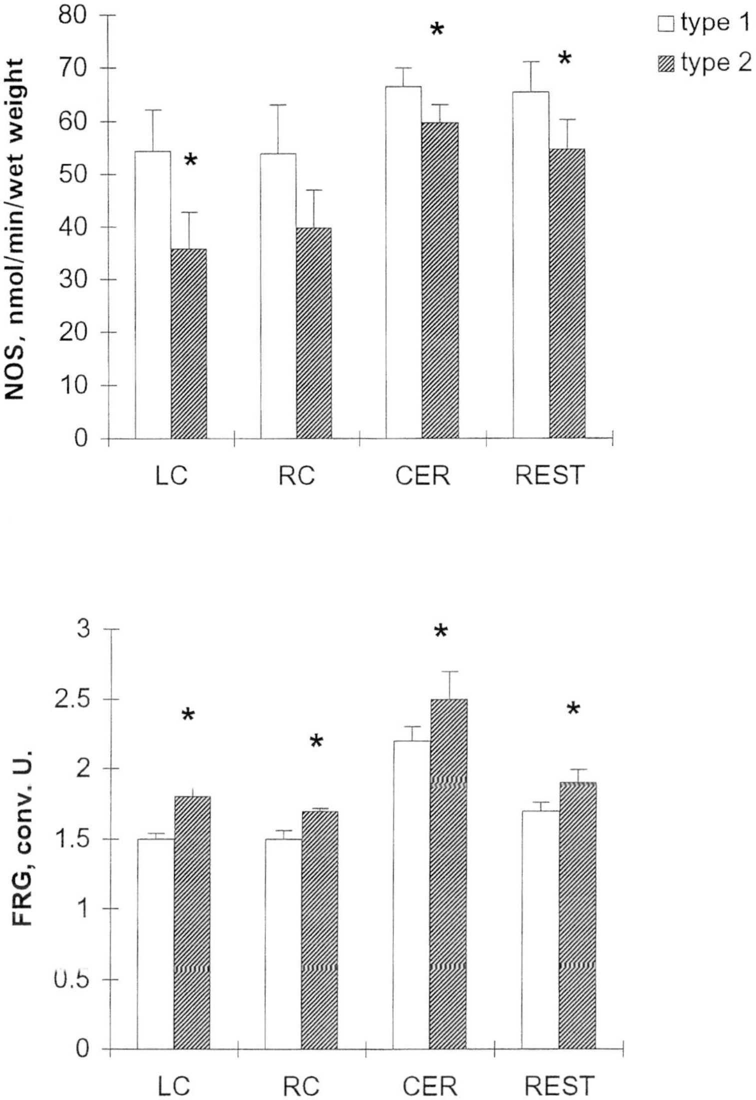

Figure 17.1. 1. Nitric oxide synthase (NOS) activity and free radical generation (FRG) in brain regions of old (23 months) rats of different behavioural types in the ER test. **LC**—left cortex, **RC**—right cortex, **CER**—cerebellum. *—significant difference between types, P < 0.05. For experimental conditions see Gulyaeva *et al.* (1994).

3) Neurochemical differences characteristic for different IB are age-dependent

The proportion of animals showing behavioural strategies typical of the ER test was different in Wistar rats at age 3 months (n = 100), 6 months (n = 162) and 20 months (n = 55) (Stepanichev, 1996). Though rats demonstrating ER comprised about 15% of the total across all age groups, the portion of nonmobile rats without exploratory behavior in the ER test apparatus (type 4) increased with age (Figure 17.2). Such rats usually sat in the

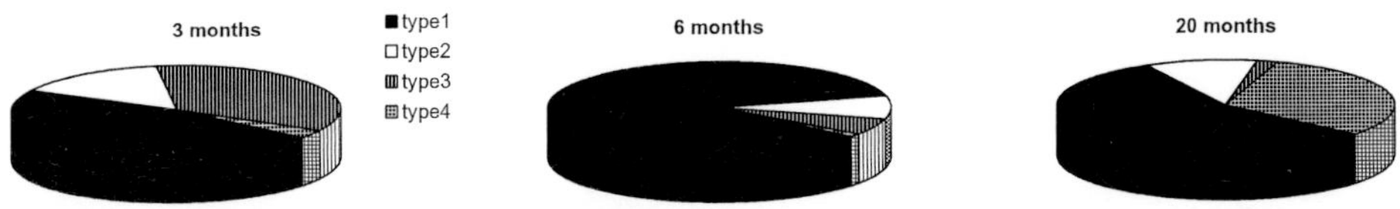

Figure 17.2. Effects of age on the emotional resonance reaction in Wistar rats. For experimental conditions see Stepanichev (1996).

corner of the light chamber and did not pass to the dark chamber. These animals made up no more than 3% among 3 month- and 6 month-old animals. However their number dramatically increased by the age of 20 months (to 31%). Novel situations might be a significant aversive stimulus for old animals of behavioural type 4, and might induce strong negative emotions preventing any activity. The age-related increase of animals that did not choose any behavioural strategy in a novel situation could be the result of a shift of emotional status towards negative emotions, which is a specific feature of aging (Rushkevich, 1989; Frolkis *et al.,* 1991). Thus, IB of rats in the ER test depended on age.

NO-synthase activity and generation of active oxygen species were both age- and IB-dependent (Onufriev *et al.,* 1995a,b).

Age-related differences were also demonstrated when rats were under a functional load or in extreme conditions. Immobilization stress-induced changes in FRMO and cerebral membrane lipids are less evident in adult rats as compared with old ones (Stepanichev, 1995, 1996). Immobilization stress produced cAMP accumulation in the hypothalamus, frontal cortex and amygdala in old rats of behavioral types 1 and 4. Only in adult rats with active exploratory behaviour in the ER-test could cAMP accumulation in the hypothalamus and amygdala be demonstrated. Statistically significant correlations of cAMP levels between symmetrical cerebral macrostructures were revealed in unstressed old rats of both behavioural types (Figure 17.3, A,C). Stress induced the appearance of additional intra- and interhemisphere correlations, their pattern depending on IB: Rats preferring the dark chamber (type 1) displayed correlations including mainly the hippocampus and hypothalamus, whereas rats that did not choose a behavioural strategy (type 4) demonstrated correlations including the frontal cortex and amygdala (Egorova *et al.*, 1995) (Figure 17.3, B,D). Correlations of cAMP content were expressed much less in adult rats, and only few of them could be revealed in rats preferring dark spaces (type 1) (Stepanichev, 1996). In many cases the neurochemical correlates of IB are expressed more in old animals compared with adult ones; however, the reasons for this phenomenon remains unclear and deserve further investigation.

4) Neurochemical IB-related differences may be global, characteristic of specific brain structures, or reflected in interhemisphere asymmetry.

Interhemispheric asymmetry of cortical phospholipids, giving higher phospholipid content in the right neocortex, was revealed in old rats preferring the dark chamber in the ER test, whereas a higher phospholipid content in the left neocortex and a higher cholesterol-to-phospholipid ratio was demonstrated in adult rats expressing the ER (Stepanichev, 1995; Stepanichev *et al.*, 1995a). Immobilization stress produced interhemispheric asymmetry of cholesterol, with dominance of the left hemisphere in the neocortex of old rats with a variety of different behaviours in the ER test.

A

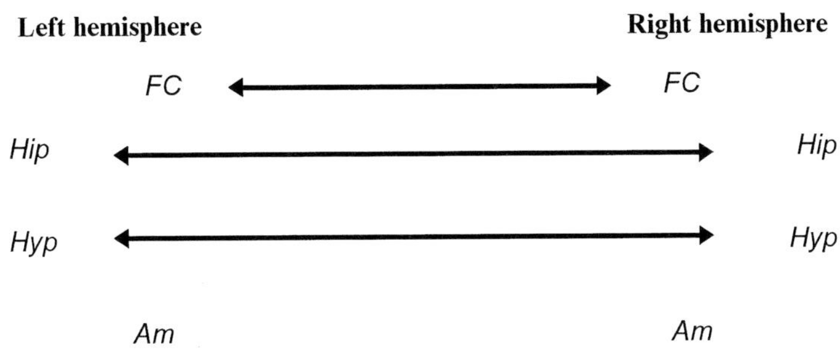

B

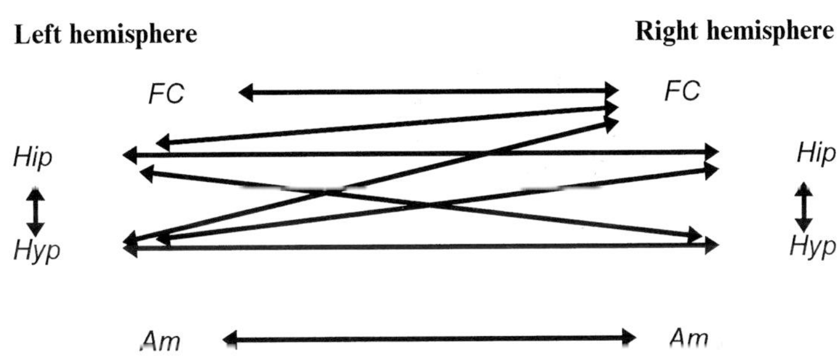

Figure 17.3. Effects of immobilization stress on inter-regional correlations of cAMP content in brains of old (20 months) rats: relation to behavioural type in the ER test. **A** and **C:** control condition; A—type 1, C—type 4. **B** and **D**: stress; B—type 1, D—type 4. **FC**—frontal cortex, **Hip**—hippocampus, **Hyp**—hypothalamus, **Am**—amygdala. Correlations with $P < 0.05$ are shown. For experimental conditions see Egorova *et al.* (1995).

Global cerebral ischemia induced by cardiac arrest (a model of postresuscitation pathology) resulted in a decrease of NO-synthase activity in the neocortex and an increased generation of active oxygen species in the neocortex and cerebellum of animals with different IB in the ER test (Stepanichev *et al.*, 1997). Rats with an active defensive behavioural strategy (type 3) demonstrated oxidative stress in the hippocampus (Figure 17.4) and rats with a passively-defensive behavioural strategy (type 1) showed NO-synthase activation in the hippocampus one week after resuscitation (Figure 17.5). As a rule, rats with a passive-defensive behavioural strategy responded to cardiac arrest by neurochemical changes in the left cortex, and rats with an active defensive behavioural strategy did so with changes in the right cortex (Figure 17.6). Multifactor analysis of variance applied to the results obtained in the postresuscitation pathology model demonstrated that changes of FRMO indices and of NO-synthase in the hippocampus depended only on IB, whereas the effects of global ischemia on these parameters in the neocortex and cerebellum was much more expressed than the effects of IB.

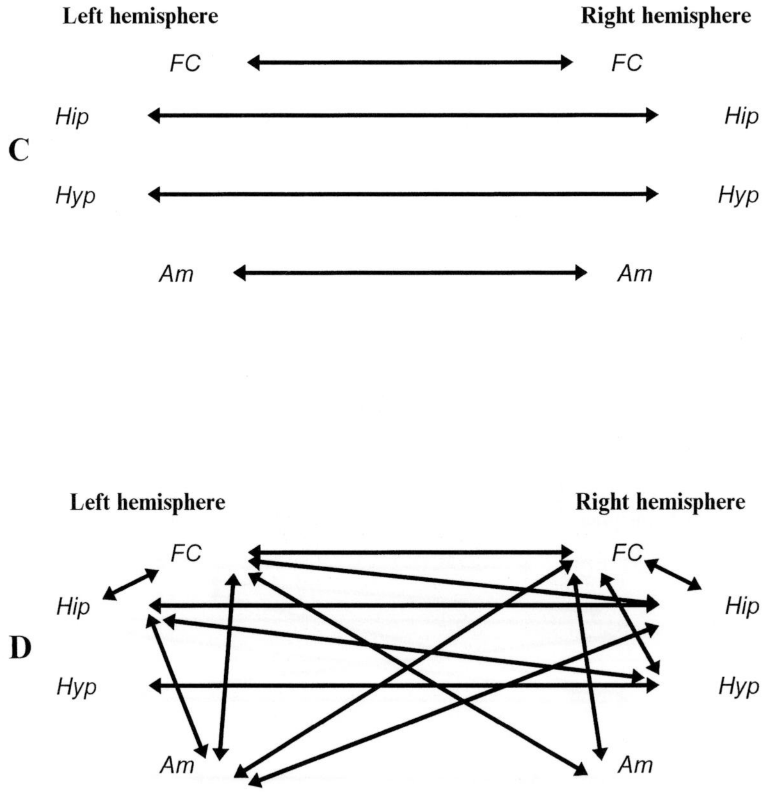

Figure 17.3. (continued). See legend on p. 309

Immobilization stress produced an accumulation of cAMP in the hypothalamus, frontal cortex and amygdala in old rats with different IB in the ER test, although when separate analysis of cAMP in the structures of the left and right hemisphere was performed, cAMP accumulation could be demonstrated only in animals without a behavioural strategy in the ER test (type 4) (Egorova *et al.*, 1995).

5) IB-dependent neurochemical differences can be revealed not only in the ER test, but also using other tests for behavioral selection of rats.

Some neurochemical differences characteristic of rats with different IB in the "open field" test were described above. Using another behavioral test (the "division of labour" or "behavioral parasitism" phenomena referred to above) we showed the usefulness of the above principles in old rats demonstrating different behavioural strategies in the situation of competition for food and food procurement (Stepanichev *et al.*, 1996). These behavioural types have different characteristic neurochemical patterns. The cholesterol-to-phospholipid ratio was higher in the left neocortex of "parasites" and of "invertors" as compared with "workers", this phenomenon resulting from differences in phospholipid but not the cholesterol content. The cAMP level was higher in the hypothalamus of

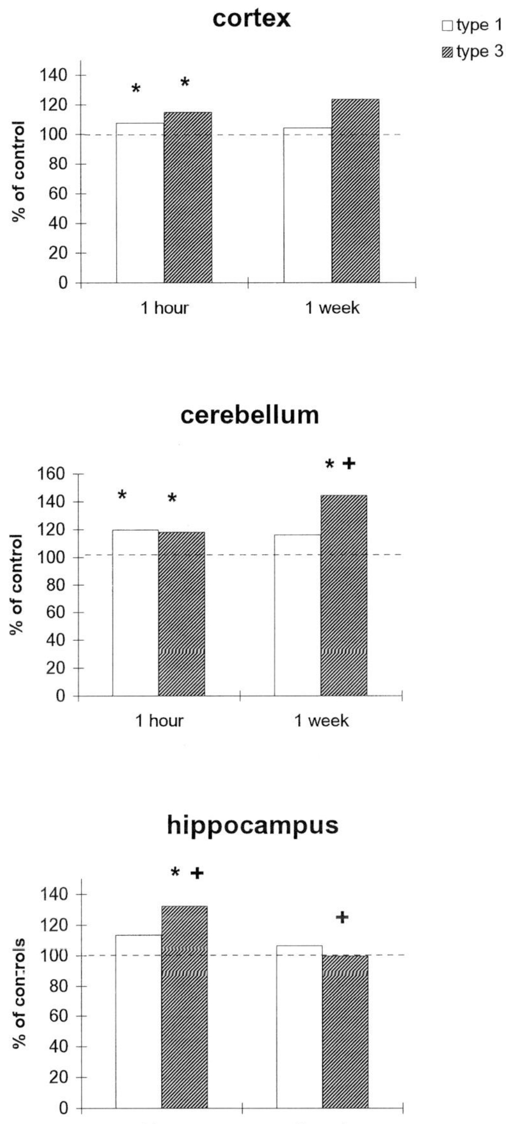

Figure 17.4. Effects of cardiac arrest (1 hour and 1 week after resuscitation) on free radical generation in brain regions of rats: relation to behavioral type in the ER test. *—significant difference from controls, +—significant difference between types, $P < 0.05$. For experimental conditions see Stepanichev *et al.* (1997).

"workers" than that of "invertors". The most striking differences were revealed in the pattern of correlations of cAMP content in brain macrostructures: "workers" demonstrated correlations only between symmetrical structures (frontal cortex, hippocampus, hypothalamus), whereas "parasites" and "invertors" displayed a rich network of intra- and inter-hemispheric correlations.

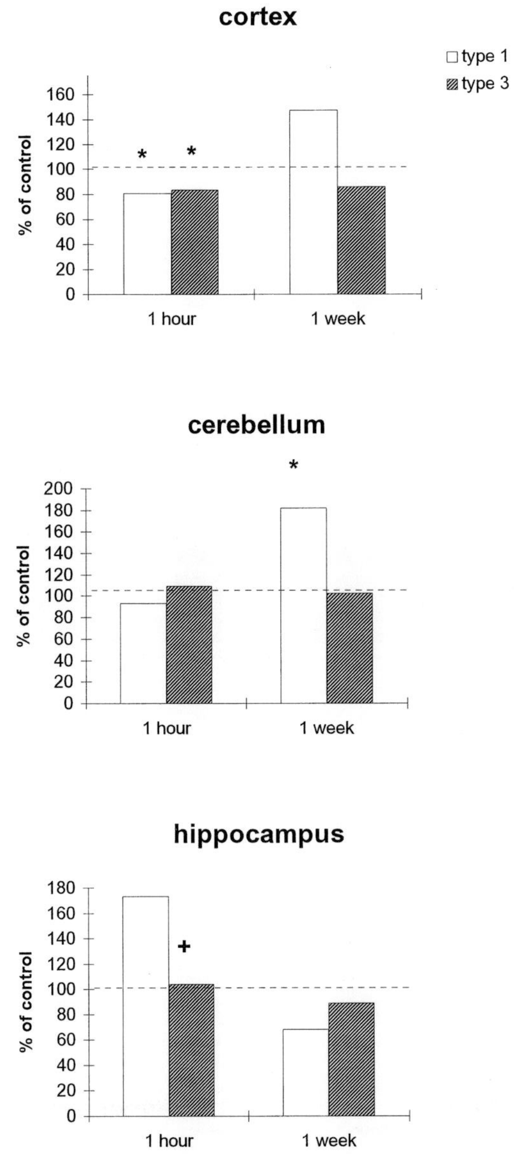

Figure 17.5. Effects of cardiac arrest (1 hour and 1 week after resuscitation) on nitric oxide synthase activity in brain regions of rats: relation to behavioral type in the ER test. *—significant difference from controls, +—significant difference between types, $P < 0.05$. For experimental conditions see Stepanichev *et al.* (1997).

These data suggest that the above conclusions based mainly on ER test experiments are of general significance, and that the methodological approaches used can be applied to study neurochemical correlates of behavior in other behavioral tests.

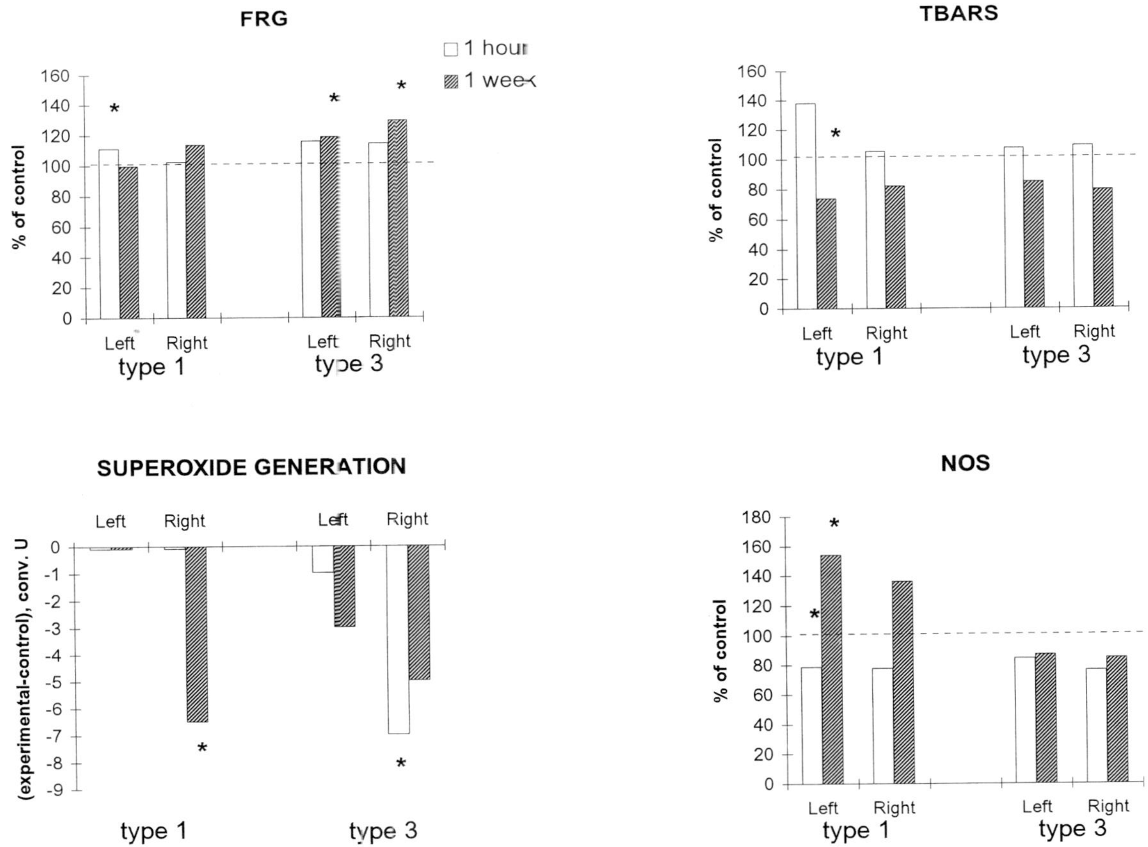

Figure 17.6. Effects of cardiac arrest (1 hour and 1 week after resuscitation) on free radical generation (FRG), thiobarbituric acid reactive substances (TBARS), superoxide generation, and nitric oxide synthase (NOS) activity in left and right cortices of rats: relation to behavioural type in the ER test. *—significant difference from controls, $P < 0.05$. For experimental conditions see in Stepanichev *et al.* (1997).

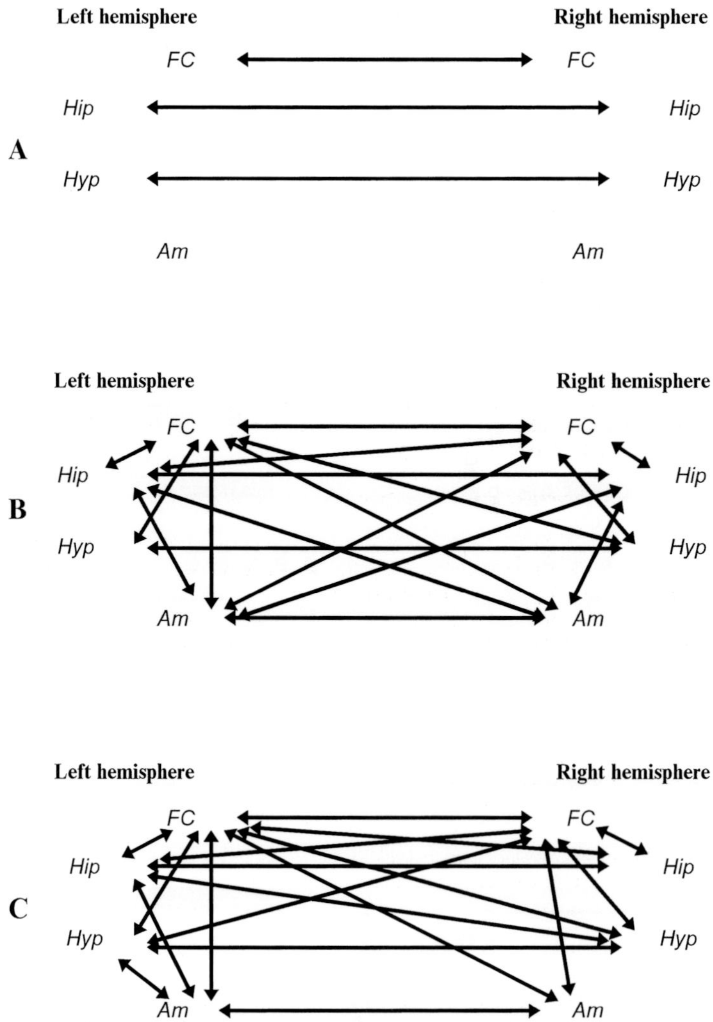

Figure 17.7. Inter-regional correlations of cAMP content in the brains of Wistar rats: relation to behavioural type in the "labor division" test ("behavioral parasitism" phenomenon). **A**—"parasites", **B—**"workers", **C**—"invertors". **FC**—frontal cortex, **Hip**—hippocampus, **Hyp**—hypothalamus, **Am**—amygdala. Correlations with $P < 0.05$ are shown. For experimental conditions see in Stepanichev *et al.* (1996).

REFERENCES

Abdulla, F.A, Abu-Bakra, M.A.J., Calaminici, M.-R, Stephenson, J.D. and Sinden, J.D. (1995) Importance of forebrain cholinergic and GABAergic systems to the age-related deficits in water maze perfomance of rats. *Neurobiology of Aging*, **16**, 41–52.

Ajrapetyanz, M.G. and Vein, A.M. (1982) *Neurosen in Experiment and Clinic* [in Russian], Nauka, Moscow.

Ajrapetyanz, M.G., Khonicheva, N.M., Mekhedova, A. Ya and Iljna Wiljar, H. (1980) Reactions to moderate functional loads in rats with different individual behavior, *Zhurnal Vysshey Nervnoy Dejatelnosty*, **30**, 994–1003.

Arons, C.D. and Shoemaker, W.J. (1992) The distribution of catecholamines and beta-endorphin in the brains of three behaviorally distinct breeds of dogs and their F1 hybrids. *Brain Research*, **594**, 31–39.

Bianki, V.I., Filippova, E.B. and Murik, S.E. (1985) Interhemisphere asymmetry of "emotional resonance" in rats. *Zhurnal Vysshey Nervnoy Dejatelnosty*, **35**, 261–266.

Bondarenko, N.A., Devjatkina, T.A., Voskresenskij, O.N. and Valdman, N.V. (1985) Effects of chronic emotional stress on lipid peroxidation in tissues and blood of emotional and non-emotional rats. *Bulletin of Experimental Biology and Medicine*, **100**, 12–14.

Bondarenko, V.A., Kamysheva, V.A., Mineeva, M.F. and Valdman, N.V. (1981) Effects of chronic stress on behavior, somatic state and tyrosine hydroxylase activity in brain of "emotional" and "non-emotional" rats. *Bulletin of Experimental Biology and Medicine*, **91**, 20–22.

Bures, J., Buresova, O. and Huston, J.P. (1983) *Techniques and Basic Experiments for the Study of Brain and Behavior.* Elsevier, Amsterdam, New York.

Burlakova, E.V. (1976) Free radical-mediated mechanism of cellular metabolism regulation and its relations with other regulatory systems. In: *Free Radical Oxidation of Lipids in Norm and Pathology* [in Russian], Nauka, Moscow, p. 18–19.

Burov, Yu.V. and Speranskaya, N.P. (1974) Analysis of neurochemical mechanisms of one of the forms of Intra-species behavior.*Zhurnal Vysshey Nervnoy Dejatelnosty*, **24**, 301–305.

Chaichenko, G.M. (1982) The dependence of learning on the general excitability of rats. *Zhurnal Vysshey Nervnoy Dejatelnosty*, **32**, 980–982.

Cutler, R.G. (1985) Dysdifferentiative hypothesis of aging: a review. In: R.S. Sohal, L.S. Birnbaum and R.G. Cutler (eds) *Molecular Biology of Aging: Gene Stability and Gene Expression.* Raven Press, New York, pp. 307–340.

Egorova, L.K., Stepanichev, M.Yu and Gulyaeva, N.V. (1995) Neurochemical peculiarities of rats with different behavior in "emotional resonance" test. Cyclic adenosine monophosphate in rat brain structures. *Zhurnal Vysshey Nervnoy Dejatelnosty*, **45**, 999–1005.

Frolkis, V.V., Bezrukov, V.V., Bogatskaya, L.N., *et al.* (1991) *Aging of Brain* [in Russian]. Nauka, Leningrad.

Gershon, D. (1988) Oxidative and peroxidative damage and its role in the aging process. In: A.Quintanilha (ed.) *Reactive Oxygen Species in Chemistry, Biology, and Medicine.* Plenum Press, New York, London, p. 211–220.

Gershtein, L.M., Kamysheva, A.S., Chebotaryeva, O.L, Sergutina, A.V. and Orlova, E.I. (1991) Morphochemical characteristics of brain of Wistar rats with different locomotor activity in the open field.*Zhurnal Vysshey Nervnoy Dejatelnosty*, **41**, 300–305.

Getsova, V.M and Orlova, N.V. (1993) Individual peculiarities of behavioral reactions an monoaminergic brain systems in rats. In: P.V. Simonov (ed.) *Individual Brain: Structural Bases of Individual Behavior*, Nauka, Moscow, pp. 68–81.

Gulyaeva, N.V. (1989) *Lipid peroxidation in the brain during the adaptation to stress.* D.Sci. Thesis, Moscow.

Gulyaeva, N.V., Onufriev, M.V. and Stepanichev, M.Yu. (1994) NO synthase and free radical generation in brain regions of old rats: correlations with individual behaviour. *NeuroReport*, **6**, 94–96.

Halliwell, B. and Gutteridge, J.M.C. (1989) *Free Radicals in Biology and Medicine.* Clarendon Press, Oxford.

Janicke, B. and Wrobel, D. (1984) Changes in motor activity with age and the effects of pharmacological treatment. *Experimental Gerontology*, **19**, 321–328.

Jeste, D.J. and Smith, J.P. (1980) Unilateral mesolimbicocortical dopamine deprivation decreases locomotion in the open-field and after amphetamine. *Pharmacology, Biochemistry and Behavior* **12**, 453–457.

Khonicheva, N.M. (1984) Specific features of instrumental food conditioning in rats with different behavior during the pain stimulation of another animal. *Zhurnal Vysshey Nervnoy Dejatelnosty*, **34**, 229–236.

Khonicheva, N.M. and Iljna Wiljar, H. (1981) Behavior in the avoidance situation as a criterion for evaluation of typological peculiarities in rats. *Zhurnal Vysshey Nervnoy Dejatelnosty*, **31**, 975–983.

Khromova, S.V. (1995) Study on behavioral "parasitism" phenomenon in rats. *Zhurnal Vysshey Nervnoy Dejatelnosty*, **45**, 479–489.

Krakovski, M.E. (1987) Activity of main redox enzymes in rabbits with different typological peculiarities. *Zhurnal Vysshey Nervnoy Dejatelnosty*, **37**, 457–461.

Krakovski, M.E., Kamenetskaja, Z.L., Tremasova, G.L. and Eshermetov, A.H. (1989) Peculiarities of some biochemical processes in the liver of rats with different behavioral types in the open field. *Zhurnal Vysshey Nervnoy Dejatelnosty*, **39**, 506–512.

Kruglikov, R.I., Getsova, V.M., Orlova, N.V., Lushkina, A.A. and Feshchenko, I.R. (1995) Changes of monoamine in the brain affects emotional resonance reaction. *Zhurnal Vysshey Nervnoy Dejatelnosty*, **45**, 551–557.

Kulagin, D.A. (1975) Emotionality and typological features of the nervous system. In: B.A. Nikitiuk (ed.) *Differential Psychophysiology and its Genetic Aspects* [in Russian], Nauka, Moscow, p. 74–77.

Kulagin, D.A. and Bolondinski, V.K. (1986) Neurochemical aspects of emotional reactivity and locomotor activity of rats in a novel situation. *Uspekhi Fiziolicheshikh Nauk*, **17**, 92–109.

Levshina, I.P. and Gulyaeva, N.V. (1991) The dependence of acute stress effect on the lateralization of lipid peroxidation products in the brain on the type of rat behavior. *Bulletin of Experimental Biology and Medicine*, **106**, 568–569.
Markel, A.L. (1981) On the evaluation of main behavioral characteristics of rats in the "open field" test. *Zhurnal Vysshey Nervnoy Dejatelnosty*, **31**, 301–307.
Masur, J. and Aparecida da Silva, V. (1977) Fighting behavior displayed during the development of a worker-parasite relationship between rats. *Behavioral Biology*, **20**, 51–59.
Meshcheryakova, O.V. (1988) Behavior of different strains of rats in the experimental model of "emotional resonance". *Zhurnal Vysshey Nervnoy Dejatelnosty*, **38**, 368–369.
Mowrer, O.H. (1940) Animal studies in the genesis of personality. *Transactions of the New York Academy of Sciences*, **3**, 8–11.
Murik, S.E. (1990) Interhemisphere asymmetry and typological peculiarities of rat nervous system. *Zhurnal Vysshey Nervnoy Dejatelnosty*, **40**, 963–967.
Onufriev, M.V., Stepanichev, M.Yu, Lazareva, N.A., Grigorian, G.L. and Guliaeva, N.V. (1995a) NO-synthase activity and radical generation in rat brain regions: dependence on age. *Bulletin of Experimental Biology and Medicine*, **120**, 54–56.
Onufriev, M.V., Stepanichev, M.Yu, Lazareva, N.A. and Gulyaeva, N.V. (1995b) NO-synthase activity and active oxygen species generation in old rat brain: correlations with individual behavior. *Bulletin of Experimental Biology and Medicine*, **120**, 145–147.
Rushkevitch, K.E. (1989) Emotional manifestation during electrical stimulation of hypothalamic nuclei of adult and old rats. *Zhurnal Vysshey Nervnoy Dejatelnosty*, **39**, 273–277.
Salomon, Y. (1991) Cellular responsiveness to hormones and neurotransmitters: Conversion of [3H] adenine to [3H]cAMP in cell monolayers, cell suspensions and tissue slices. *Methods in. Enzymology*, **195**, Academic Press, New York, London, p. 22–35.
Semagin, V.N., Zukhar, A.V. and Kulikov, M.A. (1988) *Type of the Nervous System, Stress-Resistance and Reproductive Function* [in Russian]. Nauka, Moscow.
Simonov, P.V. (1976) Conditioned reactions of emotional resonance in rats. In: *Neurophysiological Approach to the Analysis of Intraspecies Behavior* [in Russian]. Nauka, Moscow, pp. 8–36.
Simonov, P.V. (1987) *Motivated Brain* [in Russian]. Nauka, Moscow.
Stepanichev, M.Yu. (1995) Lipid component of rat cerebral membranes under stress: dependence on age and individual behavior. *Nejrokhimija*, **12(3)**, 40–46.
Stepanichev, M.Yu. (1996) *Neurochemical correlates of individual behavior in rats.* Ph.D. Thesis, Moscow.
Stepanichev, M.Yu., Onufriev, M.V., Lazareva, N.A. and Gulyaeva, N.V. (1995) Neurochemical peculiarities of rats with different behavior in "emotional resonance" test. Free radical-mediated processes and lipids in neocortex of old rats.*Zhurnal Vysshey Nervnoy Dejatelnosty*, **45**, 990–998.
Stepanichev, M.Yu., Egorova, L.K., Lazareva, N.A.,Khromova, S.V and Guliaera, N.V. (1996) Neurochemical investigation of behavioral "parasitism" phenomenon. *Zhurnal Vysshey Nervnoy Dejatelnosty*, **46**, 137–145.
Stepanichev, M.Yu, Onufiev, M.V., Lazareva, N.A., Zarzhetsky, Y.U., Mutuskina, E.A., Gurvitch,.A.M. and Gulyaeva, N.V. (1997) Post-resuscitation changes in brain free radical-mediated processes and nitric oxide synthase activity in rats: effects of individual behavior in emotional resonance test. *Neurochemical Research*, **22**, 743–753.
Strek, K.F. (1989) Manipulation of serotonin protects against an hypoxia-induced deficit of passive avoidance response in rats. *Pharmacology Biochemistry and Behavior*, **33**, 241–244.
Troshikhin, V.A., Kozlova, L.N., Kruchenko, Zh.A. and Sirotski, V.V. (1971) *Forming and Development of Main Higher Nervous Activity Characteristics in Ontogenesis* [in Russian]. Naukova Dumka, Kiev.
Yumatov, E.A. (1986) *Central neurochemical mechanisms of the resistance to emotional stress.* M.D. Thesis, Moscow.
Yumatov, E.A. and Meshcheryakova, O.V. (1990) The predictability of the resistance to emotional stress on the basis of individual behavior testing. *Zhurnal Vysshey Nervnoy Dejatelnosty*, **40**, 575–579.
Vincent, S.R. (1994) Nitric oxide: A radical neurotransmitter in the central nervous system. *Progress in Neurobiology*, **42**, 129–160.

18 Brain Serotonin in Genetically Defined Defensive Behaviour

N.K. Popova

Institute of Cytology and Genetics, Siberian Branch of Russian Academy of Sciences, Novosibirsk, Russia
e-mail: npopova@ghost.bionet.nsc.ru

Data concerning the role of brain serotonin (5-HT) and serotonergic 5-HT_{1A} receptors in the expression of a genetic predisposition to active (aggression) or passive (freezing) defensive behavior are reviewed. Significant diversity was shown in changes in serotonin metabolism, in rats bred for 40 generations for predisposition to freezing (catalepsy), and in Norway rats bred for low and high aggressiveness to man. Activity of the rate-limiting enzyme in 5-HT biosynthesis (tryptophan hydroxylase) in the midbrain, as well as the level of 5-HT and its principal metabolite (5-hydoxy-indoleacetic acid) in some brain regions were significantly lower in highly aggressive rats than in animals bred for the lack of aggressive defence reactions to man. In contrast, an increased activity of tryptophan hydroxylase was found in the striatum—i.e. the brain region involved in the control of muscular tone—of rats genetically predisposed to catalepsy. This was not found in the midbrain. A similar increase in tryptophan hydroxylase activity in the striatum, without any significant changes in the midbrain, was shown in mice with a genetic predisposition to freezing.

However, unlike 5-HT metabolism, alterations in 5-HT_{1A} receptors in animals with different defensive strategies were similar. In rats with both genetically defined active defensive reactions and in rats with passive defensive freezing, a decreased [^{3}H]8-OH DPAT specific binding in some brain structures was shown.

It was hypothesized that the density of serotonergic 5-HT_{1A} receptors participates in the mechanisms of anxiety and fear that trigger any kind of defensive response, whereas genetically defined predispositions to active and passive strategies of defence are associated with specific changes of 5-HT metabolism in midbrain and striatum.

KEYWORDS: strategy of defensive behavior, genetic predisposition to defence, serotonin, tryptophan hydroxylase, 5-HT_{1A} receptors

1. INTRODUCTION

An animal's response to a threatening situation is displayed, in most species, in three kinds of fear-induced defensive behavior—flight, fight or freezing (catalepsy). Animals having no way to flee usually exhibit fear-induced passive defensive response—freezing or an active defence—fighting. The strategy of defensive behavior depends on both environmental (Blanchard, 1989) and genetic (Kulikov and Popova, 1991; Kulikov *et al.*, 1993; Popova *et al.*, 1993) factors, defining the predisposition to a given kind of defence. Since the genetic control of brain neurotransmitters undoubtedly represents a principal element of genetic regulation of behaviour it can be suggested that the strategy of defence is conditioned to a great extent by specific, genetically defined features of brain neurotransmitter systems.

Data obtained in experiments carried out over more than 20 years in our laboratory suggest a significant role for the brain serotonergic system. A particular interest in the brain neurotransmitter serotonin arose from data indicating the involvement of the serotonergic system in regulation of both catalepsy (Kostowsky *et al.*, 1972) and some types of aggressive behavior (Lagerspetz and Lagerspetz, 1974; Popova *et al.*, 1978). Two main methodological approaches were used: 1) In order to elucidate the genetic correlations between brain serotonin metabolism and the predisposition to freezing, a number of inbred mouse strains were studied. 2) Two different rat strains were studied, selectively bred at the Institute of Cytology and Genetics, Novosibirsk, for their predisposition to catalepsy, and for their lack of defensive aggression to man, as well as silver foxes bred for more than 30 years for nonaggressive behaviour.

2. ACTIVE DEFENCE AGGRESSION

Aggression against man, in animals to whom man is not the prey, represents fear-induced active defence (Moyer, 1968). This kind of aggressive behavior has received relatively little attention. We have found (Popova *et al.*, 1975, 1991a) significant alterations in the brain serotonergic system in silver foxes selected for more than 30 years for tame behavior, displaying no defensive response towards man (Belyaev, 1979). Levels of serotonin and its metabolite 5-hydroxyindoleacetic acid (5-HIAA) in the hypothalamus, and midbrain of tame nonaggressive silver foxes were significantly higher than in nonselected "wild" animals. There were no substantial changes in serotonin in the hippocampus of silver foxes selected for the absence of a defensive responses towards man, although the level of its main metabolite in the hippocampus of tame animals was increased (Figure 18.1).

Activity of the main enzymes in serotonin metabolism, tryptophan hydroxylase and monoamine oxidase type A, was also different in highly aggressive and in tame silver foxes. A significant difference in the activity of the key enzyme in serotonin biosynthesis (tryptophan hydroxylase) was found between foxes selected for a low-level defensive response towards man and those selected for a high aggressive reaction. Tryptophan hydroxylase activity in the midbrain was 27% higher in nonaggressive foxes, and was 34% lower in the aggressive population, compared with nonselected silver foxes (Figure 18.2). Selection of silver foxes for tame behavior was also followed by a moderate decrease in type A monoamine oxidase activity in their brainstem (Popova *et al.*, 1991b). It is relevant to note that these changes were specifically related to the defensive aggression of animals, since in silver foxes selected for a high-level aggressive reaction towards man, tryptophan hydroxylase activity in the midbrain and hypothalamus was decreased, whereas in animals selected for tame behaviour, the enzyme activity was increased compared to nonselected animals (Kulikov *et al.*, 1989).

Increased activity of tryptophan hydroxylase, considered as a marker of serotonergic system functional activity (Boadle-Biber, 1982), as well as an increased serotonin level in the brain of low aggressive silver foxes are in good agreement with some data implicating serotonin as an inhibitory factor in fear-induced defensive aggression.

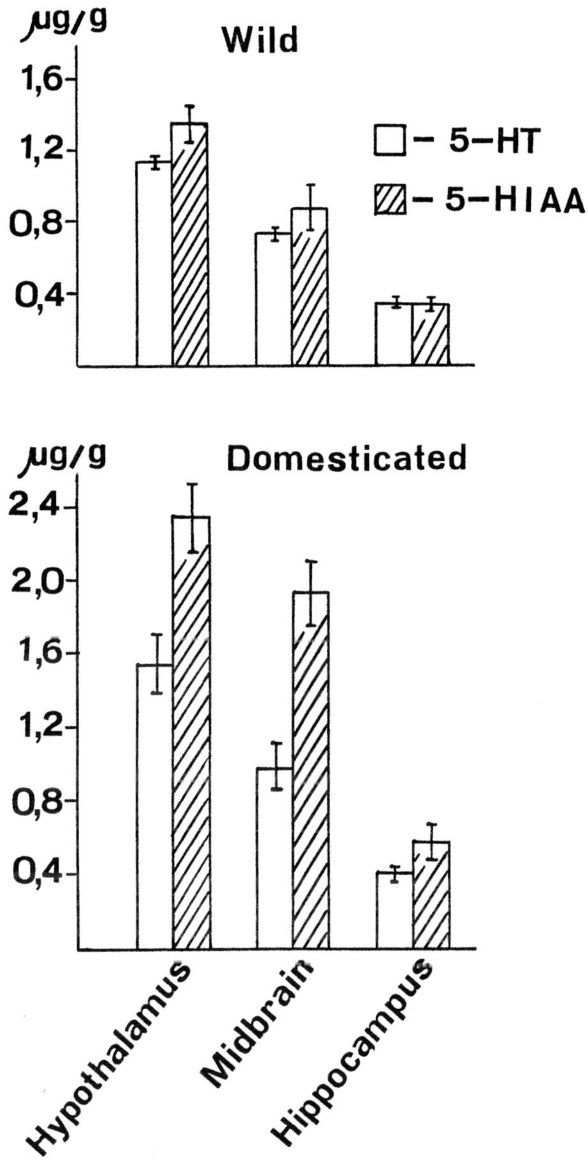

Figure 18.1. Concentrations of serotonin (5-HT) and 5-hydroxyindoleacetic acid (5-HIAA) in brain regions of domesticated and nonselected (wild) silver foxes. The difference between domesticated and wild animals in 5-HT in hypothalamus, in midbrain and in hippocampus: $p < 0.05$; in 5-HIAA in hypothalamus and in midbrain: $p < 0.001$, in hippocampus: $p = 0.05$ (from Popova *et al.*, 1991b).

We hypothesized (Popova *et al.*, 1975, 1980) that behavioural selection appears to involve the selection for specific activity levels of neurotransmitter systems regulating given behaviours. Changed serotonin metabolism in nonaggressive silver foxes compared

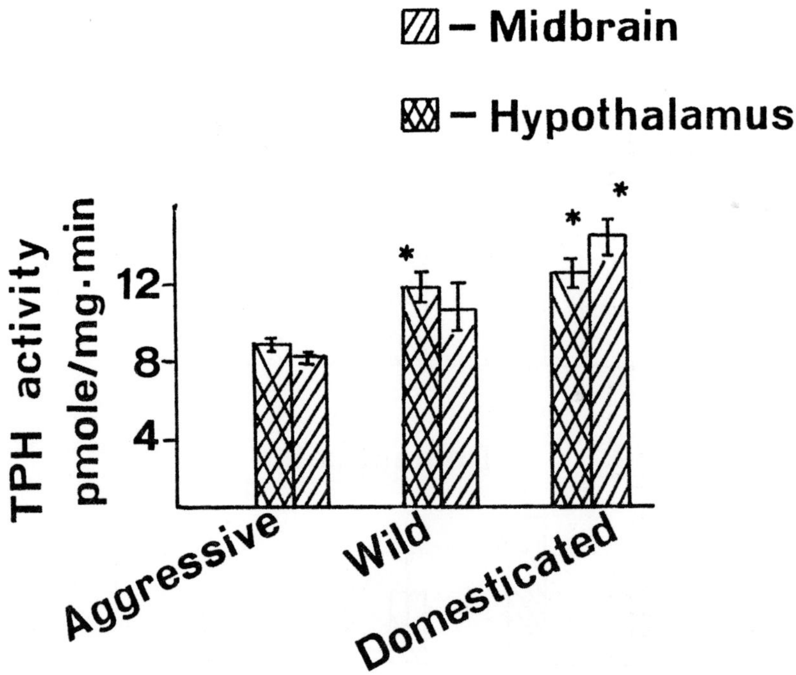

Figure 18.2. Activity of tryptophan hydroxylase (TPH) in the brain of domesticated and nonselected (wild) silver foxes, in in those selected for high aggressiveness (aggressive). *$p < 0.05$ vs aggressive animals (from Popova *et al.*, 1991b).

with their aggressive counterparts gave us a reason to suggest the involvement of the brain serotonergic system in genetically defined affective defensive aggression.

This idea was supported by findings obtained in another species, i.e. in Norway rats selected for high and low aggressiveness. As a result of long-term selection for non-aggressive behaviour towards man, a population of Norway rats was obtained quite different in their behaviour from Norway rats of the initial population. The rats selected for tame behavior not only were not afraid of people, but their reaction to other threatening stimuli was also diminished (Blanchard *et al.*, 1994). At the same time, selection for low reactivity towards man was followed by a marked change in brain serotonin metabolism (Nikulina *et al.*, 1985; Naumenko *et al.*, 1989; Popova *et al.,* 1991a).

It is essential to note that the changes in serotonin metabolism found in rats selected for nonaggressive behaviour, were similar to those shown in brains of silver foxes selected for low reactivity to man (Kulikov *et al.,* 1989). A comparison of aggressive and tame rats showed that in the 15–16th generation of selection, the content of the serotonin metabolite 5-HIAA appeared to be significantly higher in the hypothalamus of tame than in that of the aggressive animals. This difference was maintained in subsequent generations. Increases of both 5-HIAA and serotonin itself were found in the hypothalamus and midbrain in the 20th generation of selection (Figure 18.3).

Increased tryptophan hydroxylase activity was found in the midbrain and hypothalamus of rats with genetically defined low aggressiveness compared to highly aggres-

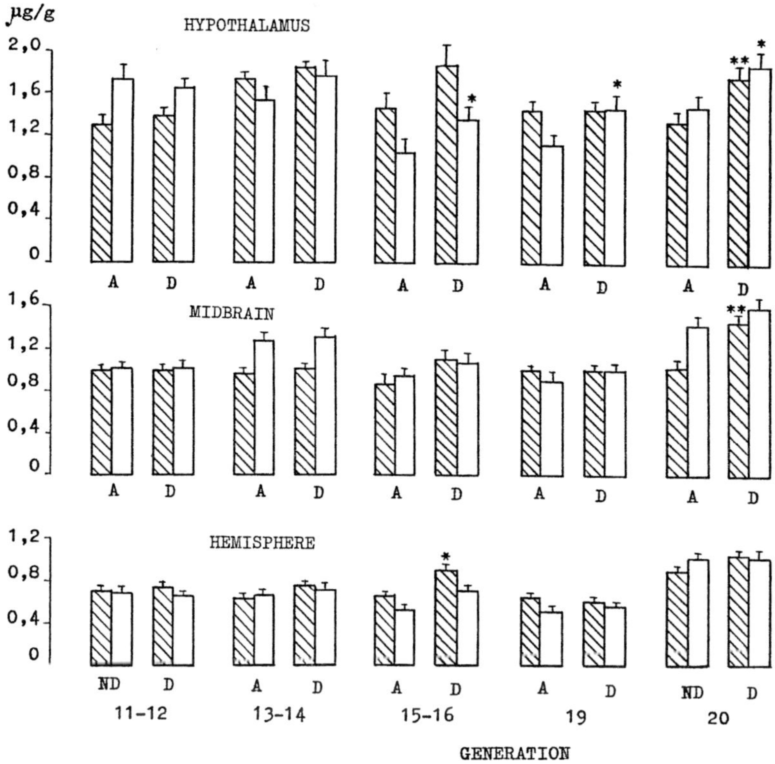

Figure 18.3 Concentrations of serotonin and 5-HIAA in brains of domesticated and aggressive Norway rats during selection for aggressive reactivity towards man. A—aggressive rats, D—nonaggressive "domesticated" rats . Striped columns: serotonin. Open columns: 5-HIAA. *p < 0.05, **p < 0.001 (from Naumenko *et al.*, 1989). Note that brains from rats in different generations were analysed at different seasons.

sive rats (Figure 18.4). It has to be noted that the changes were found in the midbrain area containing most of the tryptophan hydroxylase-synthesizing cell bodies (Popova *et al.*, 1991a). These findings were interpreted as indicating an increased activity of the brain serotonergic system in tame animals, and, subsequently, a decreased activity of this system in highly aggressive animals.

Therefore, although the kinetic characteristics of tryptophan hydroxylase in Norway rats differ significantly from those in silver foxes (Kulikov *et al.*, 1989) similar changes in the enzyme activity in animals with a hereditary lack of an aggressive reaction towards man was found. The remarkable consistency obtained in such diverse species as silver foxes and Norway rats suggests the involvement of the brain serotonergic system in control of fear-induced defensive aggression in various species. It convincingly supports our hypothesis (Popova *et al.*, 1975, 1980) that the brain serotonergic system plays an essential role in brain mechanisms converting wild aggressive animals into their tame counterparts, which, according to Belayev (1979), is the background of domestication of animals.

Our experiments also show that the distinction between genetically defined aggressive and nonaggressive animals is not limited to a difference in brain serotonin metabolism.

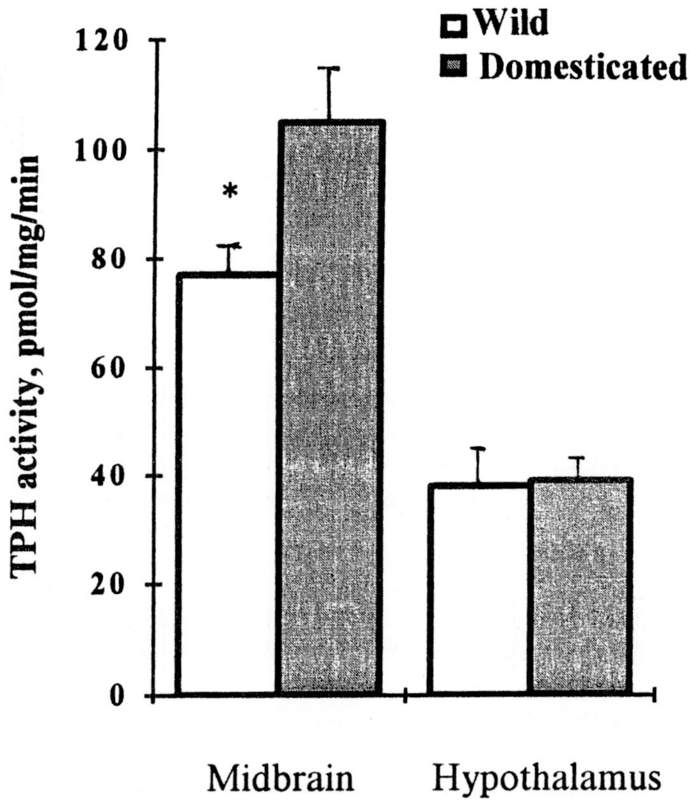

Figure 18.4. Activity of tryptophan hydroxylase (TPH) in the brain of wild and nonaggressive domesticated rats. $*p < 0.05$ *vs* domesticated animals.

Considerable changes in serotonergic 5-HT_{1A} receptors were also shown in Norway rats selected for low aggressiveness.

Among an amazing variety of cloned and identified serotonergic receptors, the 5-HT_{1A} subtype of 5-HT1 receptors attracts special attention in view of the data on anxiolytic (Nutt and Glue, 1985) and antidepressant (Kurtz, 1992; Lesh, 1992) effects of 5-HT_{1A} receptor agonists. Taking into consideration that fear appears to be the trigger of defensive behavior, it was of particular interest to study the 5-HT_{1A} receptors in the brain of animals with a hereditary predisposition to different kinds of defence.

Radioligand studies of specific [3H]8-OH DPAT binding in the brain of wild Norway rats, and of Norway rats selected for 40 generations for low aggressiveness towards man, revealed changes in the 5-HT_{1A} receptor density in some brain areas (Popova *et al.*, 1996). The clearest differences between affectively aggressive and nonaggressive rats were found in the hypothalamus, frontal cortex and in the amygdala, where the 5-HT_{1A} receptor density in wild animals was considerably lower than those with the tame phenotype. Bmax values for specific receptor binding of [3H]8-OH DPAT in tame animals was higher by 63% ($p<0.05$) in the hypothalamus, by 57% ($p<0.05$) in the frontal cortex, and by 36% ($p<0.05$) in the amygdala compared with aggressive wild rats. At the same time, in the midbrain (the region of the main concentration of presynaptic somatodendritic 5-HT_{1A} receptors) no changes in specific [3H]8-OH DPAT binding were found.

It is well known that $5HT_{1A}$ receptors are located both presynaptically and postsynaptically, and that the effect on cell body firing and on serotonin release is quite different for receptors in these different locations. Cell body somatodendritic autoreceptors belong to the 5-HT_{1A} receptor subtype in all species studied so far. These presynaptic 5-HT_{1A} receptors inhibit cell firing and serotonin release at the terminals. At the same time stimulation of postsynaptic receptors produces the effect of activation of serotonergic system.

As shown above, most differences were revealed in the frontal cortex, hypothalamus and amygdala—brain regions with a preferential postsynaptic localization of 5-HT_{1A} receptors.

Involvement of the hypothalamus in control of aggressive behaviour is firmly established, and has been confirmed by numerous data (Popova *et al.,* 1978). Increased density of serotonergic 5-HT_{1A} receptors in the hypothalamus of nonaggressive rats may account for the finding, in rats bred for low aggression, of an altered hormonal regulation, such as decreased functional activity of hypothalamic-pituitary-adrenocortical axis (Naumenko *et al.*, 1989) and hypothalamic-pituitary-testicular system (Shishkina *et al.,* 1993).

Lately, evidence has accumulated suggesting the participation of serotonin 5-HT_{1A} receptors of the amygdaloid complex in the control of emotions. However, data on their influence are rather contradictory since both facilitatory and inhibitory functions have been ascribed. On the one hand, a bilateral microinjection of the 5-HT_{1A} receptor agonists ipsapirone, buspirone and 8-OH-DPAT into the amygdala decreased the electric current-induced vocalization, which is regarded as an indicator of attenuation of anxiety (Schreiber, 1992). On the other hand, experimental data show that serotonin in the amygdala participates in the generation of anxiety and fear, thus facilitating the defensive behaviour integrated by the amygdala (Graeff, 1994).

Increase in the density of 5-HT_{1A} receptors in the limbic system of low aggressive rats found in our experiments is to some extent in accord with results obtained when studying the distribution of this type of 5-HT receptors by autoradiography (Hammer *et al.*, 1992). In these studies, carried out on Norway rats of the first generations of breeding for nonaggressive behaviour towards man, an increase of [3H]8-OH DPAT label was found in some brain structures, the maximal increase being observed in structures of the limbic system.

Taken together, the evidence reviewed above suggests that serotonergic 5-HT_{1A} receptors in the limbic system of the brain may fulfil a significant role in the expression of affective aggressiveness, and hereditary traits of high or low aggressiveness may be determined, at least partly, by the density of 5-HT_{1A} receptors in the limbic system.

However, it is noteworthy that, although 8-OH DPAT is considered to be the most selective 5-HT_{1A} agonist, recently rather high affinity of 8-OH DPAT to 5-HT_7 serotonin receptors was found (Lucas and Hen, 1995). These, currently the latest-cloned serotonin receptors, are regarded as 5-HT_{1A}-like receptors positively coupled with adenylate cyclase. However, since 5-HT_7 receptors were found in the same structures where 5-HT_{1A} receptors are localized, and where we determined the changes in [3H]8-OH DPAT binding (i.e. hippocampus, hypothalamus, frontal cortex) we can not rule out the possibility that hereditary aggressiveness is associated not only with 5-HT_{1A} receptors but also with 5-HT_{1A}-like serotonergic receptors of the 5-HT_7 type.

3. PASSIVE DEFENCE FREEZING

Changes in the brain's serotonergic system were found in animals with a genetic predisposition to another kind of defensive behavior—freezing (catalepsy). Freezing is considered as a natural fear-induced immobility response to threatening situations, for example, the appearance of a predator (Wallnau *et al.*, 1981). In this situation animals remain completely immobile—with the exception of occasional and brief movements of the vibrissae—often in a crouching posture, with eyes wide open and irregular respiration (Graeff, 1994). In humans, catalepsy in an exaggerated form is a syndrome of schizophrenia.

A significant role of the genotype in predisposing animals to catalepsy was revealed in inbred strains of mice, using pinch-induced catalepsy as the model (Kulikov *et al.*, 1993). More recently, it was shown that a genetic predisposition to catalepsy was associated with significant changes in brain serotonin metabolism, and in serotonergic receptors. It has been shown in various experimental models of freezing (i.e. pinch-induced catalepsy in mice, and catalepsy in rats selected for more than 30 generations for predisposition to catalepsy) (Kolpakov *et al.*, 1981), that the genetic predisposition to catalepsy is characterized by an increase of the tryptophan hydroxylase activity in the striatum (Kolpakov *et al.*, 1985; Popova *et al.*, 1985; Kulikov *et al.*, 1995; Popova and Kulikov, 1995) (see Figure 18.5).

It is worth noting that the increase in enzyme activity was shown selectively in the striatum, which has been implicated as a major brain structure involved in the mechanism of catalepsy (Koffer *et al.*, 1978; Di Chiara and Morelly, 1984), whereas no differences in enzyme activity were revealed in the hippocampus and the midbrain between catalepsy-prone and control rats. At the same time hereditary catalepsy does not seem to be associated with midbrain tryptophan hydroxylase. It is well known that tryptophan hydroxylase is sythesized mainly in cell bodies of serotonergic neurons in the midbrain and then transported to the forebrain regions (Meek and Neff, 1972). Activation of tryptophan hydroxylase in the striatum, side-by-side with unaltered enzyme in the midbrain of cataleptic rats and mice, suggests a local modification of the enzyme molecules in the striatum rather than an increase in the expression of mRNA encoding tryptophan hydroxylase.

It was shown that the main mechanism of increasing tryptophan hydroxylase activity is by reversible phosphorylation of the enzyme, dependent on Ca^{2+} and calmodulin (Boadle-Biber, 1982). A significant increase in endogenous phosphorylation of tryptophan hydroxylase in the striatum of cataleptic rats and mice has been revealed (Kulikov *et al.*, 1992; Kulikov and Voronova, 1995) thus indicating the mechanism of selective changes of enzyme activity in the striatum of rats with a hereditary predisposition to catalepsy.

In genetically defined catalepsy, alterations were shown, not only in serotonin metabolism, but also in serotonergic receptors. Radioligand binding of [3H]8-OH DPAT in brain regions of rats, selected over 40 generations for the predisposition to catalepsy, showed decreased $5\text{-HT}_{1A}/5\text{-HT}_{1A}$-like receptor density in 4 out of 5 brain regions studied (Popova *et al.*, 1996). The most pronounced changes in specific [3H]8-OH DPAT binding was found in brain regions where postsynaptic 5-HT_{1A} receptors are mainly localized—that is, in the frontal cortex, striatum, and in the hippocampus. At the same time, a trend towards decreased 5-HT_{1A} receptors was found in the midbrain as well, where, in the median and dorsal raphe nuclei, the bulk of presynaptic somatodendritic 5-HT_{1A} receptors are concentrated (Sharp and Hjorth, 1992).

Since stimulation of presynaptic receptors leads to a decrease of firing of serotonin neurones, and of release of serotonin from nerve terminals, while stimulation of post-synaptic receptors provokes specific serotonergic responses, involvement of both kinds of 5-HT_{1A} receptors may produce rather complicated changes in serotonergic regulation.

One may suggest that the decreased $5\text{-HT}_{1A}/5\text{-HT}_{1A}$-like receptor density in frontal cortex, in striatum and in the hippocampus are associated not only with a predisposition to catalepsy, but also with some behavioural peculiarities of rats predisposed to freezing reaction, such as a decrease of motor activity, enhancement of emotional reaction towards man, and longer immobility time in Porsolt's (forced swimming) test. This suggestion is based on data demonstrating that 5-HT_{1A} agonists influence motor activity (Rodgers *et al.*, 1994), display an anxiolytic effect (Broekkamp *et al.*, 1989) and prevent the expression of catalepsy in genetically catalepsy-prone rats (Kulikov *et al.*, 1994). Concomitantly, the 5-HT_{1A} deficit in the midbrain of cataleptic rats probably contributes to the expression of catalepsy, since it has been found that 5-HT_{1A} activation of the median raphe nucleus in the midbrain (which has the highest density of presynaptic 5-HT_{1A} receptors) decreases serotonin release in the striatum (Invernizzi *et al.,* 1989). This is the structure in which we consider enhancement of serotonergic function to be crucial for the mechanism of catalepsy (Popova and Kulikov, 1995).

4. SPECIFIC FEATURES OF BRAIN SEROTONIN METABOLISM AND 5-HT1A RECEPTORS IN ANIMALS WITH A GENETICALLY DEFINED PREDISPOSITION TO DIFFERENT KINDS OF DEFENSIVE BEHAVIOUR

In this way, notwithstanding the fact that selection was carried out for different kinds of defensive behavior—aggression towards man, or predisposition to freezing the hereditary trait to fear-induced defensive responses in both cases was associated with the changes in brain serotonin metabolism and in 5-HT_{1A} serotonergic receptors. Evidently, this finding indicates the involvement of the serotonergic system in the expression of various defensive behaviours, irrespective of their type.

However, significant diversity was found in changes in serotonin metabolism in animals genetically predisposed to active defence, or predisposed to passive defensive behaviour. First, genetically defined high aggressiveness in wild rats is associated with decreases of both level and metabolism of serotonin, compared with nonaggressive animals, whereas a predisposition to catalepsy is characterized by increased serotonin metabolism. Second, selection for low or high aggressiveness towards man affects the midbrain, where the perikarya of serotonin neurones are located (mainly in the median and dorsal raphe nuclei). This results in changes of all elements of serotonin metabolism (e.g. the key enzymes of serotonin synthesis and degradation, as well as levels of serotonin and its metabolite in some brain regions (Popova *et al.*, 1991a,b). On the other hand, changes in animals manifesting a hereditary predisposition to passive defensive freezing are quite different. An increased activity of the key enzyme in serotonin biosynthesis, tryptophan hydroxylase, which is selective for the striatum (the brain region involved in the control of muscular tone) but not in the midbrain, was found in rats selected for many generations for excessive freezing. A similar increase in tryptophan hydroxylase activity in the striatum, without any significant changes in the midbrain, was shown in mice with a genetic predisposition to catalepsy. Moreover, blockade of serotonin synthesis by inhibitors of tryptophan hydroxylase prevented the expression of hereditary catalepsy (Popova and Kulikov, 1995).

Thus selection for these two defensive behaviours appears to be associated with different hereditary changes in the brain serotonergic system.

However, unlike serotonin metabolism, alterations in 5-HT_{1A} receptors in animals with different defensive reactions were similar in highly aggressive rats and in rats genetically predisposed to freezing (Popova *et al.,* 1998). In both genetically defined fear-induced active defensive aggression towards man, and in passive defensive freezing, a decrease in the 5-HT_{1A}/5-HT_{1A}-like serotonergic receptor density in some brain structures was shown. The greatest coincidence of changes was found in the frontal cortex, where [3H]8-OH DPAT-specific binding in rats predisposed to freezing, as well as in highly aggressive rats, was half the value found in the respective control animals. Differences between animals with different types of defensive behavior were found in the hippocampus, where a considerable increase was shown in Kd, indicating diminished affinity for the ligand in rats predisposed to freezing responses, while no changes were found in rats which expressed an aggressive defence reaction.

Since the main factor inducing both kinds of defensive reaction is fear, one may hypothesize that a genetic predisposition to high expression of any kind of defence is

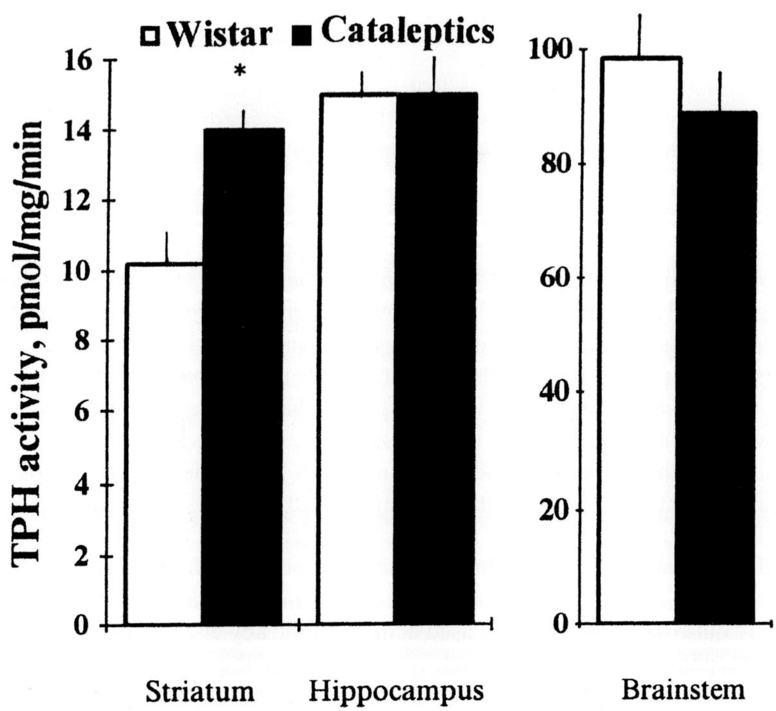

Figure 18.5. Tryptophan hydroxylase activity in the brain of rats genetically predisposed to catalepsy and control Wistar rats. *p < 0.05 vs. Wistar (from Kulikov *et al.*, 1992).

defined by a reduced density of 5-HT_{1A}/5-HT_{1A}-like serotonergic receptors, associated with anxiety and fear that trigger various forms of defensive behavior.

Therefore, the data presented give strong evidence for a crucial role of the brain serotonergic system in the mechanisms of both kinds of genetically defined defensive behavior—aggression and freezing. It is suggested that serotonergic 5-HT_{1A} receptors in the frontal cortex can participate in the mechanisms of anxiety and fear that trigger any kind of defensive response, whereas a genetically defined predisposition to an active or passive strategy of defence is associated with peculiarities of serotonin metabolism in such brain structures as the midbrain and striatum.

ACKNOWLEDGEMENT

The work was partly supported by the grant 96–04–49 960 of the Russian Foundation for Basic Researches.

REFERENCES

Belyaev, D.K. (1979) Destabilizing selection as a factor in domestication. *Journal of Heredity*, **70**, 301–308.

Blanchard, R.J. (1989) Attack and defense in rodents as ethoexperimental models for the study of emotions. *Progress in Neuro-Psychopharmacology and Biological Psychiatry*, **13**, S3–S14.

Blanchard, D.C., Popova, N.K., Plyusnina, I.F., Velichko, I.L., Campbell, D., Blanchard, R.J., Nikulina J. Nikulina, E.M. (1994) Defensive reactions of "wild-type" and "domesticated" wild rats to approach and contact by a threat stimulus. *Aggressive Behavior*, **20**, 387–397.

Boadle-Biber, M.C. (1982) Biosynthesis of serotonin. In: N.N. Osborne (ed.) *Biology of Serotonergic Transmission*. Chichester. John Wiley and Sons, New York. pp. 63–64.

Broekkamp, C.L., Berendsen, H.H., Jenck, F. and Van Delft, A. (1989) Animal models for anxiety and response to serotonergic drugs. *Psychopathology*, **22**. (Suppl. 1), 2–12.

Di Chiara, G. and Morelli, M. (1984) Output pathways mediating basal ganglia function. In: McKenzie, J.S., R.E. Kemm and L.N.Wilcock (eds). *The Basal Ganglia: Structure and Function. (Advances in Behavioral Biology, vol. 27)*, Plenum Press, New York, pp. 443–446.

Graeff, F.G. (1994) Neuroanatomy and neurotransmitter regulation of defensive behaviors and related emotions in mammals. *Brazilian Journal of Medical Biological Research*, **27**, 811–829.

Hammer, R., Hori, K., Blanchard, R. and Blanchard, D.C. (1992) Domestication alters 5-HT_{1A} receptor binding in rat brain. *Pharmacology Biochemistry and Behaviour* **42**, 25–28.

Invernizzi, R., Carli, M., DiClemente, A. and Samanin, R. (1989) Forebrain serotonin synthesis and release are regulated by serotonin $_{1A}$ receptors in the nucleus raphe dorsalis. In: R. Paoletti (ed.) *Serotonin from Cell Biology to Pharmacology and Therapeutics*. Kluwer Academic Publisher, Boston, Dordrecht, pp. 17.

Koffer, K.B., Berney, S. and Hornykiewicz, O. (1978) The role of the corpus striatum in neuroleptic and narcotic-induced catalepsy. *European Journal of Pharmacology*, **47**, 81–86.

Kolpakov, V.G., Barykina, N.N. and Chepkasov, I.L. (1981) Genetic predisposition to catatonic behavior and methylphenidate sensitivity in rats. *Behavioral Processes*, **6**, 269–281.

Kolpakov, V.G., Kulikov, A.V., Barykina, N.N., Alekhina, T.A. and Popova, N.K. (1985) Catalepsy and increased tryptophan hydroxylase activity in rat striatum. *Biogenic Amines*, **2**, 131–136.

Kostowski, W., Gumulka, W. and Czlonowski, A. (1972) Reduced cataleptogenic effects of some neuroleptics in rats with lesioned midbrain raphe and treated with p-chlorophenylalanine. *Brain Research*, **48**, 443–446.

Kulikov, A.V. and Popova, N.K. (1991) Kinds of aggressive behavior. In: *Uspekhi Sovremennoy Genetiki*. Moscow. Nauka, pp.131–151 (in Russian).

Kulikov, A.V., Kozlachkova, E.Y. and Popova, N.K. (1992) Activity of tryptophan hydroxylase in brain of hereditary predisposed to catalepsy rats. *Pharmacology Biochemistry and Behaviour*, **43**, 999–1003.

Kulikov, A.V., Kozlachkova, E.Y., Maslova, G.B. and Popova, N.K. (1993) Inheritance of predisposition to catalepsy in mice. *Behavior Genetics*, **23**, **N 4**, 378–384.

Kulikov, A.V., Kolpakov, V.G., Maslova, G.B., Kozintsev, S. and Popova, N.K. (1994) Effect of selective 5-HT_{1A} agonists and 5-HT2A antagonists on inherited catalepsy in rats. *Psychopharmacology*, **114**, 172–174.

Kulikov, A.V., Kozlachkova, E.Y., Kudryavtseva, N.N. and Popova, N.K. (1995) Correlation between tryptophan hydroxylase activity in the brain and predisposition to pinch-induced catalepsy in mice. *Pharmacology Biochemistry and Behaviour*, **50, N 3**, 431–435.

Kulikov, A.V. and Voronova, I.P. (1995) On the role of reversible phosphorylation in genetically determined polymorphism in brain tryptophan hydroxylase activity. *Bulletin of Experimental Biology and Medicine*, **91, N1**, 67–68 (in Russian).

Kulikov, A.V., Zhanaeva, E.Y. and Popova, N.K. (1989) Changes in brain tryptophan hydroxylase activity in the process of selection according behavior silver foxes and Norway rats. *Genetika*, **25**, **N2**, 346–350 (in Russian).

Kurtz, N. (1992) Efficacy of azapirones in depression. In: Stahl, S.M. *et al.* (eds.) *Serotonin$_{1A}$ Receptors in Depression and Anxiety*. Raven Press. N.Y., pp. 163–170.

Lagerspetz, K.M. and Lagerspetz, K.Y. (1974) Genetic determination of aggressive behavior. In: J. van Abeleen (ed.) *The Genetics of Behavior*. Elsevier, Amsterdam. pp. 321–346.

Lesh, K.-P. (1992) The ipsapirone/5-HT_{1A} receptor challenge in anxiety disorders and depression. In: Stahl, S.M. (eds). *Serotonin $_{1A}$ Receptors in Depression and Anxiety*. Raven Press, New York pp. 135–162.

Lucas, J. and Hen, R. (1995) New players in the 5-HT receptor field: genes and knockouts. *Trends in Pharmacological Sciences*, **16**, N7, 246–252.

Meek, J. and Neff, N.H. (1972) Tryptophan 5-hydroxylase: approximation of half-life and rate of axonal transport. *Journal of Neurochemistry*, **19**, 1519–1525.

Moyer, K.E. (1968) Kinds of aggression and their physiological basis. *Communications in Behavioral Biology*, **2**, 65–87.

Naumenko, E.V., Popova, N.K., Nikulina, E.M., Dygalo, N.N., Shishkina, G.T., Borodin, P.M. and Markel, A.L. (1989) Behavior, adrenocortical activity and brain monoamines in Norway rats selected for reduced aggressiveness towards man. *Pharmacology Biochemistry and Behaviour*, **33**, 85–92.

Nikulina, E.M., Borodin, P.M. and Popova, N.K. (1985) Changes in some kinds of aggressive behavior and monoamines level during selection of wild rats for tame behavior. *Zhurnal Vysshey Nervnoy Dejatelnosty*, **35 N4**, 703–708 (in Russian).

Nutt, D.J. and Glue, P. (1991) Clinical pharmacology of anxiolytics and antidepressants: a psychopharmacological perspective. In: S.E. File (ed.) *Psychopharmacology of Anxiolytics and Antidepressants*. Pergamon Press. New York, pp. 1–28.

Popova, N.K., Voitenko, N.N. and Trut, L.N. (1975) Changes in serotonin and 5-hydroxyindoleacetic acid content in the brain of silver foxes under selection for behavior. *Doklady Acadademii. Nauk SSSR*. **223**, 1496–1500 (in Russian).

Popova, N.K., Naumenko, E.V. and Kolpakov, V.G. (1978) *Serotonin and Behavior*. Nauka, Novosibirsk. 304 P (in Russian).

Popova, N.K., Voitenko, N.N., Pavlova, S.I., Trut, L.N., Naumenko, E.V. and Belyaev, D.K. (1980) Genetics and phenogenetics of hormonal characteristics in animals. VII. Relationships between brain serotonin and hypothalamo-pituitary-adrenal axis in emotional stress in domesticated and non-domesticated silver foxes. *Genetika*, **16**, **N10**, 1865–1870 (in Russian).

Popova, N.K., Nikulina, E.M. (1983) Inhibitory effect of serotonin on expression of predatory aggression in mice. *Zhurnal Vysshey Nervnoy Dejatelnosty*, **33**, **N6**, 1098–1102 (in Russian).

Popova, N.K., Kulikov, A.V., Kolpakov, V.G., Barykina, N.N. and Alekhina, T.A. (1985) Changes in the brain serotonergic system in the rats genetically predisposed to catalepsy. *Zhurnal Vysshey Nervnoy Dejatelnosty*, **35**, 742–746 (in Russian).

Popova, N.K. and Zhanaeva, E.Y. (1988) Brain tryptophan hydroxylase: Functional role and regulation of activity. *Neurochemia*, **7**, **N2**, 274 –287 (in Russian).

Popova, N.K., Kulikov, A.V., Nikulina, E.M., Kozlachkova, E.Y. and Maslova G.B. (1991a) Serotonin metabolism and serotonergic receptors in Norway rats selected for low aggressiveness to man. *Aggressive Behavior*, **17**, **N4**, 207–213.

Popova, N.K., Voitenko, N.N., Kulikov, A.V. and Avgustinovich, D.F. (1991b) Evidence for the involvement of central serotonin in mechanism of domestication of silver foxes. *Pharmacology Biochemistry and Behaviour*, **40**, 751–756.

Popova, N.K., Nikulina, E.M. and Kulikov, A.V. (1993) Genetic analysis of different kinds of aggressive behavior. *Behavior Genetics*, **5**, 491–497.

Popova, N.K. and Kulikov, A.V.(1995) On the role of brain serotonin in expression of genetic predisposition to catalepsy in animal models. *American Journal of Medical Genetics. Neuropsychiatric Genetics*, **60**, **N3**, 214–220.

Popova, N.K., Avgustinovich, D.F., Shigantsov, S. and Kulikov, A.V. (1996) 5-HT_{1A} receptors in brain regions of genetically predisposed to catalepsy rats. *Zhurnal Vysshey Nervnoy Dejatelnosty*, **46**, **N3**, 578–582 (in Russian).

Popova, N.K., Avgustinovich, D.F. and Plyusnina, I.F. (1997) [3H] 8-OH DPAT binding in brain regions of wild Norway rats selected for lack of aggressiveness. *Neurochemia*, **13**, 260–264 (in Russian).

Popova, N.K., Avgustinovich, D.F., Kolpakov, V.G. and Plyusnina, I. (1998) Specific [^{3}H] 8-OH DPAT binding in brain regions of rats genetically predisposed to various defense behavior strategy. *Pharmacology Biochemistry and Behaviour*, **59**, 793–797.

Rodgers, R.J., Cole, J. and Davies, A. (1994) Antianxiety and behavioral supressant actions of the novel 5-HT_{1A} receptor agonist, flesinoxan. *Pharmacology Biochemistry and Behaviour*, **48**, N3, 959–963.

Schreiber, R. (1992) *Behavioral Pharmacology of 5-HT_{1A} Receptor Ligands*. Hamburg. Proefschrift., 185P.

Sharp, T. and Hjorth, S. (1992) *In vivo* neurochemical studies of 5-HT_{1A} autoreceptor function. In: P. Bradley (ed.). *Serotonin CNS Receptors and Brain Function*. Pergamon Press, Oxford, pp. 297–323.

Shishkina, G.T., Borodin, P.M. and Naumenko, E.V. (1993) Sexual maturation and seasonal changes in plasma levels of sex steroids and fecundity of wild Norway rats selected for reduced aggressiveness towards humans. *Physiology and Behavior*, **53**, 389–393.

Wallnau, L.B., Bordash, G. and Corso, P. (1981) The effects of tryptophan and manipulations of serotonergic receptors on tonic immobility in chickens. *Pharmacology Biochemistry and Behaviour*, **14**, 463–468.

Index